AF835294

Springer Series in Statistics

Advisors:
P. Bickel, P. Diggle, S. Fienberg, U. Gather,
I. Olkin, S. Zeger

Springer Series in Statistics

For other titles published in this series, go to
http://www.springer.com/692

Michael S. Hamada
Alyson G. Wilson
C. Shane Reese
Harry F. Martz

Bayesian Reliability

Springer

Michael S. Hamada
Los Alamos National Laboratory
Los Alamos, NM 87545, USA
hamada@lanl.gov

C. Shane Reese
Department of Statistics
Brigham Young University
Provo, UT 84602, USA
reese@stat.byu.edu

Alyson G. Wilson
Los Alamos National Laboratory
Los Alamos, NM 87545, USA
agw@lanl.gov

Harry F. Martz
Los Alamos National Laboratory
Los Alamos, NM 87545, USA
martz@rt66.com

ISSN 0172-7397
ISBN 978-0-387-77948-5 e-ISBN 978-0-387-77950-8
DOI: 10.1007/978-0-387-77950-8

Library of Congress Control Number: 2008930561

© 2008 Springer Science+Business Media, LLC
All rights reserved. This work may not be translated or copied in whole or in part without the written permission of the publisher (Springer Science+Business Media, LLC, 233 Spring Street, New York, NY 10013, USA), except for brief excerpts in connection with reviews or scholarly analysis. Use in connection with any form of information storage and retrieval, electronic adaptation, computer software, or by similar or dissimilar methodology now known or hereafter developed is forbidden.
The use in this publication of trade names, trademarks, service marks, and similar terms, even if they are not identified as such, is not to be taken as an expression of opinion as to whether or not they are subject to proprietary rights.

Printed on acid-free paper

springer.com

To Jung Hee, Christina, and Alexandra - M.S.H.
To Carol Ann - H.F.M.
To Wendy, Madison, Brittany, Bryon, and Mom - C.S.R.
To Greg - A.G.W.

Preface

In this book, we present modern methods and techniques for analyzing reliability data from a Bayesian perspective. The acceptance and application of Bayesian methods in virtually all branches of science and engineering have significantly increased over the past few decades. This increase is largely due to advances in simulation-based computational tools for implementing Bayesian methods. We extensively use such tools here.

We focus our attention on assessing the reliability of components and systems with particular attention to models containing explanatory variables. Such models include failure time regression models, accelerated testing models, and degradation models. We also pay special attention to Bayesian goodness-of-fit testing, model validation, reliability test design, and assurance test planning. Throughout the book we use Markov chain Monte Carlo (MCMC) algorithms for implementing Bayesian analyses. MCMC makes the Bayesian approach to reliability computationally feasible and conceptually straightforward; this is an important advantage in complex settings where classical approaches fail or become too difficult for practical implementation.

We intend this book to be primarily a reference collection of modern Bayesian methods in reliability for use by reliability practitioners. To this end, we have included more than 70 illustrative examples. Most have a real data component, and several of the corresponding datasets have not previously been published. We note, however, that space constraints have made it impractical to fully detail model diagnostics and goodness-of-fit procedures in all examples.

This book can also be used as a textbook for an undergraduate or graduate course in reliability. Therefore, we have included more than 165 exercises to further illustrate and emphasize text material. We base many of the exercises on real data. A solution manual for the exercises that also contains code for the examples is available for instructors at `http://www.springer.com`.

As a prerequisite, readers should have a basic knowledge of probability and statistics, as presented in a first course in applied statistics. In particular, prior familiarity with probability distributions, statistical estimation, and regression analysis is useful. We present fundamental notions of reliability in Chap. 1, so prior knowledge of reliability concepts is not required. Basic calculus and matrix algebra concepts are also required.

Noteworthy highlights of the book include the following:

- Development and use of Bayesian goodness-of-fit and model selection methods,
- Introduction and use of Bayesian hierarchical models for reliability estimation,
- Consideration of a Bayesian fault tree analysis method supporting data acquisition at all levels in the tree,
- Bayesian networks in reliability analysis,
- Bayesian methods for analyzing both failure count and failure time data collected from repairable systems and the assessment of various related performance criteria,
- Estimation of reliability using information contained in explanatory variables,
- Bayesian approaches for designing and analyzing reliability improvement experiments,
- Bayesian methods for modeling and analyzing nondestructive and destructive degradation data,
- Illustration of a Bayesian approach for the optimal design of reliability experiments, and a
- Bayesian hierarchical approach to reliability assurance testing.

Of course, we have not covered all topics in reliability. For example, we have chosen not to cover topics like nonparametric methods in reliability (including hazard function and proportional hazards modeling), software and structural reliability, and certain topics related to repairable systems, such as maintenance. We also do not discuss probability plotting as a means for identifying a sampling distribution, mainly because this topic is already well covered in other books.

Chapter 1 develops the main definitions of reliability and introduces reliability and lifetime data. In Chap. 2, we cover basic concepts common to all Bayesian analyses, including the definitions and specifications of prior distributions, likelihood functions and sampling distributions, posterior distributions, and predictive distributions. Chapter 3 introduces the primary numerical, simulation-based tool for estimating these distributions: MCMC algorithms. We provide detailed examples to illustrate the two most common types of MCMC algorithms, the Gibbs sampler, and Metropolis-Hastings algorithm. We then introduce the notions of hierarchical modeling and empirical Bayes methods.

Reliability models and lifetime analyses for component-level data are presented and developed in Chap. 4. In this first applications chapter, we discuss diagnostics for addressing model fit and describe hierarchical models that facilitate the joint analyses of data collected from similar components.

In Chap. 5, we extend the models for component-level data to the system level. This extension requires us to specify logical relationships between the components in a system and how the functioning of the complete system depends on the functioning of each of its components. Probability models developed in Chap. 5 account for both dependent and independent components and multilevel data.

Chapter 6 develops a Bayesian treatment of the classical models for repairable systems: renewal processes and homogeneous and nonhomogeneous Poisson processes. We also consider some alternative models as well as Bayesian hierarchical adaptations of these common models. Several real-data examples address the reliability of highly parallel supercomputers.

Bayesian estimation methods for the standard regression models used in reliability are considered in Chap. 7. In particular, we consider linear, nonlinear, logistic, and Poisson regression models. We also present Bayesian methods for accelerated life testing models. The chapter also contains methodology for analyzing reliability improvement experiments.

Chapter 8 extends Bayesian methods to degradation data models. In addition to a general model for degradation data, we consider models that include both continuous and discrete covariates. We compare reliability estimates based on degradation data to those based on lifetime data. We also consider models for destructive degradation data, as well as an alternative stochastic process-based degradation model.

Chapter 9 presents methods for the optimal design of reliability experiments. These designs attempt to allocate resources in the most efficient way to meet specified experimental goals. These goals usually involve the quality of the inferences that can be made using experimental data.

Finally, in Chap. 10, we apply these ideas to design tests that assure, at some level of confidence, that a reliability-related quantity exceeds a specified requirement. Within the framework of Bayesian hierarchical models, we derive test plans for binomial, Poisson, and Weibull sampling distributions.

We use several existing statistical software packages for solving examples and exercises. One is the software package WinBUGS, which is a Windows-based implementation of BUGS (**B**ayesian inference **U**sing **G**ibbs **S**ampling). The package contains flexible software for analyzing complex statistical models using MCMC methods. It is available for free download at `http://www.mrc-bsu.cam.ac.uk/bugs/`. This program is relatively simple to use, and detailed examples of its implementation accompany the package.

YADAS (**Y**et **A**nother **D**ata **A**nalysis **S**ystem) is another Bayesian software system for doing MCMC calculations that is based entirely on the Metropolis and Metropolis-Hastings algorithms. It is written in Java and provides tools to implement nonstandard models. In several examples, we found it to be easier to use than WinBUGS. A detailed description of YADAS is available at `http://yadas.lanl.gov`, and it is also available for free download.

In many of the examples, we used the statistical software package R. Although it does not directly support Bayesian MCMC calculations, R is a language and environment for general statistical computing and graphics. It runs on a wide variety of platforms, including UNIX, Windows, and Mac operating systems, and is also available for free download at `http://www.r-project.org`.

We provide a list of acronyms in Appendix A. For convenient reference, Appendix B contains an extensive list of probability distributions and their properties. For each distribution, we define a standard form used throughout this book. For example, $X \sim Beta(\alpha, \beta)$ means that the random variable X has a beta distribution with parameters α and β. If we need to precisely indicate which random variable we are considering, we sometimes include it in the notation. For example, $Beta(x|\alpha, \beta)$ indicates that X is a random variable having a $Beta(\alpha, \beta)$ distribution. Throughout the book we use $\mathbf{P}(A)$ to denote the probability of the event A.

We are indebted to several people for their valuable help. Val Johnson contributed substantially throughout the writing of the book. Valerie Riedel painstakingly edited the original manuscript; and Megan Wyman, a later draft. Hazel Kutac provided invaluable word processing and editing support. Todd Graves provided support for developing YADAS code, as well as help on several of the research topics considered in Chap. 5. Brian Weaver assisted in preparing the distribution appendix and solutions manual. Finally, we thank Sallie Keller-McNulty and David Higdon for providing support and encouragement by allocating time for us to write the book.

Los Alamos, NM *Michael Hamada*
Los Alamos, NM *Harry Martz*
Provo, UT *Shane Reese*
Los Alamos, NM *Alyson Wilson*
February 2008

Contents

1

Reliability Concepts

This chapter introduces the fundamental definitions of reliability and gives examples of common types of reliability data.

1.1 Defining Reliability

There are many ways to define *reliability*. Colloquially, reliability is the property that a thing works when we want to use it. By necessity, more formal definitions of reliability must account for whether or not an item performs at or above a specified standard, how long it is able to perform at that standard, and the conditions under which it is operated. The reliability of an electrical switch, for example, may be defined as the probability that it successfully functions under a specified load and at a particular temperature. In contrast, reliability may be expressed as an explicit function of time. Defining the reliability of a pump in a nuclear power plant depends both on its environment and on its ability to provide a specified capacity over time.

As these examples illustrate, an operational definition of reliability must be specific enough to permit a clear distinction between items that are reliable and those that are not, but also must be sufficiently general to account for the complexities that arise in making this determination. In an effort to achieve both aims, the International Organization for Standardization (ISO) defines reliability as "the ability of an item to perform a required function, under given environmental and operating conditions and for a stated period of time" (ISO, 1986).

From this definition of reliability, we see that reliability analyses often involve the analysis of binary outcomes (i.e., success/failure data). However, in practice, it is often as important to analyze the time periods over which items or systems function. Such analyses are called *lifetime* or *failure time* analyses.

Lifetime analyses involve the analysis of positive, continuous-valued quantities (e.g., the length of time an item functions), and so require different statistical models than analyses based on success/failure data. There are advantages and disadvantages to each type of analysis. Much of the information contained in experimental data may be lost when it is distilled into a success/failure format. For example, failures at 1 and 99 hours are regarded as equivalent if a "reliable" component must operate for 100 hours. On the other hand, accounting for the distribution of times when items fail requires additional assumptions in the statistical models, and these additional assumptions may be difficult to validate.

The notion of randomness is inherent to both types of analyses. For example, suppose that we would like to predict whether electrical switches from a particular production lot can successfully complete 10,000 on/off cycles. We choose 100 switches from the lot and test them to see whether they complete 10,000 cycles or not. This testing will generally not allow us to predict whether any particular electrical switch from the lot will complete 10,000 cycles or not. Instead, we want to estimate the probability that a switch selected at random from the lot will complete 10,000 cycles. Similarly, if we view this problem from a lifetime analysis perspective, we want to estimate the probability that a randomly selected switch fails on or before a particular cycle. That is, we might want to estimate the probability that a switch fails before its ith cycle, for $i = 1, \ldots, 10,000$. In fact, there are a number of summaries that we might be interested in; some of these are described in Sect. 1.2.

The random nature of item failures requires us to choose a philosophy for performing statistical inference about an item's reliability or lifetime. Throughout this book, we have chosen to adopt a Bayesian approach to statistical inference. In our view, the Bayesian approach toward inference offers many advantages. Among these, it allows us to pool information obtained from related experiments into the joint estimation of quantities of interest from each, and it allows us to incorporate expert opinion and subject matter expertise into the analysis of an experiment in a coherent way. Perhaps as importantly, it provides a remarkable degree of flexibility in modeling the phenomena that contribute to reliability and lifetime. With the advent of Markov chain Monte Carlo (MCMC) algorithms, fitting complicated statistical models to data and evaluating the uncertainty in fitted values is now almost routine.

To study these ideas in greater depth, we first need to establish some definitions for describing the properties of random quantities. We then use this expanded vocabulary to discuss a variety of simple examples that illustrate the range of experiments and types of data that can be analyzed using the techniques described in this book.

1.2 Measures of Random Variation

We define a *random variable* to be a function that maps the outcome of an experiment to a real number. For example, if we cycle a switch until it fails, the number of cycles before failure is a random variable.

The *sample space* S of an experiment is the set of all possible outcomes of the experiment. In a simple electrical switch experiment, the sample space for a single cycle is the set containing the events "operates" and "fails," and a random variable for this experiment could be defined as 0 if the switch fails and 1 if it operates. Alternatively, we might perform an experiment that tests an item until it fails. The sample space of that experiment is any positive time, and we can define a random variable T as the lifetime of the item.

Much of reliability and lifetime analysis focuses on modeling the failure time distribution for an item — in other words, modeling the properties of a random variable like T. We can specify the properties of a random variable using different representations, all of which contain equivalent information. Each representation is useful in specific contexts. These representations include the *probability density* or *probability mass function*, the *reliability function*, the *cumulative distribution function*, and the *hazard function*.

For a discrete random variable X with sample space S, the *probability mass function* is a function, $m(x)$, that satisfies

$$m(x) \geq 0, \qquad x \in S,$$

and

$$\sum_{x \in S} m(x) = 1.$$

For a continuous random variable T taking values on the real line, the *probability density function* is a function, $f(t)$, that satisfies

$$f(t) \geq 0, \qquad -\infty < t < \infty,$$

and

$$\int_{-\infty}^{\infty} f(t)\mathrm{d}t = 1.$$

Thus, any nonnegative function that integrates to 1 over the real line is a probability density function. For simplicity, we refer to both probability mass functions and probability density functions as probability density functions throughout the book.

Example 1.1 Exponential probability density function. Exponential random variables are widely used for modeling lifetimes. We say that the random variable T has an exponential distribution (or is an exponential random variable), and we write $T \sim \mathit{Exponential}(\lambda)$ if the probability density function for T is

$$\begin{aligned} f(t) &= \lambda \mathrm{e}^{-\lambda t}, \qquad t > 0, \quad \lambda > 0, \\ &= 0, \qquad t \leq 0. \end{aligned} \tag{1.1}$$

This function meets the requirements for a probability density function because $f(t) \geq 0$ for all t, and $\int_{-\infty}^{\infty} f(t)\mathrm{d}t = 1$ for any value of $\lambda > 0$. Figure 1.1

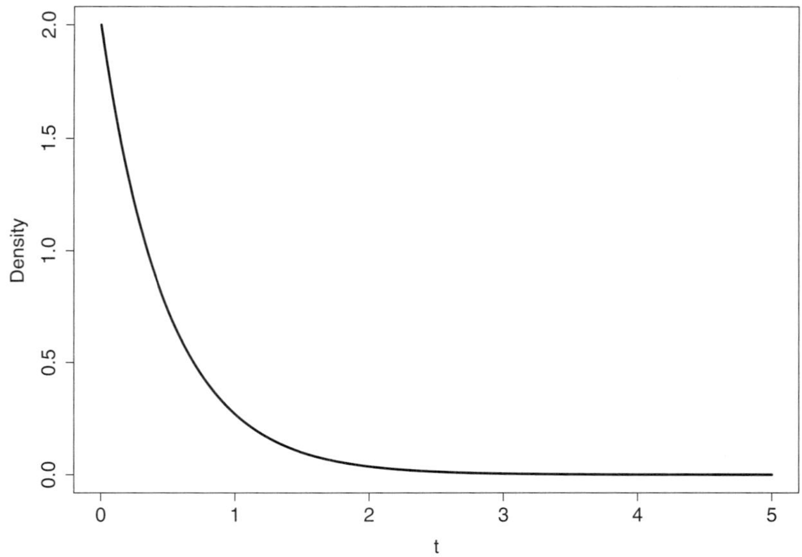

Fig. 1.1. The probability density function for an exponential random variable with $\lambda = 2$.

is a plot of the probability density function for an exponential random variable with $\lambda = 2$.

A second way to specify the properties of a random variable is through its *reliability function*, also known as the *survival function*. We define the reliability function as $R(t) = \mathbf{P}(T > t) = \int_t^\infty f(s)\mathrm{d}s$, where $f(t)$ is a probability density function. Notice that the reliability function takes values in $[0, 1]$.

A third way to specify the properties of T is the *cumulative distribution function*. The cumulative distribution function defines the probability that a random variable takes on a value less than or equal to t. The cumulative distribution function is the complement of the reliability function, so it is also called the *unreliability function*. Mathematically,

$$F(t) = \mathbf{P}(T \le t) = \int_{-\infty}^t f(s)\mathrm{d}s.$$

Example 1.2 Exponential reliability and cumulative distribution functions. The reliability function for an exponential random variable is

$$R(t) = \mathbf{P}(T > t) = \int_t^\infty f(s)\mathrm{d}s$$

$$= \int_t^{\infty} \lambda e^{-\lambda s} ds$$
$$= e^{-\lambda t}.$$

Figure 1.2 is a plot of the reliability function for an exponential random variable with $\lambda = 2$.

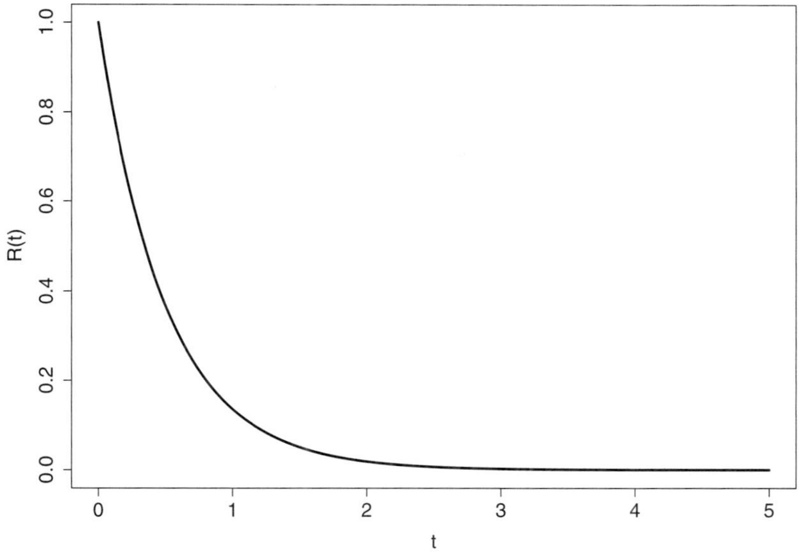

Fig. 1.2. The reliability function for an exponential random variable with $\lambda = 2$.

The cumulative distribution function for an exponential random variable is

$$F(t) = \mathbf{P}(T \leq t) = \int_{-\infty}^{t} f(s) ds$$
$$= \int_0^t \lambda e^{-\lambda s} ds$$
$$= 1 - e^{-\lambda t},$$

where $f(t)$ is the probability density function for an exponential random variable. Figure 1.3 is a plot of the cumulative distribution function for an exponential random variable with $\lambda = 2$.

Another way to specify the properties of a random variable is the *hazard function*, also called the *instantaneous failure rate function*. Suppose that we

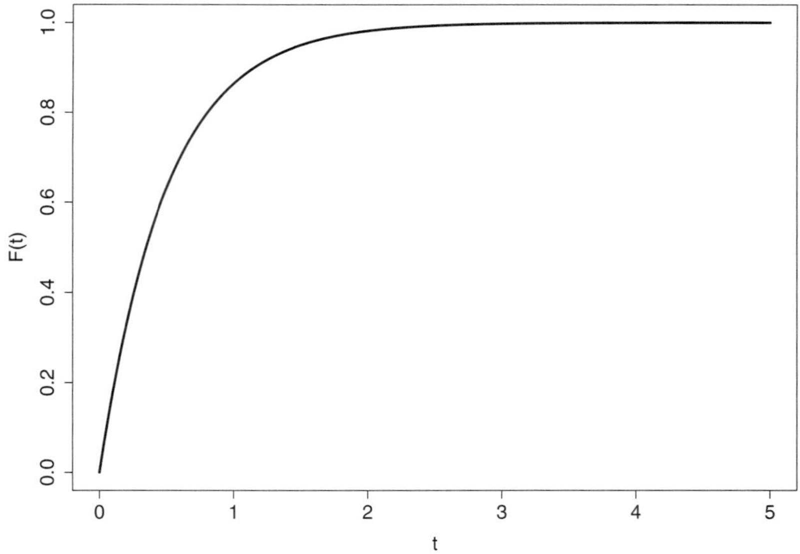

Fig. 1.3. The cumulative distribution function for an exponential random variable with $\lambda = 2$.

are interested in the probability that an item will fail in the time interval $[t, t + \Delta t]$ when we know that item is working at time t. Let $\mathbf{P}(A \mid B)$ denote the conditional probability of an event A, given that event B has occurred. From elementary probability, we know that $\mathbf{P}(A \mid B) = \mathbf{P}(A \bigcap B)/\mathbf{P}(B)$. We can write the probability that an item will fail in the time interval $[t, t + \Delta t]$, given that the item is working at time t, as

$$\mathbf{P}(t < T \leq t + \Delta t | T > t) = \frac{\mathbf{P}(t < T \leq t + \Delta t)}{P(T > t)} = \frac{F(t + \Delta t) - F(t)}{R(t)}.$$

If we want to know the failure rate, we divide by the length of the interval, Δt, and let $\Delta t \to 0$. This gives

$$\begin{aligned} h(t) &= \lim_{\Delta t \to 0} \frac{\mathbf{P}(t < T \leq t + \Delta t | T > t)}{\Delta t} \\ &= \lim_{\Delta t \to 0} \frac{F(t + \Delta t) - F(t)}{\Delta t} \frac{1}{R(t)}. \end{aligned}$$

The first term on the right-hand side is the derivative of the cumulative distribution function, $F(t)$, which is the probability density function, $f(t)$. Therefore,

$$h(t) = \frac{f(t)}{R(t)}.$$

We call $h(t)$ the *hazard function*. We can think of the hazard function as an item's propensity to fail in the next short interval of time, given that the item has survived to time t.

Figure 1.4 shows four of the most common types of hazard functions. These include:

1. *Increasing failure rate (IFR)*: the instantaneous failure rate (hazard) increases as a function of time. We expect to see an increasing number of failures for a given period of time.
2. *Decreasing failure rate (DFR)*: the instantaneous failure rate decreases as a function of time. We expect to see a decreasing number of failures for a given period of time.
3. *Bathtub failure rate (BFR)*: the instantaneous failure rate begins high because of early failures ("infant mortality" or "burn-in" failures), levels off for a period of time ("useful life"), and then increases ("wearout" or "aging" failures).
4. *Constant failure rate (CFR)*: the instantaneous failure rate is constant for the observed lifetime. We expect to see a relatively constant number of failures for a given period of time.

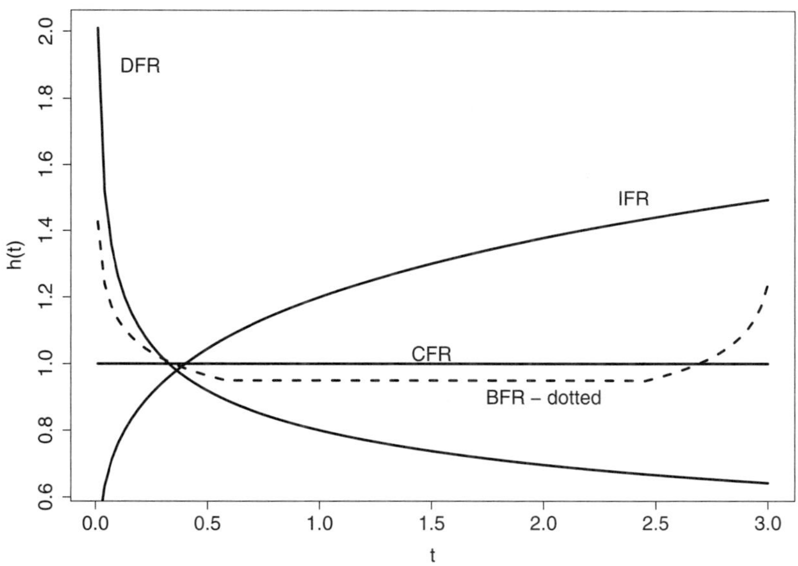

Fig. 1.4. Four different classifications of hazard functions. The dotted line represents the bathtub hazard function.

The *cumulative hazard function* is defined as $H(t) = \int_{-\infty}^{t} h(s)\mathrm{d}s$, where $h(t)$ is the hazard function. The *average hazard rate* (AHR) between times t_1 and t_2 is defined as

$$AHR(t_1, t_2) = \frac{H(t_2) - H(t_1)}{t_2 - t_1}.$$

Example 1.3 Exponential hazard and cumulative hazard functions. The hazard function for an exponential random variable is

$$h(t) = \frac{f(t)}{R(t)} = \frac{\lambda \mathrm{e}^{-\lambda t}}{\mathrm{e}^{-\lambda t}} = \lambda.$$

Notice that this hazard function is constant, which implies that an item's propensity to fail in the next small unit of time does not change as the item ages.

The cumulative hazard function for an exponential random variable is

$$H(t) = \int_{-\infty}^{t} h(s)\mathrm{d}s = \int_{0}^{t} \lambda \mathrm{d}s = \lambda t.$$

For positive random variables, Table 1.1 summarizes the mathematical relationships between the probability density function [$f(t)$], cumulative distribution function [$F(t)$], reliability function [$R(t)$], hazard function [$h(t)$], and cumulative hazard function [$H(t)$].

In applications, less complete descriptions of a random variable are also reported. Such descriptions usually involve the report of a mean, median, or variance of a random variable without specifying its complete distribution. For example, one such summary is the *mean time to failure* (MTTF), which is defined as

$$MTTF = \mathrm{E}(T) = \int_{-\infty}^{\infty} t f(t)\mathrm{d}t,$$

where $\mathrm{E}(T)$ is the *expected value* of T. The MTTF is also called the *expected life*.

Example 1.4 MTTF for an exponential random variable. The MTTF for an exponential random variable is

$$\begin{aligned} MTTF &= \int_{-\infty}^{\infty} t f(t)\mathrm{d}t \\ &= \int_{0}^{\infty} t\lambda \mathrm{e}^{-\lambda t}\mathrm{d}t \\ &= \frac{1}{\lambda}. \end{aligned}$$

Table 1.1. Relationships between the probability density function, cumulative distribution function, reliability function, hazard function, and cumulative hazard function, assuming $f(t) = 0$ for $t < 0$

	$f(t)$	$F(t)$	$R(t)$	$h(t)$	$\mathrm{H(t)}$
$f(t)$	$f(t)$	$\frac{\mathrm{d}}{\mathrm{d}t}F(t)$	$-\frac{\mathrm{d}}{\mathrm{d}t}R(t)$	$h(t)\exp[-\int_0^t h(s)\mathrm{d}s]$	$[\frac{\mathrm{d}}{\mathrm{d}t}H(t)]\exp[-H(t)]$
$F(t)$	$\int_0^t f(s)\mathrm{d}s$	$F(t)$	$1-R(t)$	$1-\exp[-\int_0^t h(s)\mathrm{d}s]$	$1-\exp[-H(t)]$
$R(t)$	$\int_t^\infty f(s)\mathrm{d}s$	$1-F(t)$	$R(t)$	$\exp[-\int_0^t h(s)\mathrm{d}s]$	$\exp[-H(t)]$
$h(t)$	$f(t)/\int_t^\infty f(s)\mathrm{d}s$	$\frac{\mathrm{d}}{\mathrm{d}t}F(t)/[1-F(t)]$	$-\frac{\mathrm{d}}{\mathrm{d}t}\log R(t)$	$h(t)$	$\frac{\mathrm{d}}{\mathrm{d}t}H(t)$
$H(t)$	$-\log[1-\int_0^t f(s)\mathrm{d}s]$	$-\log[1-F(t)]$	$-\log[R(t)]$	$\int_0^t h(s)\mathrm{d}s$	$H(t)$

Other summaries of the properties of a random variable are *quantiles* and *mean residual life*. A quantile is the inverse of the cumulative distribution function; it is the time by which a specified proportion of the population fails. Mathematically, the quantile q at which a proportion p of the population fails is the value q such that $F(q) = p$, or equivalently, $q = F^{-1}(p)$. If the cumulative distribution function is flat over some interval, we define q as the earliest time for which $F(q) = p$.

The *reliable life* is the time for which $100R\%$ of a population will survive, where R is a specified proportion between 0 and 1. For example, with $R = 0.5$, the reliable life is median of the lifetime distribution.

The *mean residual life* is the expected time to failure of a device that has survived to time t. We define the mean residual life as

$$M(t) = \frac{1}{R(t)} \int_t^{\infty} s f(t+s) \mathrm{d}s,$$

where $R(t)$ is the reliability function, and $f(t)$ is a probability density function. Notice that $M(0) = \mathrm{E}(T)$.

1.3 Examples of Reliability Data

In Sects. 1.1 and 1.2, we defined reliability and discussed various ways to summarize the distribution of the lifetime of an item. In this section, we give a few simple examples of the kinds of reliability data that are often encountered in practice. These include pass/fail, failure count, failure time, and degradation data.

1.3.1 Bernoulli Success/Failure Data

The simplest form of reliability data is "pass/fail" or *Bernoulli trial* data. This data can arise from simple "pass/fail" testing. In addition, it can be derived from lifetime data by letting the random variable $X(t) = 1$ ("pass" or "success") if the item is functioning at time t, and $X(t) = 0$ ("fail") if the item has failed by time t.

Table 1.2 contains the outcomes from a set of Bernoulli trials. These data are the launch outcomes of new aerospace vehicles conducted by "new" companies during the period 1980–2002. A total of 11 launches occurred; 3 were successes and 8 were failures. Reliability is the probability of a successful launch. We analyze these data in Chap. 2.

1.3.2 Failure Count Data

Bernoulli data can also be recorded as a function of time. *Failure count data* represent the number of failures that occur over a period of time. For example,

Table 1.2. New launch vehicle outcomes (Johnson et al., 2005)

Vehicle	Outcome
Pegasus	Success
Percheron	Failure
AMROC	Failure
Conestoga	Failure
Ariane 1	Success
India SLV-3	Failure
India ASLV	Failure
India PSLV	Failure
Shavit	Success
Taepodong	Failure
Brazil VLS	Failure

Gaver and O'Muircheartaigh (1987) provides the data shown in Table 1.3 on the number of pump failures x_i observed in t_i thousands of operating hours for 10 different systems at the Farley 1 United States commercial nuclear power plant. The random variable is the number of pump failures, X_i, and the reliability is the probability that no pump failures occur in a given period of time.

Table 1.3. Pump failure count data from Farley 1 U.S. nuclear power plant (number of failures x in t thousands of operating hours) (Gaver and O'Muircheartaigh, 1987)

System	x_i (failures)	t_i (thousand hours)
1	5	94.320
2	1	15.720
3	5	62.880
4	14	125.760
5	3	5.240
6	19	31.440
7	1	1.048
8	1	1.048
9	4	2.096
10	22	10.480

We consider this dataset in Chap. 10; we present a general discussion of failure count data in Chap. 4.

1.3.3 Lifetime/Failure Time Data

Table 1.4 presents an example of *lifetime data*. This dataset comprises 11 observed failure times and 55 times when a test was suspended (censored) before item failure occurred (*) for a 4.5 roller bearing in a set of J-52 engines from EA-6B Prowler aircraft (Muller, 2003). Degradation of the 4.5 roller bearing in the J-52 engine has caused in-flight engine failures. The random variable is the time to failure T, and reliability is the probability that a roller bearing does not fail before time t. We analyze these data in Chap. 4.

Table 1.4. Roller bearing lifetime data (in operating hours) for the Prowler attack aircraft (Muller, 2003)

Failure Times (operating hours)		
1,085*	1,795*	100*
1,500*	1,890	1,628
1,390*	1,145*	759*
152*	1,380*	246*
971*	61*	861*
966*	1,165*	462*
997*	437*	1,079*
887*	1,152*	1,199*
977*	159*	424*
1,022*	3,428*	763*
2,087*	555*	1,297*
646	727*	2,238*
820*	2,294*	1,388
897	663*	1,153*
810*	1,427*	2,892*
80*	951	2,153*
1,167	767*	853*
711	546*	911*
1,203	736*	2,181
85*	917*	1,042*
1,070*	2,871*	799*
719*	1,231*	750

1.3.4 Degradation Data

In some applications, it is useful to measure the *degradation* of an item rather than its lifetime. This is particularly true when items have relatively long lifetimes, which makes experimentation difficult. Many item failures can be traced to an underlying degradation mechanism; for example, when degradation reaches a certain critical threshold, the item fails. Studying degradation mechanisms can often tell us how to improve the reliability of an item.

Table 1.5 presents an example of degradation data adapted from Chow and Shao (1991). One aspect of developing a new drug is to determine its shelf life. Since the potency of a drug degrades over time, its lifetime or shelf life is defined to be the time when its potency reaches 90% of its stated potency.

The random variable is the potency of the drug, and reliability at time t is the probability that the drug has not degraded past 90% of its stated potency by t. We analyze these data in Chap. 8.

Table 1.5. Drug potency degradation data (in percent of stated potency)

	Time (months)					Time (months)			
Batch	0	12	24	36	Batch	0	12	24	36
1	99.9	98.9	95.9	92.9	13	99.8	98.8	93.8	89.8
2	101.1	97.1	94.1	91.1	14	100.1	99.1	93.1	90.1
3	100.3	98.3	95.3	92.3	15	100.7	98.7	93.7	91.7
4	100.8	96.8	94.8	90.8	16	100.3	98.3	96.3	93.3
5	100.0	98.0	96.0	92.0	17	100.2	98.2	97.2	94.2
6	100.1	98.1	98.1	95.1	18	99.8	97.8	95.8	90.8
7	99.6	98.6	96.6	92.6	19	100.8	98.8	95.8	94.8
8	100.4	99.4	96.4	95.4	20	100.0	98.0	96.0	92.0
9	100.9	98.9	96.9	96.9	21	99.6	99.6	92.6	88.6
10	100.5	99.5	94.5	93.5	22	100.2	98.2	97.2	94.2
11	101.1	98.1	93.1	91.1	23	99.8	97.8	95.8	90.8
12	100.9	97.9	95.9	93.9	24	100.0	99.0	95.0	92.0

1.4 Censoring

One of the features of reliability data is the presence of *censoring.* Lifetime data are censored when the exact failure time for a specific item is unknown. There are several types of censoring, including left, right, interval, time, and failure censoring.

Left censoring occurs when an item fails before the first inspection. For example, suppose that an experiment tests the lifetime of a new battery. A set of 500 batteries are tested at 8:00 a.m. every day for 90 days to determine whether each battery is still usable. Suppose the test starts at midnight. The data for any battery that fails before 8:00 a.m. on the first day are *left censored.*

Right censoring occurs when an item has not failed by the last inspection. Consider our battery example. Data for any battery that has not failed by 8 a.m. on the 90th day are *right censored.*

Both left and right censoring are special cases of *interval censoring*. Interval censoring occurs when an item's failure time is only known to be in an interval, (t_i, t_{i+1}). If an observation is left censored at t, then its failure time is in $(0, t)$. If an observation is right censored at t, then its failure time is in (t, ∞). In our battery example, the failure times are *interval censored* because they can only be determined to within a 24-hour interval.

Other categories of censoring describe the cause of the censoring. *Type I censoring* or *time censoring* occurs when we remove unfailed items from testing at a prespecified time; in other words, the test ends after a fixed amount of time. In our battery example, data from any battery that has not failed before 8 a.m. on the 90th day are *time censored.*

Type II censoring (also called *failure censoring* or *item censoring*) occurs when a test ends after a specific number of failures have occurred. Suppose that we are testing a set of 150 air conditioners. We decide that the test will end after 100 failures. The data for the 50 air conditioners that do not fail are *failure censored.*

Type III censoring combines Type I (time) and Type II (failure) censoring. Type III censoring occurs when we set both time and failure criteria and end the experiment when either the time or failure criterion (whichever comes first) has been met. For example, suppose that the test of our 150 air conditioners must end after 1 year or 100 failures, whichever comes first. Data for any air conditioners that have not failed when the experiment ends are *Type III censored.*

Systematic multiple censoring (also called *Type IV censoring*) occurs when items enter into an experiment over a period of time. For example, suppose that our air conditioners are entered into the study as they come off the production line over a six-month period. Suppose that the study ends after 50 air conditioners have failed. The data from any air conditioner still working after the trial ends are subject to *systematic multiple censoring.*

Random right censoring occurs when we remove an item from a test because of a failure that is not of interest. For example, suppose that we are testing our air conditioners for wearout failures, but partway through the test, one falls off the test stand and breaks. The air conditioner's datum for the time to failure is *random right censored.*

We usually assume that the censoring and survival times are independent. This is known as *independent censoring* or *noninformative censoring.* For example, if an air conditioner is removed from our study because it is sold to a customer, then the censoring of its failure time is independent of its survival time.

Because Bayesian modeling uses the observed lifetime in its analyses, censoring mechanisms are easily addressed. Suppose that we observe that an item has failed before time t_L, and its lifetime data are left censored. We know that its lifetime is in $[0, t_L]$. The probability of observing a failure in this interval is

$$\begin{aligned}\mathbf{P}(T \leq t_L) &= \int_0^{t_L} f(t)\mathrm{d}t \\ &= F(t_L).\end{aligned}$$

As we will see later, $F(t_L)$ represents this item's contribution to the likelihood function for estimating the parameters of $f(\cdot)$, and the cause of the censoring does not matter. Similar derivations lead to the expressions displayed in Table 1.6. These probabilities are central to Bayesian and likelihood-based analyses and represent all information provided by the censored data.

Table 1.6. The probability of observing a failure in censored and uncensored data

Type of Observation	Failure Time	Contribution
Uncensored	$T = t$	$f(t)$
Left censored	$T \leq t_L$	$F(t_L)$
Interval censored	$t_L < T \leq t_R$	$F(t_R) - F(t_L)$
Right censored	$T > t_R$	$1 - F(t_R)$

1.5 Bayesian Reliability Analysis

The acceptance and applicability of Bayesian methods have increased in recent years. Today, with advances in computation and methodology, researchers are using Bayesian methods to solve an increasing variety of complex problems. In many applications, Bayesian methods provide important computational and methodological advantages over classical techniques.

This book focuses on Bayesian reliability analysis, which includes the topics of modeling, computation, sensitivity analysis, and model checking. While reading the examples in the book, it is important to keep the following in mind:

> Every attempt to use mathematics to study some real phenomena must begin with building a mathematical model of these phenomena. Of necessity, the model simplifies matters to a greater or lesser extent and a number of details are ignored. The success depends on whether or not the details ignored are really unimportant in the development of the phenomena studied. The solution of the mathematical problem may be correct and yet it may be in violent conflict with realities simply because the original assumptions of the mathematical model diverge essentially from the conditions of the practical problem considered. Beforehand, it is impossible to predict with certainty whether or not a given mathematical model is adequate. To find this out, it is necessary to deduce a number of consequences of the model and to compare them with observation. (Neyman, 1949, p. 22)

In statistical analyses, we must make trade-offs between selecting a model that is sufficiently simple to be readily interpretable and selecting one that is sufficiently rich to capture the essential features of the problem. In Bayesian reliability analysis, the statistical model consists of two parts: the *likelihood function* and the *prior distribution*. The likelihood function is typically constructed from the *sampling distribution* of the data, defined by the probability density function assumed for the data. For example, Eq. 1.1 could be a sampling distribution for lifetime data. The sampling distribution usually contains unknown parameters, such as λ in Eq. 1.1. Once we perform the experiment and observe its outcome, we regard the sampling distribution as a function of the unknown parameters. This function (or any function proportional to it) is called the *likelihood function*. Bayesian inference is the only framework for statistical inference that consistently obeys the *likelihood principle*. Simply put, the likelihood principle states that all information contained in experimental data is contained in the sampling density of the observed data.

In Bayesian analysis, the parameters in the likelihood function are treated as unknown quantities, and we use a probability density function to describe our uncertainty about them. Before analyzing experimental data, we call the distribution that represents our knowledge about these parameters the *prior distribution*. In Bayesian analysis, the likelihood function and the prior distribution are the basis for parameter estimation and inference. Details of Bayesian inference are discussed in Chap. 2.

Bayesian analysis differs from *classical frequency-based* analysis in several key ways. One major philosophical difference is the notion of probability. Classical methods are rooted in the notion of probability as the limiting relative frequency of an event in a repeated series of identical trials. In contrast, the cornerstone of Bayesian methods is the notion of *subjective probability*. Bayesian methods consider probability to be a subjective assessment of the state of knowledge (also called degree of belief) about model parameters of interest, given all available evidence.

As a direct consequence of its use of subjective probability, Bayesian methods permit us to incorporate and use information beyond that contained in experimental data. Whether a reliability analyst does or does not have such test data available, he will often have other relevant information about the value of the unknown reliability parameters. Such relevant information is an extremely useful and powerful component in the Bayesian approach, and thoughtful Bayesian parameter estimates reflect this knowledge. This relevant information is often derived from combinations of such sources as physical/chemical theory, engineering and qualification test results, generic industrywide reliability data, computational analysis, past experience with similar devices, previous test results obtained from a process development program, and the subjective judgment of experienced personnel.

After the test data have been obtained, the *posterior distribution* fully describes the uncertainty associated with the parameter. We calculate the posterior distribution via Bayes' Theorem using the likelihood function and the prior distribution. The logical sequence of likelihood function, prior distribution, Bayes' Theorem, and posterior distribution makes Bayesian reliability methods easy to describe and the derived estimates easy to interpret and use. For example, a Bayesian interval estimate may be directly interpreted as a probability statement about a parameter. In contrast, a corresponding frequency-based confidence interval has no such direct interpretation.

Ignoring the interpretation of probability statements, the differences between Bayesian and classical inferences often become negligible as sample sizes become large. However, when test data are scarce, these differences are often significant, and Bayesian interval estimates based on informative prior distributions are often narrower than classical confidence intervals.

An advantage of the Bayesian approach is illustrated in binomial or Poisson sampling models when no failures have occurred during an experiment. In this case, the classical maximum likelihood estimator (MLE) of the binomial failure probability or Poisson failure rate is zero, which is clearly too optimistic (see Example 1.5). Although there are various ad hoc classical methods to rectify this shortcoming, Bayesian point estimates are naturally nonzero.

Because Bayesian posterior distributions are true probability statements about unknown parameters, they may be easily propagated through complex system models, such as fault trees, event trees, and other logic models. Except in the simplest cases, it is difficult or impossible to propagate classical confidence intervals through such models. Features and nuisances of real-world reliability problems, such as complex censoring and random hierarchical effects, can easily be accommodated and modeled by Bayesian methods. Such considerations are often either difficult or impossible to consider when using classical methods.

When analyzing censored data, Bayesian methods have an important advantage over classical methods. From a classical perspective, confidence intervals and other inferential statements must be made with respect to repeated sampling of the data. From a Bayesian perspective, only the observed censoring pattern is relevant.

Table 1.6 contains the contribution to the likelihood function from censored and uncensored observations. The contribution of Type I, Type II, Type III, Type IV, and random right-censored data are all described by the right-censored row. Each of these types of censoring describes a different reason why an item was removed from a test, but in each case, the item was observed for a period of time without failing, which means that its datum is right censored.

Bayesian hierarchical computations, which until a decade or so ago were essentially impossible to perform, are now straightforward using modern computer software like WinBUGS (Gilks et al., 1994; Spiegelhalter et al., 2003), R (Venables et al., 2006), or YADAS (Graves, 2007a,b). The availability of such software permits the reliability analyst to concentrate on modeling the distinguishing features of the problem, without worrying about its numerical solution.

Example 1.5 High reliability estimation using Bayesian methods. Dastrup (2005) considers the reliability of field programmable gate arrays (FPGAs). These highly flexible microchips allow reprogramming after deployment, making them ideally suited for use in various spacecraft applications. One drawback of the space applications is that the FPGAs experience an increased exposure to radiation, causing the FPGAs to malfunction. While the FPGA can be repaired, the failures must be monitored to determine when reprogramming is required due to radiation exposure.

Testing of FPGAs is accomplished by placing them in a proton accelerator and bombarding them with a proton beam. The number of bits that are upset (n) and the number of FPGA failures as a result of the upset (Y) are recorded.

During one test scenario, the number of upsets observed was $n = 62$ and the number of FPGA failures caused by the 62 upsets was $y = 0$. A simplistic analysis of these data using the standard classical MLE might suggest that the probability that an upset results in an FPGA failure (π) is $\widehat{\pi} = x/n = 0/62 = 0$, which implies that the FPGA is completely reliable. Clearly, this is an unsatisfactory estimate of the actual failure rate.

In addition to the data from the test, engineers have developed a simulation program to assess the probability of an FPGA failure as a function of upsets. For the scenario above, the simulation procedure suggests a failure probability of 0.08 with an associated standard deviation of 0.05.

Let π be the probability that an upset results in an FPGA failure, or $\pi = \mathbf{P}(X_i = 1)$, where X_i is the ith upset, and $Y = \sum_{i=1}^{n} X_i$. One simple model for the sampling distribution of Y is the binomial distribution, written

$$Y \sim Binomial(n, \pi).$$

We use the information from the simulation program to specify a prior distribution for π. A beta distribution with parameters $\alpha = 2.4$ and $\beta = 27.6$, written

$$\pi \sim Beta(2.4, 27.6),$$

matches the mean and standard deviation from the simulation.

Using Bayes' Theorem, we find the posterior distribution of π is

$$\begin{aligned} \pi \,|\, Y &\sim Beta(a + y, b + (n - y)) \\ &\sim Beta(2.4 + 0, 27.6 + 62). \end{aligned}$$

Figure 1.5 shows both the prior and posterior distributions for the FPGA example. We note that the 95% credible interval for π is (0.004, 0.067). This interval is scientifically justifiable and consistent with all of our available information.

Many of the underlying themes presented in Example 1.5 appear repeatedly throughout this book.

1.6 Related Reading

There is extensive literature on reliability analysis. Barlow and Proschan (1965) develops the theory of mathematical reliability. Meeker and Escobar (1998) presents reliability from a classical perspective for engineers and statisticians. Lewis (2001), Blischke and Murthy (2000), and Tobias and Trindade (1995) introduce reliability with an applied focus. Rausand and Høyland (2003) concentrates on system reliability, but also presents component models and qualitative system analysis. Martz and Waller (1982) is one of the few

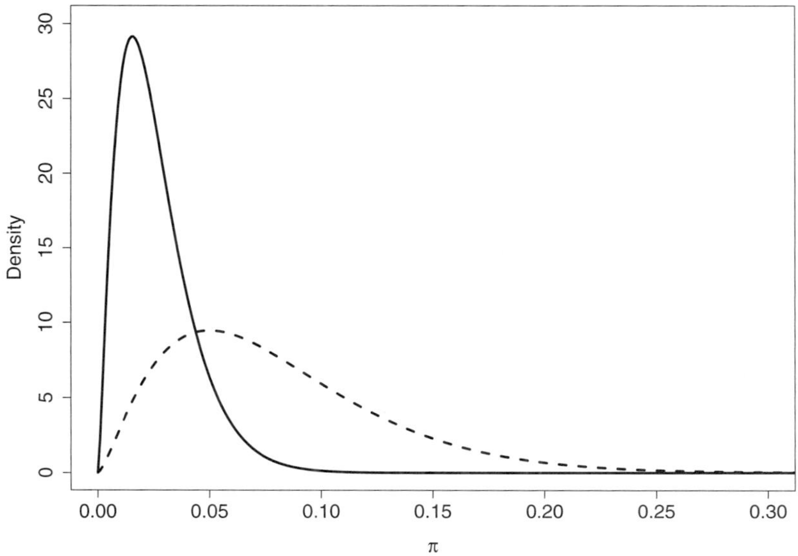

Fig. 1.5. Prior (- - -) density and posterior (—) density for the FPGA example.

surveys of reliability from a Bayesian perspective. Although written before the advent of MCMC and other recent computational advances, the book is a systematic collection of Bayesian reliability techniques.

1.7 Exercises for Chapter 1

1.1 Use the ISO definition of reliability ("the ability of an item to perform a required function, under given environmental and operating conditions and for a stated period of time") to develop explicit descriptions of reliability for three everyday items.

1.2 Consider an experiment where we test a light bulb for 1,000 hours. Define a sample space for the outcomes of the experiment. Define a random variable on this sample space.

1.3 The probability density function for the gamma distribution is

$$f(t \mid \alpha, \lambda) = \frac{\lambda^\alpha}{\Gamma(\alpha)} t^{\alpha-1} \exp(-\lambda t).$$

What is the MTTF for the gamma distribution?

1.4 The probability density function for the Weibull distribution is

$$f(t \mid \lambda, \beta, \theta) = \lambda\beta(t-\theta)^{\beta-1} \exp\left[-\lambda(t-\theta)^\beta\right], \qquad 0 \leq \theta < t, \lambda > 0, \beta > 0.$$

What is the reliability function for the Weibull distribution?

1.5 The probability density function for an exponential random variable is

$$f(t) = \lambda e^{-\lambda t}, \qquad t > 0, \qquad \lambda > 0.$$

What is the average hazard rate for an exponential random variable?

1.6 Find studies in the reliability literature that analyze pass/fail data, failure count data, lifetime data, and degradation data. What is the sampling distribution assumed for the data in each of these studies?

1.7 Suppose that we are using the exponential distribution to model an item's lifetime.

a) We observe that the item failed at 6 hours. What is the likelihood function for this observation?

b) We observe that the item failed at some time between 5 and 10 hours. What is the likelihood function for this observation?

c) We observe the item for 20 hours, and it does not fail. What is the likelihood function for this observation?

2

Bayesian Inference

In this chapter we review the fundamental concepts of Bayesian and likelihood-based inference in reliability. We explore prior distributions, sampling distributions, posterior distributions, and the relation between the three quantities as specified through Bayes' Theorem. We also provide examples of inference in both discrete and continuous settings.

2.1 Introductory Concepts

The unifying theme of this book is the application of Bayesian statistical methods to problems in reliability and lifetime analysis. Because coverage of Bayesian methods in introductory statistical courses is typically sparse, in this chapter we review the basic principles of Bayesian analysis. Readers familiar with fundamentals of Bayesian inference may proceed to Chap. 3.

A primary goal of Bayesian inference is summarizing available information about unknown parameters that define statistical models through the specification of probability density functions. "Unknown parameters that define statistical models" refers to things like failure probabilities or mean system lifetimes; they are the parameters of interest. "Available information" normally comes in the form of test data, experience with related systems, and engineering judgment. "Probability density functions" occur in four flavors: prior densities, sampling densities or likelihood functions, posterior densities, and predictive densities.

To illustrate these concepts more concretely, we consider the following problem. Johnson et al. (2005) presents data for estimating the failure probabilities of launch vehicles used to place satellites in orbit. Because estimates of these failure probabilities play a prominent role in prelaunch risk assessments, they have a significant impact on both public safety and the ability of aerospace manufacturers to develop and field new rocket systems. The Federal Aviation Administration (FAA) and United States Air Force (USAF) were particularly interested in estimating the failure probability for new rockets fielded by companies that had limited design experience.

Table 2.1 displays historical data for launches of new rockets conducted by "new" companies during the period 1980–2002. A total of 11 launches were performed; 3 were successes and 8 were failures. Our goal in presenting these data is to specify a statistical model that can be used for predicting the future success of new rocket systems. Because a launch outcome can be regarded as either a success or failure, we can model launch outcome as Bernoulli data.

Table 2.1. Outcomes for 11 launches of new vehicles performed by companies with limited launch-vehicle design experience, 1980–2002 (Johnson et al., 2005)

Vehicle	Outcome
Pegasus	Success
Percheron	Failure
AMROC	Failure
Conestoga	Failure
Ariane 1	Success
India SLV-3	Failure
India ASLV	Failure
India PSLV	Failure
Shavit	Success
Taepodong	Failure
Brazil VLS	Failure

When we use a Bernoulli model for success/failure data, the basic assumption we make is that the success or failure of each experimental unit is conditionally independent of the success or failure of other units, assuming that we know the probability of success for the population of items. Two events A and B are *independent* if $\mathbf{P}(A \bigcap B) = \mathbf{P}(A)\mathbf{P}(B)$. Two events A and B are *conditionally independent* given an event C if $\mathbf{P}(A \bigcap B \mid C) = \mathbf{P}(A \mid C)\mathbf{P}(B \mid C)$. In Bayesian analyses, outcomes of experiments are usually not independent unless the values of the parameters underlying their distributions are known. For this reason, we often say that two observations are conditionally independent given the values of the parameters that determine their distributions.

Operationally, once we assume that the success or failure of each experimental unit is conditionally independent of the success or failure of other units, we can multiply probabilities of success and failure together to obtain the probability of observing a given sequence of successes and failures. If we let π denote the probability that a new launch vehicle selected at random from the population of new launch vehicles designed by new companies or agencies succeeds, then we can express the probability of observing the sequence of successes and failures reported in Table 2.1 as

$$\pi(1-\pi)(1-\pi)(1-\pi)\pi(1-\pi)(1-\pi)(1-\pi)\pi(1-\pi)(1-\pi) = \pi^3(1-\pi)^8. \quad (2.1)$$

Generalizing Eq. 2.1 to the situation in which we observe y successes in n trials leads to the binomial probability density function, which we can write as

$$f(y \mid n, \pi) = \binom{n}{y} \pi^y (1 - \pi)^{n-y}. \tag{2.2}$$

The quantity $\binom{n}{y}$ accounts for the number of ways that y successes can occur in n trials. The vertical bar in $f(y \mid n, \pi)$ denotes a conditional relationship and is read y "given" n and π. In the launch vehicle example, $y = 3$ and $n = 11$. We denote binomial probability distribution (or density) functions by $Binomial(y \mid n, \pi)$ or, when it is clear that the random variable is Y, simply as $Binomial(n, \pi)$.

Because the binomial probability density function $f(y \mid n, \pi)$ specifies the probability of observing an outcome of a future experiment conducted on a sample of items drawn from the population of interest, we call it a *sampling distribution*. Identifying an appropriate sampling distribution is a major component in specifying a statistical model. In Chap. 4, we explore the use of alternative sampling distributions, including the Poisson, normal, gamma, Weibull, and exponential distributions.

To avoid confusion over notation, we denote an arbitrary sampling distribution by $f(\mathbf{y} \mid \boldsymbol{\theta})$, where $\mathbf{y}$ denotes the (possibly vector-valued) random variable that constitutes the data, and $\boldsymbol{\theta}$ denotes the (possibly vector-valued) parameter that indexes the family of densities. In the binomial example, $\boldsymbol{\theta} = \pi$, and $\mathbf{y} = y$ is simply the scalar-valued number of successes observed.

Of course, once an experiment has been conducted or data have been collected, the value of the random variable, in this case, y, is known. It then makes sense to regard the sampling distribution as a function of the unknown model parameter, in this case, π. When we do, the sampling distribution (or any function proportional to it) is called the *likelihood function*. The likelihood function contains all information in the data that is relevant for estimating unknown model parameters.

Although the likelihood function is known to be a sufficient statistic, many classical statisticians prefer not to base inference solely on it. In so doing, they implicitly reject either the *conditionality principle*, which states that only evidence collected from the experiment actually performed (rather than experiments that might have been performed) is relevant for parameter estimation, or the *sufficiency principle*, which states that all information about the unknown parameter is conveyed through the sufficient statistic. Together, the conditionality principle and the sufficiency principle imply what is known as the *likelihood principle* (Birnbaum, 1962). In contrast, Bayesians and adherents of other forms of likelihood-based inference accept both the conditionality principle and the sufficiency principle, and base their inference instead on the likelihood function and information known about parameters prior to the conduct of an experiment.

2.1.1 Maximum Likelihood Estimation

The simplest way to use the likelihood function for estimating model parameters is to find the value of the parameters that maximize the value of the likelihood function. Estimates obtained in this way are called *maximum likelihood estimates* (MLEs). MLEs make the observed data as "likely" as possible.

For computational reasons, it is often more convenient to maximize the logarithm of the likelihood function rather than the likelihood function itself; the same value maximizes both functions. Not surprisingly, we call the logarithm of the likelihood function the *log-likelihood function*. When observations are conditionally independent, the log-likelihood function is mathematically easier to handle than the likelihood function because it takes the form of a sum rather than of a product. The log-likelihood function is the sum of the logarithm of the density values evaluated at each observation, whereas the likelihood function is the product of the sampling density evaluated at each observation.

Returning to the binomial likelihood function for the launch vehicle data given in Eq. 2.2, it follows that the log-likelihood function for these data is proportional to

$$\log[f(y \mid n, \pi)] \propto y \log(\pi) + (n - y) \log(1 - \pi), \tag{2.3}$$

where $y = 3$ and $n = 11$. Taking the first derivative of the log-likelihood function with respect to π and setting the result equal to 0, we see that the MLE must satisfy

$$0 = \frac{\mathrm{d} \log f(y \mid n, \pi)}{\mathrm{d}\pi} = \frac{y}{\pi} - \frac{n - y}{1 - \pi}. \tag{2.4}$$

Solving for π implies that the MLE of π, say $\widehat{\pi}$, is given by

$$\widehat{\pi} = \frac{y}{n} = \frac{3}{11}.$$

In other words, the MLE of the success probability π in a binomial model is simply the observed proportion of successes. The log-likelihood function for these data is displayed in Fig. 2.1.

In many cases, it is relatively easy to calculate the MLE of model parameters. But ease of calculation does not, by itself, make the MLE a good choice for an estimator. It is more important that estimators be as close as possible to the true value of the parameter. We also want estimators that converge to the true parameter value as the number of observations available for estimating that parameter becomes large. In statistical terms, these requirements can be summarized by saying that we want our estimator to be *efficient* and *consistent*. A nice feature of the MLE in regular statistical models is that it is both efficient and consistent as the sample size becomes large.

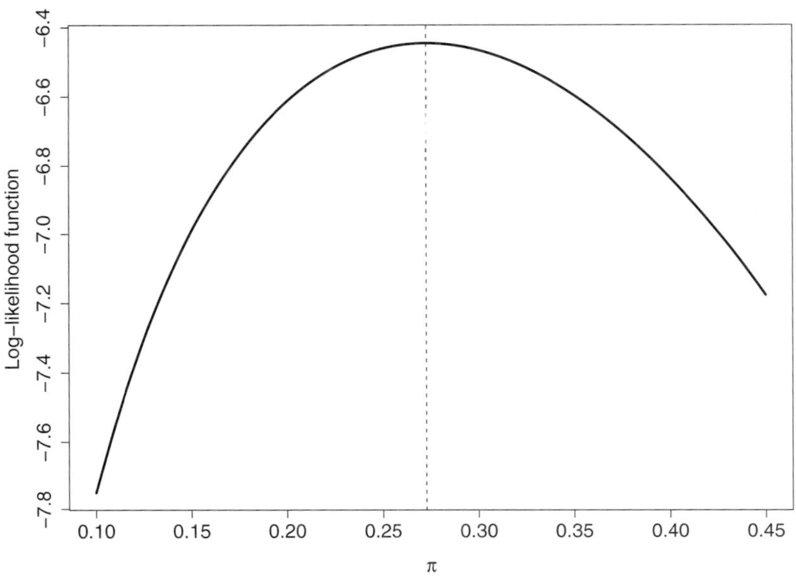

Fig. 2.1. Log-likelihood function for binomial data with three successes and one failure. The vertical line is at the maximum value of the log-likelihood function, the MLE $\widehat{\pi} = 3/11$.

From a classical perspective, inference concerning the value of a parameter is based on the sampling distribution of an estimator. The sampling distribution of an estimator refers to the variation in the estimator when similar samples are repeatedly drawn from the population of interest. In the binomial setting, the sampling distribution of $\widehat{\pi}$ is the distribution of this estimator when repeated binomial samples of size n are drawn from a population with probability of success π. In the context of new launch vehicles manufactured by companies with limited design experience, applying classical inference procedures requires that we imagine ourselves repeatedly identifying 11 new rocket manufacturers, asking each of these manufacturers to design a new rocket, and then testing the 11 new rockets so obtained. For each test of the 11 new rockets, we would calculate the MLE of π and, based on these repeated samples of the MLE, we would estimate its sampling distribution.

In simple statistical models, the sampling distributions of estimators can sometimes be derived analytically. For example, we know that the sample mean of n draws from a normal population with mean μ and standard deviation σ has a normal distribution with mean μ and standard deviation $\sigma/\sqrt{n}$.

Unfortunately, in many situations the sampling distribution of an estimator cannot be derived analytically. In such circumstances, classical inference relies on asymptotic results. These results approximate the sampling distribution of the estimator when the sample size n is "large." In the case of the MLE, there is a convenient asymptotic approximation to the sampling distribution that pertains in most applied settings. For large n, the MLE of a scalar parameter θ, say $\widehat{\theta}$, is approximately normally distributed with mean θ and variance equal to the negative reciprocal of the second derivative of the log-likelihood evaluated at the MLE. We call the value of the second derivative of the log-likelihood evaluated at the MLE the *observed information*, $I(\theta)$. More specifically, we define the observed information of a model and data to be

$$-\left[\frac{\mathrm{d}^2 \log f(\mathbf{y} \mid \theta)}{\mathrm{d}\theta^2}\right]_{\theta=\widehat{\theta}}, \tag{2.5}$$

evaluated at $\theta = \widehat{\theta}$.

In the case of a binomial probability, the second derivative of the log-likelihood is

$$\frac{\mathrm{d}^2 \log f(y \mid n, \pi)}{\mathrm{d}\pi^2} = -\frac{y}{\pi^2} - \frac{n-y}{(1-\pi)^2}.$$

Substituting the MLE $\widehat{\pi}$ into this expression and taking the reciprocal and square root, we estimate the standard deviation of the MLE as

$$\sqrt{\frac{\widehat{\pi}(1-\widehat{\pi})}{n}}.$$

For binomial data, the MLE $\widehat{\pi}$ is asymptotically normally distributed with mean π and standard deviation $\sqrt{\widehat{\pi}(1-\widehat{\pi})/n}$. The standard deviation of an estimator is usually called its standard error (se), although this distinction is often not made when conducting a Bayesian analysis.

2.1.2 Classical Point and Interval Estimation for a Proportion

In the last section, we described the large sample properties of the MLE. Let us now examine how these properties can be used to obtain both a point estimate of a binomial probability π and an interval surrounding this estimate that will contain the true value of the probability of a specified proportion of the time in repeated sampling. Such an interval is called a *confidence interval*.

Example 2.1 Calculating a confidence interval for the launch vehicle data. For the launch vehicle data in Table 2.1, a point estimate of the failure probability of a new launch system developed by an inexperienced manufacturer is provided by the MLE:

$$\widehat{\pi} = \frac{y}{n} = \frac{3}{11} = 0.272.$$

The standard error for this estimate is

$$se(\widehat{\pi}) = \sqrt{\frac{\widehat{\pi}(1-\widehat{\pi})}{n}} = \sqrt{\frac{0.272(1-0.272)}{11}} = 0.134.$$

An interval estimate for the population proportion π can be obtained using the asymptotic normal sampling distribution of the MLE $\widehat{\pi}$. In this case, the standardized statistic

$$\frac{\widehat{\pi}-\pi}{se(\widehat{\pi})}=\frac{\widehat{\pi}-\pi}{\sqrt{\frac{\widehat{\pi}(1-\widehat{\pi})}{n}}}$$

is approximately distributed as a standard normal. It follows that an approximate $(1-\alpha)\times 100\%$ confidence interval for π is given by

$$(\widehat{\pi}-z_{\alpha/2}\,se(\widehat{\pi}),\widehat{\pi}+z_{\alpha/2}\,se(\widehat{\pi})),$$

where $z_{\alpha/2}$ is the $\alpha/2$ quantile of the standard normal distribution. In this example, if $\alpha=0.10$, then the approximate 90% confidence interval for π is

$$(0.272-1.645\times 0.134, 0.272+1.645\times 0.134)=(0.052, 0.492).$$

The "confidence" of this interval is a reflection of the initial probability statement about the sampling distribution of $\widehat{\pi}$. In repeated sampling, we expect the random interval $[\widehat{\pi}-z_{\alpha/2}\,se(\widehat{\pi}),\widehat{\pi}+z_{\alpha/2}\,se(\widehat{\pi})]$ to include the unknown parameter π with probability close to $(1-\alpha)$.

In this simple setting, the exact confidence interval can also be calculated by finding values of π for which more than two successes or fewer than four successes would be observed in 5% of samples, respectively. This leads to an exact 90% confidence interval of (0.079, 0.564).

Consistency and efficiency are properties of estimators most relevant for estimation when sample sizes are large. However, because sample sizes available for estimation are never infinite, inference for small-to-moderate sample sizes is also important. In this regard, the large sample properties of the MLE do not pertain in more complicated settings. For example, the MLE is not appropriate in hierarchical models (Chap. 3), or when parameter values fall close to the boundary of the parameter space. Finally, deriving analytic expressions for the MLE is often difficult in high-dimensional settings, and obtaining an analytic expression for the information matrix required for inference may not be feasible. The magnitude of this problem is indicated by the large volume of literature devoted to addressing these problems (see, for example, Meeker and Escobar (1998)). As we demonstrate throughout the remainder of this book, many of these difficulties can be avoided or eliminated by adopting a Bayesian approach toward inference.

2.2 Fundamentals of Bayesian Inference

Bayesian inference is based on the subjective view of probability. Rather than specifying a rule for constructing an interval that will contain the true value of a parameter a specified proportion of the time in an infinite sequence of repetitions of the same experiment, Bayesians combine knowledge of a parameter available before data are analyzed with information collected during an experiment to update their belief about the value of the parameter after the experiment has been completed. Bayesians summarize knowledge of the parameter after seeing the results of an experiment using a probability density function. The use of a probability density function to summarize uncertainty about the value of a parameter does not mean that we believe that values of unknown parameters are random; it only means that our knowledge of a parameter's value is uncertain, and that our uncertainty about this value can be represented using an appropriate probability density function.

The mechanism for updating probability density functions that represent uncertainty about the value of a parameter is Bayes' Theorem. Mathematically, Bayes' Theorem can be expressed

$$p(\boldsymbol{\theta} \mid \mathbf{y}) = \frac{f(\mathbf{y} \mid \boldsymbol{\theta}) p(\boldsymbol{\theta})}{m(\mathbf{y})}, \tag{2.6}$$

where

$$m(\mathbf{y}) = \int f(\mathbf{y} \mid \boldsymbol{\theta}) p(\boldsymbol{\theta}) \, d\boldsymbol{\theta}.$$

The function $p(\boldsymbol{\theta} \mid \mathbf{y})$ is called the *posterior density*, $p(\boldsymbol{\theta})$ is called the *prior density*, $m(\mathbf{y})$ is the *marginal density* of the data, and, as already noted, $f(\mathbf{y} \mid \boldsymbol{\theta})$ is the sampling density of the data. Notationally, we use $p(\cdot)$ and $p(\cdot \mid \cdot)$ to denote a generic density or conditional density function; the arguments supplied to p determine the particular density function to which reference is made.

We now explore each of the components of Bayes' Theorem in more detail, beginning with the prior density.

2.2.1 The Prior Distribution

In the launch vehicle example, the parameter of interest is the value of the success probability π. Conceivably, π could be any value in the interval (0,1). Within the Bayesian paradigm, we specify prior information regarding the value of this parameter (information that is available before analyzing the experimental data) by using a probability density function on the unit interval. This probability density is called the prior density, since it reflects information about π prior to observing experimental data. In practice, the distribution used to reflect prior information may be dispersed, reflecting the fact that little is known about the parameter, or it may be concentrated in a particular region of the parameter space, reflecting the fact that more specific information is available. In the former case, the prior distribution is sometimes called *diffuse*, *noninformative*, or *vague*; in the latter, it is called *informative*.

To illustrate the role of the prior distribution, suppose that comparatively little information about π is available before data are collected. A priori, we might then suppose that all values of π between 0 and 1 are equally plausible. We might summarize this type of information by assuming that the prior distribution for π is uniform on the unit interval, that is,

$$p(\pi) = 1, \qquad 0 < \pi < 1,$$

as graphed in Fig. 2.2. We can use this prior distribution to compute the prior probabilities that π falls in any subinterval of (0,1). For example, $\mathbf{P}(\pi < 0.25) = \mathbf{P}(\pi > 0.75) = 0.25$. The uniform prior distribution reflects the prior belief that the unknown proportion is as likely to be small (for example, less than 0.25) as large (greater than 0.75). This prior distribution is an example of a diffuse prior since it reflects a lack of precise prior information about the true value of the proportion.

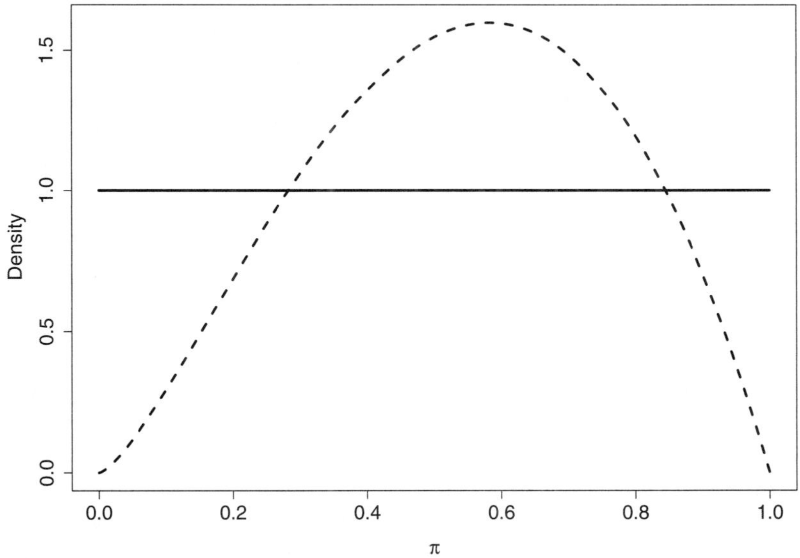

Fig. 2.2. Two prior densities for a proportion: uniform density (solid line) that reflects diffuse prior beliefs and $Beta(2.4, 2)$ prior density (dashed line) reflecting prior information that the true proportion is likely to be close to 0.55, but not arbitrarily close to either 0 or 1.

Alternatively, we might use experience from vehicles launched prior to 1980 to specify an informative prior distribution for the success probabilities of post-1980 launch vehicles. Although engineering practice has advanced rapidly since early launches, so too has the complexity and size of launch platforms. As a consequence of this balance between increased complexity and improved engineering practice, the success of early launch vehicles might still provide a useful baseline for the success of the more modern vehicles.

With these considerations in mind, we could reasonably allow a prior distribution for the success of new launch vehicles based on data collected prior to 1980 to concentrate its mass around 0.55, which is not too far from the observed proportion of successful launches observed before 1980. Figure 2.2 displays one prior density that we might use to represent this prior information. The distribution depicted in this plot is a beta density with parameters $\alpha = 2.4$ and $\beta = 2$. The probability density function of a beta distribution with parameters α and β, denoted by $Beta(\alpha, \beta)$, is

$$p(\pi \mid \alpha, \beta) = \frac{\Gamma(\alpha + \beta)}{\Gamma(\alpha)\Gamma(\beta)} \pi^{\alpha-1}(1-\pi)^{\beta-1}, \qquad 0 \leq \pi \leq 1, \quad \alpha, \beta > 0.$$

Note that most of the mass of this prior distribution lies between 0.2 and 0.8, and the prior distribution has a mean of $\alpha/(\alpha + \beta) = 0.545$ and a mode of $(\alpha - 1)/(\alpha + \beta - 2) = 0.583$. The probabilities that π falls in the intervals (0, 0.25) and (0.75, 1) for this prior density are 0.10 and 0.20, respectively. This prior distribution concentrates more of its mass around values of π near 0.5, and assigns a prior weight of 0.7 to the central interval $(0.25, 0.75)$. By comparison, the uniform prior distribution assigns mass 0.5 to this interval. Finally, note that the uniform distribution is also a beta distribution, but with parameters $\alpha = \beta = 1$.

2.2.2 Combining Data with Prior Information

The prior distribution $p(\theta)$ reflects knowledge of a parameter before data are analyzed. Once data are obtained, the prior distribution is updated using the new information. The updated probability distribution on the parameter of interest is called the *posterior distribution*, because it reflects probability beliefs posterior to, or after, seeing the data.

According to Bayes' Theorem, the posterior distribution is computed (up to a proportionality constant) by multiplying the likelihood function by the prior distribution. We can reexpress Eq. 2.6 to obtain the following general updating strategy:

$$\text{posterior} \propto \text{likelihood} \times \text{prior}.$$

In this context, the proportionality constant absorbs the term $m(\mathbf{y})$ of Eq. 2.6 as it does not depend on the model parameter θ. In terms of probability density functions then,

$$p(\theta \mid \mathbf{y}) \propto f(\mathbf{y} \mid \theta)\, p(\theta). \tag{2.7}$$

Example 2.2 Calculating posterior distributions for the launch vehicle failure data. We discussed two prior distributions in the previous section: a uniform $Beta(1, 1)$ prior distribution and a more informative $Beta(2.4, 2)$ prior distribution. Substituting the uniform prior distribution for $p(\pi)$ in

Eq. 2.7 leads to a posterior distribution for the launch vehicle success probability that is proportional to

$$\begin{aligned} p(\pi \mid y) &\propto \pi^3(1-\pi)^8 \times \pi^{1-1}(1-\pi)^{1-1} \\ &\propto \pi^{4-1}(1-\pi)^{9-1}. \end{aligned}$$

Examining the last expression in this formula, we see that the posterior distribution is proportional to yet another beta distribution. Because the posterior density is a beta distribution, we can analytically determine the constant of proportionality (also called the normalizing constant), which in this case is equal to $\Gamma(13)/[\Gamma(4)\Gamma(9)]$. Figure 2.3 provides a plot of this (normalized) posterior distribution, based on data in Table 2.1 and a uniform prior distribution.

We can apply a similar procedure to obtain the posterior distribution based on the same launch failure data and the informative $Beta(2.4, 2)$ prior distribution. Again applying Eq. 2.7, we obtain

$$\begin{aligned} p(\pi \mid y) &\propto \pi^3(1-\pi)^8 \times \pi^{2.4-1}(1-\pi)^{2-1} \\ &= \pi^{5.4-1}(1-\pi)^{10-1}. \end{aligned}$$

Once again, the posterior distribution is proportional to a beta distribution, so we can analytically determine its normalizing constant. In this case, it is equal to $\Gamma(15.4)/[\Gamma(5.4)\Gamma(10)]$. Figure 2.3 depicts a plot of this normalized posterior distribution.

Both posterior distributions in Example 2.2 turned out to be beta distributions. The prior distribution is a beta distribution. While the sampling distribution is binomial, the corresponding likelihood function is proportional to a beta distribution. This model is called the *beta-binomial model*. Prior distributions that take the same functional form as the posterior distribution are called *conjugate prior distributions*. In simple problems, conjugate prior distributions can make posterior analysis easy because they eliminate the need to numerically determine normalizing constants. Of course, prior distributions should not be specified simply for computational convenience, and if a conjugate prior distribution that provides an adequate representation of information available before the conduct of an experiment cannot be found, nonconjugate prior distributions should be used instead. We explore numerical techniques for handling nonconjugate prior distributions, or conjugate prior distributions that do not lead to posterior distributions of a form that admit simple analytical treatments, in Chap. 3.

Returning to the analysis of the launch failure probabilities, we found that the posterior distributions depicted in Fig. 2.3 represent all available information about π after both prior information and experimental data are combined. As Bayesians, we base all inferences about the success probability π on such distributions. For example, if we want to determine the probability that π falls in any particular interval, we compute the area under the

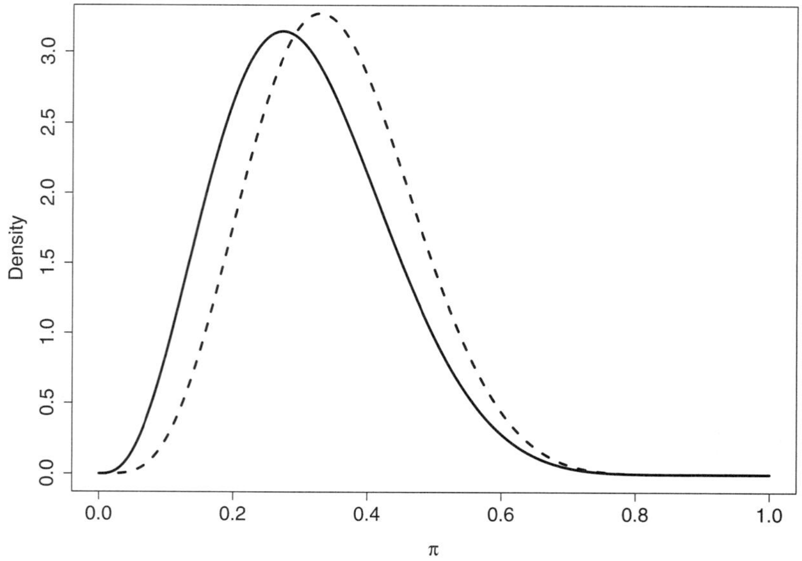

Fig. 2.3. Posterior distributions for launch vehicle success probabilities: the posterior distribution (solid line) resulting from the observation of three successes and eight failures after the assumption of a uniform prior distribution, and the posterior distribution (dashed line) based on the same data but under the assumption of a $Beta(2.4, 2)$ prior distribution.

posterior distribution within that interval. For example, if the uniform prior distribution represents available prior knowledge, then we obtain the posterior probability that π falls in the interval $(0.2, 0.5)$ by integrating a $Beta(4, 9)$ distribution over this range. The result is 0.72. If we judge that an informative $Beta(2.4, 2)$ distribution adequately represents our prior knowledge, then we obtain the posterior probability assigned to the interval $(0.2, 0.5)$ by integrating a $Beta(5.4, 10)$ distribution over this region, which yields 0.79.

These intervals are the Bayesian analogues of classical confidence intervals and are called *posterior probability intervals* or *posterior credible intervals*. Frequently, we summarize the posterior distribution using a central $(1-\alpha)\times 100\%$ interval, which is a range of values having $(\alpha/2)\times 100\%$ of the posterior probability above and below the endpoints. For our example, with a $Beta(5.4, 10)$ posterior distribution, the central 95% posterior credible interval is $(0.14, 0.60)$. Another alternative is the $(1-\alpha)\times 100\%$ highest posterior density interval. This interval both contains $(1-\alpha)\times 100\%$ of the posterior probability and has the property that the density within the region is never lower than the density outside. For our example, with a $Beta(5.4, 10)$ posterior distribution, the 95% highest posterior density interval is $(0.13, 0.58)$.

There are three Bayesian analogues of the MLE, which is a classical point estimator. The first is the *maximum a posteriori* estimate, or MAP estimate. This estimate corresponds to the point in the parameter space at which the posterior density function achieves its maximum. The MAP estimates corresponding to the two choices of prior distributions above are 0.272 and 0.328. The second and most commonly reported Bayesian point estimator is the *posterior mean*, determined as the first moment of the posterior distribution. Mathematically, the posterior mean of the binomial success probability can be expressed as

$$\mathrm{E}(\pi \mid y) = \int_0^1 \pi \, p(\pi \mid y) \, \mathrm{d}\pi.$$

Because the mean of a $Beta(\alpha, \beta)$ distribution is $\alpha/(\alpha+\beta)$, the posterior means corresponding to the two choices of prior distributions above are 0.307 and 0.351, respectively. The third popular Bayesian point estimator is the *posterior median*. The posterior median, $\tilde{\pi}$, satisfies the equation

$$\int_0^{\tilde{\pi}} p(\pi \mid y) \mathrm{d}\pi = 0.5.$$

The posterior medians corresponding to the two choices of prior distributions above are 0.298 and 0.344.

One way we can understand the combination of information from the prior distribution and the data is through the notion of shrinkage. The mean of a beta distribution with parameters α and β is

$$\frac{\alpha}{\alpha+\beta}.$$

Based on y successes and $n-y$ failures, the posterior mean is thus

$$\mathrm{E}(\pi \mid y) = \frac{\alpha+y}{\alpha+\beta+n},$$

which we can reexpress as

$$\mathrm{E}(\pi \mid y) = w\frac{\alpha}{\alpha+\beta} + (1-w)\frac{y}{n},$$

where $\frac{\alpha}{\alpha+\beta}$ is the prior mean for the proportion of successes, y/n is the proportion of successes in the sample, and

$$w = \frac{\alpha+\beta}{\alpha+\beta+n} \tag{2.8}$$

is a fraction between 0 and 1. The posterior mean can be called a *shrinkage estimate* because it moves the observed proportion of successes y/n toward the prior mean $\alpha/(\alpha+\beta)$. The degree of shrinkage is controlled by the fraction w. The value of this fraction depends on the relative size of $(\alpha+\beta)$ to the sample size n. For this reason, we can think of $\alpha+\beta$ as a *prior sample size*, or the number of observations afforded to the prior distribution in determining the posterior mean.

Equation 2.8 also provides insight into the large sample properties of the posterior mean. From this equation, we see that the weight given to the prior mean decreases to 0 as the number n of experimental units becomes large. That is, for large samples, the prior mean loses its impact on the posterior mean. Similar comments apply also to the posterior distribution. For binomial data with a $Beta(\alpha, \beta)$ prior distribution, the posterior distribution is $Beta(y + \alpha, n - y + \beta)$. When y and $n - y$ are large, the difference between this distribution and a $Beta(y, n - y)$ distribution becomes small, and so the influence of the prior distribution on the posterior distribution diminishes. Furthermore, for large values of y and $n - y$, a $Beta(y, n - y)$ distribution looks very much like a normal distribution. Figure 2.4 illustrates the similarity between a $Beta(30, 80)$ distribution (with 10 times more successes and failures than observed in the launch vehicle data) and an approximating normal distribution with the same mean and variance.

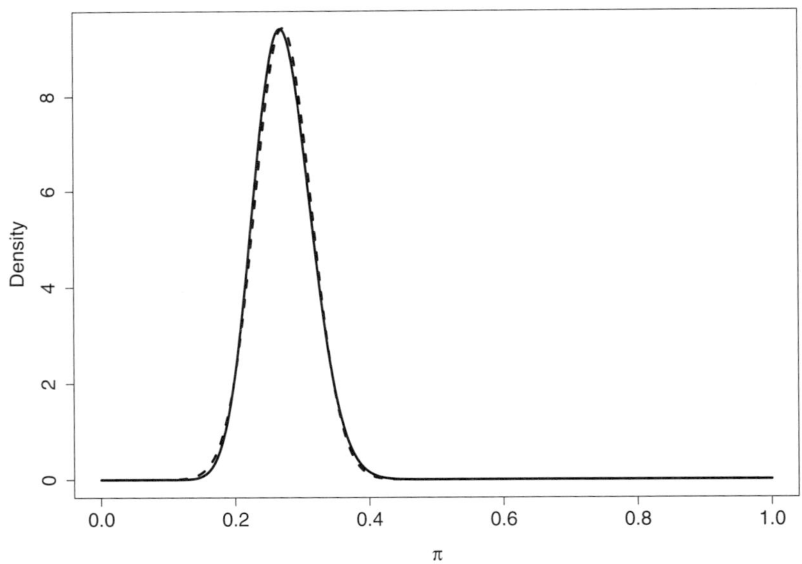

Fig. 2.4. Beta distribution (solid line) when number of failures and successes is large versus approximating normal distribution (dashed line).

The similarities displayed in Fig. 2.4 between the exact posterior distribution and the approximating normal distribution typify the large sample properties of posterior distributions. Loosely speaking, if we choose prior distributions so that they assign nonnegligible mass to the region surrounding the true value of a parameter, then the posterior distribution will converge to a normal (Gaussian) distribution centered on the MLE. Similarly, the variance of the posterior distribution converges to the inverse of the information matrix. In this sense, the asymptotic (large sample) properties of the MLE and the posterior distribution are similar. Of course, their probabilistic interpretations are different.

2.3 Prediction

So far, we have concentrated on inference regarding the true value of a parameter. However, in many situations, the goal of an analysis is to predict values of a future sample. For example, in the case of failure probabilities of launch vehicles, the FAA and USAF actually need to estimate the number of new launch vehicles that will succeed in, say, m future launches scheduled during the next calendar year.

If we knew the success probability for the launch of a new vehicle, π, this problem would be simple; the corresponding binomial distribution would exactly predict the distribution on the number of failures in future launches. The probability of observing $m-z$ failures in m future launches of new vehicles would be equal to the probability of observing z successes, or

$$f(z \mid \pi) = \binom{m}{z} \pi^z (1-\pi)^{m-z}.$$

In practice, however, we do not know the value of π. We only know its posterior distribution. In this case, the *predictive probability* of z (for a future sample of size m), given a posterior distribution on π based on past data y, is given by the integral

$$p(z \mid y) = \int_0^1 f(z \mid \pi) p(\pi \mid y) \mathrm{d}\pi, \qquad z = 0, \ldots, m. \tag{2.9}$$

In essence, by integrating the sampling distribution $f(z \mid \pi)$ over the posterior distribution on the parameter π, we average over the uncertainty in this parameter. The predictive distribution provides a full account for the uncertainty in the unknown parameter, in this case π.

Example 2.3 Calculating the predictive distribution for the number of successes of a new launch vehicle. Assume that a uniform distribution is used to model the prior distribution on the launch vehicle success probability. With this prior distribution, we previously found that the posterior distribution on π was a $Beta(4, 9)$ distribution. Now suppose that five additional launches of new vehicles are scheduled over the next year.

The posterior mean of π under these model assumptions is $4/13 = 0.307$. If we were to assume that the true value of π was exactly equal to this value, then the probabilities of observing z successes, for $z = 0, \ldots, 5$, are as displayed in the first row of Table 2.2. From this table, we see that the most likely number of successes to be observed is 1, and this event is predicted to occur with probability 0.354, or a little greater than 1/3.

Table 2.2. Comparison of the plug-in predictions for the number of successes in five future launches of new launch vehicles to the corresponding formal estimates of the predictive probabilities obtained by integrating over the posterior distribution of π

	0	1	2	3	4	5
Plug-in: $\pi = 0.307$	0.160	0.354	0.314	0.139	0.031	0.003
Predictive	0.208	0.320	0.267	0.145	0.051	0.009

In contrast, the second row of Table 2.2 displays the predicted probabilities of observing each number of successes obtained using Eq. 2.9. As expected, the predictive distribution is more dispersed in that the modal values of 1 and 2 successes are assigned less probability by the predictive distribution than they are using the plug-in estimate. On the other hand, more probability is distributed in the tails of the distribution, reflecting the fact that more extreme numbers of successes are more likely when the success probability π is not assumed to be known exactly.

Equation 2.9 extends directly to more general settings. If $\boldsymbol{\theta}$ denotes a generic parameter, $p(\boldsymbol{\theta} \,|\, \mathbf{y})$ its posterior distribution based on the data vector $\mathbf{y}$, and $f(\mathbf{z} \,|\, \boldsymbol{\theta})$ the sampling distribution of $\mathbf{z}$ given $\boldsymbol{\theta}$, then the predictive distribution for $\mathbf{z}$ may be expressed as

$$p(\mathbf{z} \,|\, \mathbf{y}) = \int_{\Theta} f(\mathbf{z} \,|\, \boldsymbol{\theta})\, p(\boldsymbol{\theta} \,|\, \mathbf{y}) d\boldsymbol{\theta}.$$

In most cases, the areas under predictive densities must be evaluated numerically, but, as we demonstrate in Chap. 3, this typically presents little difficulty when using modern Markov chain Monte Carlo (MCMC) algorithms.

2.4 The Marginal Distribution of the Data and Bayes' Factors

When only inference regarding the value of an unknown parameter is performed, the denominator of the term on the right-hand side of Eq. 2.6 — the marginal distribution of the data — is often overlooked. After all, the numerator in this expression represents an unnormalized density, so for purposes of inference the denominator, the marginal distribution, represents little more than a normalizing constant.

The situation changes when we turn our attention from parameter estimation and prediction and instead focus on model selection. In this setting, the marginal distribution $m(\mathbf{y})$ plays a critical role in Bayesian inference. To see why, suppose that two probability models, say M_1 and M_2, with sampling and prior distributions $f_1(\mathbf{y} \mid \boldsymbol{\theta}_1, M_1), p_1(\boldsymbol{\theta}_1 \mid M_1)$ and $f_2(\mathbf{y} \mid \boldsymbol{\theta}_2, M_2), p_2(\boldsymbol{\theta}_2 \mid M_2)$, respectively, are entertained as approximations to the process underlying an observed set of data $\mathbf{y}$. (Note that $\boldsymbol{\theta}_1$ and $\boldsymbol{\theta}_2$ need not be defined on the same parameter space.) Let $\mathbf{P}(M_1)$ denote the prior probability assigned to the first model, and let $\mathbf{P}(M_2) = 1 - \mathbf{P}(M_1)$ denote the prior probability assigned to the second model. Then the *posterior odds* that model M_1 is true are

$$\begin{aligned}\frac{\mathbf{P}[M_1 \mid \mathbf{y}]}{\mathbf{P}[M_2 \mid \mathbf{y}]} &= \frac{\mathbf{P}(\mathbf{y} \mid M_1)\mathbf{P}(M_1)}{\mathbf{P}(\mathbf{y} \mid M_2)\mathbf{P}(M_2)} \\ &= \frac{\mathbf{P}(M_1) \int_{\Theta_1} f_1(\mathbf{y} \mid \boldsymbol{\theta}_1, M_1) p(\boldsymbol{\theta}_1 \mid M_1) \mathrm{d}\theta_1}{\mathbf{P}(M_2) \int_{\Theta_2} f_2(\mathbf{y} \mid \boldsymbol{\theta}_2, M_2) p(\boldsymbol{\theta}_2 \mid M_2) \mathrm{d}\theta_2} \\ &= \frac{\mathbf{P}(M_1)}{\mathbf{P}(M_2)} \times \frac{m_1(\mathbf{y} \mid M_1)}{m_2(\mathbf{y} \mid M_2)}. \end{aligned} \tag{2.10}$$

That is,

$$\text{Posterior odds} = \text{Prior odds} \times \text{Bayes' factor},$$

where the *Bayes' factor* is defined to be the ratio of the marginal densities of the data under the two models considered.

As is apparent from Eq. 2.10, Bayesian model selection and testing differs drastically from classical significance testing. In classical testing, a test statistic is chosen and a p-value or significance level is defined by computing the probability that a test statistic is more extreme than the value observed. In contrast, Bayesians compute the actual odds that a model is true.

Example 2.4 Calculating the Bayes' factor between the uniform and informative prior models for the launch vehicle data. Consider again the two prior distributions assumed for the launch vehicle failure data. Under the uniform prior distribution (M1), the marginal probability of observing 3 successes in 11 launches can be computed as

$$\begin{aligned} m_1(3 \mid M_1) &= \int_0^1 \binom{11}{3} \pi^3 (1-\pi)^8 \mathrm{d}\pi \\ &= \binom{11}{3} \frac{\Gamma(4)\Gamma(9)}{\Gamma(13)} \int_0^1 \frac{\Gamma(13)}{\Gamma(4)\Gamma(9)} \pi^{4-1} (1-\pi)^{9-1} \mathrm{d}\pi \\ &= \frac{11!}{3!\,8!} \times \frac{3!\,8!}{12!} \times 1 \\ &= \frac{1}{12} \\ &= 0.08333. \end{aligned}$$

As it happens, the marginal probability of all 12 possible outcomes from the 11 launches (i.e., 0, 1, 2, ..., 11 successes) are equally likely under the uniform prior, and so are all assigned a marginal probability of 1/12.

Under the second model (M2), we assumed a *Beta*(2.4, 2) prior distribution. It follows that the marginal probability of observing 3 successes in 11 launches under that model is

$$\begin{aligned} m_2(3 \mid M_2) &= \int_0^1 \binom{11}{3} \pi^3 (1-\pi)^8 \, \pi^{2.4-1} (1-\pi)^{2-1} \mathrm{d}\pi \\ &= \binom{11}{3} \frac{\Gamma(5.4)\Gamma(10)}{\Gamma(15.4)} \int_0^1 \frac{\Gamma(15.4)}{\Gamma(5.4)\Gamma(10)} \pi^{5.4-1} (1-\pi)^{10-1} \mathrm{d}\pi \\ &= \frac{11!}{3!\,8!} \times \frac{\Gamma(5.4) 9!}{\Gamma(15.4)} \times 1 \\ &= 0.01045. \end{aligned}$$

Thus, the Bayes' factor in favor of the first model is 0.08333/0.01045 = 7.97, or about 8 to 1.

If both models are given equal weight a priori, (i.e., $\mathbf{P}(M_1) = \mathbf{P}(M_2) = 0.5$), then the posterior probability that the first model is true is

$$\begin{aligned} \mathbf{P}[M_1 \mid \mathbf{y}] &= \frac{\mathbf{P}(\mathbf{y} \mid M_1)\mathbf{P}(M_1)}{\mathbf{P}(\mathbf{y} \mid M_1)\mathbf{P}(M_1) + \mathbf{P}(\mathbf{y} \mid M_2)\mathbf{P}(M_2)} \\ &= \frac{\mathbf{P}(M_1) m_1(\mathbf{y} \mid M_1)}{\mathbf{P}(M_1) m_1(\mathbf{y} \mid M_1) + \mathbf{P}(M_2) m_2(\mathbf{y} \mid M_2)} \\ &= \frac{(0.5)(0.08333)}{(0.5)(0.08333) + (0.5)(0.01045)} \\ &= 0.89. \end{aligned}$$

Practitioners accustomed to classical testing procedures should take special note of the interpretation of the Bayes' factors derived in Example 2.4. The probability 0.89 is the posterior probability that the model employing the uniform prior distribution is true, given that one of the two models is true and assuming that both models are given equal prior weight. This statement differs at a fundamental, philosophical level from statements made in classical hypothesis testing. Under the classical paradigm, probabilities regarding the truth of a model are not cited. Instead, probability statements made from within the classical paradigm refer only to the probability of observing a test statistic *more* extreme than the one actually observed. Such statements do not directly address the question of whether a particular model is true.

There is an important proviso regarding the use of Bayes' factors for model testing: Bayes' factors are only defined when proper prior distributions are used. A *proper prior* integrates to one. Both of the prior distributions that we used for launch vehicle success probabilities were proper because they were both beta densities. The next section provides an example of a model that uses an improper prior distribution.

2.5 A Lognormal Example

We now turn to a model that involves continuous-valued random variables. The particular dataset we consider represents viscosity breakdown times for 50 samples of a lubricant. Table 2.3 lists individual values of the breakdown times. Figure 2.5 depicts a histogram summary of these values with an overlaid kernel density estimate of the same data. A *kernel density estimate* is a smoothed version of a histogram. Heuristically, it is computed by placing a bubble, or kernel, on top of each of the n data points, and then adding up the bubbles to calculate the height of the density estimate at each point on the horizontal axis. Typically, the kernels are versions of a probability density function that are rescaled to give each point $1/n$ mass.

Table 2.3. Viscosity breakdown times, in thousands of hours, for 50 samples of a lubricating fluid

5.45	16.46	15.70	10.39	6.71	3.77	7.42	6.89	9.45	5.89
7.39	5.61	16.55	12.63	8.18	10.44	6.03	13.96	5.19	10.96
14.73	6.21	5.69	8.18	4.49	3.71	5.84	10.97	6.81	10.16
4.34	9.81	4.30	8.91	10.07	5.85	4.95	7.30	4.81	8.44
6.56	9.40	11.29	12.04	1.24	3.45	11.28	6.64	5.74	6.79

Figure 2.5 clearly shows that the right-hand tail of the data distribution extends farther away from the center of the distribution than does the left-hand tail. That is, there are more large extreme values than there are small extreme values. We call this tendency for data values to be more spread out on the right right-skewness; it is a common feature of data that assumes only positive values.

Many statistical analyses become easier when data are not skewed, for then a normal, or Gaussian distribution, can serve as an appropriate model. To remove skewness from positive data, it is common to transform data to the logarithmic scale. If t_i denotes the ith fluid breakdown time, we can define the transformed variables y_i as

$$y_i = \log(t_i), \qquad i = 1, \ldots, 50.$$

Figure 2.6 displays the histogram and density estimates of the transformed variables y_i. Note the more symmetrical, bell-shaped appearance of these plots. On this scale, the data can reasonably be modeled as arising from a normal distribution.

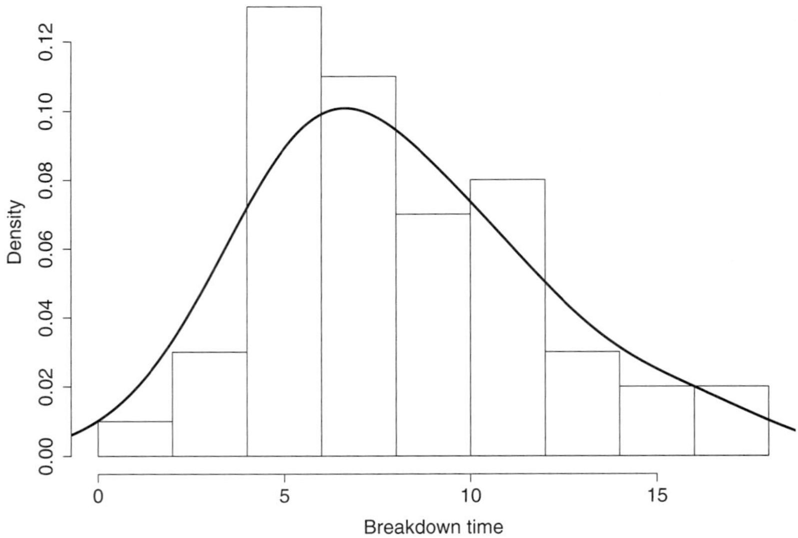

Fig. 2.5. Histogram and density estimate of fluid breakdown times.

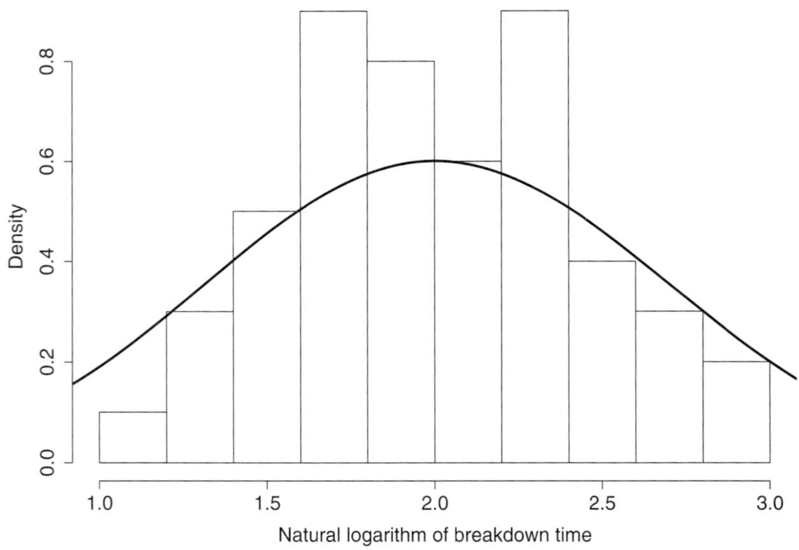

Fig. 2.6. Histogram and density estimate of the natural logarithm of the fluid breakdown times.

The probability density function for a normally distributed random variable is

$$f(y \mid \mu, \sigma^2) = \frac{1}{\sqrt{2\pi\sigma^2}} \exp\left[-\frac{1}{2\sigma^2}(y-\mu)^2\right],$$

where the expected value of Y is μ and the variance of Y is σ^2. Notice that this implies that $T_i \sim LogNormal(\mu, \sigma^2)$.

If we assume that the logarithms of the fluid breakdown times are conditionally independent given μ and σ^2, then the likelihood function for the natural logarithm of the $n = 50$ values displayed in Table 2.3 is given by

$$\prod_{i=1}^{n} \frac{1}{\sqrt{2\pi\sigma^2}} \exp\left[-\frac{1}{2\sigma^2}(y_i-\mu)^2\right]. \tag{2.11}$$

To obtain a joint posterior distribution for the values of μ and σ^2, we must combine this sampling distribution with a prior distribution via Bayes' Theorem. Three general choices of prior distribution are available for this purpose: the standard noninformative prior distribution; an informative, conjugate normal-inverse gamma prior distribution; or a more general informative prior distribution. We consider the first two choices in detail in this section. Discussion of the third is left to the reader as Exercise 2.3.

Analysis with a Noninformative Prior Distribution

We specify the default, noninformative prior distribution for modeling an unknown mean and variance for normally distributed data as being proportional to

$$p(\mu, \sigma^2) \propto \frac{1}{\sigma^2}. \tag{2.12}$$

In specifying this prior distribution, we regard σ^2, rather than σ, as the second model parameter.

Justifying this prior distribution falls beyond our immediate scope. In general, however, noninformative prior distributions are obtained by determining a scale for a parameter on which the likelihood is approximately "data translated," and then taking a uniform prior distribution on that scale. Holding σ^2 fixed, the normal likelihood in Eq. 2.11 maintains the same shape as $\bar{y}$, the sufficient statistic for μ, is shifted. Thus, the noninformative prior distribution for μ is uniform on the real line. If μ is regarded as known, the likelihood function maintains the same shape on the logarithmic scale as $\sum_i (y_i - \mu)^2$, the sufficient statistic for σ^2, is varied. Transforming from a uniform prior distribution on the $\log(\sigma^2)$ scale back to the original scale yields the prior distribution given in Eq. 2.12. Further details on the specification and motivation for noninformative prior distributions may be found in Box and Tiao (1973).

It is important to note that the noninformative prior distribution that we specified in Eq. 2.12 is not integrable on the positive real line. Because it is not integrable, it cannot be normalized to have unit area, and so it is not proportional to any real probability density function. It is thus called an *improper prior*. Improper prior distributions can be regarded as convenient approximations to real (proper) prior distributions, provided that they yield (when multiplied by a likelihood function) a posterior distribution that is integrable. To avoid nonintegrable posterior distributions, we recommend the use of proper prior distributions.

Multiplying the likelihood function specified in Eq. 2.11 by the prior distribution given in Eq. 2.12 results in a joint posterior distribution for (μ, σ^2) that is proportional to

$$p(\mu, \sigma^2 \,|\, \mathbf{y}) \propto (\sigma^2)^{-n/2-1} \exp\left[\sum_{i=1}^{n} -\frac{1}{2\sigma^2}(y_i - \mu)^2\right], \tag{2.13}$$

which is a proper density function.

In contrast to the beta-binomial model discussed for a success probability in the last section, the posterior distribution in Eq. 2.13 contains two unknown parameters. The existence of a second parameter complicates matters and makes inference for either parameter more difficult, regardless of which parameter is of interest. In general, we call parameters that are not of direct interest, but which appear in the posterior distribution, *nuisance* parameters. If, for example, we are interested in performing inference on the mean of the logarithmic breakdown times μ, then σ^2 would be considered a nuisance parameter.

From the Bayesian perspective, nuisance parameters can be handled in a straightforward way. They are simply integrated out of the posterior distribution to obtain the *marginal posterior distribution* on the parameters of interest.

Suppose then that μ is the parameter of interest in Eq. 2.13. To obtain the marginal posterior distribution on μ, we must therefore integrate out σ^2. That is, we must compute

$$p(\mu \,|\, \mathbf{y}) \propto \int_0^\infty (\sigma^2)^{-n/2-1} \exp\left[\sum_{i=1}^{n} -\frac{1}{2\sigma^2}(y_i - \mu)^2\right] d\sigma^2. \tag{2.14}$$

Examining the integrand in Eq. 2.14 and regarding it as a function of σ^2, we see that the integrand takes the form of an (unnormalized) inverse gamma probability density function. Because the integral of an inverse gamma probability density function is one, we can thus rewrite this equation as

$$\begin{aligned} p(\mu \,|\, \mathbf{y}) \propto \frac{\Gamma(n/2)}{\left(\sum_{i=1}^{n}(y_i - \mu)^2/2\right)^{n/2}} \int_0^\infty \frac{\left(\sum_{i=1}^{n}(y_i - \mu)^2/2\right)^{n/2}}{\Gamma(n/2)} (\sigma^2)^{-n/2-1} \\ \times \exp\left[\sum_{i=1}^{n} -\frac{1}{2\sigma^2}(y_i - \mu)^2\right] d\sigma^2 \end{aligned}$$

$$= \frac{\Gamma(n/2)}{\left(\sum_{i=1}^{n}(y_i-\mu)^2/2\right)^{n/2}}.$$

Noting that $\sum_1^n(y_i-\mu)^2 = n(\bar{y}-\mu)^2+\sum_1^n(y_i-\bar{y})^2$ and rearranging terms, we find that

$$p(\mu\,|\,\mathbf{y}) \propto \left[1+\frac{(\mu-\bar{y})^2}{(n-1)\frac{s^2}{n}}\right]^{-n/2}, \tag{2.15}$$

where $\bar{y} = \frac{1}{n}\sum_{i=1}^n y_i$ and $s^2 = \sum_{i=1}^n(y_i-\bar{y})^2/(n-1)$. Thus, the marginal posterior distribution of μ has the form of a Student's t distribution having $n-1$ degrees of freedom with mean $\bar{y}$ $(n>1)$ and scale parameter s^2/n.

This result is similar to the result obtained from a classical statistics perspective, except for its interpretation. From the classical perspective, the sampling distribution of $\bar{y}$ has a Student's t distribution centered on μ, and confidence intervals are determined by forming a pivotal statistic $\sqrt{n}(\bar{y}-\mu)/s$ whose distribution is independent of the parameter values.

From the Bayesian perspective, the marginal posterior distribution of μ has a Student's t distribution centered on $\bar{y}$, and inferences concerning μ are based entirely on this posterior distribution. For example, we can calculate a 95% posterior probability interval for μ from Student's t distribution tables as

$$(\bar{y}+t_{n-1,0.025}s/\sqrt{n}, \bar{y}+t_{n-1,0.975}s/\sqrt{n}),$$

where $t_{n-1,\alpha}$ denotes the αth quantile from a Student's t distribution having $n-1$ degrees of freedom. In this example, $n=50, \bar{y}=2.01$, and $s^2=0.178$. It follows that a 95% posterior probability interval for μ for these data is $(1.89, 2.13)$.

Suppose now that the parameter of interest in Eq. 2.13 is σ^2, and that μ is the nuisance parameter. To obtain the marginal posterior distribution of σ^2, we must integrate out μ. Fortunately, the same trick we used previously to integrate out σ^2 also works for integrating out μ, except that the part of the joint distribution in Eq. 2.13 that depends on μ has the form of a normal distribution rather than the form of an inverse gamma distribution. More specifically, we can write

$$\begin{aligned}
p(\sigma^2\,|\,\mathbf{y}) &\propto \int_{-\infty}^{\infty}(\sigma^2)^{-n/2-1}\exp\left[-\frac{1}{2\sigma^2}\sum_1^n(y_i-\mu)^2\right]d\mu \\
&= \int_{-\infty}^{\infty}(\sigma^2)^{-n/2-1}\exp\left[-\frac{(\bar{y}-\mu)^2}{2\sigma^2/n}-\frac{(n-1)s^2}{2\sigma^2}\right]d\mu \\
&\propto (\sigma^2)^{-n/2-1}\exp\left[-\frac{(n-1)s^2}{2\sigma^2}\right]\sqrt{2\pi\sigma^2/n} \\
&\times \int_{-\infty}^{\infty}\frac{1}{\sqrt{2\pi\sigma^2/n}}\exp\left[-\frac{(\bar{y}-\mu)^2}{2\sigma^2/n}\right]d\mu \\
&\propto (\sigma^2)^{-(n-1)/2-1}\exp\left[-\frac{(n-1)s^2}{2\sigma^2}\right].
\end{aligned} \tag{2.16}$$

The function displayed in Eq. 2.16 is proportional to an inverse gamma probability density function with shape and scale parameters $(n-1)/2$ and $(n-1)s^2/2$, respectively. The marginal distribution of the data is not defined in this model because we used an improper prior distribution. However, the proportionality constant follows from the recognizable form of the probability density function. As was the case for μ, all inferences for σ^2 follow directly from this distribution. Note also the similarity with classical inference, which is based on the fact that the sampling distribution of $(n-1)s^2/\sigma^2$ is $ChiSquared(n-1)$.

Analysis with Conjugate Inverse-Gamma Prior Distributions

We can extend the noninformative analysis described in the preceding section to incorporate informative prior distributions. Again examining Eq. 2.11 and regarding μ and $\mathbf{y}$ as being fixed, we see that the likelihood function for σ^2 has the form of an inverse gamma probability density function. If we were to multiply Eq. 2.11 by an inverse gamma prior distribution, the result would also take the form of an inverse gamma distribution. Thus, an inverse gamma prior distribution of the form

$$p(\sigma^2 \mid \alpha, \lambda) = \frac{\lambda^\alpha}{\Gamma(\alpha)}(\sigma^2)^{-\alpha-1} \exp\left(-\frac{\lambda}{\sigma^2}\right) \tag{2.17}$$

is the conjugate prior distribution for σ^2. Of course, this does not mean that an inverse gamma prior distribution necessarily represents our available prior knowledge of σ^2. However, if by varying the parameters α and λ of an inverse gamma distribution, we can find a distribution that approximately represents our prior belief about σ^2, then posterior inferences are easier from a computational viewpoint.

Turning now to the specification of a prior distribution on μ, we can re-express the likelihood function for μ given σ^2 as

$$\begin{aligned} L(\mu \mid \mathbf{y}, \sigma^2) &\propto \frac{1}{(\sigma^2)^{n/2}} \exp\left[-\frac{1}{2\sigma^2}\left(\sum_i (y_i - \bar{y})^2 + n(\bar{y}-\mu)^2\right)\right], \\ &\propto \exp\left[-\frac{n}{2\sigma^2}(\mu - \bar{y})^2\right]. \end{aligned} \tag{2.18}$$

Viewed as a function of μ, Eq. 2.18 has the form of a normal distribution, in this case with mean $\bar{y}$ and variance σ^2/n.

To complete the specification of the prior distribution for μ and σ^2, we need to specify $p(\mu \mid \sigma^2)$. From Eq. 2.18, $p(\mu \mid \sigma^2)$ must be normal with variance proportional to σ^2. We set the variance equal to σ^2/κ. This form explicitly relates the variance of μ to the variance of the observations. This is not to say that we need to calculate σ^2 or observe data to specify $p(\mu \mid \sigma^2)$. Instead, κ controls how diffuse the prior distribution of μ given σ^2 is with respect to the data. Intuitively, κ plays the role of the prior sample size. The full specification of the prior distribution for μ given σ^2 is

$$p(\mu \,|\, \sigma^2, \kappa, \delta) = \frac{\sqrt{\kappa}}{\sqrt{2\pi\sigma^2}} \exp\left[-\frac{\kappa}{2\sigma^2}(\mu - \delta)^2\right]. \tag{2.19}$$

This corresponds to a prior distribution for μ conditional on σ^2 that is normal with mean δ and variance σ^2/κ. We can, of course, specify other prior distributions when available prior knowledge of μ warrants their use, but in the absence of such prior information, Eq. 2.19 provides a computationally convenient and relatively flexible mechanism for incorporating prior information.

Taken together, Eqs. 2.17 and 2.19 lead to a joint prior distribution for μ and σ^2 that is proportional to

$$p(\mu, \sigma^2 \,|\, \kappa, \delta, \alpha, \lambda) \propto (\sigma^2)^{-\alpha-3/2} \exp\left[-\frac{\kappa}{2\sigma^2}(\mu - \delta)^2 - \frac{\lambda}{\sigma^2}\right].$$

Multiplying this times the likelihood function in Eq. 2.11 and simplifying yields a joint posterior distribution that is proportional to

$$\begin{aligned} &p(\mu, \sigma^2 \,|\, \mathbf{y}) \propto \\ &\quad (\sigma^2)^{-(n+1)/2-\alpha-1} \times \\ &\quad \exp\left[-\frac{2\lambda + (n-1)s^2 + (n+\kappa)(\mu - b)^2 + \frac{n\kappa(\bar{y}-\delta)^2}{n+\kappa}}{2\sigma^2}\right], \end{aligned} \tag{2.20}$$

where

$$b = \mathrm{E}(\mu \,|\, \mathbf{y}) = \frac{n\bar{y} + \kappa\delta}{n + \kappa}. \tag{2.21}$$

Equation 2.20 has the same general form as the joint posterior distribution obtained in the noninformative case (cf. Eq. 2.13). With μ held fixed, Eq. 2.20 is proportional to an inverse gamma distribution on σ^2; with σ^2 held fixed, it has the form of a normal distribution on μ. As a result, we can derive both marginal distributions in closed form. The marginal posterior distributions for μ and σ^2 are given by

$$\mu \,|\, \mathbf{y} \sim t\left(n + 2\alpha, b, \frac{2\lambda + (n-1)s^2 + n\kappa(\bar{y} - \delta)^2/(n+\kappa)}{(n+\kappa)(n+2\alpha)}\right) \tag{2.22}$$

and

$$\sigma^2 \,|\, \mathbf{y} \sim \mathit{InverseGamma}\left(\alpha + \frac{n}{2}, \lambda + \frac{(n-1)s^2}{2} + \frac{n\kappa(\bar{y} - \delta)^2}{2(n+\kappa)}\right).$$

Note that the posterior mean b (Eq. 2.21) is a weighted average of the data mean $\bar{y}$ and prior mean δ. While the previous examples illustrated Bayesian inference closed-form posterior distributions, the next chapter discusses computational techniques for approximating posterior distributions that do not have closed-form expressions.

2.6 More on Prior Distributions

A prior distribution captures all of the information known about the parameters $\boldsymbol{\theta}$ before we collect data. Along with the likelihood function, it is one of the two key components of a Bayesian model. In this chapter, we have introduced a number of strategies for choosing a prior distribution, including the development of informative priors (Sect. 2.2.1), conjugate priors (Sects. 2.2.2 and 2.5), and noninformative or diffuse priors (Sects. 2.2.1 and 2.5). In the following sections, we provide more detail for each of these strategies.

2.6.1 Noninformative and Diffuse Prior Distributions

We use noninformative or diffuse prior distributions when we feel that we have very little prior knowledge about the model parameters. In this book, we tend to call prior distributions that express a "lack of knowledge" developed using some formalism *noninformative*, and dispersed but proper prior distributions *diffuse*, but these categorizations overlap, so we use the terms somewhat interchangeably.

While there are a number of formalisms for developing noninformative prior distributions, one of the most common uses *Jeffreys' rule*, which results in a distribution often called a *Jeffreys' prior*. (See Sect. 2.5 for an example of a different formalism.) Suppose that we have a one-to-one transformation of our parameter $\phi = h(\theta)$. There are two ways we can think about determining a prior distribution for ϕ. One is to use a rule to determine a prior distribution $p(\theta)$ for θ and to use the *change of variables* technique (see Example 3.2) to determine the implied prior distribution for ϕ; the second is to use the same rule to directly determine a prior distribution for ϕ. Jeffreys' rule states that any rule for determining a prior distribution should yield the same prior distribution for ϕ whether we transform from a prior on θ or determine a prior directly for ϕ.

Define the *expected Fisher information* as

$$I(\theta) = -E_\theta \left[\frac{\mathrm{d}^2 \log(p(y \mid \theta))}{\mathrm{d}\theta^2} \right].$$

Jeffreys' rule defines a noninformative prior as $p(\theta) \propto [I(\theta)]^{1/2}$.

Table 2.4 summarizes common choices for noninformative prior distributions. Some of the prior distributions in Table 2.4 are proper, and others are improper. For a more detailed discussion, see Box and Tiao (1973).

Table 2.4. Common choices for noninformative prior distributions

Parameters	Noninformative Prior
Binomial (π)	$Beta(0.5, 0.5)$
Multinomial ($\boldsymbol{\pi}$)	$Dirichlet(0.5, 0.5, \ldots, 0.5)$
Poisson (λ)	$\lambda^{-1/2}$
Normal (μ, σ known)	constant k
Normal (σ, μ known)	σ^{-1}

We often use a diffuse prior distribution when there are a range of parameter values about which we are relatively indifferent. In this case, we would typically specify a prior distribution that is at least approximately uniform over the range of indifference. Examples of these include $Beta(\alpha, \beta)$ distributions with α and β small or $InverseGamma(\alpha, \lambda)$ distributions with α and λ small. Note that WinBUGS requires the use of proper prior distributions. Also, as noted in Sect. 2.5, posterior distributions obtained using improper prior distributions are not always proper. To avoid nonintegrable posterior distributions, we recommend the use of proper prior distributions.

2.6.2 Conjugate Prior Distributions

Intuitively, a conjugate prior distribution $p(\theta)$ for a given sampling distribution $f(x \mid \theta)$ is one where the prior distribution $p(\theta)$ and the posterior distribution $p(\theta \mid x)$ have the same functional form. Historically, conjugate prior distributions were useful because they provided tractable analytical results. Of course, prior distributions should not be specified simply for computational convenience, and if a conjugate prior distribution that provides an adequate representation of information available before the conduct of an experiment cannot be found, we use a nonconjugate prior distribution and the analytical techniques described in Chap. 3. Table 2.5 summarizes many common conjugate priors for a variety of sampling distributions.

2.6.3 Informative Prior Distributions

We use informative prior distributions when we have information about the parameters of our model before we collect data. In reliability problems, there are six broad sources of information for constructing informative prior distributions:

1. physical/chemical theory,
2. computational analysis,
3. previous engineering and qualification test results from a process development program,
4. industrywide generic reliability data,
5. past experience with similar devices, and

Table 2.5. Common conjugate priors

Sampling Distribution (Parameter)	Conjugate Prior
Binomial (π)	Beta
Exponential (λ)	Gamma
Gamma (λ)	Gamma
Multinomial ($\boldsymbol{\pi}$)	Dirichlet
Multivariate Normal ($\boldsymbol{\mu}, \Sigma$)	Normal Inverse Wishart
Negative Binomial (π)	Beta
Normal (μ, σ^2 known)	Normal
Normal (σ^2, μ known)	Inverse Gamma
Normal (μ, σ^2)	Normal Inverse Gamma
Pareto (β)	Gamma
Poisson (λ)	Gamma
Uniform$(0, \beta)$	Pareto

6. expert opinion.

There are numerous industrywide generic sources of reliability data reported in a variety of media, such as reliability databases or reliability data handbooks. These sources report the results of analyses performed on actual failure or maintenance event data or, in some cases, are based on expert opinion. They usually contain component failure probabilities, failure rates, and, in some cases, initiating event frequencies.

Expert judgment is often used in assessing a prior distribution. In assessing probability distributions based on expert opinion there are many potential biases that have been identified that either must be minimized (or altogether avoided) or, at the very least, accounted for when assessing prior distributions (Fischoff, 1982). The procedures for formally eliciting probability distributions are well established as described by Winterfeldt and Edwards (1986), Morgan and Henrion (1991), and Meyer and Booker (2001). Appendix C.5 of U. S. Nuclear Regulatory Commission (1994) also contains an excellent summary on the use of expert judgment in eliciting prior probability distributions.

Siu and Kelly (1998) presents several heuristics in connection with developing an informative prior distribution that are worth summarizing here:

1. Beware of zero values. If the prior distribution says that a value of the parameter is impossible, then no quantity of data can overcome this.
2. When using expert opinion, beware of cognitive biases caused by the way people think.
3. Beware of generating overly narrow prior distributions.
4. Ensure that the information used to generate the prior distribution is relevant to the problem at hand.
5. Be careful when assessing prior distributions on parameters that are not directly observable.
6. Beware of conservatism. Realism is the desired ideal, not conservatism.

2.7 Related Reading

Readers interested in a more expanded introduction to Bayesian statistics may consult a variety of texts, including Lee (1997), Congdon (2001), Robert (2001), Gill (2003), and Gelman et al. (2004), or for those seeking a more theoretical treatment Bernardo and Smith (1994) and Berger (1985).

2.8 Exercises for Chapter 2

2.1 Suppose that we want to develop an informative prior distribution for the probability of observing heads when we flip a coin. Suppose that we think that the most likely probability of heads is 0.5 and that 0.75 would be "extreme." Find the parameters of a beta density so that the median is approximately 0.5 and the 0.9 quantile is 0.75.

2.2 Suppose that we are going to flip a coin 20 times.
  a) Using a beta distribution, write down a prior density that describes your uncertainty about the probability of "heads."
  b) Flip a coin 20 times and record the outcomes. Write down the likelihood function for the observed data.
  c) Calculate the maximum likelihood estimate for the probability of "heads" and a 95% confidence interval.
  d) Calculate the posterior distribution for the probability of "heads" and a 95% credible interval.
  e) Plot the log-likelihood function.
  f) Plot the prior density.
  g) Plot the posterior density.
  h) Calculate the Bayes' factor comparing a uniform prior density to your informative prior density.

2.3 Consider again the fluid breakdown times introduced in Sect. 2.5. Two models were proposed for these data. The first incorporated a normal likelihood function and a noninformative prior distribution; the second a normal likelihood function and a conjugate inverse-gamma/normal prior distribution. Now suppose that the properties of the manufacturing process were controlled when these samples of lubricant were produced so that it is known that the true mean of the sample values must lie between 6.0 and 7.4 (on the original measurement scale). No further information is available concerning the value of the variance parameter σ^2.
  a) Assume that the joint prior distribution for (μ, σ^2) is proportional to $1/\sigma^2$ whenever $\mu \in (\log(6.0), \log(7.4))$, and is 0 otherwise. Find an expression for a function that is proportional to the joint posterior distribution.
  b) Find a function that is proportional to the marginal posterior distribution of μ.

c) Find a function that is proportional to the marginal posterior distribution of σ^2.

2.4 Show that the beta distribution is the conjugate prior distribution for the binomial likelihood.

2.5 Show that the gamma distribution is the conjugate prior distribution for the mean of a Poisson likelihood.

2.6 Show that the gamma distribution is the conjugate prior distribution for the exponential likelihood.

2.7 Derive the mean and variance for the lognormal distribution.

2.8 Suppose we are using an Exponential(λ) distribution to model the lifetimes of n items.

a) Find the maximum likelihood estimator of λ.

b) Assume n is large and find the standard error of $\hat{\lambda}$.

c) Suppose that we observed $n = 50$ items and that $\sum_{i=1}^{50} t_i = 25$. Find a 90% confidence interval for λ.

d) Suppose that $\lambda \sim Gamma(1, 2)$. Find the posterior distribution for λ.

e) Suppose that we observed $n = 50$ items and that $\sum_{i=1}^{50} t_i = 25$. What is the posterior probability that λ falls in the 90% confidence interval found in (c)?

3

Advanced Bayesian Modeling and Computational Methods

We extend the model structures described in the previous chapter using Bayesian hierarchical models. Because we generally cannot write the posterior distributions that result from these more complicated models in closed form, we begin this chapter with a description of Markov chain Monte Carlo algorithms that can be used to generate samples from intractable posterior distributions. These samples provide the basis for subsequent model inference. We also discuss empirical Bayes' methods. Finally, we describe techniques for assessing the sensitivity of model inferences to prior assumptions and a broadly applicable model diagnostic.

3.1 Introduction to Markov Chain Monte Carlo (MCMC)

As the final example of the previous chapter suggests, analytically deriving marginal posterior densities by integrating out nuisance parameters can be a chore. In more complicated models, analytically integrating out parameters from a joint posterior distribution, or even determining the normalizing constant of the posterior distribution, is generally not possible. Furthermore, calculating the posterior distribution of functions of parameters is difficult. For many years, the difficulty associated with performing such marginalizations, as well as many other Bayesian inference tasks that required high-dimensional integration, prevented practitioners from applying Bayesian modeling techniques to real-world problems. That situation changed in the late 1980s and early 1990s with the advent of MCMC algorithms.

MCMC algorithms are a general class of computational methods used to produce samples from posterior distributions. They are often easy to implement and, at least in principle, can be used to simulate from very high-dimensional posterior distributions. Since their introduction in the 1990s, they have been successfully applied to literally thousands of applications.

The basic goal of an MCMC algorithm is to simulate values (also called samples or draws) from the posterior distribution of a parameter vector. Inference about likely parameter values, or functions of parameter values, is then based on these simulated values. Letting the jth value in such a sequence of draws of the parameter vector $\boldsymbol{\theta}$ be denoted by $\boldsymbol{\theta}^{(j)}$, MCMC algorithms have the property that the distribution of the jth iterate in the sequence of sampled values converges to a random sample drawn from the posterior distribution as j becomes large. In general, successive draws from the posterior are correlated, but this correlation tends to die out as the interval between draws increases. Thus, if a large number of sample updates are performed, the last group of sampled values in the sequence, say $\boldsymbol{\theta}^{(m)}, \boldsymbol{\theta}^{(m+1)}, \ldots, \boldsymbol{\theta}^{(m+K)}$, represents a (dependent) sample from the posterior distribution of interest. The iterations, $\boldsymbol{\theta}^{(1)}, \ldots, \boldsymbol{\theta}^{(m-1)}$, are known as *burn-in* and do not represent samples from the posterior distribution.

Viewed from a slightly more general perspective, MCMC algorithms produce random walks over a probability distribution. By taking a sufficient number of steps in this random walk, the MCMC simulation algorithm visits various regions of the parameter space in proportion to their posterior probabilities. We can, for inferential purposes, summarize the iterates obtained in these random walks much as we would summarize an independent sample from the posterior distribution.

We consider two general categories of MCMC algorithms: Metropolis-Hastings algorithms and Gibbs samplers. We begin with Metropolis-Hastings algorithms.

3.1.1 Metropolis-Hastings Algorithms

Metropolis-Hastings algorithms provide a simple, generic prescription for obtaining draws from a posterior distribution. The basic steps of a Metropolis-Hastings algorithm follow. For ease of description, we assume that $\boldsymbol{\theta}$ is a q-dimensional, real-valued parameter vector. If we were to apply the Metropolis-Hastings algorithm to the lognormal example in Sect. 2.5 (which had two unknown parameters, μ and σ^2), then $\boldsymbol{\theta}$ would be (μ, σ^2) and q would be 2.

The first step in Metropolis-Hastings algorithms is to generate a candidate point, denoted here by $\boldsymbol{\theta}^*$. Often, the candidate point differs from the current value of the parameter in only one or two components; for example, in the normal example, we may alternate between updating the value of μ and the value of σ^2. A common method for generating the candidate value $\boldsymbol{\theta}^*$ is to add a mean-zero normal deviate to a single component of $\boldsymbol{\theta}^{(j-1)}$, say $\theta_i^{(j-1)}$. This means that we can express the candidate value $\boldsymbol{\theta}^*$ as the vector with components

$$\begin{aligned} \theta_i^* &= \theta_i^{(j-1)} + sZ, \\ \theta_k^* &= \theta_k^{(j-1)} \qquad \text{for } k \neq i, \end{aligned} \tag{3.1}$$

where Z is a standard normal deviate and s is an arbitrary constant. For continuous-valued components of the parameter vector, let $f(\boldsymbol{\theta}^*|\boldsymbol{\theta}^{(j-1)})$ denote the *proposal density* used to generate $\boldsymbol{\theta}^*$ from $\boldsymbol{\theta}^{(j-1)}$. For example, in Eq. 3.1, the proposal density $f(\cdot)$ represents a normal distribution with mean $\theta_i^{(j-1)}$ and standard deviation s. For discrete-valued components of the parameter vector, $f(\cdot)$ represents the probability mass function used to generate candidate points. The probability of moving from the candidate point back to the original value is denoted, in a similar way, by $f(\boldsymbol{\theta}^{(j-1)}|\boldsymbol{\theta}^*)$.

In theory, any density or mass function can serve as the proposal density as long as it satisfies three conditions. First, the proposal density must allow us to move from any subset of the parameter space to any other subset of the parameter space in a finite number of moves. Second, the proposal density cannot be periodic. Informally this means that, in the long run, moves to any subset of the parameter space can occur at any time. Finally, we require that the rule used to specify the proposal density satisfies

$$0 < \frac{f(\boldsymbol{\theta}^*|\boldsymbol{\theta}^{(j-1)})}{f(\boldsymbol{\theta}^{(j-1)}|\boldsymbol{\theta}^*)} < \infty,$$

for all values $\boldsymbol{\theta}^{(j-1)}$ and $\boldsymbol{\theta}^*$.

Having generated a candidate point $\boldsymbol{\theta}^*$, we perform the second step in a Metropolis-Hastings algorithm; we compute the probability that the candidate value will be accepted as the next simulated value in the sequence. We call this quantity the *acceptance probability* and denote its value by r. With this notation, the acceptance probability r is defined as

$$r = \min\left(1, \frac{p(\boldsymbol{\theta}^*|\text{data})}{p(\boldsymbol{\theta}^{(j-1)}|\text{data})} \frac{f(\boldsymbol{\theta}^{(j-1)}|\boldsymbol{\theta}^*)}{f(\boldsymbol{\theta}^*|\boldsymbol{\theta}^{(j-1)})}\right).$$

In this formula, the acceptance probability represents the product of the ratio of the posterior density evaluated at the candidate and current parameter values, $p(\boldsymbol{\theta}^*|\text{data})/p(\boldsymbol{\theta}^{(j-1)}|\text{data})$, and the ratio of the proposal densities of the current and candidate point, $f(\boldsymbol{\theta}^{(j-1)}|\boldsymbol{\theta}^*)/f(\boldsymbol{\theta}^*|\boldsymbol{\theta}^{(j-1)})$. The first ratio encourages the algorithm to move to parameter values that have high posterior probability, and the second ratio accounts for the fact that the proposal density might favor some values of the parameter over others. Note that if the proposal density is symmetric — that is, if $f(\boldsymbol{\theta}^{(j-1)}|\boldsymbol{\theta}^*) = f(\boldsymbol{\theta}^*|\boldsymbol{\theta}^{(j-1)})$ — this second ratio is 1 and can be omitted from the formula for the acceptance probability.

Having computed an acceptance probability, we perform the third step in a Metropolis-Hastings algorithm. We accept or reject the candidate point with probability equal to r. To do so, we draw a $Uniform(0,1)$ random variable, say u, and compare u to r. If $u \leq r$, then we accept the candidate value and set $\boldsymbol{\theta}^{(j)} = \boldsymbol{\theta}^*$. On the other hand, if $u > r$, then we reject the candidate value and set $\boldsymbol{\theta}^{(j)} = \boldsymbol{\theta}^{(j-1)}$ (that is, we keep the same value). This process is repeated for each component of $\boldsymbol{\theta}$.

To illustrate the Metropolis-Hastings algorithm, consider again the launch vehicle failure data discussed in Chap. 2. In that example, 3 successes out of 11 tests were observed. Assuming a binomial model for these data and letting π denote the success probability, we know that the likelihood function is proportional to

$$\pi^3(1-\pi)^8.$$

In our previous discussion of these data, we assumed that the prior distribution for π took the form of a conjugate beta density. Now suppose that the rocket scientists tell us that past data and their engineering expertise require that the prior distribution be uniform on the interval $(0.1, 0.9)$. That is, the prior for π is taken to be proportional to

$$p(\pi) \propto \begin{cases} 1 \text{ if } 0.1 < \pi < 0.9 \\ 0 \text{ otherwise.} \end{cases}$$

Multiplying the likelihood function and prior density together, we find that the posterior density is proportional to

$$p(\pi \mid \text{data}) \propto \begin{cases} \pi^3(1-\pi)^8 & \text{if } 0.1 < \pi < 0.9 \\ 0 & \text{otherwise.} \end{cases}$$

This distribution does not have the form of a standard beta density because it is not defined on $0 < \pi < 1$. To determine the normalizing constant for it, we would have to resort to tables of incomplete beta densities or numerically evaluate the posterior distribution. For purposes of illustration, we construct a Metropolis-Hastings algorithm to evaluate the posterior distribution. This is illustrated in Fig. 3.1.

In this example, we choose the proposal density to be the prior distribution. Because the proposal density in this algorithm does not depend on $\pi^{(j-1)}$, this Metropolis-Hastings algorithm is called an *independence sampler*. Figure 3.2 presents a *trace plot* of the first 500 values drawn from this chain. The posterior mean and variance of the values depicted in this plot are 0.31 and 0.015, respectively.

In general, independence samplers work well when the proposal density represents a reasonable approximation to the posterior density. They work poorly when the proposal density assigns negligible mass near the region of the parameter space where the posterior density is most concentrated. The following example illustrates the type of problem that can occur when the proposal density used to define an independence sampler is not "close" to the posterior distribution.

Example 3.1 Convergence properties of a Metropolis-Hastings sampler for a binomial success probability. Suppose for the moment that in the launch vehicle data we had observed 300 successes and 800 failures instead of the 3 successes and 8 failures that we did. Repeating the above algorithm with r now defined as

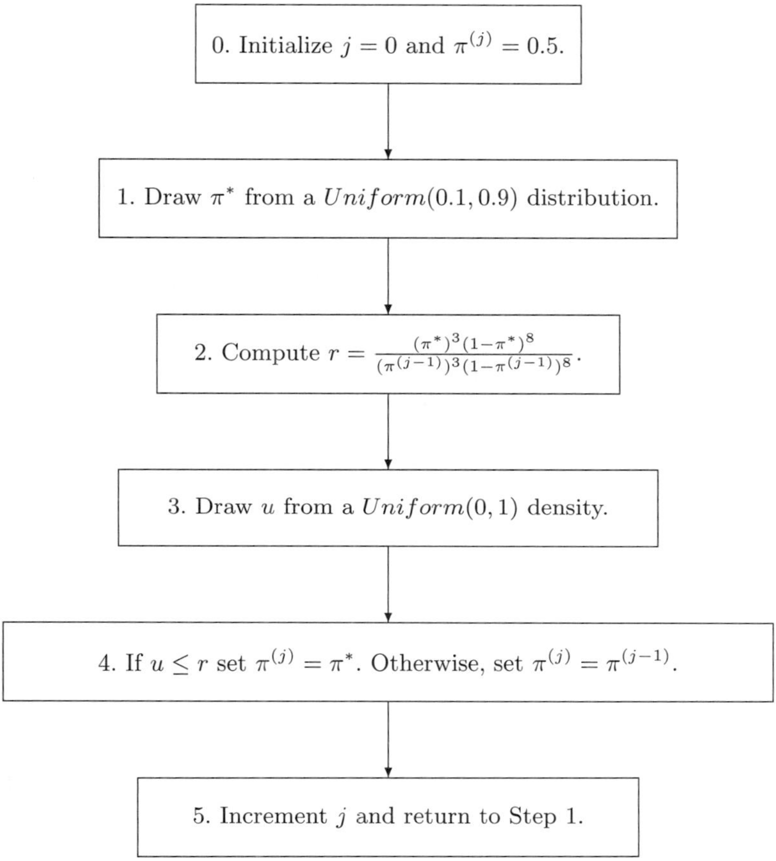

Fig. 3.1. Metropolis-Hastings algorithm for sampling from posterior distribution on a binomial success probability using a truncated prior density.

$$r = \frac{(\pi^*)^{300}(1-\pi^*)^{800}}{(\pi^{(j-1)})^{300}(1-\pi^{(j-1)})^{800}},$$

we obtain the trace plot depicted in Fig. 3.3.

The problem with this sampler is evident from the plot. Once a candidate value of π^* close to $3/11$ is drawn and accepted, it is retained as the current value of π until another point that is close to $3/11$ is proposed. In high-dimensional problems, this property of independence samplers has two implications. First, it means that a sampler may take a long time for the proposal density to "find" a value near the mass of the posterior distribution. Thus, the algorithm can take a long time to "burn-in." Second, once

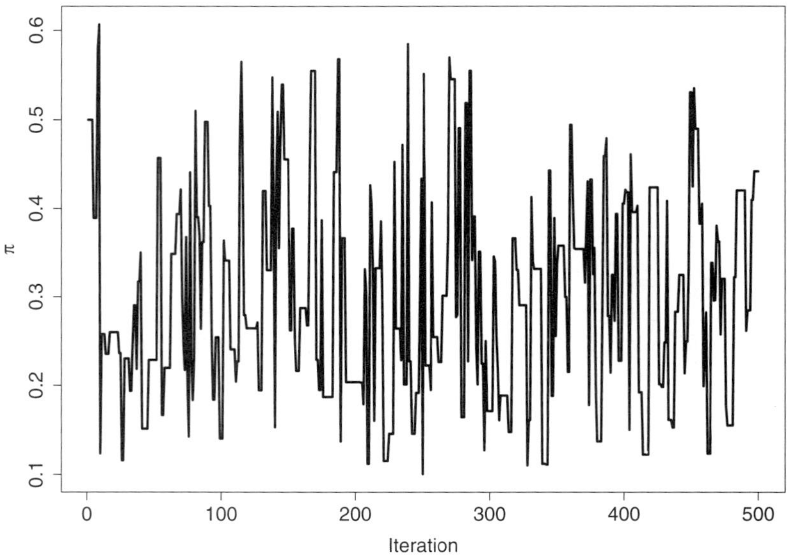

Fig. 3.2. Plot of the successive values of π generated in 500 updates of the independence sampler.

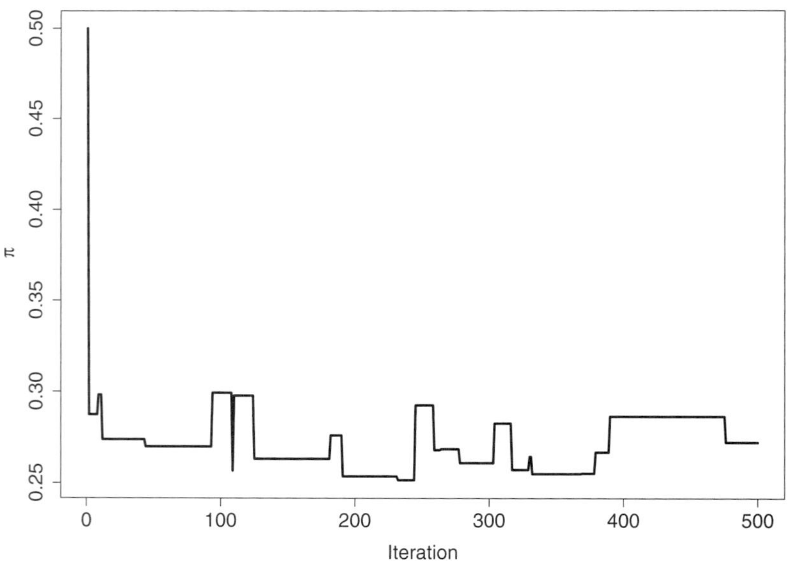

Fig. 3.3. Plot of the successive values of π generated in 500 updates of the independence sampler, now assuming 300 successes and 800 failures.

a point near the mass of the posterior has been drawn, many iterations can pass before it is replaced with another point that also has relatively high posterior probability. Indeed, in the updates of π depicted in Fig. 3.3, nearly 94% of the candidate draws for π were rejected. Successive iterates in the Metropolis-Hastings algorithm were thus highly correlated, which means that a large number of iterates would have to be drawn before, for example, the sample mean of the sequence $\{\pi^{(j)}\}$ would provide an accurate estimate of the posterior mean of π.

Because of the difficulty in specifying an appropriate proposal density for an independence sampler, *random-walk* Metropolis-Hastings algorithms are usually used to draw samples from more complicated models. To illustrate these schemes, consider again the lognormal lifetime data discussed in Sect. 2.5. There, the data model contained two unknown parameters, μ and σ^2.

As before, we assume that the likelihood function for the logarithm of the breakdown times has the form specified in Eq. 2.11, and that a noninformative prior distribution on (μ, σ^2) proportional to $1/\sigma^2$ is employed, so that the posterior distribution is proportional to Eq. 2.13.

In specifying a random-walk Metropolis-Hastings algorithm, we can update both μ and σ^2 simultaneously, or we can alternate between updates of μ and σ^2. The simplest method is to alternate, and so we take that approach here. In general, however, it is often better to update correlated parameters simultaneously using a proposal density that approximately matches the posterior correlation of the parameters. Often, we can base such a proposal on the asymptotic covariance matrix estimated from the inverse of the information matrix.

One proposal density for generating a candidate draw μ^* for μ may be defined as being normally distributed with mean $\mu^{(j-1)}$ and variance s_1^2.

Choosing a proposal density for σ^2 is slightly more difficult because we want the proposed values to be positive, and a random-walk Metropolis-Hastings algorithm using normally distributed updates can generate negative candidate values. One way of simulating positive candidate values is to generate candidates on the logarithmic scale and then transform them to the original scale. That is, we might define candidate draws for σ^2 according to

$$\log\left((\sigma^2)^*\right) = \log\left((\sigma^2)^{(j-1)}\right) + \nu, \qquad \text{where} \qquad \nu \sim Normal(0, s_2^2),$$

for a given variance parameter s_2^2. Transforming back to the σ^2 scale, it follows that the proposal density for $(\sigma^2)^*$ is given by

$$f\left((\sigma^2)^* \mid (\sigma^2)^{(j-1)}\right) = \frac{1}{(\sigma^2)^*}\frac{1}{\sqrt{2\pi}s_2} \times \\ \exp\left\{-\frac{1}{2s_2^2}\left[\log\left((\sigma^2)^*\right) - \log\left((\sigma^2)^{(j-1)}\right)\right]^2\right\},$$

where $(\sigma^2)^* > 0$. The ratio of the proposal densities that appears in the Metropolis-Hastings acceptance probability thus simplifies to

$$\frac{f\left((\sigma^2)^{(j-1)} \mid (\sigma^2)^*\right)}{f\left((\sigma^2)^* \mid (\sigma^{(j-1)})^*\right)} = \frac{(\sigma^2)^*}{(\sigma^2)^{(j-1)}}.$$

With these proposal densities defined, we can specify the random-walk Metropolis-Hastings scheme to generate posterior samples of μ and σ^2 illustrated in Fig. 3.4. Note that in Step 6 of this algorithm, the prior distribution specified for σ^2 has canceled the contribution from the ratio of the proposal densities.

Figure 3.5 depicts trace plots for the first 5,000 iterates obtained from the algorithm above with $s_1^2 = 0.5$ and $s_2^2 = 1.0$. Figure 3.6 shows histograms of the sampled values of μ and σ^2 obtained after iteration 50 (i.e., after burn-in). For comparison, Fig. 3.6 also displays the exact marginal posterior density functions computed in Eqs. 2.15 and 2.16.

In examining the trace plots of Fig. 3.5, no long-term trend appears evident. This suggests that the simulated values of μ and σ^2 represent approximate draws from their marginal posterior densities.

Example 3.2 Change of variables. Suppose that we have a random variable $X \sim Gamma(\alpha, \lambda)$ and that we are interested in learning about the distribution of $Y = g(X) = 1/X$. There are two ways to approach this problem. The first is to use the *change of variables* technique. Mathematically,

$$f_Y(y) = f_X(g^{-1}(y))|\frac{\mathrm{d}}{\mathrm{d}y}g^{-1}(y)|.$$

For the gamma distribution,

$$f_X(x) = \frac{\lambda^\alpha}{\Gamma(\alpha)}x^{\alpha-1}\exp(-\lambda x)$$

and

$$\frac{\mathrm{d}}{\mathrm{d}y}g^{-1}(y) = \frac{\mathrm{d}}{\mathrm{d}y}1/y = -1/y^2.$$

Substituting, we obtain

$$\begin{aligned} f_Y(y) &= f_X(g^{-1}(y))|\frac{\mathrm{d}}{\mathrm{d}y}g^{-1}(y)| \\ &= \frac{\lambda^\alpha}{\Gamma(\alpha)}(1/y)^{\alpha-1}\exp(-\lambda/y)\frac{1}{y^2} \\ &= \frac{\lambda^\alpha}{\Gamma(\alpha)}y^{-(\alpha+1)}\exp(-\lambda/y). \end{aligned}$$

This is the density function for an inverse gamma distribution with parameters α and λ. See Casella and Berger (1990) for a more detailed description of both the univariate and multivariate cases of the change of variables technique.

0. Initialize $j = 0, \mu^{(j)} = 1, (\sigma^2)^{(j)} = 0.5$.

1. Draw μ^* from a $Normal(\mu^{(j-1)}, s_1^2)$ distribution.

2. Compute $r = \frac{\exp(-\sum(y_i-\mu^*)^2/2(\sigma^2)^{(j-1)})}{\exp(-\sum(y_i-\mu^{(j-1)})^2/2(\sigma^2)^{(j-1)})}$.

3. Draw u from a $Uniform(0, 1)$ density.

4. If $u \leq r$ set $\mu^{(j)} = \mu^*$. Otherwise, set $\mu^{(j)} = \mu^{(j-1)}$.

5. Draw ν from a $Normal(0, s_2^2)$ distribution and set $(\sigma^2)^* = (\sigma^2)^{(j-1)} \exp(\nu)$.

6. Compute $r = \frac{(\sigma^{-n})^* \exp(-\sum(y_i-\mu^{(j)})^2/2(\sigma^2)^*)}{(\sigma^{-n})^{(j-1)} \exp(-\sum(y_i-\mu^{(j)})^2/2(\sigma^2)^{(j-1)})}$.

7. Draw u from a $Uniform(0, 1)$ density.

8. If $u \leq r$ set $(\sigma^2)^{(j)} = (\sigma^2)^*$. Otherwise, set $(\sigma^2)^{(j)} = (\sigma^2)^{(j-1)}$.

9. Increment j and return to Step 1.

Fig. 3.4. Metropolis-Hastings algorithm for sampling from posterior distribution on a normal mean and variance parameter.

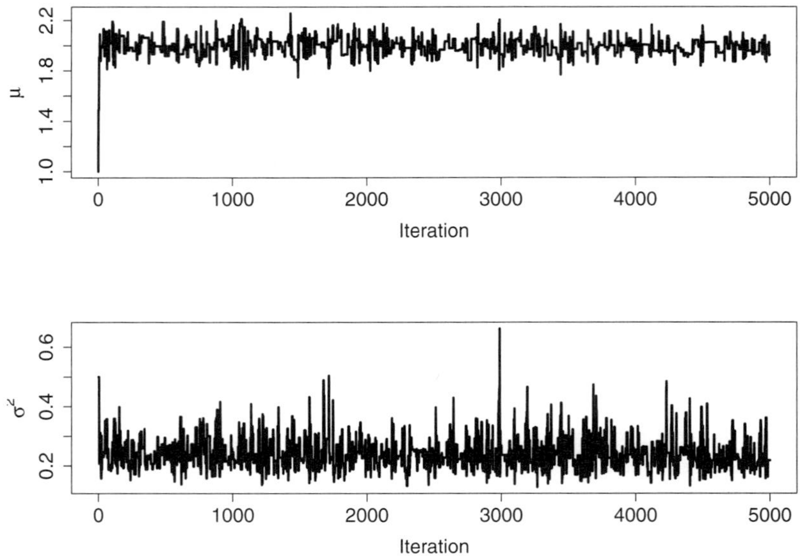

Fig. 3.5. Trace plots of successive values of μ and σ^2 generated in 5,000 updates from a random-walk Metropolis-Hastings sampler.

As a second approach, suppose that we have a random sample $X_1, \ldots, X_n$ from a $Gamma(\alpha, \beta)$ distribution. Suppose that we set $Y_i = 1/X_i$ for $i = 1, \ldots, n$. While we do not know the functional form of the probability density function of $Y = 1/X$, we can use the random sample Y_i to draw a kernel density estimate of the probability density function of Y, estimate the mean, variance, and quantiles of Y — in short, we can estimate (increasingly well as n gets large) many of the quantities of interest about Y.

This approach is particularly useful when we have draws from the posterior distribution of a set of parameters $\boldsymbol{\theta}$. Suppose that we are interested in learning about a function $g(\boldsymbol{\theta})$. We may not be able to analytically calculate the probability density function of $g(\boldsymbol{\theta})$, but we can always plug in the MCMC draws to get $g(\boldsymbol{\theta}^{(j)})$ and use this random sample to estimate quantities of interest.

3.1.2 Gibbs Sampler

Metropolis-Hastings algorithms often provide effective methods for simulating from a posterior distribution of an unfamiliar form. However, the success of these methods depends upon determining reasonable choices of proposal densities, which, in some cases, can be difficult. When poor proposal densities are selected, one of two problems may arise. First, if the proposal density is

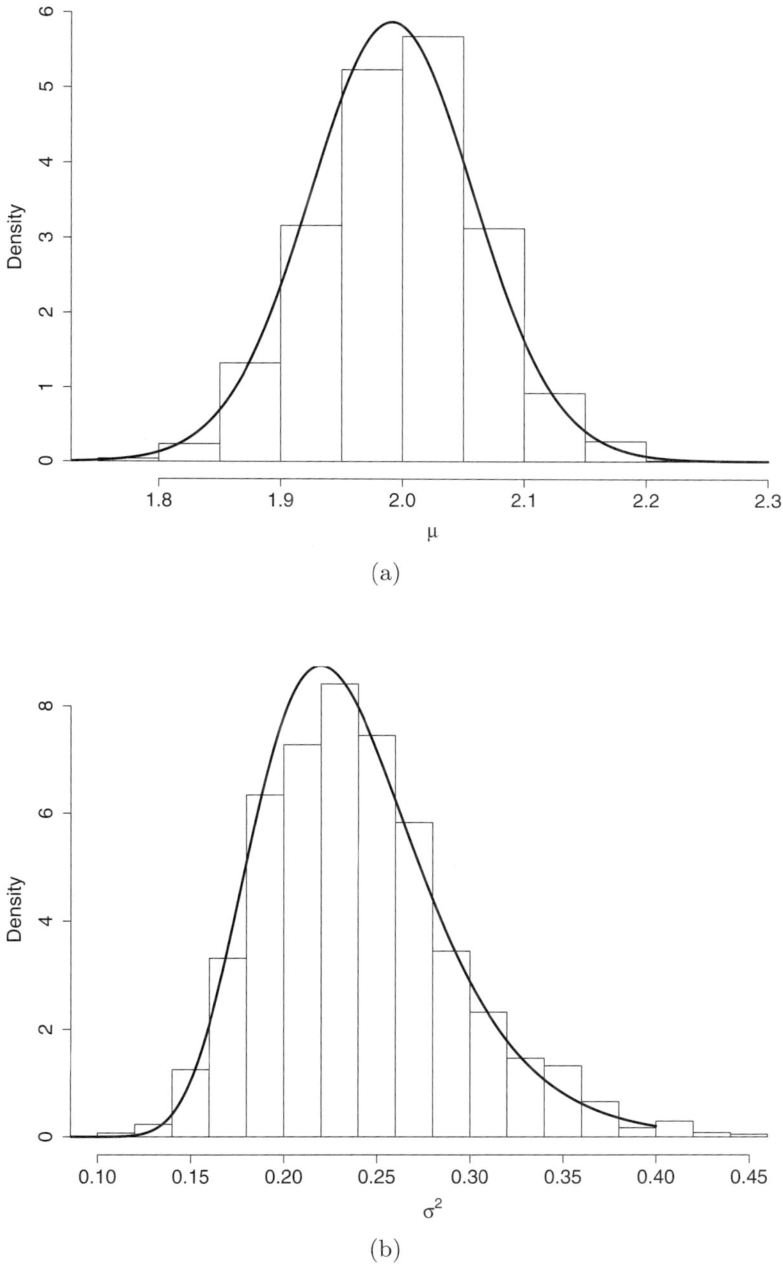

Fig. 3.6. Histogram of draws from the posterior distributions of μ and σ^2 obtained using the last 4,950 updates in the chain shown in Fig. 3.5. For comparison, the exact marginal posterior distributions are depicted as solid lines.

too narrow, the Metropolis-Hastings algorithm may spend all of its time in a limited region of the parameter space and may not be able to visit "distant" modes of the posterior distribution in any reasonable number of updates. In addition, proposal densities that are too narrow lead to high correlations between iterates of the chain, making the effective independent sample size quite small. On the other hand, if the proposal density is too broad, the chain may freeze in a single state for hundreds or even thousands of iterations, generating very few unique values.

Generally speaking, a better MCMC method for sampling from a posterior distribution can be obtained by replacing generic proposal densities in Metropolis-Hastings algorithms by the conditional distribution of the parameter component that is being sampled. MCMC algorithms that employ this strategy are referred to as Gibbs samplers.

To describe a Gibbs sampler, suppose that the parameter vector $\boldsymbol{\theta}$ can be partitioned into q components $\boldsymbol{\theta} = (\theta_1, \theta_2, \ldots, \theta_q)$, and denote the conditional posterior distributions (*full conditional distributions*) by

$$p_1(\theta_1|\theta_2, \ldots, \theta_q, \text{data})$$

$$p_2(\theta_2|\theta_1, \theta_3, \ldots, \theta_q, \text{data})$$

$$\vdots$$

$$p_q(\theta_q|\theta_1, \ldots, \theta_{q-1}, \text{data}).$$

In the simple case of $q = 2$, the probability density p_1 represents the posterior density of component vector θ_1, conditional on the value of component θ_2. Likewise, p_2 is the conditional posterior density of θ_2 given θ_1.

In many models, it is difficult to simulate directly from the full parameter vector $\boldsymbol{\theta}$. However, it is often possible to generate simulated values from each of the full conditional densities $p_1, p_2, \ldots, p_q$. When this is the case, a Gibbs sampler may be defined according to the algorithm depicted in Fig. 3.7.

Example 3.3 Illustration of a Gibbs sampler. Consider again the fluid breakdown time data from Table 2.3 and the joint posterior density that results from a normal model assumption and a noninformative prior on μ and σ^2:

$$p(\mu, \sigma^2 \,|\, \mathbf{y}) \propto (\sigma^2)^{-n/2-1} \exp\left[\sum_{i=1}^{n} -\frac{1}{2\sigma^2}(y_i - \mu)^2\right].$$

To apply the Gibbs sampler in this setting, we must identify conditional distributions for μ and σ^2. Holding μ fixed, we see that the conditional distribution of σ^2 is an inverse gamma distribution with parameters $n/2$ and $\sum(y_i - \mu)^2/2$. If we hold σ^2 constant and recall that $\sum(y_i - \mu)^2 = (n-1)s^2 + n(\mu - \bar{y})^2$, then the conditional distribution of μ is normal with mean $\bar{y}$ and variance σ^2/n.

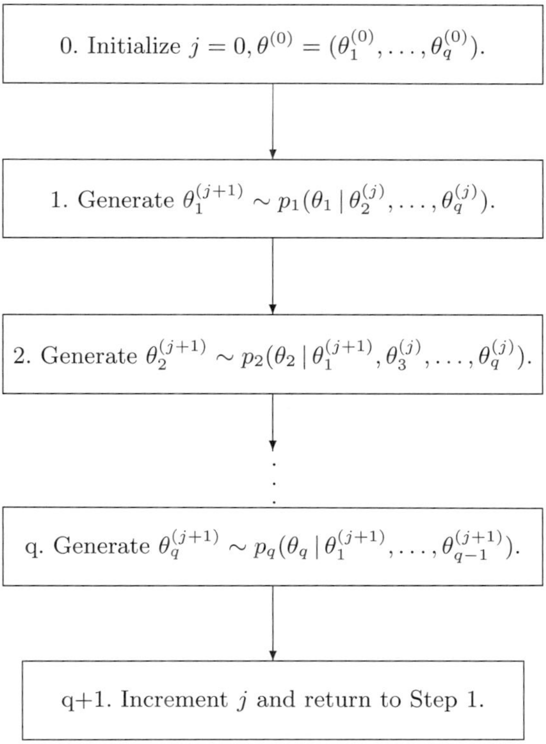

Fig. 3.7. Gibbs sampling algorithm for sampling from posterior distribution of a q-dimensional parameter vector $\boldsymbol{\theta}$.

A Gibbs sampler for this model thus reduces to simply alternating between sampling σ^2 from its conditional inverse gamma distribution given the most recent sampled value of μ, and sampling μ from its conditional normal distribution given the last sampled value of σ^2. Figures 3.8 and 3.9 provide trace plots and histograms of the marginal posterior distributions of μ and σ^2 based on 1,000 Gibbs updates.

A comparison of the trace plots obtained using the Gibbs sampler (as seen in Fig. 3.8) and the random-walk Metropolis-Hastings algorithm (as seen in Fig. 3.5) suggests that the Gibbs sampler "mixes" more efficiently; it reaches the target distribution faster and the successive values in the chain are less highly correlated. Of course, we could tune the Metropolis-Hastings algorithm to nearly mimic the performance of the Gibbs sampler by judicious choice of s_1^2 and s_2^2, but that would require additional simulation and analyst study. The Gibbs sampler achieves nearly optimal performance without the selection of

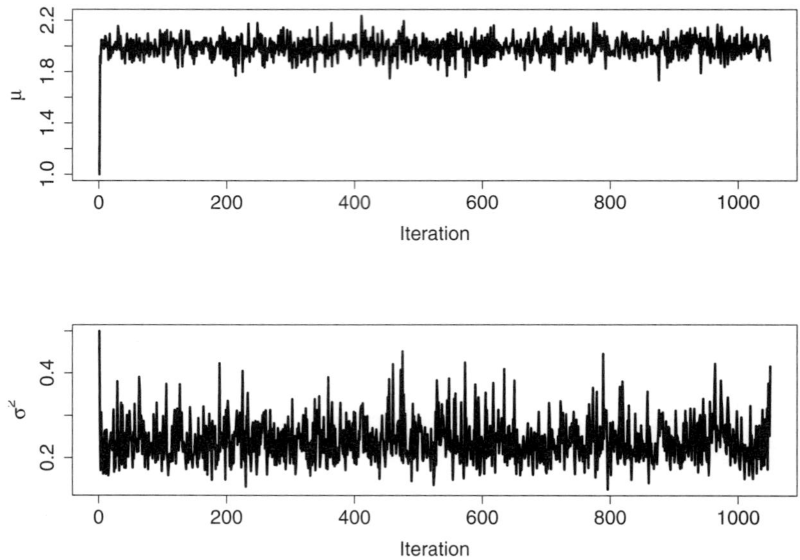

Fig. 3.8. Trace plots of successive values of μ and σ^2 generated in 1,000 updates from a Gibbs sampler.

"good" proposal densities. The downside of the Gibbs sampler is that we must derive full conditional distributions for all parameters or parameter vectors.

In many applications, full conditional distributions for several parameters will be recognizable, while for others they will not. Hybrid Gibbs/Metropolis-Hastings algorithms are then commonly used, with Gibbs updates of parameters for which full conditional distributions are available in closed form and easily simulated, and Metropolis-Hastings with convenience proposal densities used for updates of parameters that are not.

3.1.3 Output Analysis

Under rather general conditions, the Gibbs and Metropolis-Hastings MCMC algorithms described in previous sections of this chapter produce parameter values that, after a large number of updates, represent a sample from the posterior distribution. Unfortunately, this theoretical result provides little practical guidance for determining how many updates must be performed to obtain an adequate sample size for accurate posterior inferences. Let us now briefly explore some of the important issues that arise in interpreting MCMC output and describe graphical and numerical diagnostics for assessing convergence.

The first issue in understanding MCMC output involves determining the *burn-in* period. Because chains are typically initialized with values that are

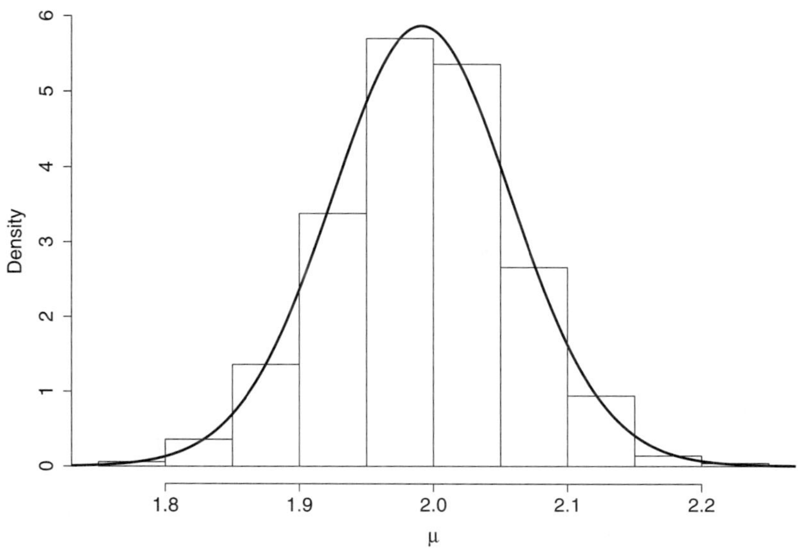

(a)

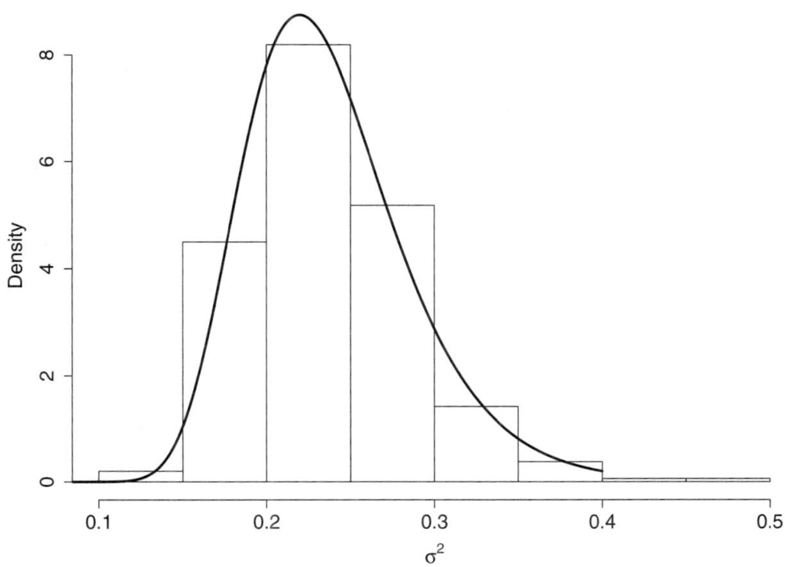

(b)

Fig. 3.9. Histogram of the marginal posterior distributions of μ and σ^2 based on the sampled values depicted in the trace plots of Fig. 3.8. For comparison, the exact marginal posterior densities are depicted as solid lines.

not actually drawn from the posterior distribution, simulated values of the parameter $\boldsymbol{\theta}$ obtained at the beginning of an MCMC run are not distributed from the posterior distribution. However, after some number of iterations have been performed (i.e., at the end of the burn-in period), the effect of the initial values wears off and the distribution of the new iterates approaches the posterior distribution. One way to estimate the length of this burn-in period is to examine the trace plots of simulated values of a component (or some other function) of $\boldsymbol{\theta}$ against the iteration number. Figures 3.5 and 3.8 provide examples of trace plots. When MCMC algorithms are initialized with parameter values that happen to fall far from the center of the posterior distribution, updates obtained early in the chain exhibit a systematic drift toward the region of the parameter space where the posterior distribution is concentrated. An increasing or decreasing trend in the parameter values in the trace plot therefore indicates that the burn-in period is not over.

Aside from burn-in, a second concern that we must address in analyzing output from MCMC algorithms is the degree of *autocorrelation* in the sampled values. For both the Metropolis-Hastings and Gibbs sampling algorithms, the simulated value of $\boldsymbol{\theta}$ at the $(j+1)$st iteration is correlated with the simulated value obtained at the jth iteration. If this correlation is strong, then consecutive values in the chain provide only marginally more information about the posterior distribution than does a single simulated value. In such cases, we say that the algorithm displays poor *mixing*.

A standard statistic for measuring the degree of dependence between successive draws in an MCMC chain is the autocorrelation function. As its name suggests, the autocorrelation function measures the correlation between sets of simulated values $\{\theta_i^{(j)}\}$ and $\{\theta_i^{(j+L)}\}$, where L is the lag or number of iterates separating the two sets of values. For a particular component or function of $\boldsymbol{\theta}$, one can compute the autocorrelation function as a function of differing values of the lag L. The mean of values in the simulated sample gives an estimate of the posterior mean:

$$\bar{\theta}_i = \frac{1}{M} \sum_{j=1}^{M} \theta_i^{(j)}.$$

For component i of the random variable $\boldsymbol{\theta}$, the lag L autocorrelation may be estimated by

$$\rho_{iL} = \frac{M}{M-L} \frac{\sum_{j=1}^{M-L} \left(\theta_i^{(j)} - \bar{\theta}_i\right)\left(\theta_i^{(j+L)} - \bar{\theta}_i\right)}{\sum_{i=1}^{M} \left(\theta_i^{(j)} - \bar{\theta}_i\right)^2},$$

where M is the number of post-burn-in samples. The value of the autocorrelation at lag 1 is generally positive for the Metropolis-Hastings and Gibbs sampling algorithms but decreases to zero as the lag value is increased.

Aside from autocorrelation, another issue that we must address in analyzing output from MCMC algorithms is the choice of the simulated sample size and the resulting accuracy of calculated posterior summaries. Because iterates in an MCMC algorithm are not independent, it is difficult to compute the standard errors of MCMC-based estimators. To obtain estimates of these simulation errors (which are not to be confused with the uncertainty inherent in the posterior distribution), several procedures are commonly used. Perhaps the simplest of these is the method of *batch means*, which we now describe.

To illustrate the estimation of MCMC-sample uncertainty using batch means, suppose that we are interested in computing the posterior mean of a component of $\boldsymbol{\theta}$, say θ_i. To compute the MCMC standard error for this estimate, we subdivide the stream of simulated values $\theta_i^{(1)}, \ldots, \theta_i^{(M)}$ into b batches, each batch of size v, where $M = bv$. For each batch, we compute a sample mean; let's call the set of sample means $\bar{\theta}_{i,1}, \ldots, \bar{\theta}_{i,b}$. Suppose that the size of the batch v has been chosen to be large enough so that the lag 1 autocorrelation in the sequence of batch means is small, say under 0.1. Then the standard error of the estimate $\bar{\theta}_i$ can be approximated by the sample standard deviation of the batch means divided by the square root of the number of batches:

$$s_{\bar{\theta}_i}^B = \sqrt{\frac{\sum_{l=1}^b (\bar{\theta}_{i,l} - \bar{\theta}_i)^2}{(b-1)b}}.$$

This standard error is useful for determining the accuracy of posterior means that are computed in the simulation run. If the MCMC standard error is too large, rerun the MCMC algorithm using a larger number of iterations.

A more sophisticated approach for monitoring the convergence of an MCMC algorithm requires that several chains, each started from different starting values, be run. Gelman and Rubin (1992) advocates this approach; it is also described by Gelman (1996). In principle, starting values of the separate chains should be widely dispersed and should in some sense "surround" the region of the parameter space where the posterior distribution is thought to concentrate.

Based on the output from several chains, we can estimate the posterior variance of any particular component i of the parameter vector in an unbiased way by the formula

$$V_i = \frac{M-1}{M} W_i + \frac{1}{M} B_i,$$

where W and B represent within- and between-chain estimates of the variance. Specifically, if K chains each of length M have been run, and sampled values for the kth chain are denoted with a second subscript, then

$$W_i = \frac{1}{K} \sum_{k=1}^{K} s_k^2,$$

where

$$s_k^2 = \frac{1}{M-1} \sum_{j=1}^{M} (\theta_{i,k}^{(j)} - \bar{\theta}_{i,k})^2,$$

and

$$B_i = \frac{M}{K-1} \sum_{k=1}^{K} (\bar{\theta}_{i,k} - \bar{\bar{\theta}}_i)^2,$$

where

$$\bar{\theta}_{i,k} = \frac{1}{M} \sum_{j=1}^{M} \theta_{i,k}^{(j)}$$

and

$$\bar{\bar{\theta}}_i = \frac{1}{K} \sum_{k=1}^{K} \bar{\theta}_{i,k}.$$

If each of the chains has reached and adequately probed the posterior distribution, the within-chain estimate of the variance W_i should approximately equal V_i, the combined estimate of the variance using both within- and between-chain variation. We can thus base a diagnostic on the ratio

$$G_i = \frac{V_i}{W_i}.$$

When $\sqrt{G_i}$ is close to 1 for each component of $\boldsymbol{\theta}$, say less than 1.05, it is reasonable to assume that an adequate number of updates have been performed. Gelman (1996) refers to $\sqrt{G_i}$ as the "potential scale reduction" as it approximately represents the decrease in the estimate of the posterior variance of a parameter that might result from running an MCMC algorithm longer. Values of $\sqrt{G_i}$ greater than about 1.1 suggest that additional updates should be performed.

Aside from batch means and the multichain diagnostics proposed by Gelman and Rubin (1992), many other proposals for monitoring convergence of MCMC chains now appear in the statistical literature. Indeed, most software packages that produce output from MCMC chains offer their own convergence diagnostics, and most diagnostics arrive at essentially the same conclusion for a given chain or chains. But regardless of which diagnostic one chooses to use, it is important to examine the convergence of an algorithm. Otherwise, conclusions from an analysis can be seriously flawed and even misleading.

3.2 Hierarchical Models

Many statistical applications require the specification of models that are more complex than those considered in the previous chapter. Often, more realistic models require the introduction of numerous parameters, many of which must be linked according to an underlying structure determined by application-specific constraints. The Bayesian paradigm provides a logical framework for dealing with such models.

We can best explain the notion of incorporating structural relationships between parameters through an example. To this end, consider again the launch vehicle success data in Table 2.1. As it happens, many of the rockets listed in Table 2.1 were launched more than once. Table 3.1 provides a more complete record of these vehicles' launch experience.

Table 3.1. Launch vehicle outcomes. The second column provides the number of successful launches and total number of launches for launch vehicles developed after 1980 (Johnson et al., 2005)

Vehicle	Outcome
Pegasus	9/10
Percheron	0/1
AMROC	0/1
Conestoga	0/1
Ariane 1	9/11
India SLV-3	3/4
India ASLV	2/4
India PSLV	6/7
Shavit	2/4
Taepodong	0/1
Brazil VLS	0/2

In the simple binomial model that we initially considered for these data, we used a single success probability π to model the probability that a particular vehicle was successfully launched. Examination of Table 3.1 calls this assumption into question. For example, while only 3 of the 11 initial launches of each vehicle were successful, the Pegasus rocket was subsequently launched a total of 10 times with 9 successful launches. As it happens, the Pegasus was designed and manufactured under a contract for the United States government. In contrast, several of the other vehicles were launched only once and were unsuccessful on that launch. These rockets tended to be designed by commercial manufacturers who, in most cases, lacked the design experience and financial resources of the other manufacturers. Because of these considerations, it seems somewhat unreasonable for us to assume that each of these launch vehicles would have the same probability of success on either initial or subsequent launches.

To more accurately model these launch data, it makes sense for us to introduce parameters π_i that denote the long-term probability that the launch of the ith vehicle is successful. Of course, the values of π_i for distinct launch vehicles (both in the present and for the future) must be linked if they are to be useful for predicting the success of launch vehicles that have yet to be launched. One way to make these connections is to assume that the success probabilities π_i are themselves drawn independently from a common distribution. For example, we might assume that, given parameters K and δ, the π_is are drawn from a beta distribution with parameters $K\delta$ and $K(1-\delta)$. That is, we model the population distribution of the π_is according to

$$\pi_i \,|\, K, \delta \sim Beta(K\delta, K[1-\delta]). \tag{3.2}$$

In this parameterization, δ represents the prior mean of each π_i. K controls the dispersion of the beta prior distribution. More specifically, the prior variance of the π_is is $\delta(1-\delta)/(K+1)$.

If K and δ are fixed, then this model reduces to the beta-binomial model described in Chap. 2. If, however, K and δ are regarded as parameters, then they too can be estimated from the data just as the π_is are. In this case, they are called *hyperparameters* because they are parameters that do not appear in the likelihood function. We adopt the convention that *first-stage parameters* are those model parameters that appear in the likelihood function, *second-stage parameters* represent those parameters that appear in the prior distributions of the first-stage parameters, and so on. Second- and higher-stage parameters collectively comprise the hyperparameters of a model.

Because hyperparameters do not appear in the likelihood function, we must take some care in specifying prior distributions for them. In particular, when prior densities on hyperparameters can become arbitrarily large for particular values of the hyperparameters, they often will if they are not constrained otherwise. Fortunately, such degeneracies usually occur only at special points in the parameter space, and we can avoid these points by choosing suitable prior densities for the hyperparameters.

Based on the binomial likelihood function and Eq. 3.2, it follows that the joint posterior density is proportional to

$$\begin{aligned} p(\boldsymbol{\pi}, K, \delta \,|\, \mathbf{y}) \propto & \left[\prod_{i=1}^{n} \pi_i^{y_i + K\delta - 1} (1-\pi_i)^{m_i - y_i + K - K\delta - 1} \right] \\ & \times \left[\frac{\Gamma(K)}{\Gamma(K\delta)\Gamma(K(1-\delta))} \right]^n p(K, \delta), \end{aligned} \tag{3.3}$$

where $\boldsymbol{\pi} = (\pi_1, \ldots, \pi_n)$, m_i and y_i are the number of launches and successful launches of the ith vehicle, $n = 11$ is the number of launch vehicles in the dataset, and $p(K, \delta)$ is the joint prior distribution for K and δ.

It is worth noting here that we have reparameterized the beta prior distribution on the π_is in terms of the prior mean δ and a dispersion parameter K. This facilitates the modeling of the mean of the π_is that follows in subsequent analyses. More generally, it often makes sense to parameterize statistical models in terms of meaningful parameters. This makes prior specifications more natural.

From Eq. 3.3, we see that the parameter K enters the joint posterior distribution much like the binomial denominators m_i do. It thus represents something like a *prior sample size*, describing the equivalent number of observations given to the prior distribution on the π_is. We use a $Gamma(\alpha, \lambda)$ distribution as the prior distribution for K.

The parameter δ represents the expected value of the success probabilities before observing any data. Because it represents a probability, we naturally assume a beta prior distribution with parameters we will call η and ν. The joint prior distribution for (K, δ) thus takes the form

$$p(K, \delta \mid \alpha, \lambda, \eta, \nu) \propto K^{\alpha-1} \exp(-\lambda K) \delta^{\eta-1} (1-\delta)^{\nu-1} \qquad (3.4)$$
$$K > 0,\ 0 < \delta < 1\,.$$

We obtain the joint posterior distribution for $(\boldsymbol{\pi}, K, \delta)$ by substituting Eq. 3.4 into Eq. 3.3. We specify values for the third-stage parameters of $\alpha = 5, \lambda = 1, \eta = 0.5$, and $\nu = 0.5$. Making these substitutions leads to a functional form that does not correspond to any recognizable joint distribution function. As a consequence, we must estimate the marginal posterior density functions by implementing a suitable MCMC algorithm. We describe such an algorithm in the next section.

3.2.1 MCMC Estimation of Hierarchical Model Parameters

To perform inference in the model specified in Eq. 3.3, we exploit the fact that the conditional posterior density of each π_i has a beta distribution. This permits us to specify a hybrid Gibbs/Metropolis-Hastings algorithm through the steps depicted in Fig. 3.10.

The sequence of variables generated by algorithm Fig. 3.10 converges to a sample from the posterior distribution of $(\boldsymbol{\pi}, K, \delta)$. We can monitor convergence of the algorithm using the diagnostics described earlier in this chapter or by visually examining trace plots.

3.2.2 Inference for Launch Vehicle Probabilities

We are primarily interested in the launch vehicle data of Table 3.1 because of its utility in estimating the probability that a newly developed vehicle succeeds on one of its early launches. For this reason, the parameters of interest in Eq. 3.3 are K and δ; these are the parameters that determine the distribution from which a future success probability, say π_f, is assumed to be drawn.

Example 3.4 Inference for launch vehicle success probabilities under a hierarchical model. Figure 3.11 displays histograms of the marginal posterior densities of K and δ obtained by running the algorithm described in the preceding section for 5,000 iterations. The third-stage prior densities for each of these hyperparameters are plotted on top of these histograms for reference. As we can see from the figure, the marginal posterior distribution

0. Initialize $j = 0, K^{(0)} = \alpha/\lambda = 5, \delta^{(0)} = \eta/(\eta + \nu) = 0.5$.

1. For $i = 1, \ldots n$, generate
$\pi_i^{(j)} \sim Beta(y_i + K^{(j-1)}\delta^{(j-1)}, m_i - y_i + K^{(j-1)} - K^{(j-1)}\delta^{(j-1)})$.

2. Draw z from a $Normal(0, 1)$ distribution and set $K^* = K^{(j-1)} \exp(z)$.

3. Compute $r = \frac{p(\boldsymbol{\pi}^{(j)}, K^*, \delta^{(j-1)} \mid \mathbf{y}) K^*}{p(\boldsymbol{\pi}^{(j)}, K^{(j-1)}, \delta^{(j-1)} \mid \mathbf{y}) K^{(j-1)}}$.

4. Draw u from a $Uniform(0, 1)$ density.

5. If $u \leq r$ set $K^{(j)} = K^*$. Otherwise, set $K^{(j)} = K^{(j-1)}$.

6. Generate $\delta^* \sim Beta(K^{(j)}a, K^{(j)}[1 - a])$, where $a = \frac{1}{n}\sum_{i=1}^{n} \pi_i^{(j)}$.

7. Compute $r = \frac{p(\boldsymbol{\pi}^{(j)}, K^{(j)}, \delta^* \mid \mathbf{y})(\delta^{(j-1)}/\delta^*)^{K^{(j)}a-1}}{p(\boldsymbol{\pi}^{(j)}, K^{(j)}, \delta^{(j-1)} \mid \mathbf{y})[(1-\delta^*)/(1-\delta^{(j-1)})]^{K^{(j)}(1-a)-1}}$.

8. Draw u from a $Uniform(0, 1)$ density.

9. If $u \leq r$ set $\delta^{(j)} = \delta^*$. Otherwise, set $\delta^{(j)} = \delta^{(j-1)}$.

10. Increment j and return to Step 1.

Fig. 3.10. MCMC algorithm for generating draws from the posterior distribution of the success probabilities in the hierarchical model for launch vehicle successes.

of δ is reasonably concentrated around its posterior mean of 0.58. Furthermore, the marginal posterior distribution has a decidedly different shape than its prior distribution. Apparently, the data shifted the posterior distribution away from the prior distribution, which means that the likelihood function played a dominant role in determining the posterior distribution. Because we assumed a noninformative prior distribution for δ, the posterior distribution on this parameter is likely to be fairly robust against variations in the prior distribution assumed for it.

In contrast, the marginal posterior distribution of K has a close resemblance to its prior distribution. This implies that the data were not informative in determining this parameter. Because we selected the prior distribution for K in a rather ad hoc fashion, we should assess the sensitivity of inferences drawn from the joint posterior distribution specified in Eq. 3.3 as the prior distribution for K is varied. We might accomplish this by refitting the model with different values of α and λ and examining how inferences concerning the marginal distribution K and other model parameters change. Performing this *sensitivity analysis* is left as Exercise 3.6.

Finally, we can estimate the posterior predictive distribution on π_f, the success probability for a future launch vehicle, directly from the MCMC algorithm described in the previous section. To make this estimate, we draw a predefined number of samples of π_f from the beta densities defined from the sampled values of $K^{(j)}$ and $\delta^{(j)}$. For example, Fig. 3.12 provides a histogram and kernel density estimate of the posterior predictive distribution of π_f that we obtained by drawing one value of π_f from each $Beta(K^{(j)}\delta^{(j)}, K^{(j)}[1-\delta^{(j)}])$ density obtained in the MCMC algorithm. Assuming that the prior distribution assumed for K is adequate, Fig. 3.12 indicates that there is substantial uncertainty in the likely values of the long-term success probabilities of future vehicles. The 90% credible interval for a single value of π_f extends from 0.16 to 0.93.

3.3 Empirical Bayes

Before the advent of MCMC algorithms, fitting hierarchical models of the type described in the previous section was not feasible. Instead, practitioners were forced to utilize models that approximated the type of structure that is now routinely incorporated into hierarchical models by estimating hyperparameters "off-line," or outside of the formal model framework. The resulting *empirical Bayes'* approach bases estimates of hyperparameters on data already incorporated into the likelihood function, which causes something of a philosophical dilemma and, of course, formally invalidates the use of Bayes' Theorem for purposes of inference. Nonetheless, as the number of units (e.g., launch vehicles) in the first stage of a hierarchical model becomes large, this

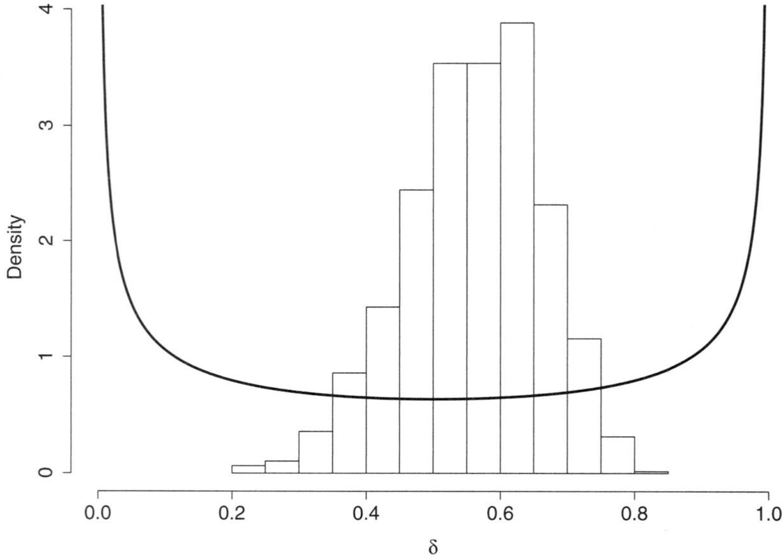

(a)

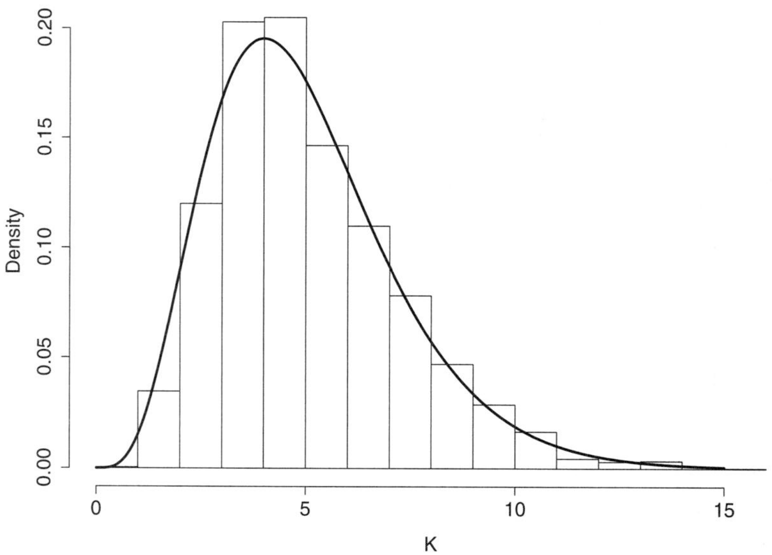

(b)

Fig. 3.11. Histogram of the marginal posterior distributions of δ and K. The prior densities chosen for these parameters are depicted as solid lines.

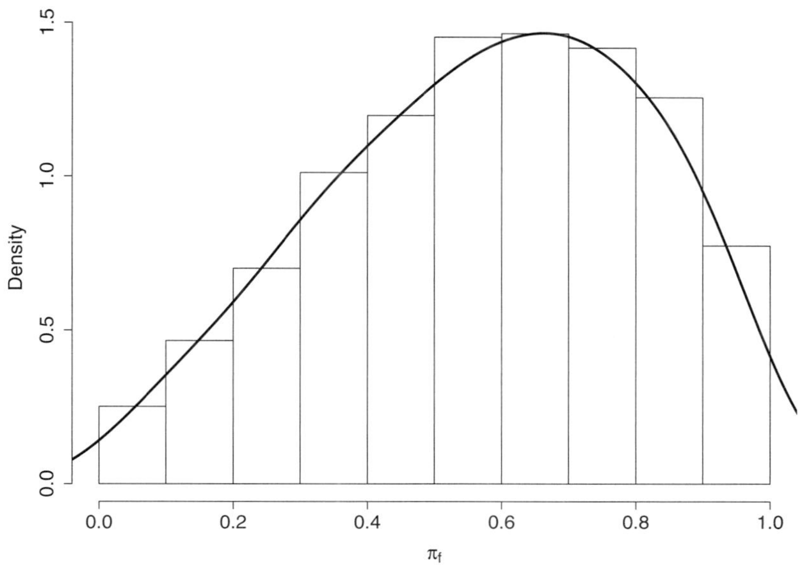

Fig. 3.12. Histogram and posterior predictive distribution of π_f.

double counting of data to estimate hyperparameters at higher stages in a model would seem to be only a minor violation of the paradigm. It also alleviates the problem of having to specify prior distributions on hyperparameters for which interpretation may be difficult anyway.

Although we do not consider empirical Bayes' methods in the chapters that follow, we illustrate them here for completeness. Returning to the model of the previous section, consider again the prior distribution specified in Eq. 3.2. In this equation, δ represents the population mean of the π_is, while K represents a dispersion parameter related to the variance of the π_is around δ. Specifically, the prior variance of the π_is is

$$\text{Var}(\pi_i) = \frac{\delta(1-\delta)}{K+1}.$$

To apply empirical Bayes' methodology in this model, we must obtain data-based estimates of the hyperparameters δ and K. If the values of π were known a priori, then we could use moment-based estimators of δ and K. Unfortunately, none of the π_is are actually known, and so we must instead use estimates of the π_is for this purpose. In this context, the maximum likelihood estimates (MLE) $\widehat{\pi}_i = y_i/m_i$ are an obvious choice of estimators. Ignoring the fact that several of the $\widehat{\pi}_i$ are based on only one or two observations, using them to obtain empirical Bayes' estimates of δ and K, we get

$$\widehat{\delta} = \frac{1}{n} \sum_{i=1}^{n} \widehat{\pi}_i$$

and

$$\widehat{K} = \frac{\widehat{\delta}(1-\widehat{\delta})}{\widehat{\text{Var}}(\widehat{\pi}_i)} - 1,$$

where

$$\widehat{\text{Var}}(\widehat{\pi}_i) = \frac{1}{n-1} \sum_{i=1}^{n} (\widehat{\pi}_i - \widehat{\delta})^2.$$

We can substitute these estimated values δ and K into Eq. 3.2 to obtain closed-form expressions for the posterior distributions on each of the π_is, or a predictive distribution for future values of π_f.

The advantages of the empirical Bayes' approach in this context are clear: it eliminates the need to implement numerical algorithms to estimate marginal posterior densities for first-stage model parameters, and it provides a reasonable method for estimating the hyperparameters δ and K in the absence of actual expert opinion regarding their values.

A disadvantage of this empirical Bayes' procedure is that it can lead to misleadingly precise estimates of model parameters. The estimates $\widehat{\delta}$ and $\widehat{K}$ are based on the assumption that the values of π_i are known; in fact, the uncertainty in the individual values of π_i is quite high. Furthermore, even if the values of π_i were known exactly, there were only 11 launch vehicles available for this analysis. The moment estimates of δ and K would thus still be subject to substantial sample variability. Predictions based on this empirical Bayes' model would account for neither of these sources of variation.

3.4 Goodness of Fit

No statistical analysis is complete without testing the adequacy of the models upon which the analysis is based. In this section, we describe a general-purpose, goodness-of-fit diagnostic that can be applied to most of the models described in subsequent chapters.

The model diagnostic described below is motivated by Pearson's chi-squared goodness-of-fit diagnostic (Pearson, 1900), which, for a completely specified model (i.e., one in which there are no unknown parameters), may be defined as follows.

Let $y_1, \ldots, y_n$ denote an independent and identically distributed sample drawn from distribution $f(y \mid \boldsymbol{\theta})$, with corresponding cumulative distribution function $F(y \mid \boldsymbol{\theta})$, but now suppose that the value of $\boldsymbol{\theta}$ is known. Let $0 = a_0 < a_1 < \ldots < a_{K-1} < a_K = 1$ denote predetermined quantiles from a uniform distribution, and define $p_j = a_j - a_{j-1}$. Finally, let m_j denote the number of observations y_i for which $a_{j-1} < F(y_i \mid \boldsymbol{\theta}) \leq a_j$. Then Pearson's chi-squared test statistic, say R^0, may be defined as

$$R^0 = \sum_{k=1}^{K} \frac{(m_k - np_k)^2}{np_k} . \tag{3.5}$$

In this setting, Pearson (1900) shows that R^0 follows a chi-squared distribution having $K - 1$ degrees of freedom when the specified model is true. We can thus perform goodness-of-fit tests by comparing the observed value of R^0 to its nominal chi-squared distribution.

The assumptions required for R^0 to achieve its nominal chi-squared distribution preclude its use in more complicated settings. In particular, rarely is $\boldsymbol{\theta}$ known, and data are often not identically distributed. Cramér (1946) and Chernoff and Lehmann (1954) address the first issue by extending the chi-squared statistic to cases in which the value of $\boldsymbol{\theta}$ is estimated by either the minimum chi-squared method or grouped maximum likelihood estimation (Cramér, 1946) or by standard maximum likelihood procedures (Chernoff and Lehmann, 1954).

In Cramér (1946), the statistic has a chi-squared distribution having $K - q - 1$ degrees of freedom in large samples, where q is the dimension of the parameter $\boldsymbol{\theta}$. But because minimum chi-squared or grouped maximum likelihood estimation of model parameters is seldom appropriate in practice, this version of the chi-squared statistic is generally not useful. The exception occurs in the case of contingency tables, where grouped maximum likelihood estimation and standard maximum likelihood estimation are equivalent.

Chernoff and Lehmann (1954) shows that the distribution of R^0 falls stochastically between chi-squared distributions having $K - 1 - q$ and $K - 1$ degrees of freedom when $\widehat{p}_k$, the estimate of p_k based on the MLE, is substituted for p_k in Eq. 3.5. However, in complicated models the difference between $K - 1 - q$ and $K - 1$ can be substantial, rendering this version of the goodness-of-fit test ineffective for high-dimensional parameters.

We can define a Bayesian version of Pearson's goodness-of-fit statistic by loosening the standard assumptions in two ways. First, rather than using an optimal value (like the MLE) of $\boldsymbol{\theta}$ to estimate p_k in Eq. 3.5, we use a randomly sampled draw of $\boldsymbol{\theta}$, say $\tilde{\boldsymbol{\theta}}$, from the posterior distribution. Second, by redefining the bin counts m_j according to the number of observations y_i for which $a_{j-1} < F_i(y_i \,|\, \tilde{\boldsymbol{\theta}}) \leq a_j$, we eliminate the restriction that the data be identically distributed. Here, $F_i(y_i \,|\, \tilde{\boldsymbol{\theta}})$ denotes the conditional distribution function of the ith observation given the sampled value of $\tilde{\boldsymbol{\theta}}$. Letting $m_k(\tilde{\boldsymbol{\theta}})$ denote the bin counts based on $\tilde{\boldsymbol{\theta}}$, a Bayesian version of the chi-squared test statistics for goodness of fit may be defined as

$$R^B(\tilde{\boldsymbol{\theta}}) = \sum_{k=1}^{K} \frac{(m_k[\tilde{\boldsymbol{\theta}}) - np_k]^2}{np_k} . \tag{3.6}$$

We refer to the use of this statistic throughout the book as the *Bayesian* χ^2 *goodness-of-fit test*. In general, $K \approx n^{0.4}$, where n is the sample size, often represents a good choice for K. Interestingly, for large n, the distribution of R^B is chi-squared on $K-1$ degrees of freedom, independently of the dimension of $\boldsymbol{\theta}$ (Johnson, 2004). This is an important feature of the statistic because no adjustment must be made for the dimension of the parameter vector $\boldsymbol{\theta}$. Also, by allowing for nonidentically distributed data (i.e., different sampling densities for the individual observations y_i, given the single, sampled parameter vector $\tilde{\boldsymbol{\theta}}$), this diagnostic can be extended to a much broader range of models than can the classical chi-squared statistic. For example, it readily extends to many random effects and spatial models.

In principle we would prefer to base our goodness-of-fit statistic on more than a single sampled value from the posterior distribution. To do so, we might consider taking the average value of R^B, averaged with respect to the posterior distribution on the parameter space. That is, we could repeatedly sample values of $\boldsymbol{\theta}$ from the posterior distribution, and then average the values of R^B. Unfortunately, the sample mean of the R^B values obtained from this procedure does not have a known reference distribution — in particular, this sample mean does not have $ChiSquared(K-1)$ distribution. A simple strategy that we can use to overcome this problem is simply to report the proportion of R^B values that exceed a specified critical value from their known $ChiSquared(K-1)$ reference distribution. Thus, we might report that, say, 50% of the R^B values calculated from draws of $\boldsymbol{\theta}$ from the posterior distribution exceeded the 0.95 quantile from the reference chi-squared distribution. Such a finding would clearly suggest a problem with the model fit. More sophisticated methods for assessing the significance of a sample of R^B values can be found in Johnson (2007).

Example 3.5 Goodness of fit in a lognormal random effects model. Consider the fluid breakdown data of Table 2.3, and suppose that each column of that table represents the viscosity breakdown time of a sample measured on a distinct testing device. That is, suppose that 10 devices were used to test samples from this batch, and that the columns in Table 2.3 record breakdown times measured on the separate devices.

As most experimentalists know, when two supposedly "identical" testing devices are used to take measurements on the same item, the measurements recorded on the two devices will usually not be exactly identical. The difference in the measurements can often be nonnegligible, and so in many applications it is important to account for *measurement errors* that can be attributed to the measuring device. In this case, we can account for the fact that different measurement devices were used to measure the fluid breakdown times by revising our model so that it includes *random effects* for the testing devices. Statistically, we can accomplish this by assuming that the measured breakdown times are generated from a model of the form

$$Y_{ij} = \mu + \beta_j + \epsilon_{ij},$$

where Y_{ij} is the logarithm of the breakdown time of the ith sample tested on the jth device, μ is the overall viscosity breakdown time for the batch (assumed to be the same for the entire batch), β_j is the random effect (that is, measurement error) attributable to the jth device, and $\epsilon_{ij} \sim Normal(0, \sigma^2)$ is the measurement repeatability. (Note that the breakdown times T_{ij} have a $Lognormal(\mu+\beta_j, \sigma^2)$ distribution, where $Y_{ij} = \log(T_{ij})$.) To finish specifying the model, we further assume the following prior distributions for the first-stage model parameters:

$$\begin{aligned}
\beta_j \mid \sigma^2, \kappa &\sim Normal(0, \kappa\sigma^2), \\
\kappa &\sim InverseGamma(3.5, 0.25), \\
\sigma^2 &\propto \frac{1}{\sigma^2}, \\
\mu &\propto \text{constant}.
\end{aligned}$$

Notice that we use a hierarchical model for the random effects β_j. We have introduced the parameter κ as a mechanism for modeling the fact that the bias attributable to the jth measuring device is likely to be a fraction of the size of the variability between samples. We assume that the prior mean for κ is 1/10.

With these prior distributions, we can use a Gibbs sampler to sample from the posterior distribution on the parameter space using the implied full conditional densities, which may be specified as

$$p(\mu \mid \beta, \sigma^2, \kappa, \mathbf{y}) \sim Normal\left(\frac{1}{50}\sum_{i=1}^{5}\sum_{j=1}^{10}(y_{ij} - \beta_j), \frac{\sigma^2}{50}\right),$$

$$p(\beta_j \mid \beta_{\mathbf{i \neq j}}, \mu, \sigma^2, \kappa) \sim Normal\left(\frac{\sum_{i=1}^{5}(y_{ij} - \mu)/\sigma^2}{5/\sigma^2 + 1/(\kappa\sigma^2)}, \frac{1}{5/\sigma^2 + 1/(\kappa\sigma^2)}\right),$$

$$p(\sigma^2 \mid \beta, \mu, \kappa, \mathbf{y}) \sim$$
$$InverseGamma\left(30, \frac{1}{2}\sum_{i=1}^{5}\sum_{j=1}^{10}(y_{ij} - \beta_j - \mu)^2 + \frac{1}{2}\sum_{j=1}^{10}\frac{\beta_j^2}{\kappa}\right),$$

and

$$p(\kappa \mid \beta, \mu, \sigma^2, \mathbf{y}) \sim InverseGamma\left(8.5, 0.25 + 0.5\sum_{j=1}^{10}\beta_j^2/\sigma^2\right).$$

The derivation of these full conditional distributions is left as Exercise 3.7.

To apply the Bayesian χ^2 goodness-of-fit test to these data, a Gibbs sampler using these conditional distributions was burned-in for 1,000 iterations. The value of the parameter vector sampled on the 1,001st iteration was

$$\boldsymbol{\beta}^T = (0.116, -0.039, 0.224, 0.186, 0.107, -0.194, -0.062, 0.128, -0.049, -0.027)$$

$$\mu = 2.035, \qquad \sigma^2 = 0.125, \qquad \kappa = 0.076.$$

We can use these values of the parameter vector to compute the Bayesian χ^2 goodness-of-fit test as follows.

First, we select quantile values that will be used to determine the bins in our chi-squared test. For purposes of illustration, we will use five equally spaced bins corresponding to $\mathbf{a} = (0.0, 0.2, 0.4, 0.6, 0.8, 1.0)$.

Second, we calculate the value of the normal cumulative distribution function for each observation at its conditional mean and variance based on the parameter values listed above; that is, we compute

$$\Phi(y_{ij} \,|\, \mu + \beta_j, \sigma^2).$$

The value of this distribution function for the first observation ($y_{11} = \log(5.45) = 1.696$) is $\Phi(1.696 \,|\, 2.035 + 0.116, 0.125) = 0.099$, so we assign the first observation to the first bin. Repeating this calculation for each observation, we find that 12 counts are assigned to the first bin, 14 counts are assigned to the second, 5 to the third, 9 to the fourth, and 10 to the fifth. Consequently, $\mathbf{m} = (12, 14, 5, 9, 10)$.

Finally, we compute the value of the test statistic for this sampled parameter value according to Eq. 3.6 as

$$R^B = \sum_{k=1}^{5} \frac{(m_k - 10)^2}{10} = 4.6.$$

Comparing this value to a $ChiSquared(4)$ distribution, we find that 4.6 does not exceed $\chi^2_{4,0.95} = 9.49$. Thus, at least for this sampled value of $(\mu, \beta, \sigma^2, \kappa)$, there is little evidence to suggest that the random effects model does not provide an adequate fit to the data.

When we repeat this computation for each posterior draw, 3% of R^B values exceeded the 0.95 quantile of a $ChiSquared(4)$ distribution. Because we expect around 5% of the values to exceed this value in repeated sampling of data and R^B values from this model, the observed value of 3% indicates that the model seems to fit well, at least with regard to this test criterion.

In the case of discrete data, we can modify the Bayesian χ^2 goodness-of-fit test in one of two ways in order to account for the fact that the probability mass function assigned to an observation can "straddle" the quantiles $\mathbf{a} = (a_0, \ldots, a_K)$. The first modification involves simple randomization to a bin cell. Specifically, drawing

$$g_i \sim Uniform[F_i(y_i - 1 \,|\, \tilde{\boldsymbol{\theta}}), F_i(y_i \,|\, \tilde{\boldsymbol{\theta}})]\,, \tag{3.7}$$

we may then redefine the bin counts $m_j(\tilde{\boldsymbol{\theta}})$ according to whether $a_{j-1} < g_i \leq a_j$. With this randomization, the distribution of R^B retains its $ChiSquared(K-1)$ distribution.

The discrete-data adjustment described in the last paragraph is recommended for data that take on a relatively broad range of values. For example, this adjustment might be applied for Poisson count data with means ranging above 2 or 3. However, for binary or multinomial data, it is usually better to fix the bins and bin counts (as is normally done), and to instead regard the bin probabilities p_j as being functions of the sampled value $\tilde{\boldsymbol{\theta}}$. That is, for each observation, we compute the probability that y_i takes one of the discrete values that fall into the kth bin according to $F_i(y_i \mid \tilde{\boldsymbol{\theta}})$. The definition of $R^B(\tilde{\boldsymbol{\theta}})$ then becomes

$$R^B(\tilde{\boldsymbol{\theta}}) = \sum_{k=1}^{K} \frac{[m_k - np_k(\tilde{\boldsymbol{\theta}})]^2}{np_k(\tilde{\boldsymbol{\theta}})}.$$

Example 3.6 Goodness of fit in a lifetime data model. In Chap. 4, we analyze the lifetimes of $n = 31$ liquid crystal display (LCD) projector lamps with data in Table 4.2. Here, we briefly summarize the analysis, and then use the Bayesian χ^2 goodness-of-fit test to assess model fit.

In Example 4.3, we assume that the lifetimes of the LCD projector lamps have an $Exponential(\lambda)$ distribution and that λ has a $Gamma(1.7, 2550)$ prior distribution. Since the gamma distribution is the conjugate prior density for λ, we can analytically obtain a $Gamma(32.7, 20457)$ posterior distribution for λ. See Example 4.3 for more details.

In applying the Bayesian χ^2 goodness-of-fit test, we use the suggested $K = 31^{0.4} \approx 4)$ number of equal probability bins so that $\mathbf{a} = (0, 0.25, 0.5, 0.75, 1)$. We make a draw from the $Gamma(32.7, 20457)$ posterior distribution, say $\tilde{\lambda} = 0.00195$. For the first lifetime of 387 hours, $F_1(387|\tilde{\lambda}) = 0.530$, where $F_1(t|\lambda)$ is the exponential cumulative distribution function. Consequently, the first lifetime belongs to the third bin.

Processing the rest of the lifetimes yields the bin counts $\mathbf{m} = (5, 9, 9, 8)$ and evaluating R^B in Eq. 3.6 with expected bin counts $n\mathbf{p} = (7.75, 7.75, 7.75, 7.75)$, we obtain $R^B = 1.387M\chi^2_{3,0.95} = 7.81$. We repeat this calculation for 100,000 draws from the posterior distribution and find that only 0.1% of the distribution falls below 0.05, again suggesting no lack of fit.

In the case of censored data, we can modify the Bayesian χ^2 goodness-of-fit test to account for the fact that we are uncertain into which bin a censored observation falls. Specifically, for interval-censored observation (a_i, b_i), we draw

$$g_i \sim Uniform[F_i(a_i \mid \tilde{\boldsymbol{\theta}}), F_i(b_i \mid \tilde{\boldsymbol{\theta}})].$$

We then redefine the bin counts $m_j(\tilde{\boldsymbol{\theta}})$ according to whether $a_{j-1} < g_i \leq a_j$. With this randomization, the distribution of R^B retains its $ChiSquared(K-1)$ distribution. Note that for right-censored data, $b_i = \infty$, so that

$$g_i \sim Uniform[F_i(a_i \,|\, \tilde{\boldsymbol{\theta}}), 1]\,, \tag{3.8}$$

and for left-censored data, $a_i = 0$, so that

$$g_i \sim Uniform[0, F_i(b_i \,|\, \tilde{\boldsymbol{\theta}})]\,.$$

3.5 Related Reading

Additional introductory material about the Gibbs sampling and Metropolis-Hastings algorithms can be found in Casella and George (1992) and Chib and Greenberg (1995). Readers interested in an expanded treatment of MCMC methods may consult a variety of texts, including Gilks et al. (1996), Gamerman (1997), Robert and Casella (2004), and Marin and Robert (2007). Hierarchical models are introduced in Lindley and Smith (1972). More information about empirical Bayes' methods can be found in Carlin and Louis (1996) and Kass and Steffey (1989).

3.6 Exercises for Chapter 3

3.1 Suppose that $X \sim Normal(0, 1)$ and $Y = \exp(X)$.
  a) Use the change of variables technique to calculate the probability density function, mean, and variance of Y.
  b) Draw a random sample $X_1, \ldots, X_{10,000}$ from a $Normal(0, 1)$ distribution. Set $Y_i = \exp(X_i)$. Draw a histogram of the Y_i and overlay a plot of the probability density function of Y.
  c) Estimate the probability density function, mean, and variance of Y using the random sample.

3.2 Suppose we perform an experiment where the data have a $Poisson(\lambda)$ sampling density. We describe our uncertainty about λ using a *Gamma* prior density with parameters α and β. We also describe our uncertainty about α and β using independent *Gamma* prior densities.
  a) Simulate 50 observations from a Poisson distribution with parameter $\lambda = 5$.
  b) Choose diffuse prior densities for α and β.
  c) Implement an MCMC algorithm to calculate posterior densities for λ, α, and β.
  d) Is $\lambda = 5$ contained in a 90% posterior credible interval for λ?

3.3 Consider again the fluid breakdown times introduced in Sect. 2.5. Two models were proposed for these data. The first incorporated a normal likelihood function and a noninformative prior density; the second, a normal likelihood function and a conjugate inverse-gamma/normal prior density. Now suppose that the properties of the manufacturing process were controlled when these samples of lubricant were produced so that it is known that the true mean of the sample values must lie between 6.0 and 7.4 (on the original measurement scale). No further information is available concerning the value of the variance parameter σ^2. Assume that the joint prior density for (μ, σ^2) is proportional to $1/\sigma^2$ whenever $\mu \in (\log(6.0), \log(7.4))$, and is 0 otherwise.

a) Find an expression for a function that is proportional to the joint posterior density.
b) Describe a hybrid Gibbs/Metropolis-Hastings algorithm for sampling from the joint posterior density.

3.4 Implement the algorithm in Fig. 3.1. Use batch means to compute the simulation error.

3.5 Implement the algorithm in Fig. 3.4. Calculate the autocorrelation for the chain.

3.6 In the analysis of the launch vehicle success probabilities described in Example 3.4, the hyperparameters α and λ were assigned values of 5 and 1, respectively.

a) Perform a sensitivity analysis for α and λ by varying their values over a suitable range.
b) Report how changes in the values assumed for K and α impact the posterior means of other model parameters.

3.7 Derive the conditional densities described in Example 3.5 for the random effects model.

4

Component Reliability

This chapter presents models for various types of component reliability data, which consist of sampling and prior distributions. Several examples with real data, including some for which the data are censored, illustrate the use of these models in assessing component reliability. The complexity of some of these examples requires the use of hierarchical models. This chapter also introduces methods for model selection.

4.1 Introduction

Component reliability is the foundation of reliability assessment and refers to the reliability of a single item; this item may be an integrated system or a minor component within a large system. Component reliability data can be discrete, as are success/failure or failure count data, or continuous, as are failure time data. In this chapter, we consider success/failure, failure count, and failure time data. See Chap. 8, which assesses component reliability with degradation data.

Analyzing reliability data begins by choosing an appropriate probability model, which in the Bayesian approach, includes a sampling distribution for the data and a prior distribution for the parameters on which the sampling distribution depends. This chapter introduces several sampling distributions for discrete and continuous data. To complete the model, we discuss appropriate prior distributions for each sampling distribution's parameters.

Often, failure time data are censored, which means that the analysts do not know the failure times exactly. In this chapter, we will see how the Bayesian approach provides a unified framework to handle various types of censored data as well as exact failure time data.

In assessing component reliability, the situation can be more complicated and warrant more complex models like the hierarchical models introduced in Sect. 3.2. Finally, in any reliability data analysis, we must consider model selection. In this chapter, model selection means choosing a distribution that is appropriate for the reliability data.

This chapter discusses such topics as sampling distributions, prior distributions, censored data, hierarchical models, and model selection.

4.2 Discrete Data Models for Reliability

In this section, we present models for discrete reliability data. In turn, we consider success/failure data and failure count data.

4.2.1 Success/Failure Data

In certain situations, analysts may use *success/failure data* in reliability assessments, which capture the component's success or failure to perform its intended function. For example, testers may try an emergency diesel generator (EDG) to see if it will start on demand, and record whether the EDG starts or not. Another example is a missile system, for which testers record whether it executes its mission successfully or not when launched. We model such data with the *binomial distribution*. The binomial distribution is appropriate for a fixed number of tested components n, where the tests are assumed to be conditionally independent given the success probability π. The binomial probability density function is

$$f(x|n, \pi) = \binom{n}{x} \pi^x (1 - \pi)^{n-x} \, , \; x = 0, \ldots, n \, , \tag{4.1}$$

where π is the success probability and $0 \leq \pi \leq 1$. Note that the binomial distribution is not an appropriate model if the tests are dependent, and that it only applies if all the items have the same success probability. Also, the *Bernoulli distribution* is a special case of the binomial distribution for $n = 1$ test.

As mentioned in Chap. 1, sometimes analysts will treat failure time data as success/failure data with respect to a specified time t; that is, x is the number of failures before t of n tested items. The primary reason for doing this is to avoid having to specify a failure time model. However, much information is lost, so that we do not recommend this as a general practice.

For the binomial model, the success probability π is the unknown model parameter that the analyst wants to estimate. In a model of success/failure data, Eq. 4.1, viewed as a function of π for observed number of successes x, is the likelihood function for binomial data. If there are m binomial datasets, say $x_1, \ldots, x_m$ successes out of $n_1, \ldots, n_m$ tests, then under conditional independence and constant success probability π, the likelihood function consists of the product of the m individual likelihood functions specified by Eq. 4.1. To complete the model, the analyst must specify a prior distribution for π. In the next section, we present a commonly used prior distribution for π.

A Prior Distribution for Binomial Data

A convenient choice for a prior distribution for π is one that is conjugate. Recall that Chap. 2 defines a conjugate prior distribution as one that has the same form as the posterior distribution. The conjugate prior distribution for binomial data is the beta distribution:

$$p(\pi|\alpha,\beta) = \frac{\Gamma(\alpha+\beta)}{\Gamma(\alpha)\Gamma(\beta)}\pi^{\alpha-1}(1-\pi)^{\beta-1}\,, \quad 0 \le x \le 1\,, \quad \alpha > 0\,, \quad \beta > 0\,,$$

where we interpret α as the prior number of successful component tests and β as the prior number of failed component tests; that is, $\alpha + \beta$ is like a prior sample size. See Appendix B for details of the gamma function, denoted by $\Gamma(\cdot)$. Note that the beta distribution is a natural choice as a prior distribution for π, because the support of a beta distribution is the interval (0,1). In Chap. 2, we showed that the posterior distribution of π (conditioned on x observed successes out of n tests) has the form

$$p(\pi|x) \propto \pi^{\alpha+x-1}(1-\pi)^{\beta+n-x-1}\,.$$

That is, the posterior distribution of π given x is

$$\pi|x \sim Beta(\alpha+x, \beta+n-x)\,. \tag{4.2}$$

Example 4.1 Binomial model for EDG demand data. Martz et al. (1996) presents demand data for EDGs in U.S. nuclear power plants. Consider the 1988–1991 data for plant number 63, which consists of $x = 212$ successful starts in $n = 212$ demands on the EDGs. As an illustration, we use plant 62's data to develop a $Beta(\alpha = (273/278)27.8, \beta = (1 - 273/278)27.8)$ prior distribution for the probability of successful start π; plant number 62 had 273 successful starts in 278 demands so we use 273/278 as a point estimate for π, but treat the binomial data as coming from a sample size of 27.8 or 10% of the plant 62's sample size. Consequently, from Eq. 4.2, π for plant number 63 has a $Beta(\alpha + x, \beta + n - x) = Beta(239.3 = (273/278)28.3 + 212, 0.5 = (1 - 5/278)28.3 + 0)$ posterior distribution. Figure 4.1 presents the prior and posterior distributions for the successful start probability π, which shows that the posterior distribution (solid line) is more peaked than the prior distribution (dashed line), and that the demand data provide more evidence in favor of a high successful start probability π.

4.2.2 Failure Count Data

Failure count data record the number of times that a component fails in a specified period of time, where we can either repair the component immediately and put it back on test or replace it with another component. Such data

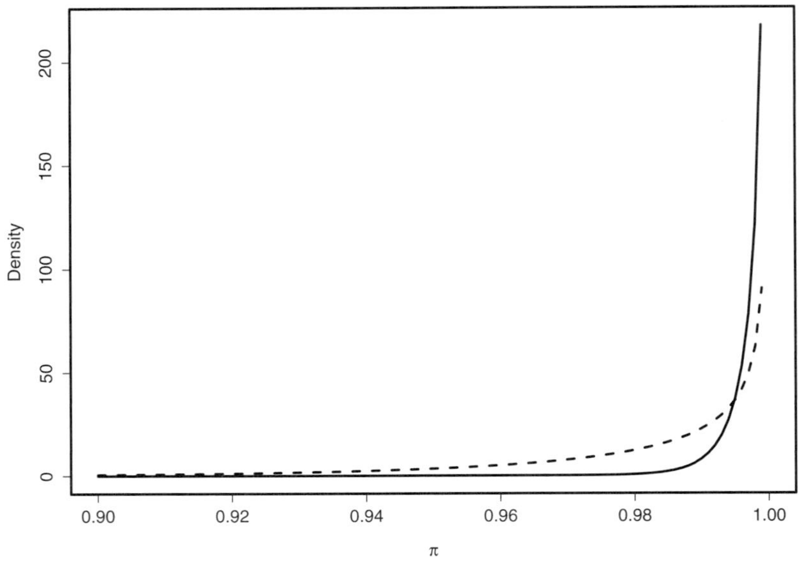

Fig. 4.1. Prior (dashed line) and posterior (solid line) distributions of the successful start probability π for the EDG example.

may arise because of limitations of the data capture system or the way that data are reported, e.g., a system may only keep track of the monthly number of failures and report them. The basic model for failure count data is the *Poisson distribution*, which is appropriate when the probability of events occurring in disjoint time intervals is independent, and when the probability of two events occurring in a short time period is small. The Poisson probability density function is

$$f(y|\lambda) = \frac{(\lambda t)^y \exp(-\lambda t)}{y!}, \tag{4.3}$$

where the observed number of failures $y = 0, 1, 2, \ldots, \lambda > 0$ is the mean number of failures per unit time, and t is the length of the specified time period. For repairable components, Chap. 6 refers to λ as the intensity. Note that the equal mean and variance (here, λt) is the most limiting characteristic of the Poisson distribution.

For the Poisson model, the mean number of failures per unit time λ is the unknown model parameter the analyst wants to estimate. In a model of failure count data, Eq. 4.3 viewed as a function of λ given the observed failure count y is the appropriate likelihood function for failure count data. If there are n failure counts, say $y_1, \ldots, y_n$ in time periods of lengths $t_1, \ldots, t_n$, then under conditional independence and constant λ, the likelihood function consists of the product of the m individual likelihood functions specified by Eq. 4.3.

To complete the model, the analyst must specify a prior distribution for λ. In the next section, we consider a prior distribution for λ.

A Prior Distribution for Poisson Data

A commonly used prior distribution for the mean number of failures per unit time λ of the Poisson distribution given in Eq. 4.3 is the gamma distribution. A major reason for its use is that it is the conjugate prior distribution for λ, as well as it having positive support. That is, the gamma prior distribution and the Poisson likelihood function have the same form, so that the resulting posterior distribution for λ is also a gamma distribution. Notationally, if

$$Y_i \sim Poisson(\lambda t_i)\,,\; i = 1, \ldots, n\,,$$

where $y_1, \ldots, y_n$ are n observed failure counts, and the prior distribution for λ is

$$\lambda \sim Gamma(\alpha, \beta)\,,$$

then the posterior distribution of λ is

$$\lambda|\mathbf{y} \sim Gamma(\alpha + \sum_{i=1}^{n} y_i, \beta + \sum_{i=1}^{n} t_i)\,, \tag{4.4}$$

where $\mathbf{y} = (y_1, \ldots, y_n)$. By inspecting Eq. 4.4, we interpret β as a prior sample size in contrast with the data sample size $\sum_{i=1}^{n} t_i$ and α as the prior number of failures in contrast with the observed number of failures $\sum_{i=1}^{n} y_i$. We leave the derivation of Eq. 4.4 as Exercise 4.3.

Example 4.2 Poisson model for supercomputer failure count data. Consider modeling the monthly number of failures of the Los Alamos National Laboratory Blue Mountain supercomputer components (shared memory processors or SMPs) by a Poisson distribution. The supercomputer consists of 47 "identical" SMPs and Table 4.1 presents their monthly number of failures for the first month of operation. Let $y_1, \ldots, y_{47}$ denote the monthly number of failures recorded for the SMPs. With $t_i = 1$ month, we model the failure count data by the Poisson distribution given in Eq. 4.3 as

$$Y_i \sim Poisson(\lambda t) = Poisson(\lambda)\,,\; i = 1, \ldots, 47\,,$$

where λ is the mean monthly number of failures.

The supercomputer engineers expect that there should be no more than 10 failures for each component in the first month of operation. One way to represent this prior information is to assume a gamma prior distribution for λ with a mean of five. As discussed above, we can express this prior information by

Table 4.1. Monthly number of failures for 47 supercomputer components

1	5	1	4	2	3	1	3	6	4	4	4
2	3	2	2	4	5	5	2	5	3	2	2
3	1	1	2	5	1	4	1	1	1	2	1
3	2	5	3	5	2	5	1	1	5	2	

$$\lambda \sim Gamma(\alpha = 5, \beta = 1) .$$

Note that for the $Gamma(5, 1)$ prior distribution, the probability that λ exceeds 10 is 0.03. Using Eq. 4.4, the posterior distribution of λ given the failure data $\mathbf{y} = (y_1, \ldots, y_{47})$ is

$$\begin{aligned} \lambda|\mathbf{y} &\sim Gamma(\alpha + \textstyle\sum_{i=1}^{n} y_i, \beta + \sum_{i=1}^{n} t_i) \\ &= Gamma(5 + 132, 1 + 47) = Gamma(137, 48) . \end{aligned}$$

Figure 4.2 presents the prior and posterior distributions for the mean monthly number of failures λ. Note the relatively diffuse prior (dashed line) distribution and the very peaked posterior (solid line) distribution, which indicates that the failure count data provide substantial evidence for a lower mean monthly number of failures than the engineers expected. The posterior mean monthly number of failures is

$$\mathrm{E}(\lambda \,|\, \mathbf{y}) = \alpha^* / \beta^* = 137/48 = 2.85 ,$$

the posterior standard deviation is

$$\sqrt{\mathrm{Var}(\lambda \,|\, \mathbf{y})} = \sqrt{\alpha^* / \beta^{*2}} = 0.24 ,$$

and a 95% credible interval is (2.40, 3.35) monthly failures.

To support the claim that a Poisson distribution models the supercomputer failure count data well, we can apply a Bayesian χ^2 goodness-of-fit test. Remember to use the modification given in Eq. 3.7 for discrete data. Based on $K = 5$ (approximately $47^{0.4}$) equal probability bins, repeatedly make draws from the $Gamma(137, 48)$ posterior distribution of λ, calculate the R^B test statistic, and compare it against the 0.95 quantile of the $ChiSquared(4)$ reference distribution. We find that 3.9% of the R^B values exceed this 0.95 quantile, which shows no lack of fit.

4.3 Failure Time Data Models for Reliability

Perhaps the most commonly used data to assess component reliability are *failure time data*, which record the continuous time to failure of the components. Other examples of failure time data are "time to death" used in survival analysis and "time to interrupt" that arises in software reliability. In

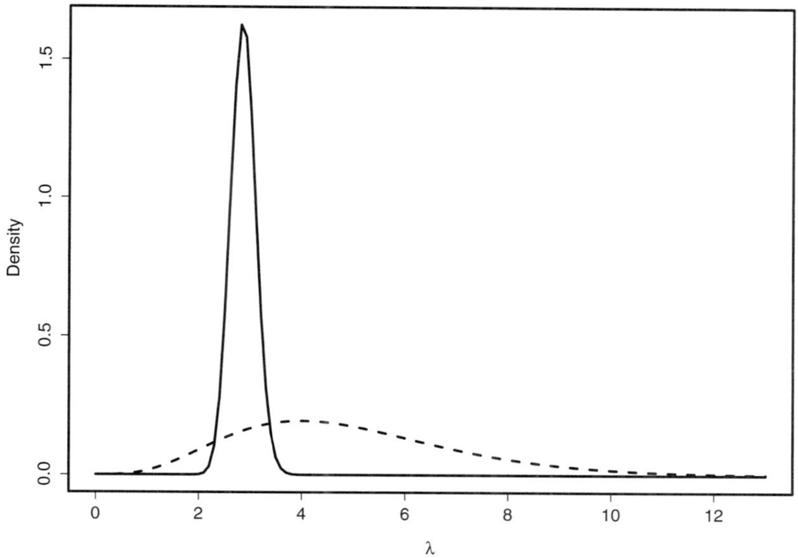

Fig. 4.2. Prior (dashed line) and posterior (solid line) distributions of the mean monthly number of failures λ for the supercomputer example.

general, failure time data record "time to some event." The reliability literature also refers to failure time data as lifetime data, and we use both terms interchangeably throughout this book.

This section presents several standard failure time models. These models differ in the number of parameters, which reflect shape, location, and scale, and for a particular application, provide a variety of hazard functions to choose from. In turn, we consider the exponential, Weibull, lognormal, gamma, inverse Gaussian, and normal failure time models. To complete these models, we also present some useful prior distributions for their model parameters.

4.3.1 Exponential Failure Times

We begin with the *exponential distribution* as a model for failure time data. Historically, reliability analysts have widely used the exponential distribution because of its simplicity and tractability. The probability density function for an exponential failure time t is

$$f(t|\lambda, \mu) = \lambda \exp[-\lambda(t - \mu)], \tag{4.5}$$

where $\mu > 0$ represents one aspect of the distribution location because $t > \mu$, and $\lambda > 0$ governs both the distribution location and scale; inspecting the exponential mean $\mu + 1/\lambda$ and standard deviation $1/\lambda$ shows the roles of these parameters. Note that λ is a rate, the mean number of failures per unit time, and is called a rate parameter. An alternative parameterization expresses the probability density function in terms of the mean time to failure (MTTF), which equals $\frac{1}{\lambda}$. In most reliability applications, $\mu = 0$, which assumes that failures can occur at any time after the start of the test ($t = 0$). For $\mu = 0$, both the mean and standard deviation of the failure time distribution are $\frac{1}{\lambda}$.

We can express the hazard function and reliability function as

$$h(t) = \lambda \quad \text{and}$$
$$R(t) = \exp[-\lambda(t - \mu)]\,.$$

An important and unique feature of the exponential distribution is that its hazard function $h(t) = \lambda$ is constant. In other words, the probability of a component's failure in the next instant of time, given it has survived to the current time, does not depend on the component's age. The constant hazard function limits the usefulness of the exponential distribution as a failure time model, however. For example, the exponential distribution often adequately models failure times of many electronic components and other components designed to last beyond their anticipated technological lives, i.e., the times when new technologies make them obsolete. However, the exponential distribution poorly models failure times of components that experience early failures or wear-out failures during their technological life. There is also a connection between the exponential distribution and the Poisson distribution, presented in Sect. 4.2. When the times between failures (i.e., interfailure times) follow an exponential distribution and are independent, the number of failures in a specified period of time has a Poisson distribution. See also Chap. 6 on repairable system reliability, where the times between repairs are called interfailure times.

Some Prior Distributions for Exponential Failure Times

This section considers prior distributions for the exponential distribution parameters λ and μ. If μ is known, a commonly used prior distribution for λ is the gamma distribution. The gamma distribution has positive support, is the conjugate prior distribution, and has probability density function given by

$$p(\lambda|\alpha, \beta) = \frac{\beta^\alpha}{\Gamma(\alpha)} \lambda^{\alpha-1} \exp(-\beta\lambda)\,,$$

where $\alpha > 0$ is a shape parameter and $\beta > 0$ is a scale parameter, as seen by inspecting $\mathrm{E}(\lambda) = \alpha/\beta$ and $\mathrm{Var}(\lambda) = \alpha/\beta^2$.

Suppose that we observe n conditionally independent component failure times $t_1, \ldots, t_n$, which follow an exponential distribution (with $\mu = 0$). Using a gamma prior distribution for λ, the model for these failure time data is

$$T_i \sim \mathit{Exponential}(\lambda)\,,\ i = 1, \ldots, n\,,\ \text{and}$$
$$\lambda \sim \mathit{Gamma}(\alpha, \beta)\,.$$

Applying Bayes' Theorem, with an exponential likelihood function based on Eq. 4.5 (i.e., the product of exponential likelihood functions for $t_1, \ldots, t_n$), yields the following posterior distribution for λ:

$$\lambda | \mathbf{t} \sim Gamma(\alpha + n, \beta + \sum_{i=1}^{n} t_i)\,, \tag{4.6}$$

where $\mathbf{t} = (t_1, \ldots, t_n)$.

If both λ and μ are unknown, Martz and Waller (1982) discuss a natural conjugate prior. This prior, while conjugate, does not have a common distributional form and is not substantially easier to work with than a nonconjugate prior. However, Markov chain Monte Carlo (MCMC) allows the analyst to use either the natural conjugate or nonconjugate prior distribution for (λ, μ). In fact, because functions of both parameters are usually of interest, it is easier to work with draws from the joint posterior distribution of (λ, μ). Consequently, we can employ MCMC no matter what prior distribution we use.

If no prior knowledge or expertise suggest a correlation structure for the parameters, a common approach for specifying a nonconjugate prior distribution is to assume independent prior distributions for λ and μ; the joint prior distribution for (λ, μ), then takes the form

$$p(\lambda, \mu) = p(\lambda)p(\mu)\,.$$

When specifying prior distributions for λ and μ, note that the support of both parameters is positive.

Example 4.3 Exponential model for projector lamp failure times. In business and educational settings, computer presentations use liquid crystal display (LCD) projectors. The most common failure mode of these projectors is the failure of the lamp. Many manufacturers include the "expected" lamp life in their technical specification documents, and one manufacturer claims that users can expect 1,500 hours of projection time from each lamp used under "normal operating conditions." To test this claim, a large private university placed identical lamps in three projector models for a total of 31 projectors. The university staff recorded the number of projection hours (as measured by the projector) when each lamp burned out. See Table 4.2, which presents the failure times (in projection hours) for the 31 lamps.

Assuming a constant hazard rate, consider modeling the lamp failure times by an exponential distribution with rate parameter λ and use the conjugate prior distribution for λ,

$$\lambda \sim Gamma(\alpha, \beta)\,.$$

One way to choose values for α and β is to use the manufacturer claimed specification as a best guess for the MTTF ($1/\lambda$) and use a large standard deviation around this specification. Here, we interpret the manufacturer's "about 1,500

Table 4.2. LCD projector lamp failure time data

LCD Model	Projection Hours	LCD Model	Projection Hours
1	387	3	1895
1	182	2	158
1	244	1	974
1	600	2	345
1	627	1	1755
2	332	3	1752
2	418	1	473
2	300	2	81
1	798	1	954
2	584	2	1407
1	660	1	230
3	39	1	464
3	274	2	380
2	174	2	131
2	50	2	1205
3	34		

hours" of lamp life as a prior mean lamp life of 1,500 hours, and also interpret "about" as not being very certain and let the prior standard deviation be 2,000 hours. In order for λ to have a $Gamma(\alpha, \beta)$ prior distribution, the MTTF must have an $InverseGamma(\alpha, \beta)$ prior distribution. By *moment matching* (i.e., equating these values to the MTTF prior mean and standard deviation, respectively), we can solve for α and β. From the inverse gamma moments given in Appendix B, the MTTF prior mean and standard deviation are

$$\frac{\beta}{\alpha-1} = 1,500 \text{ and}$$
$$\sqrt{\frac{\beta^2}{(\alpha-1)^2(\alpha-2)}} = 2,000\,. \tag{4.7}$$

Solving for α and β in Eq. 4.7 yields approximately $\alpha = 2.5$ and $\beta = 2,350$.

From Eq. 4.6, the posterior distribution for λ is

$$\begin{aligned}\lambda|\mathbf{y} &\sim Gamma(\alpha + n, \beta + \textstyle\sum_{i=1}^{n} t_i) \\ &= Gamma(2.5 + 31; 2,350 + 17,907) = Gamma(33.5; 20,257)\,.\end{aligned}$$

Figure 4.3 displays the prior (dashed line) and posterior (solid line) distributions for λ, which shows that the failure time data indicate a rate much worse than claimed. Figure 4.4 also displays the posterior for MTTF ($\frac{1}{\lambda}$), which confirms that the MTTF is much smaller than the claimed 1,500 hours; based on 100,000 draws of the MTTF posterior distribution, the probability of exceeding 1,500 hours is 0.00001, and the probability of even exceeding 1,000 hours is only 0.004. We obtain draws from the MTTF posterior distribution by making draws from the λ posterior distribution and taking reciprocals.

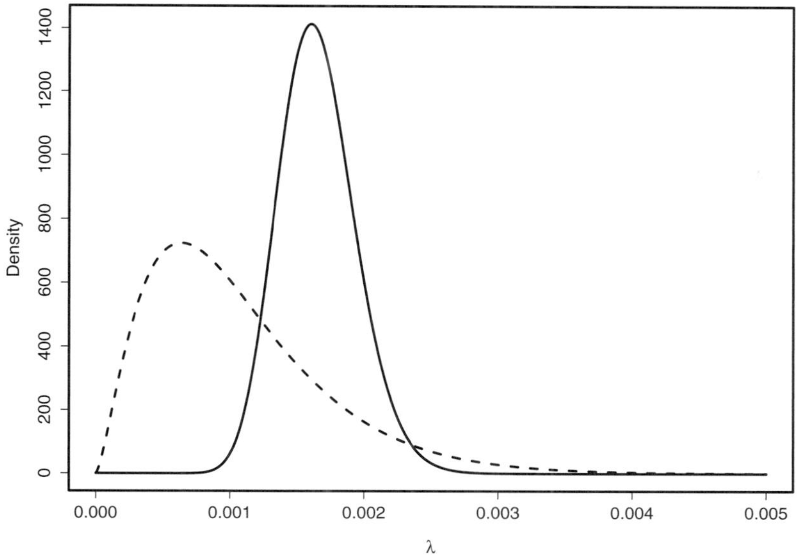

Fig. 4.3. The prior (dashed line) and posterior (solid line) distributions of the exponential rate parameter λ for the LCD projector example.

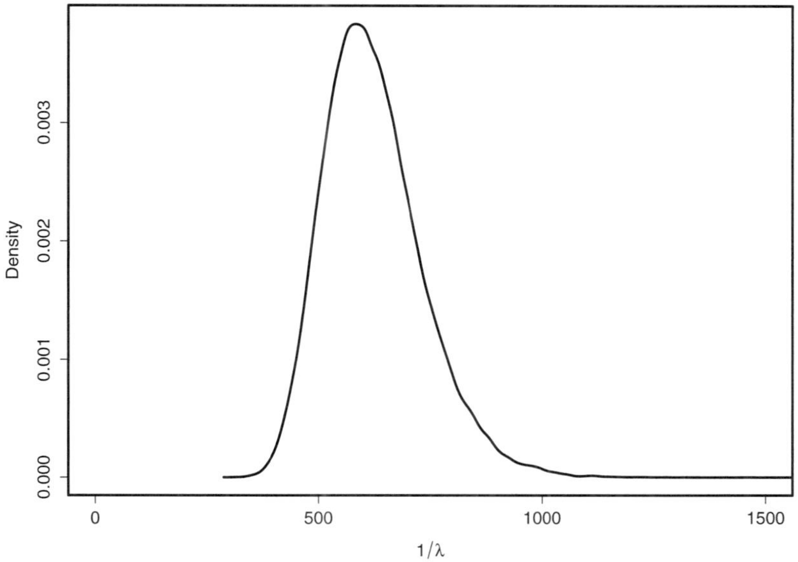

Fig. 4.4. The posterior distribution for the $MTTF = \frac{1}{\lambda}$ (in hours) of the LCD projector example under the exponential distribution.

We can also approximate the posterior distribution of the reliability function $R(t)$ over time t by making draws from the λ posterior distribution and evaluating the reliability $R(t)$ given by

$$R(t) = \exp(-\lambda t),$$

to obtain draws from the $R(t)$ posterior distribution. Figure 4.5 presents the posterior median reliability function as the solid line and the corresponding 90% credible intervals as dashed lines. At 1,000 hours, note that the median posterior reliability is 0.194 and the reliability is between 0.117 and 0.298 with 0.90 probability. One reason for providing 90% credible intervals is that they provide 95% credible lower or upper bounds on reliability; that is, the 95% credible upper bound on reliability is 0.298. When the reliability is much higher, the analyst may want to report a 95% credible lower bound on reliability.

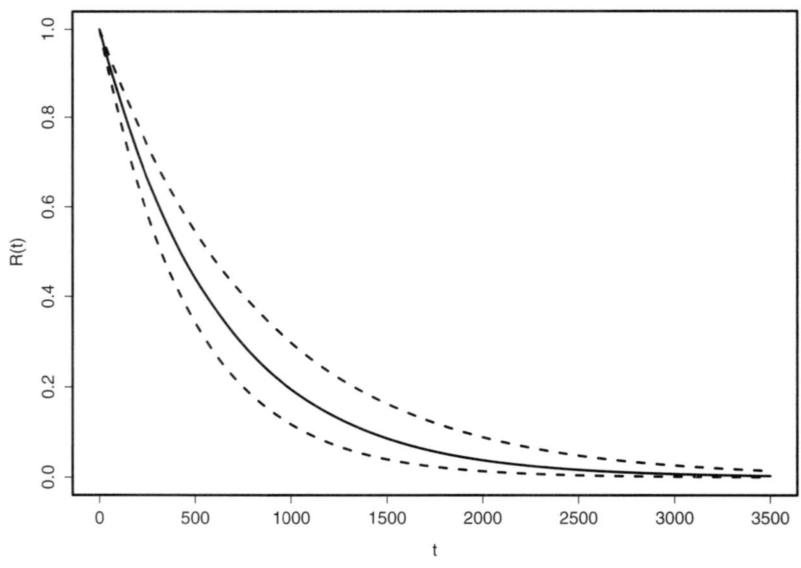

Fig. 4.5. The posterior medians (solid line) with 90% credible intervals (dashed lines) for the LCD projector lamp reliability over time t (in hours) under the exponential distribution.

To assess whether an exponential distribution models the lamp failure times well, we can apply a Bayesian χ^2 goodness-of-fit test. Based on $K = 4$ ($\approx 31^{0.4}$) equal probability bins, repeatedly make draws from the λ posterior distribution, calculate the R^B test statistic, and compare it against the 0.95 quantile of the $ChiSquared(3)$ reference distribution. We find that 0.1% of the R^B values exceed this 0.95 quantile, which supports the claim that the exponential model fits the data well.

When the hazard function is not constant, the analyst may model the failure time data with a generalized form of the exponential distribution, called the Weibull distribution. We consider the Weibull distribution next.

4.3.2 Weibull Failure Times

Consider the *Weibull distribution* as a model for failure time data. Historically, reliability analysts have also widely used the Weibull distribution because of its tractability and flexibility; as for the exponential distribution, many software packages implement classical statistical methods for the Weibull distribution. One motivation for the Weibull distribution is that it is the asymptotic distribution of the scaled minimum of i.i.d. random variables meeting certain conditions; that is, the Weibull distribution arises when the weakest of many factors causes failure. (See Lawless (1982), Exercise 1.11, for more details.) The probability density function for a Weibull failure time t is

$$\begin{aligned} f(t|\lambda, \beta, \theta) &= \lambda\beta(t-\theta)^{\beta-1}\exp[-\lambda(t-\theta)^{\beta}], \\ &\qquad t > \theta, \lambda > 0, \beta > 0, \theta > 0, \end{aligned} \tag{4.8}$$

where θ determines the location, λ the scale, and β the shape of the distribution. In most reliability applications, $\theta = 0$, which assumes that failures can occur at any time after the start of the test ($t = 0$). Also, when performing reliability analyses, the analyst must make sure which of the many Weibull parameterizations a software package is implementing; see Appendix B, which presents three standard parameterizations.

We can express the hazard function and reliability function for a Weibull failure time model by

$$\begin{aligned} h(t) &= \lambda\beta(t-\theta)^{\beta-1} \quad \text{and} \\ R(t) &= \exp[-\lambda(t-\theta)^{\beta}]. \end{aligned}$$

Notice that the hazard function is decreasing for $\beta < 1$, which applies to components in the "infant mortality" phase of the failure time distribution, and is increasing for $\beta > 1$, which applies to components in the wear-out phase. When $\beta = 1$, the Weibull distribution reduces to the exponential distribution, which has a constant hazard function. In practice, the Weibull distribution is an attractive model choice because it allows for increasing, decreasing, or constant failure rates (i.e., IFR, DFR, or CFR). Furthermore, within the Bayesian framework, the analyst can answer the question "What is the probability that the failure time distribution has an increasing failure rate?" by evaluating the posterior probability that $\beta > 1$.

No natural conjugate prior distribution exists if both the shape and scale parameters of the Weibull distribution are assumed to be unknown. In specifying a prior distribution for (λ, β, θ), note that all these parameters have positive support.

We revisit the LCD projector example next to illustrate the use of the Weibull distribution.

Example 4.4 Weibull model for projector lamp failure times. Returning to the LCD projector lamp failure time data, presented in Example 4.3, consider modeling these failure times by a Weibull distribution with $\theta = 0$. That is, the model for the $n = 31$ lamp failure times $t_1, \ldots, t_{31}$ given in Table 4.2 is

$$T_i \sim Weibull(\lambda, \beta)\,,\; i = 1, \ldots, 31\,.$$

The Weibull likelihood function based on Eq. 4.8 is the product of Weibull likelihood functions for $t_1, \ldots, t_{31}$.

To complete the model, we then need to choose a "comparable" joint prior distribution for λ and β. Making this choice is a challenge because the exponential distribution has only one parameter λ and is a special case of the Weibull distribution. We can assess the comparability of the prior distribution choices by inspecting their *prior predictive distributions*. To approximate the prior predictive distribution, make draws from the model parameters' prior distribution and then make draws from the data or sampling distribution given the drawn model parameter values. For example, for the exponential model, make gamma draws for λ and then make $Exponential(\lambda)$ draws. Similarly, for the Weibull model, make draws from the (λ, β) prior distribution and then make $Weibull(\lambda, \beta)$ draws.

Because both λ and β have positive support, here we use separate and independent gamma distributions as prior distributions. Notationally, these choices are

$$\lambda \sim Gamma(\alpha_\lambda, \theta_\lambda) \quad \text{and} \tag{4.9}$$

$$\beta \sim Gamma(\alpha_\beta\, \theta_\beta)\,, \tag{4.10}$$

where α_λ and α_β are the respective shape parameters, and θ_λ and θ_β are the respective scale parameters. By assuming independence, their joint prior density function is the product of the two prior density functions associated with Eqs. 4.9 and 4.10.

Now what remains is to specify values for $\alpha_\lambda, \theta_\lambda, \alpha_\beta$, and θ_β. In this case, we use the same prior distribution for λ as used for the exponential model, i.e., $\lambda \sim Gamma(2.5, 2350)$. For β, we use $\beta \sim Gamma(1, 1)$, because it is centered at 1 (i.e., the exponential model), but allows for either a decreasing or increasing hazard function. Note that the exponential and Weibull prior predictive distributions cannot be exactly the same (unless $\beta = 1$). But it is worth inspecting their prior predictive distributions as displayed in Fig. 4.6, where the solid line is the exponential prior predictive distribution and the dashed line is that for the Weibull distribution. Figure 4.6 shows that the Weibull prior predictive distribution is somewhat spread out more than the exponential prior predictive distribution.

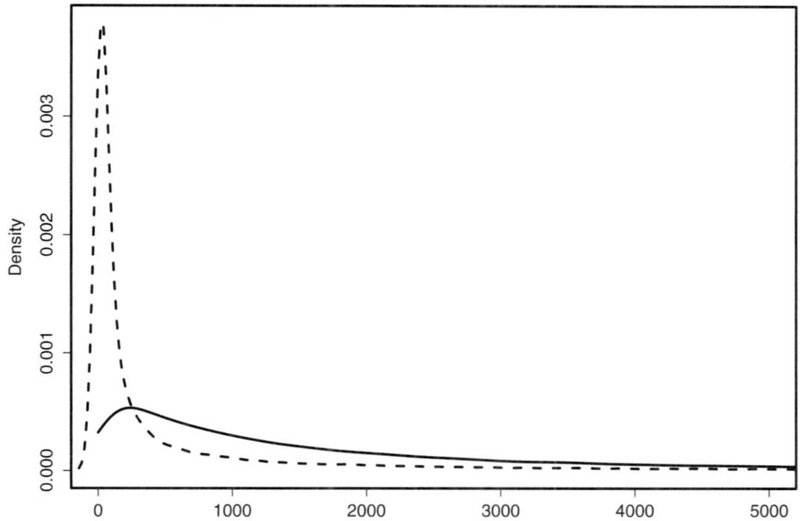

Fig. 4.6. Prior predictive failure time (in hours) distributions for the exponential (solid line) distribution and for the Weibull (dashed line) distribution for the LCD projector example.

To analyze the lamp failure time data, we use MCMC to obtain draws from the joint posterior distribution of (λ, β). See Fig. 4.7, which displays the marginal posterior distributions for λ and β. We approximate the posterior distribution of MTTF, which takes the form

$$MTTF = \lambda^{-1/\beta} \Gamma(\frac{\beta+1}{\beta}), \tag{4.11}$$

by evaluating Eq. 4.11 with the (λ, β) posterior draws. Using the MTTF posterior distribution, the claimed 1,500-hour lamp lifetime is also suspect under the Weibull failure time model. In fact, the posterior probability that MTTF even exceeds 1,000 hours is only 0.002. By inspecting the posterior distribution for β in Fig. 4.7, there is weak evidence for an increasing hazard function; moreover, the posterior probability that $\beta > 1$ is 0.866.

We can also approximate the posterior distribution of the reliability function $R(t)$ over time t by making the draws from the (λ, β) posterior

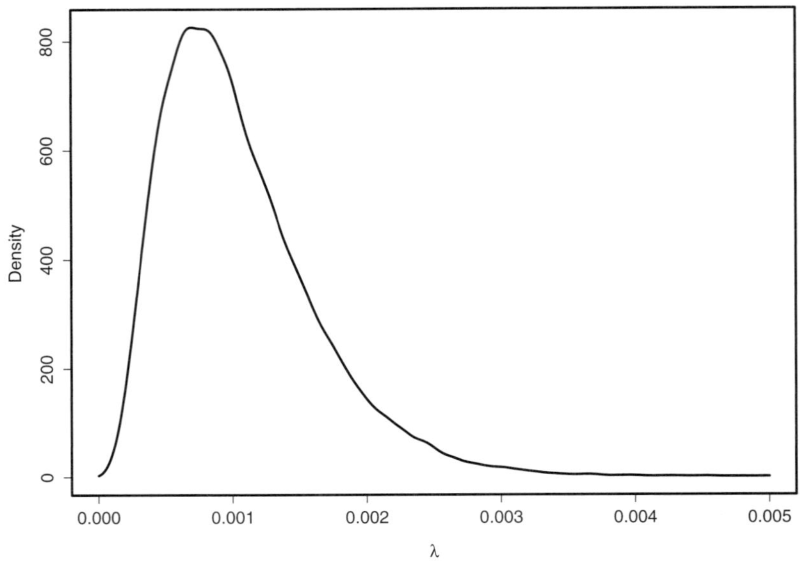

(a)

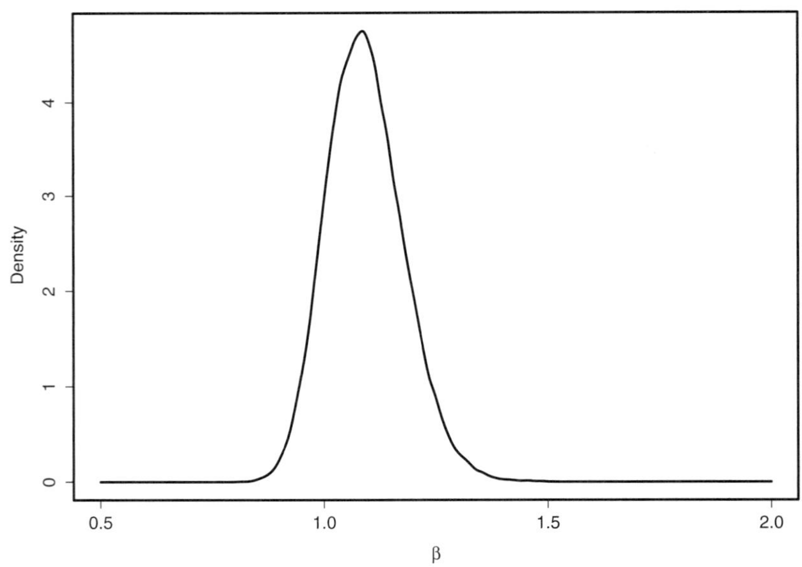

(b)

Fig. 4.7. The posterior distributions of the Weibull distribution parameters for the LCD projector example: (a) scale λ and (b) shape β.

distribution and evaluating the reliability $R(t)$ given by

$$R(t) = \exp[-\lambda t^{\beta}],$$

to obtain draws from the $R(t)$ posterior distribution. Figure 4.8 presents the posterior median reliability function as the solid line and the corresponding 90% credible intervals as dashed lines. At 1,000 hours, note that the median posterior reliability is 0.176 and the reliability is between 0.100 and 0.280 with 0.90 probability.

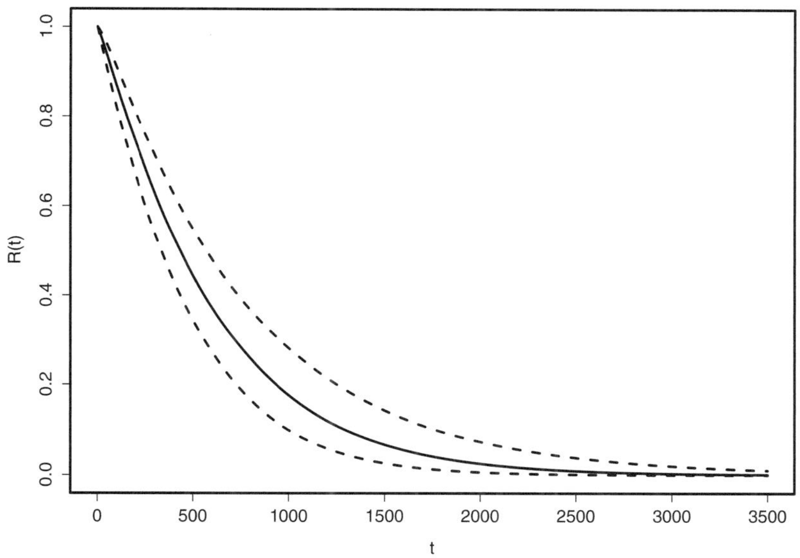

Fig. 4.8. The posterior medians (solid line) with 90% credible intervals (dashed lines) for the LCD projector lamp reliability over time t (in hours) under the Weibull distribution.

To assess whether a Weibull distribution models the lamp failure times well, we can apply a Bayesian χ^2 goodness-of-fit test. Based on $K = 4$ ($\approx 31^{0.4}$) equal probability bins, repeatedly take draws from the (λ, β) posterior distribution, calculate the R^B test statistic, and compare it against the 0.95 quantile of the $ChiSquared(3)$ reference distribution. We find that 0.2% of the R^B values exceed this 0.95 quantile, which suggests that the Weibull model fits the data well. This result does not surprise us, because Example 4.3 showed that the exponential distribution, a special case of the Weibull distribution, fit the data well.

4.3.3 Lognormal Failure Times

Consider the *lognormal distribution* as a model for failure time data. The lognormal distribution's connection with the normal distribution follows from: if X has a normal distribution, then $T = \exp(X)$ has a lognormal distribution. Whereas the normal distribution is symmetric about its mean, the lognormal distribution is skewed, which makes it a potential model for failure times that often exhibit a skewed distribution. The probability density function for a lognormal failure time t is

$$f(t|\mu, \sigma) = \frac{1}{t\sqrt{2\pi\sigma^2}} \exp\left(\frac{-1}{2\sigma^2}[\log(t) - \mu]^2\right), \tag{4.12}$$

where μ and σ are the mean and standard deviation of the distribution of the log failure time $x = \log(t)$. We can express the hazard function and reliability function for the lognormal distribution as

$$\begin{aligned} h(t) &= \tfrac{f(t)}{R(t)} \quad \text{and} \\ R(t) &= \textstyle\int_t^\infty f(x)dx = 1 - \phi\{[\log(t) - \mu]/\sigma\}, \end{aligned} \tag{4.13}$$

where $f(x)$ is the lognormal probability density function given in Eq. 4.12 and $\Phi(\cdot)$ is the standard normal cumulative distribution function. Note that neither the hazard nor the reliability functions have closed forms. Historically, the lack of closed form functions is a major reason why reliability analysts did not regularly use the lognormal distribution. Today, however, software packages routinely evaluate these functions using numerical algorithms. One feature of the lognormal distribution is its unique hazard function; the lognormal hazard function increases initially and then decreases and approaches zero at very long times. Despite a distribution with decreasing hazard function at long times being untenable, the lognormal distribution has been useful in many applications.

Some Prior Distributions for Lognormal Failure Times

If there is no available information about a joint distribution for (μ, σ^2), we can use independent and separate prior distributions for μ and σ^2. Recall the interpretations of μ and σ^2 as the mean and variance of the logged failure times, which have support on the real line and positive real line, respectively. One choice of prior distributions with the same supports is

$$\begin{aligned} \mu &\sim Normal(\theta, \tau^2) \quad \text{and} \\ \sigma^2 &\sim InverseGamma(\alpha, \beta). \end{aligned} \tag{4.14}$$

Take care in specifying the hyperparameters θ, τ^2, α, and β because they must be interpreted in terms of logged failure times. We can check their specification by inspecting the prior predictive distribution, obtained by making draws for μ and σ^2 using Eq. 4.14 and then making $Lognormal(\mu, \sigma^2)$ draws.

One commonly used form of the joint prior distribution for μ and σ^2 is

$$\begin{aligned} \mu &\sim Normal(\theta, \kappa\sigma^2) \quad \text{and} \\ \sigma^2 &\sim InverseGamma(\alpha, \beta)\,. \end{aligned} \tag{4.15}$$

Use this prior distribution if μ is thought to have more uncertainty when σ^2 is larger.

We illustrate the use of the lognormal distribution as a failure time model for the LCD lamp failure times next.

Example 4.5 Lognormal model for projector lamp failure times. Returning to the LCD projector lamp failure time data, displayed in Table 4.2, assume that the failure times follow a lognormal distribution. Notationally, for the 31 observed failure times $t_1, \ldots, t_{31}$,

$$T_i \sim LogNormal(\mu, \sigma^2)\,,\; i = 1, \ldots, 31\,.$$

The lognormal likelihood function based on Eq. 4.12 is the product of lognormal likelihood functions for $t_1, \ldots, t_{31}$. To complete the model, we use the prior distributions for μ and σ^2 given in Eq. 4.14, where the chosen hyperparameter values provide a comparable prior distribution using the prior predictive distribution approach. That is, the chosen hyperparameter values should yield comparable lognormal and Weibull prior predictive distributions. By letting $\theta = 6, \tau^2 = 25, \alpha = 6.5$, and $\beta = 23.5$, the prior predictive distributions are comparable as demonstrated in Fig. 4.9.

To analyze the failure time data, we use MCMC to obtain draws from the (μ, σ^2) joint posterior distribution. We can approximate the posterior distribution of reliability $R(t)$ over time t by evaluating $R(t)$ given in Eq. 4.13 with these (μ, σ^2) draws. Figure 4.10 displays the median posterior reliability function as a solid line and the corresponding 90% credible intervals as dashed lines. At 1,000 hours, note that the median posterior reliability is 0.234 and that the reliability is between 0.147 and 0.342 with 0.90 probability. Like the previous analyses, the manufacturer's claimed expected lamp lifetime of 1,500 hours is suspect.

To assess whether a lognormal distribution models the lamp failure times well, we can apply a Bayesian χ^2 goodness-of-fit test. Based on $K = 4$ ($\approx 31^{0.4}$) equal probability bins, repeatedly take draws from the (μ, σ^2) posterior distribution, calculate the R^B test statistic, and compare it against the 0.95 quantile of the $ChiSquared(3)$ reference distribution. We find that 6.0% of the R^B values exceed this 0.95 quantile, which suggests that the lognormal model fits the data, but not as well as the exponential and Weibull models.

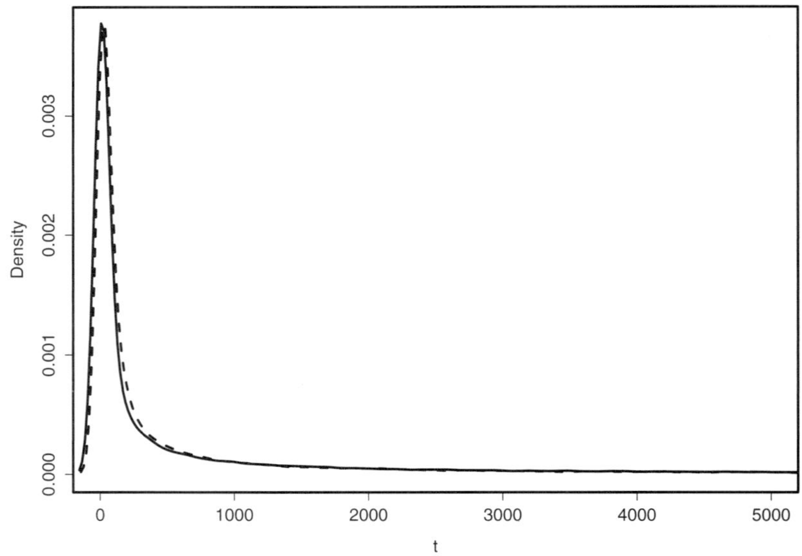

Fig. 4.9. Prior predictive failure time (in hours) distributions for the Weibull (solid line) distribution and for the lognormal (dashed line) distribution.

4.3.4 Gamma Failure Times

Consider the *gamma distribution* as a model for failure time data. The probability density function for a gamma failure time t is

$$f(t|\alpha, \lambda) = \frac{\lambda^\alpha}{\Gamma(\alpha)} t^{\alpha-1} \exp(-\lambda t), \tag{4.16}$$

where $\alpha > 0$ determines the shape and $\lambda > 0$ the scale of the distribution. We can express the hazard function and reliability function as

$$\begin{aligned} h(t) &= \frac{\lambda^\alpha}{\Gamma(\alpha, \lambda t)} t^{\alpha-1} \exp(-\lambda t) \quad \text{and} \\ R(t) &= \int_t^\infty f(x)dx = \Gamma(\alpha, \lambda t), \end{aligned}$$

where $\Gamma(\cdot, \cdot)$ is the upper incomplete gamma function defined in Appendix B. The exponential distribution is a special case of the gamma distribution when $\alpha = 1$. Consequently, when $\alpha = 1$, the hazard function is constant; the hazard function is decreasing when $\alpha < 1$ and increasing when $\alpha > 1$.

Like the Weibull distribution, the gamma distribution is flexible, but used less often in practice. One reason for its little use is that the hazard and

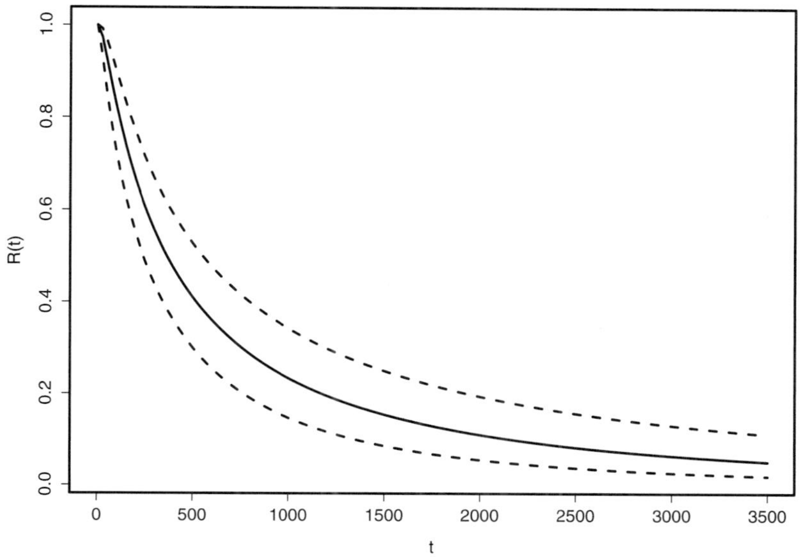

Fig. 4.10. The posterior median (solid line) with 90% credible intervals (dashed lines) for the LCD projector lamp reliability over time t (in hours) under the lognormal distribution.

reliability functions do not have a simple and closed form, and, therefore, are not as easy to use as those for the Weibull distribution.

We leave the analysis of the LCD projector lamp failure time data in Table 4.2 using the gamma failure time model as Exercise 4.6.

4.3.5 Inverse Gaussian Failure Times

Consider the *inverse Gaussian distribution* as a model for failure time data. The probability density function for an inverse Gaussian failure time t is

$$f(t \mid \mu, \lambda) = \sqrt{\frac{\lambda}{2\pi t^3}} \exp\left(\frac{-\lambda(t-\mu)^2}{2t\mu^2}\right) , \tag{4.17}$$

where $\mu > 0$ determines the location and $\lambda > 0$ the shape of the distribution. The reliability function takes the form

$$R(t) = \Phi\left[\sqrt{\frac{\lambda}{t}}\left(1 - \frac{t}{\mu}\right)\right] - \exp(2\lambda/\mu)\Phi\left[-\sqrt{\frac{\lambda}{t}}\left(1 - \frac{t}{\mu}\right)\right] , \tag{4.18}$$

where $\Phi(\cdot)$ is the standard normal cumulative distribution function. To obtain the inverse Gaussian hazard function, we can use the standard hazard function definition

$$h(t) = \frac{f(t)}{R(t)}$$

with definitions for $f(t)$ and $R(t)$ given in Eqs. 4.17 and 4.18, respectively.

There is a relationship between the inverse Gaussian distribution and the degradation models presented in Sect. 8.6. If the degradation follows a Wiener process, the first time that the degradation crosses a specified threshold (i.e., first crossing time) has an inverse Gaussian distribution. Consequently, identifying an underlying degradation process motivates the modeling of the failure times by an inverse Gaussian distribution.

4.3.6 Normal Failure Times

Consider the normal distribution as a model for failure time data. The probability density function for a normal failure time t is

$$f(t|\mu, \sigma^2) = \frac{1}{\sqrt{2\pi\sigma^2}} \exp\left(-\frac{1}{2\sigma^2}(t-\mu)^2\right), \tag{4.19}$$

where μ and σ are the mean and standard deviation of the distribution. The hazard and reliability functions take the following forms

$$\begin{aligned} h(t) &= \tfrac{f(t)}{R(t)} \quad \text{and} \\ R(t) &= \textstyle\int_t^\infty f(x)dx = 1 - \Phi[(t-\mu)/\sigma], \end{aligned} \tag{4.20}$$

where $\Phi(\cdot)$ is the standard normal cumulative distribution function.

Reliability analysts have seldom used the normal distribution, perhaps because its support is the real line; the normal distribution is also symmetric, whereas failure times tend to exhibit a skewed distribution. Nevertheless, Martz and Waller (1982) notes its applicability when μ is large relative to σ, so that the probability below 0 is negligible. Meeker and Escobar (1998), Sect. 4.5, also notes several applications that employed the normal distribution in reliability assessments. We discuss the choice of prior distributions at length in Chap. 2.

The normal distribution also plays an important role in hierarchical models for reliability data. We have already seen its use in Example 3.5 to model machine-to-machine differences (i.e., random effects) for machines that are similar but not identical. Throughout the remainder of the book, hierarchical models often arise, which employ the normal distribution to model random effects.

In the next section, we address the situation when some components placed on test have not failed when we stop testing.

4.4 Censored Data

A unique feature of reliability data, especially failure time data, is that some of the data may be *censored*. When collection stops and a unit has not failed, its failure time is censored and the exact failure time is unknown. For example, in Table 4.4, all the roller bearings with asterisks were working so that their failure times are censored; their failure times exceed the times when they were inspected. Such data are right-censored data. Right-censored data also arise from a Type I- or time-censoring scheme, where testing stops at a specified time; all components still working at the end of the test have right-censored failure times. As discussed in Chap. 1, there are left- and interval-censored data. Left censoring arises when the component fails before the first inspection. For example, suppose that we put a batch of new batteries on test for 90 days and check them at 8 a.m. every day to see whether they are still working. The failure times for the batteries that fail during the first day are left censored; also, the batteries that fail between the first and second days, and so on, before the 90th day, are interval censored. The Type II- or failure-censoring scheme stops testing components after a specified number of failures occur; the components that are still operating have right-censored failure times. Finally, a component may need to be removed from the test; for example, the experimenter may damage it accidentally, so that it can no longer be tested. This is an example of random censoring, which for this example, produces a right-censored failure time. See also Chap. 1, which discusses censoring in more detail.

In a model for censored data, the analyst needs to understand the censored data's contribution to the likelihood function. Thus far, we have dealt with failure time data that are uncensored or *complete*, where the times are known exactly. Recall that for a complete failure time t, its contribution to the likelihood is its probability density function $f(t)$. Now, for a censored failure time, represented by an interval, its contribution to the likelihood function is the probability of the failure time occurring within the interval. For example, we can represent a right-censored failure time by (t_R, ∞), which has probability $F(\infty) - F(t_R) = 1 - F(t_R)$; this probability is the censored failure time's contribution to the likelihood function. By assuming that all the data are independent, the likelihood function is the product of all their contributions. For the other censored data types, see Table 4.3 for their likelihood function contributions. As noted in Chap. 1, an advantage of the Bayesian approach is that only the censoring pattern, e.g., a right-censored failure time, is relevant, not which censoring scheme, such as Type I, Type II, or random censoring, produced it.

Notationally, the likelihood function takes the following form. Let $\mathbf{t}_{\text{full}}$ denote all the data, both uncensored and censored, $\mathbf{t}_{\text{complete}}$ denote the uncensored or complete data, and $\mathbf{t}_{\text{censored}}$ denote the censored data. Then, the likelihood function for all the data $\mathbf{t}_{\text{full}} = (\mathbf{t}_{\text{complete}}, \mathbf{t}_{\text{censored}})$ is

Table 4.3. The likelihood function contributions of uncensored and censored data

Type of Observation	Failure Time	Contribution
Uncensored	$T = t$	$f(t)$
Left censored	$T \leq t_L$	$F(t_L)$
Interval censored	$t_L < T \leq t_R$	$F(t_R) - F(t_L)$
Right censored	$T > t_R$	$1 - F(t_R)$

$$f(\mathbf{t}_{\text{full}}|\boldsymbol{\theta}) = f(\mathbf{t}_{\text{complete}}|\boldsymbol{\theta}) \times f(\mathbf{t}_{\text{censored}}|\boldsymbol{\theta}), \tag{4.21}$$

where $\boldsymbol{\theta}$ is the vector of model parameters. Note that Eq. 4.21 assumes that the failure times are conditionally independent (given $\boldsymbol{\theta}$) and that the censoring scheme is independent of the failure times.

For example, if only complete (uncensored) and right-censored data are available, as for the roller bearing example in Table 4.4, the likelihood function takes the form

$$f(\mathbf{t}_{\text{full}}|\boldsymbol{\theta}) = \prod_{i=1}^{n_{\text{complete}}} f(\mathbf{t}_{\text{complete},i}|\boldsymbol{\theta}) \times \prod_{j=1}^{n_{\text{censored}}} [1 - F(\mathbf{t}_{\text{censored},j}|\boldsymbol{\theta})],$$

where n_{complete} and n_{censored} are the number of complete and censored data, respectively, and $\mathbf{t}_{\text{complete},i}$ and $\mathbf{t}_{\text{censored},j}$ are the ith complete and jth censored data, respectively.

An appealing feature of the Bayesian approach is that now with the likelihood function specified, we need only choose appropriate prior distributions for the model parameters $\boldsymbol{\theta}$ and approximate their joint posterior distribution using MCMC. Unlike classical statistical methods, which have different methods for each censored data type, the Bayesian approach provides a common framework to analyze all censored data types.

Finally, we discuss the likelihood function for failure time data collected under a Type II-censoring or failure-censoring scheme, which tests n components and stops after the rth failure. Under this scheme, the data consist of $t_1 < \ldots < t_r$ uncensored failure times and $n - r$ censored failure times at t_r. The joint probability density function for these data has the form

$$\frac{n!}{(n-r)!} \left(\prod_{j=1}^{r} f(t_j) \right) R(t_r)^{n-r}, \tag{4.22}$$

where $f(\cdot)$ and $R(\cdot)$ are the appropriate probability density and reliability functions, respectively. Equation 4.22 is the likelihood function for these data. But this exposition is unnecessary, because there are r failure times $t_1 < \ldots < t_r$ that contribute $f(t_i), \ldots, f(t_r)$ to the likelihood function and $n - r$ right-censored failure times (t_r) that together contribute $R(t_r)^{n-r}$ to the likelihood function. In other words, these failure time data's contribution to the likelihood is

$$\left(\prod_{j=1}^{r} f(t_j)\right) R(t_r)^{n-r},$$

which is proportional to Eq. 4.22. This is another demonstration that under the Bayesian approach, the censoring scheme (in this case, Type II censoring) need not be accounted for, only the censoring patterns of the failure time data.

Example 4.6 Lognormal distribution with censored data for the roller bearing failure time data. The U.S. military uses an aircraft called the Prowler to combat opposition air defenses. While the Prowler is aging, it has unique capabilities that make it indispensable to many military missions. In November 2001, the Prowler aircraft had two engine failures in the same week, and an ensuing investigation determined that 4.5 roller bearings caused the engine failures. Military analysts were interested in assessing the roller bearing reliability, not only because it is one of the engine's most frequent failure modes, but also because replacing the roller bearing costs up to 100 times less than replacing an entire engine. Consequently, a reliability assessment of the roller bearings would not only provide a better understanding of aircraft reliability, but potentially might provide guidance for reducing maintenance costs for the aging Prowler aircraft. Muller (2003) presents the original failure time data, as displayed in Table 4.4. Table 4.4 presents the failure time data of the 4.5 roller bearings from 66 Prowler attack aircraft. Some of these data (11 of 66) are recorded failure times. The remaining data (55 of 66) are the ages of the roller bearings at the last engine inspection; that is, these data are right censored and indicated by asterisks.

To illustrate a model of the roller bearing failure time data, assume that the failure times follow a lognormal distribution. Group the complete data together, denoted by $\mathbf{t}_F = (t_1, \ldots, t_{11})$. Similarly, group the censored data together, denoted by $\mathbf{t}_S = (t_1, \ldots, t_{55})$. The likelihood function then takes the form

$$f(\mathbf{t}_F, \mathbf{t}_S|\mu, \sigma^2) = \prod_{i=1}^{11} f(\mathbf{t}_{F,i}|\mu, \sigma^2) \prod_{j=1}^{55} (1 - F(\mathbf{t}_{S,j}|\mu, \sigma^2)),$$

where $f(t)$ and $F(t)$ are the lognormal probability density and cumulative distribution functions, respectively. The choice made for prior distributions for μ and σ^2 is

$$\begin{aligned} \mu &\sim Normal(6.5, 25) \quad \text{and} \\ \sigma^2 &\sim InverseGamma(6.5, 23.5)\,. \end{aligned} \tag{4.23}$$

We based these prior distributions on the Prowler engineers' expectation that the roller bearings would last 1,000 flight hours, but where they have much uncertainty. These prior distributions yield a lognormal prior predictive distribution that reflects the engineers' input.

Table 4.4. Roller bearing failure time data (in operating hours) for the Prowler attack aircraft (Muller, 2003). An asterisk indicates a right-censored failure time

Failure Time (operating hours)		
1,085*	1,795*	100*
1,500*	1,890	1,628
1,390*	1,145*	759*
152*	1,380*	246*
971*	61*	861*
966*	1,165*	462*
997*	437*	1,079*
887*	1,152*	1,199*
977*	159*	424*
1,022*	3,428*	763*
2,087*	555*	1,297*
646	727*	2,238*
820*	2,294*	1,388
897	663*	1,153*
810*	1,427*	2,892*
80*	951	2,153*
1,167	767*	853*
711	546*	911*
1,203	736*	2,181
85*	917*	1,042*
1,070*	2,871*	799*
719*	1,231*	750

Using MCMC, we obtain draws from the (μ, σ^2) joint posterior distribution and approximate the posterior distribution of reliability $R(t)$ over time t by taking the draws from the (μ, σ^2) joint posterior distribution and evaluating $R(t)$ given in Eq. 4.13. Figure 4.11 displays the median posterior reliability function as a solid line and the corresponding 90% credible intervals as dashed lines. At 1,000 hours, note that the median posterior reliability is 0.875 and that the reliability is between 0.799 and 0.834 with 0.930 probability.

To assess whether a lognormal distribution models the roller bearing failure times well, we can apply a Bayesian χ^2 goodness-of-fit test. Based on $K = 5$ $(\approx 66^{0.4})$ equal probability bins, repeatedly take draws from the (μ, σ^2) posterior distribution, calculate the R^B test statistic, and compare it against the 0.95 quantile of the $ChiSquared(4)$ reference distribution. We find that about 8% of the R^B values exceed this 0.95 quantile, which suggests that the lognormal model fits the data well. Note that the calculation of the R^B test statistic used the modification for Type I- or right-censored data given in Eq. 3.8.

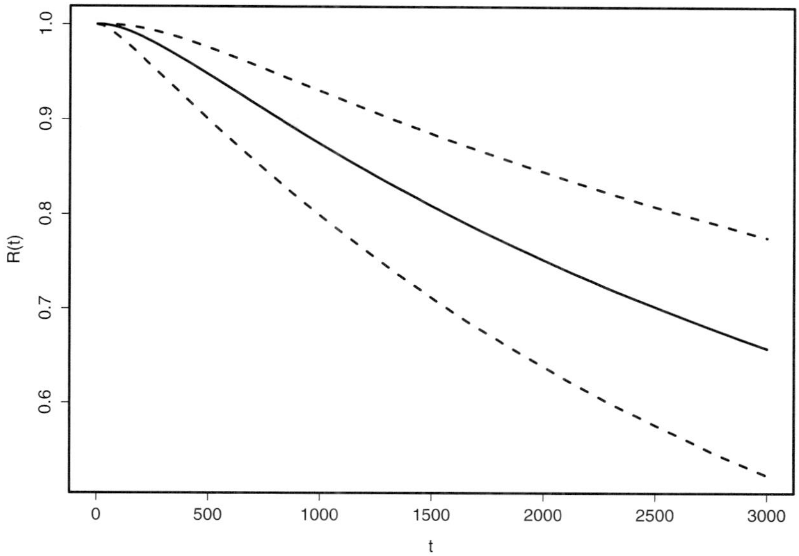

Fig. 4.11. The posterior medians (solid line) with 90% credible intervals (dashed lines) for the Prowler aircraft 4.5 roller bearing reliability over time t (in hours) based on a lognormal distribution.

4.5 Multiple Units and Hierarchical Modeling

Section 3.2 introduced *hierarchical models* and showed their use in handling the complexity often found in reliability applications. One such situation arises when there is an underlying stochastic structure that links the model parameters together. In this section, we consider how hierarchical models can be used in analyzing component reliability data.

In Example 4.2, a $Poisson(\lambda)$ distribution modeled the monthly supercomputer failure count data, presented in Table 4.1. By assuming a common failure rate λ, we treated the 47 components (shared memory processors or SMPs) as identical. But the actual situation is more complicated, because each SMP experiences a different usage (or load), although there is no record of the usage. We can allow for different loads by assuming individual failure rates, however. That is,

$$Y_i \sim Poisson(\lambda_i)\,, \tag{4.24}$$

where λ_i is the individual monthly failure rate for the ith SMP. Further, because the same scientists use the supercomputer, the λ_i are likely related, which

$$\lambda_i|\nu,\kappa \sim Gamma(\nu,\kappa) \tag{4.25}$$

captures. It is Eq. 4.25 that stochastically links the individual λ_i together; also assume that the λ_i are conditionally independent. To complete the model, note that from Eq. 4.25 prior distributions for ν and κ need specification. Because both parameters have positive support, one choice for prior distributions is

$$\begin{aligned} \nu &\sim Gamma(a_\nu, 1) \quad \text{and} \\ \kappa &\sim Gamma(a_\kappa, 1)\,, \end{aligned} \tag{4.26}$$

where a_ν and a_κ are the prior means of ν and κ.

Example 4.7 Hierarchical Poisson model for supercomputer failure count data. Consider the analysis of the supercomputer failure count data in Table 4.1 with a hierarchical Poisson model. Equations 4.24, 4.25, and 4.26 with $a_\nu = 5$ and $a_\kappa = 1$ specify the hierarchical Poisson model, motivated by the available information that the supercomputer engineers provided in Example 4.2. Note that the likelihood function is based on Eq. 4.24 for the 47 observed failure counts $y_i, i = 1, \ldots, 47$.

To analyze these failure count data, we use MCMC to obtain draws from the joint posterior distribution of $(\lambda_i, i = 1, \ldots, 47), \nu$, and κ. Figure 4.12 displays the prior and posterior predictive $Gamma(\nu, \kappa)$ distributions for λ as solid and dashed lines; the figure shows how flat the prior predictive distribution is. The posterior predictive distribution has a median of 2.687 monthly failures with a 95% credible interval of (1.005, 5.564) monthly failures and shows the impact of the different loads on the failure rates λ.

To assess how well the hierarchical Poisson model fits the supercomputer failure count data, we can apply a Bayesian χ^2 goodness-of-fit test. Recall the modification for discrete data given in Eq. 3.7. Based on $K = 5$ $(\approx 47^{0.4})$ equal probability bins, repeatedly make draws from the $(\lambda_i, i = 1, \ldots, 47)$ joint posterior distribution, calculate the R^B test statistic, and compare it against the 0.95 quantile of the $ChiSquared(4)$ reference distribution. We find that about 11% of the R^B values exceed this 0.95 quantile, which suggests that this model fits the data well. Note that the calculation of the R^B test statistic uses the joint posterior draws of $(\lambda_i, i = 1, \ldots, 47)$.

In assessing an individual-specific parameter, an important advantage of a hierarchical model is its ability to "borrow strength" from all the data (i.e., from all the individuals) to improve the estimation of the individual-specific parameter. For example, for the first SMP, based on the analysis using the hierarchical Poisson model just discussed, a 95% credible interval for λ_1 is (0.872, 4.284) monthly failures. Using only the SMP 1 data (i.e., $y_1 = 1$) and using the same prior distribution, we obtain a 95% credible interval for λ_1 of (0.288, 10.130), which is much wider than the one obtained using all the SMP data.

In the next example, we consider a hierarchical model for failure time data.

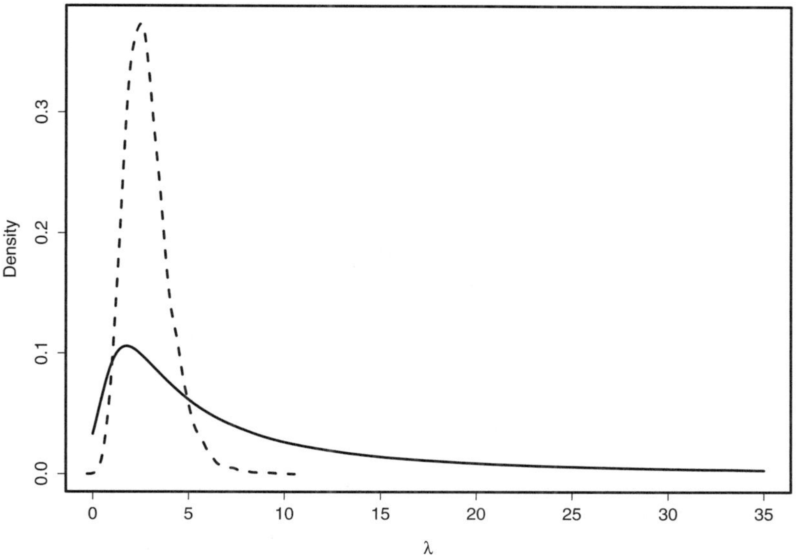

Fig. 4.12. The prior (solid line) and posterior (dashed line) predictive distributions of failure rate λ under the hierarchical Poisson model for the supercomputer example.

Example 4.8 Hierarchical Weibull model for bearing failure time data. Ku et al. (1972) reports on fatigue testing of bearings used with a particular lubricant and assumes that the failure times follow a Weibull distribution. The experimenters used 10 testers, bench-type rigs, and found that the testers impacted the measured failure times. See Table 4.5, which presents the bearing failure time data (in hours) that they collected when they used an aviation gas turbine lubricant O-64-2. The experimenters want to determine the bearing failure time distribution when they use the bearings with O-64-2, by removing the tester effect. In analyzing these data, we can account for the tester-to-tester differences by specifying

$$Y_{ij} \sim Weibull(\alpha_i, \beta)\,, \tag{4.27}$$

where y_{ij} is the jth observed failure time from the ith tester. Note that this model specification uses the first parameterization of the Weibull distribution given in Appendix B. Further, we model the logged scale parameter α_i (i.e., the ith tester effect) by

$$\begin{aligned} \log(\alpha_i) &= \mu + \gamma_{0,i}\,, \\ \gamma_{0,i} &\sim Normal(0, \sigma^2)\,. \end{aligned} \tag{4.28}$$

In other words, the α_i are assumed to have a $LogNormal(\mu, \sigma^2)$ distribution and be conditionally independent given μ and σ^2. Note that σ^2 characterizes the tester-to-tester variation.

Table 4.5. Bearing fatigue failure times (in hours) for lubricant O-64-2 (Ku et al., 1972)

Tester	Failure Time									
1	130.3	135.2	152.4	161.7	74.0	155.0	141.2	167.8	137.2	110.1
2	243.6	242.1	239.0	202.1	190.5	159.8	275.5	192.4	183.8	203.7
3	71.3	137.8	101.2	75.3	164.5	113.9	5 4.7	224.0	171.7	226.5
4	183.4	276.9	210.3	262.8	115.3	242.2	293.5	221.3	108.9	191.5
5	132.9	74.0	169.2	126.4	79.9	139.7	139.0	104.3	100.2	108.2
6	117.9	168.4	153.7	174.7	65.8	158.4	115.7	133.4	171.4	203.0
7	208.5	135.2	217.7	158.5	215.7	136.6	223.3	188.2	190.3	159.8
8	167.5	164.6	215.6	118.3	151.1	166.5	162.6	215.6	171.6	207.6
9	94.2	113.0	180.2	90.4	118.0	101.8	97.8	104.6	154.9	181.3
10	138.0	134.4	200.8	202.7	181.6	126.9	80.0	152.6	173.1	169.5

Let us now analyze the bearing failure time data. From Eq. 4.27, the likelihood function is based on Eq. 4.8 and consists of the product of individual Weibull likelihood functions for the observed failure times $y_{ij}, i = 1, \ldots, 10, j = 1, \ldots, 10$. We multiply the likelihood function by the contributions from the tester effects as specified by Eq. 4.28 and expressed as

$$\prod_{i=1}^{10} \frac{1}{\alpha_i \sqrt{2\pi\sigma^2}} \exp\left(\frac{-1}{2\sigma^2} [\log(\alpha_i) - \mu]^2 \right) .$$

To complete the model, we specify the following diffuse priors distributions for μ, σ^2, and β:

$$\begin{aligned} &\mu \sim Normal(0, 1000) , \\ &\sigma^2 \sim InverseGamma(0.001, 0.001) , \text{ and} \\ &\beta \sim Gamma(1.5, 0.5) . \end{aligned}$$

We approximate the joint posterior distribution of ($(\gamma_{0,i}, i = 1, \ldots, 10)$, μ, σ^2, β) by making draws using MCMC. Table 4.6 presents marginal posterior summaries for these model parameters. The results for σ (obtained by taking the square root of the σ^2 posterior draws) show some differences in the testers (i.e., with a posterior median of 0.82), although there is much overlap in the posterior distributions for $\gamma_{0,1}, \ldots, \gamma_{0,10}$, the individual tester effects.

Recall that the experimenters want to determine the bearing failure time distribution with lubricant O-64-2, by removing the tester effect, i.e., the bearing failure times have a $Weibull[\exp(\mu), \beta]$ distribution. We can approximate the posterior distribution for reliability over time t by taking the (μ, β) posterior draws and evaluating

$$R(t) = \exp[-\exp(\mu)t^{\beta}],$$

to obtain $R(t)$ posterior draws. Table 4.6 presents summaries for the posterior reliability at 50, 100, 150, and 200 hours. For example, at 150 hours, the median posterior reliability is 0.578 with a 95% credible interval of (0.369, 0.751).

Table 4.6. Posterior summaries of hierarchical Weibull model parameters for the bearing example

			Quantiles				
Parameter	Mean	Std Dev	0.025	0.050	0.500	0.950	0.975
$\exp(\mu)$	7.723E-10	2.927E-9	2.345E-12	4.829E-12	1.534E-10	3.122E-9	5.526E-9
μ	−22.67	1.964	−26.78	−26.06	−22.60	−19.58	−19.01
σ	0.8713	0.2823	0.4654	0.5099	0.8239	1.3760	1.5520
$\gamma_{0,1}$	-22.02	1.886	-25.95	−25.26	−21.95	−19.03	−18.54
$\gamma_{0,2}$	−23.69	2.062	−27.94	−27.20	−23.63	−20.41	−19.85
$\gamma_{0,3}$	−22.62	1.990	−26.76	−26.05	−22.54	−19.49	−18.93
$\gamma_{0,4}$	−23.87	2.093	−28.20	−27.48	−23.80	−20.57	−19.96
$\gamma_{0,5}$	−21.59	1.852	−25.45	−24.81	−21.53	−18.66	−18.16
$\gamma_{0,6}$	−22.40	1.932	−26.44	−25.76	−22.34	−19.34	−18.83
$\gamma_{0,7}$	−23.11	2.004	−27.26	−26.58	−23.04	−19.93	−19.43
$\gamma_{0,8}$	−22.92	1.976	−27.04	−26.32	−22.84	−19.82	−19.24
$\gamma_{0,9}$	−21.91	1.898	−25.86	−25.15	−21.85	−18.93	−18.41
$\gamma_{0,10}$	−22.60	1.948	−26.67	−25.97	−22.54	−19.56	−19.00
β	4.403	0.3718	3.714	3.823	4.391	5.045	5.174
$R(50)$	0.9950	0.0029	0.9876	0.9895	0.9956	0.9984	0.9987
$R(100)$	0.9075	0.0334	0.8320	0.8472	0.9117	0.9534	0.9597
$R(150)$	0.5745	0.0969	0.3689	0.4088	0.5783	0.7224	0.7507
$R(200)$	0.1570	0.0831	0.0277	0.0421	0.1468	0.3059	0.3493

To assess how well the hierarchical Weibull model fits the bearing failure time data, we can apply a Bayesian χ^2 goodness-of-fit test. Based on $K = 5$ $(\approx 100^{0.4})$ equal probability bins, repeatedly make draws from $[(\gamma_{0,i}, i = 1, \ldots, 10), \beta]$ posterior distribution, calculate the R^B test statistic, and compare it against the 0.95 quantile of the $ChiSquared(4)$ reference distribution. We find that about 4% of the R^B values exceed this 0.95 quantile, which suggests that this model fits the data well.

4.6 Model Selection

Model selection encompasses many aspects. In this chapter, we have introduced a number of distributions useful for modeling reliability data. For example, for analyzing failure times, most applications choose from the exponential, Weibull, or lognormal distributions. Consequently, one aspect of model selection is choosing a distribution for the reliability data. Examples 4.2 and 4.7, which analyzed the supercomputer failure count data by a Poisson and hierarchical Poisson models, respectively, suggest another aspect: do the reliability data require a hierarchical model or not?

In Sect. 2.4, we presented Bayes' factors as a powerful model selection method. Bayes' factors involve multidimensional integrals, which require computationally difficult numerical approximation and, consequently, limit their use.

In this section, we present three general model selection methods, which are

- Bayesian information criterion (BIC),
- deviance information criterion (DIC), and
- Akaike information criterion (AIC).

The BIC, DIC, and AIC are all information-based criteria and have the same basic form, which is

$$IC = -2\log[f(\mathbf{t}|\widehat{\boldsymbol{\theta}})] + g(k)\,, \tag{4.29}$$

where $\widehat{\boldsymbol{\theta}}$ is an estimate of the vector of model parameters $\boldsymbol{\theta}$ based on its posterior distribution (e.g., posterior median, mean, or mode), k is the dimension of $\boldsymbol{\theta}, g(\cdot)$ is a function that changes depending on which information criterion is being used, and IC denotes an information criterion.

All of the information-based criteria choose between models i and j by calculating *information criterion differences*, or

$$\Delta_{ij} = IC_i - IC_j\,, \tag{4.30}$$

where IC_i represents the model information criterion for model i, and IC_j represents the model information criterion for model j.

While we use BIC as the primary method for model selection, DIC is highly useful when the models are hierarchical as discussed in Sects. 3.2 and 4.5. The AIC is a general-purpose model selection procedure used in many classical analyses.

4.6.1 Bayesian Information Criterion

The *Bayesian information criterion* (*BIC*) has the same basic form as Eq. 4.29, where

$$g(k) = \log(n)k, \tag{4.31}$$

n is the number of observations, and k is the dimension of $\boldsymbol{\theta}$. Because $g(k)$ is positive and a lower BIC is better, the implication of using this criterion is that the penalty factor for using k parameters is $\log(n)$. Motivated by each potential model being equiprobable, Schwarz (1978) first developed BIC as a criterion for model selection. Assuming diffuse prior distributions on all the parameters, the criterion penalizes models with increasing complexity as demonstrated by Eq. 4.31. When comparing several models, the model with the lowest BIC fits best.

We noted previously that Bayes' factors, while being a powerful model selection method, are hard to implement. However, DiCiccio et al. (1997) discusses how BIC is an approximation to Bayes' factors, which provides another justification for using BIC.

Example 4.9 Illustration of BIC. Consider modeling the failure times (3.12, 5.13, 1.01, 4.17, 3.08, 1.44, 2.39, 2.44, 6.48, 3.33, 2.65, 3.36, 0.36, 3.16, 0.39, 4.55, 3.30, 4.74, 1.83, 1.51, 1.05, 5.70, 1.42, 2.24, 5.49) by exponential and Weibull distributions. (We actually simulated these failure times from a Weibull distribution with scale $\lambda = 0.1$ and shape $\beta = 2$.) To complete the model, let us use independent $InverseGamma(0.1, 0.1)$ and $Gamma(2, 2)$ prior distributions for λ and β, respectively.

For the exponential model fit with $n = 25$ and $k = 1$, BIC $= 107.71$, and for the Weibull model fit with $k = 2$, BIC $= 100.76$. Not surprisingly, the Weibull model has the lower BIC and fits the data better.

In the next example, we consider the use of BIC in the analyses of the LCD projector failure time data.

Example 4.10 Comparison of distributions for the LCD projector example. Examples 4.3, 4.4, and 4.5 considered exponential, Weibull, and lognormal distributions, respectively, for analyzing the LCD project lamp bearing failure times. For the exponential model fit with $n = 31$ and $k = 1$, BIC $= 459.753$, for the Weibull model fit with $k = 2$, BIC $= 463.083$, and for the lognormal model fit with $k = 2$, BIC $= 468.2956$. Between the two parameter models, the Weibull model fits better. Based on BIC alone, the results suggest that the exponential model fits better. Recall that the exponential distribution is a special case of the Weibull distribution with shape parameter $\beta = 1$; also, the analysis results using the Weibull model in Example 4.4 weakly suggest that β exceeds 1.

4.6.2 Deviance Information Criterion

The BIC method requires the analyst to specify the number of parameters exactly. For many data models, however, the number of estimated parameters is not clearly defined. Hierarchical models arising from data on multiple units provide one clear example where specifying the number of parameters is difficult. Consider the following model for failure times $\mathbf{t}$:

$$\begin{aligned} t_{ij}|\theta_i, \sigma^2 &\sim LogNormal(\mu + \theta_i, \sigma^2),\ i = 1, \ldots, n, j = 1, \ldots, m\,, \\ \theta_i|\tau^2 &\sim Normal(0, \tau^2),\ i = 1, \ldots, n\,, \\ \mu &\sim Normal(0, 10^6)\,, \\ \tau^2 &\sim InverseGamma(0.001, 0.001)\,,\ \text{and} \\ \sigma^2 &\sim InverseGamma((0.001, 0.001)\,. \end{aligned} \tag{4.32}$$

A very strict view suggests that the dimension of the parameter space is $n+3$ (one dimension for $\theta_1, \ldots, \theta_n, \mu, \tau^2$, and σ^2). Another view suggests that the *actual* parameters are the subset (μ, τ^2, σ^2), so that the dimension is 3. A more realistic and justifiable answer lies somewhere between 3 and $n + 3$.

The *deviance information criterion* (*DIC*) solves the problem of a poorly defined parameter space. Before defining DIC, several other quantities need definitions. First, *model deviance* is

$$D(\boldsymbol{\theta}) = -2\log[f(\mathbf{t}|\boldsymbol{\theta})]\,,$$

where $f(\mathbf{t}|\boldsymbol{\theta})$ is the likelihood function, $\mathbf{t}$ is the vector of failure times, and $\boldsymbol{\theta}$ is the vector of unknown model parameters. One measure of the quality of a particular model's fit is the expected deviance, or

$$\bar{D} = E_{\boldsymbol{\theta}}[D]\,,$$

where the expectation is over the posterior distribution of $\boldsymbol{\theta}$. As with BIC, a penalty term addresses the model complexity (or the number of estimated parameters). As such, the definition for the number of estimated parameters p_D is

$$\begin{aligned} p_D &= E_{\boldsymbol{\theta}}[D] - D[E_{\boldsymbol{\theta}}(\boldsymbol{\theta})] \\ &= \bar{D} - D(\bar{\boldsymbol{\theta}})\,, \end{aligned}$$

which we interpret as the effective number of estimated parameters; instead of using $\bar{\boldsymbol{\theta}}$, the marginal posterior mean, use $\tilde{\boldsymbol{\theta}}$, the marginal posterior median, because of its stability. Now, formal definition for DIC is

$$DIC = \bar{D} + p_D\,.$$

This definition allows for a penalty that treats the number of parameters somewhere between the two extremes (in the example above, 3 is the minimum and $n + 3$ is the maximum). We recommend using DIC for model selection when the proposed models include a hierarchical specification. As with BIC, models with lower values of DIC are preferred.

Example 4.11 Illustration of DIC. Consider the calculation of DIC for the failure time model given in Eq. 4.32 using the failure times presented in Table 4.7. These are simulated data using $\mu = 0, \sigma^2 = 1$, and $\tau^2 = 1$. Note that there are two failure times associated with each of the $n = 25$ θ_is. We use a large number of draws (e.g., 10,000) from the joint posterior distribution for μ, σ, and $\theta_i, i = 1, \ldots, 25$, to evaluate $\bar{D}$ (using the lognormal probability density function given in Eq. 4.12). That is, approximate $\bar{D}$ by the average of D over these draws and $\tilde{\boldsymbol{\theta}}$ by the marginal posterior median over these draws. Consequently, calculating $\bar{D}, D(\tilde{\boldsymbol{\theta}}), p_D$, and DIC yields 139.38, 117.05, 22.33, and 161.71, respectively. Contrast these results with those from fitting a common mean model, i.e., with no θ_is, which has $\bar{D}, D(\tilde{\boldsymbol{\theta}}), p_D$, and DIC values of 203.11, 201.08, 2.03, and 205.14, respectively. Not surprisingly, the DIC of the true model that generated these data is smaller than that of the common mean model.

Table 4.7. Failure times for DIC illustration

i	$\mathbf{t_i}$
1	32.512, 40.429
2	0.994, 2.602
3	0.965, 8.703
4	4.088, 8.917
5	13.205, 2.607
6	0.730, 0.791
7	0.043, 0.024
8	0.166, 0.257
9	0.063, 0.233
10	0.195, 0.889
11	0.687, 0.162
12	0.289, 0.115
13	5.247, 4.406
14	2.137, 0.928
15	11.943, 0.491
16	0.700, 0.320
17	1.309, 0.391
18	0.016, 0.088
19	1.252, 1.050
20	1.834, 0.480
21	34.747, 6.062
22	3.345, 2.143
23	1.054, 4.368
24	0.206, 1.531
25	1.404, 1.576

4.6.3 Akaike Information Criterion

Akaike information criterion (AIC) is an approximation to the Kullback-Leibler distance, a measure of the difference between two probability density functions. Burnham and Anderson (2002) presents a detailed discussion of AIC and related issues. Using the general information criterion specification in Eq. 4.29, the definition for AIC is

$$g(k) = k\,,$$

where k is the number of estimated parameters. Classical analysts mostly use AIC from among the information criterion-based model selection methods. Analysts can also use a Bayesian version of AIC by using the posterior mode or mean for $\widehat{\boldsymbol{\theta}}$. The AIC method, like other information criterion-based methods, prefers models with lower values of AIC.

4.7 Related Reading

Meeker and Escobar (1998) extensively examines component reliability from a classical statistical viewpoint. See also Nelson (1982), which provides many graphical methods for identifying an appropriate failure time distribution. From a Bayesian viewpoint, Martz and Waller (1982) compiles many closed form results for component reliability for a variety of data types and prior distributions. Martz and Waller (1982) preceded the advent of MCMC, but provides an extensive collection of situations that a reliability analyst may face and offers advice on prior distribution selection.

4.8 Exercises for Chapter 4

4.1 Reanalyze the plant 63 EDG demand data in Example 4.1 using a uniform prior distribution for the probability of successful start π. Compare the results with those obtained in Example 4.1 in terms of the point estimate, credible interval, and length of the credible interval.

4.2 Tetrahedron Inc. (1996) presents success/failure data for tests on blowout prevention (BOP) systems, which prevent uncontrolled releases of reservoir fluids in the drilling of oil and gas fields. A test is a success if no piece of equipment in the BOP system needs repair. The initial experiment consisted of 44 unused BOP systems tested with 17 successes. Estimate the initial BOP system reliability and provide a credible interval.

4.3 Derive the posterior distribution for a Poisson sampling distribution with gamma prior distribution.

4.4 Reanalyze the supercomputer data in Example 4.2 using a uniform prior distribution on the interval $(0, 20)$ for the mean monthly number of failures λ. Compare the results with those obtained in Example 4.2 in terms of the point estimate, credible interval, and length of the credible interval.

4.5 Like Example 4.2, analyze the failure count data for SMP 21 in Table 6.3 assuming a Poisson distribution. Separately analyze the failure count data from SMP 1 and compare the mean number of failures λ for these two SMPs.

4.6 Analyze the LCD projector lamp failure time data in Table 4.2 using the gamma failure time model. Choose hyperparameters for the α and λ prior distributions so that the prior predictive distribution is similar to that for the Weibull failure time model used in Example 4.4. Evaluate the posterior distribution of the reliability function $R(t)$ at 1,000 hours. Assess how well the gamma distribution fits these data. Compute BIC and compare with BIC calculated for the Weibull and lognormal distribution fits in Examples 4.4 and 4.5.

4.7 Consider the following failure times (in 1,000 hours) for a particular component of an anti-aircraft missile system: 14.4, 2.1, 0.4, 18.6, 1.2, 2.6, 11.5, 18.4, 14.0, 2.8, 7.6, 2.7, 35.4, 10.4, 19.8, 11.3, 2.6, 0.8, 11.3, 5.4. Using the BIC model selection method in Sect. 4.6, determine which distribution among the exponential, Weibull, lognormal, or gamma distributions best fits the data. Assess the goodness of fit for these distributions using a Bayesian χ^2 goodness-of-fit test.

4.8 Table 4.8 presents failure times in hours for the model 7835 power amplifier vacuum tube used in the Linac accelerator at Fermi National Accelerator Laboratory. See McCrory (2006) for more details. Using the BIC model selection method in Sect. 4.6, determine which distribution among the exponential, Weibull, lognormal, or gamma distributions best fits the data. Assess the goodness of fit for these distributions using a Bayesian χ^2 goodness-of-fit test.

Table 4.8. Failure times (in hours) for power amplifier vacuum tube, model 7835 (McCrory, 2006)

25	13414	15943	22269	10265	10452	16640	16766	13736	13779
20340	21983	10634	22874	13853	15843	13594	8568	16942	8404
8947	5929	10830	10615	10039	13540	11184	10567	12920	8589
8050	11795	10378	10388	11448	9022	10051	15363	15996	15018
18102	9515	13513	15299	11466	8912	13084	16752	14174	13852
15346	9773	10687	7845	16547	12258	12824	7190	19433	14004
15002	14145	16214	19117	19203	22699	17163	14492	10064	13282
10216	20939	13444	14741	10322	19598	9896	12350	14438	21094
15981	11674	8729	16987	18632	14868	14791	48449	17833	22910
21587	12484	20054	16852	12651	10116	17948	10914	16284	11914
7828	15599	6845	26466	7867	12027	40859	9319	13264	11687
15812	31882	25773	8063	11896	10880	10880	5469	11305	15870
14190	2664	17973							

4.9 Coit and Jin (2000) presents the reliability data for an airplane indicator light, as displayed in Table 4.9. Note that the actual failure times are not available. Assuming that the failure times follow a gamma distribution, use the result that a sum of independent gamma random variables also has gamma distribution to write down the likelihood for these data. (That is, if $T_j \sim Gamma(\alpha, \lambda), j = 1, \ldots, n_i$, then $\sum_{j=1}^{n_i} T_j \sim Gamma(n_i\alpha, \lambda)$.) Analyze these data using diffuse prior distributions for α and λ. Plot the median posterior reliability function with corresponding 90% credible intervals from 0 to 25,000 hours.

Table 4.9. Airplane indicator light reliability data (Coit and Jin, 2000)

Number of Failures	Cumulative Operating Time (hours)
2	51000
9	194900
8	45300
8	112400
6	104000
5	44800

4.10 Cox and Oakes (1984) presents failure times of springs (in 1,000 cycles) under repeated loadings at 700 N/mm^2: 3402, 9417, 1802, 4326, 11520*, 7152, 2969, 3012, 1550, 11211, where the fourth failure time marked by an asterisk is right censored.
   a) Analyze these data assuming a Weibull distribution.
   b) Provide a plot of the median posterior reliability function over a suitable time period with corresponding 90% credible intervals.

4.11 Show that the likelihood for observed failure times $t_1, \ldots, t_n$, which have an inverse Gaussian distribution, depend on the failure time data through $\bar{t} = \sum_{i=1}^{n} t_i/m$ and $\overline{t^{-1}} = \sum_{i=1}^{n} t_i^{-1}/n$. Nádas (1973) summarizes the failure time data for an unspecified electronic device based on $n = 10$ devices as $\bar{t} = 1.352$ and $\overline{t^{-1}} = 0.948$. Analyze these failure time data summaries using an inverse Gaussian distribution.

4.12 Chhikara and Folks (1977) presents repair times in hours of an airborne communication transceiver as follows: 0.2, 0.3, 0.5, 0.5, 0.5,0.5, 0.6, 0.6, 0.7, 0.7, 0.7, 0.8, 0.8, 1.0, 1.0, 1.0, 1.0, 1.1, 1.3, 1.5, 1.5, 1.5, 1.5, 2.0, 2.0, 2.2, 2.5, 2.7, 3.0, 3.0, 3.3, 3.3, 4.0, 4.0, 4.5, 4.7, 5.0, 5.4, 5.4, 7.0, 7.5, 8.8, 9.0, 10.3, 22.0, 24.5.
   a) Fit these repair times using an inverse Gaussian distribution.
   b) Assess how well the inverse Gaussian distribution fits these data using a Bayesian χ^2 goodness-of-fit test.
   c) Plot the median posterior reliability function from 0 to 30 hours with corresponding 90% credible intervals.

4.13 Example 10.2 analyzes success/failure data from EDGs from 63 nuclear plants as reported by Martz et al. (1996) and displayed in Table 10.1. See Example 10.2 for more details, which uses a hierarchical binomial model. Does the hierarchical binomial model fit the data well based on a Bayesian χ^2 goodness-of-fit test? Using the DIC model selection method, is a hierarchical model necessary? In other words, can we assume that the failure probability is constant across the plants?

4.14 Consider the AFW demand data in Table 10.8. See Exercise 10.7 for more details. Analyze these data using a hierarchical binomial model. Does the hierarchical binomial model fit the data well based on a Bayesian χ^2 goodness-of-fit test? Using the DIC model selection method, is a hierarchical model necessary? In other words, can we assume that the failure probability is constant across the plants?

4.15 Consider the number of minor defectives in successive MIL-STD-105B samples of some material, as presented in Table 10.9. See Exercise 10.9 for more details. Analyze these data using a hierarchical binomial model. Does the hierarchical binomial model fit the data well based on a Bayesian χ^2 goodness-of-fit test? Using the DIC model selection method, is a hierarchical model necessary? In other words, can we assume that the failure probability is constant across the samples?

4.16 As in Example 4.2, analyze the supercomputer failure count data for the last month (e.g., month 15) in Table 6.3 using a hierarchical model. Drop the monthly failure count data for SMP 21, which is different than the other 47 SMPs. Is there a difference in supercomputer performance between the first and last month? One measure of performance is the predictive distribution for an SMP. Also, assess how well the Poisson model fits these data using a Bayesian χ^2 goodness-of-fit test.

4.17 Example 10.4 analyzes pump failure count data for the Farley 1 nuclear power plant given in Table 10.3. See Example 10.4 for more details, which uses a hierarchical Poisson model. Does the hierarchical Poisson model fit the data well based on a Bayesian χ^2 goodness-of-fit test? Using the DIC model selection method, is a hierarchical model necessary? In other words, can we assume that the failure rate is constant across the pump systems?

4.18 In Example 4.8, suppose that reliability using a randomly chosen tester is of interest. Assume that the 10 testers in Example 4.8 represent a random sample from a population of testers; that is, their effects are conditionally independent. Evaluate the posterior distribution of reliability $R(t)$ for a randomly chosen tester.

4.19 Similar to Example 4.8, Exercise 7.9 reports on failure time data in Table 7.31 for a different lubricant O-67-22. Analyze these data with the same model used in Example 4.8. How do the two lubricants compare?

4.20 Example 10.5 analyzes the pressure vessel failure time data collected at 23.4 MPa given in Table 10.5. See Example 10.5 for more details, which uses a hierarchical Weibull model. Does the hierarchical Weibull model fit the data well based on a Bayesian χ^2 goodness-of-fit test? Using the DIC model selection method, is a hierarchical model necessary? In other words, can we assume that the Weibull scale parameter is constant across the spools?

5

System Reliability

This chapter extends the models for component data to systems. This extension requires us to specify logical relationships between the components in a system and how the functioning of the complete system depends on the functioning (or not) of each of its components. We consider models for both independent and dependent component failures.

5.1 System Structure

To this point, we have considered only single components. Now suppose that we want to assess the reliability of a system that is composed of multiple components. To assess systems, we must first understand the *structural properties* of the system. In this chapter, we consider several ways to represent the relationships among the components that comprise a system, including structure functions, minimal cut and path sets, fault trees, reliability block diagrams, and Bayesian networks (BN). We must also understand the *probabilistic properties* of a system. We explore a variety of models for expressing the reliability of a system. While most of our models assume that components fail independently, we also consider models for dependent failures. This chapter focuses on the first failure of the system, while Chap. 6 discusses repairable systems.

A common strategy in many assessment problems is to break the large or complex problem into several small problems, perform assessments on the smaller problems, and then aggregate these assessments into an estimate for the overall quantity of interest. This process has two steps: *modeling* and *information* (Mosleh and Bier, 1992). For system assessments, *modeling* is the process of defining the overall system in terms of its basic elements, typically components, and defining the relationships among those components. *Information*, perhaps from many sources, is then available to assess the reliability of the basic elements and combinations of the basic elements.

5.1.1 Reliability Block Diagrams

In reliability analysis, we often model systems graphically. This provides a visual representation of the components and how they are configured to form a system. One of the most commonly used system representations in risk and reliability analysis is the *reliability block diagram*. Figure 5.1 shows how components are represented in a reliability block diagram. In this diagram, "connection" through a block implies that the component is working and that a failure has not occurred.

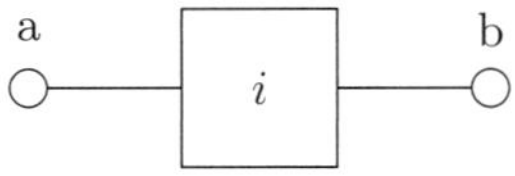

Fig. 5.1. Component i in a reliability block diagram.

A system that functions if and only if all of its n components are functioning is a *series system*. Figure 5.2 shows the reliability block diagram for a series system. For the system to be functioning, there has to be a functioning path from point (a) to point (b); in Fig. 5.2, all n components must be functioning.

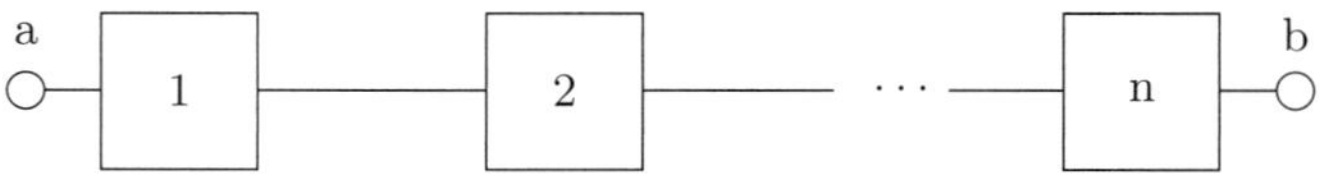

Fig. 5.2. Series system reliability block diagram.

A system that functions if at least one of its n components is functioning is a *parallel system*. Figure 5.3 shows the reliability block diagram for a parallel system. For the system to be functioning, there has to be a functioning path from point (a) to point (b); in Fig. 5.3, at least one of the n components must be functioning.

Series and parallel systems are special cases of k-of-n systems. A *k-of-n system* functions if at least k of its n components are functioning. If $k = n$, we have a series system; if $k = 1$, we have a parallel system. Figure 5.4 shows the reliability block diagram for a k-of-n system with $k = 2$ and $n = 3$.

5.1.2 Structure Functions

Structure functions provide another way to summarize the relationships between components in a system. Consider a system with n components. For the ith component and time t, define a random variable $X_i(t)$ so that

$$X_i(t) = x_i = \begin{cases} 1 \text{ if the } i\text{th component is functioning at time } t \\ 0 \text{ if the } i\text{th component has failed prior to time } t. \end{cases}$$

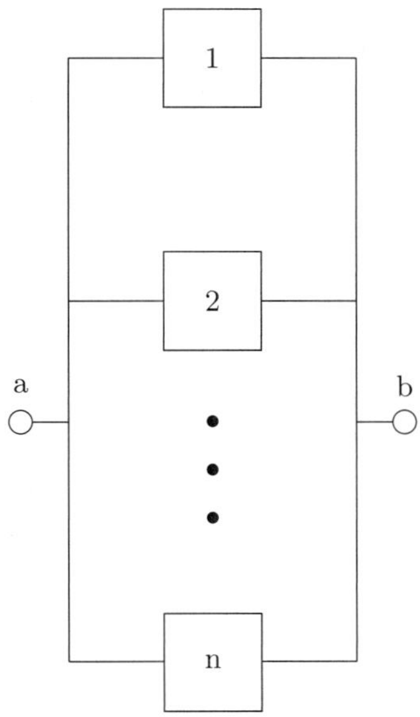

Fig. 5.3. Parallel system reliability block diagram.

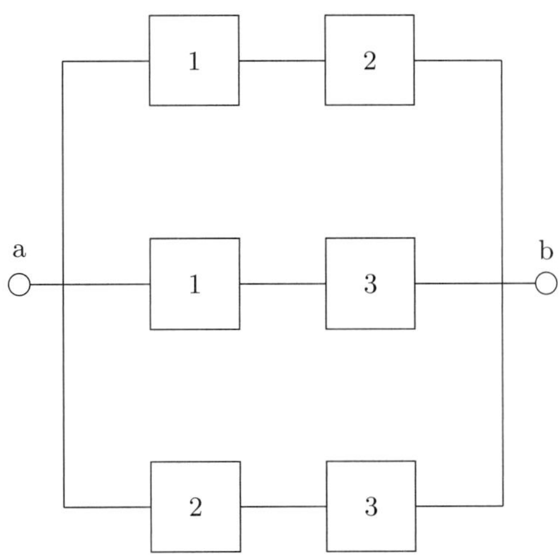

Fig. 5.4. 2-of-3 system reliability block diagram.

We can summarize the state of all of the components by a vector $\mathbf{x} = (x_1, x_2, \ldots, x_n)$. Some of the 2^n states correspond to a functioning system; some correspond to a failed system. The state of the system is thus a function of $\mathbf{x}$. We call this function the *structure function* and define it as

$$\phi(\mathbf{x}) = \begin{cases} 1 \text{ if the system is functioning} \\ 0 \text{ if the system has failed.} \end{cases}$$

Consider a series system, which functions if and only if all of its n components are functioning. Thus, $\phi(\mathbf{x}) = 1$ if $x_1 = x_2 = \ldots = x_n = 1$, and is 0 otherwise. We can write the following three equivalent expressions:

$$\begin{aligned} \phi(\mathbf{x}) &= \begin{cases} 1 \text{ if } x_i = 1 \text{ for all } i \\ 0 \text{ if } x_i = 0 \text{ for any } i, \end{cases} \\ &= \min(x_1, x_2, \ldots, x_n), \\ &= \prod_{i=1}^{n} x_i. \end{aligned}$$

A parallel system functions if at least one of its components is functioning. Thus, $\phi(\mathbf{x}) = 0$ if $x_1 = x_2 = \ldots = x_n = 0$, and is 1 otherwise. We can write the following three equivalent expressions:

$$\begin{aligned} \phi(\mathbf{x}) &= \begin{cases} 1 \text{ if } x_i = 1 \text{ for any } i \\ 0 \text{ if } x_i = 0 \text{ for all } i, \end{cases} \\ &= \max(x_1, x_2, \ldots, x_n), \\ &= 1 - \prod_{i=1}^{n}(1 - x_i). \end{aligned}$$

A k-of-n system functions if k or more of its components function. We can write

$$\begin{aligned} \phi(\mathbf{x}) &= \begin{cases} 1 \text{ if } \sum_{i=1}^{n} x_i \geq k \\ 0 \text{ if } \sum_{i=1}^{n} x_i < k, \end{cases} \\ &= \sum_j (\prod_{i \in A_j} x_i)[\prod_{i \in A_j^c} (1 - x_i)], \end{aligned} \tag{5.1}$$

where A_j is any subset of $\{1, 2, \ldots, n\}$ with at least k elements, and the sum is over all such subsets. For example, the structure function for a 2-of-3 system is

$$\begin{aligned} \phi(\mathbf{x}) &= \sum_j (\prod_{i \in A_j} x_i)[\prod_{i \in A_j^c} (1 - x_i)] \\ &= x_1 x_2 (1 - x_3) + x_1 x_3 (1 - x_2) + x_2 x_3 (1 - x_1) + x_1 x_2 x_3 \\ &= x_1 x_2 + x_1 x_3 + x_2 x_3 - 2 x_1 x_2 x_3. \end{aligned}$$

Of particular interest in reliability is the set of *coherent systems*. A system is *coherent* if its structure function satisfies the following conditions:

1. $\phi(0, 0, \ldots, 0) = 0$,
2. $\phi(1, 1, \ldots, 1) = 1$,
3. $\phi(\mathbf{x})$ is nondecreasing in each argument.

We can summarize these conditions as follows. If every component in the system has failed, the system has failed; if every component in the system is functioning, the system is functioning. The third condition implies that if the system is functioning, and a failed component is restored to a functioning state, then the system is still functioning.

Let $\phi(\mathbf{x})$ be the structure function of a coherent system. Then

$$\prod_{i=1}^{n} x_i \leq \phi(\mathbf{x}) \leq 1 - \prod_{i=1}^{n}(1 - x_i). \tag{5.2}$$

Equation 5.2 indicates that any coherent system functions at least as well as a system in which the same n components are connected in series, and functions no better than a system in which the same n components are connected in parallel.

5.1.3 Minimal Path and Cut Sets

In addition to reliability block diagrams and structure functions, we can use minimal path and cut sets to represent the structure of a system. We call any $\mathbf{x}$ for which $\phi(\mathbf{x}) = 1$ a *path vector* for the system, and any $\mathbf{x}$ for which $\phi(\mathbf{x}) = 0$ a *cut vector* for the structure. The set of component indices corresponding to the functioning (failed) components of a path vector (cut vector) is a *path set* (*cut set*).

Define $\mathbf{y} < \mathbf{x}$ if for all i, $y_i \leq x_i$, and for some i, $y_i < x_i, i = 1, \ldots n$. A path vector, $\mathbf{x}$, is a *minimal path vector* if for every $\mathbf{y} < \mathbf{x}$, $\phi(\mathbf{y}) = 0$. The *minimal path set* is the set of components in a minimal path vector that are functioning; that is, a minimal set of components such that if they are all functioning, the system is functioning, but if one of them fails (and all of the components outside the set have failed), then the system fails.

A cut vector, $\mathbf{x}$, is a *minimal cut vector* if for every $\mathbf{y} > \mathbf{x}$, $\phi(\mathbf{y}) = 1$. The *minimal cut set* is the set of components in a minimal cut vector that are failed; that is, a minimal set of components such that if they have all failed, the system has failed, but if one of them is functioning (and all of the components outside the set are functioning), then the system is functioning.

We can determine the structure function of a coherent system from either its minimal path sets or its minimal cut sets. Suppose that $\{a_1, a_2, \ldots, a_m\}$ is the collection of all minimal path sets of a coherent system, with x_i being the state variable of the ith component. The system is functioning if all of the components in one or more path sets are functioning. We can think of this as a parallel arrangement of m sets of components in series. In terms of the minimal path sets, the structure function of the system is

$$\phi(\mathbf{x}) = 1 - \prod_{j=1}^{m}(1 - \prod_{i \in a_j} x_i). \tag{5.3}$$

A similar result holds for cut sets. Let $\{b_1, b_2, \ldots, b_k\}$ be the collection of all minimal cut sets of a coherent system, with x_i being the state variable of the ith component. The system fails if all of the components in one or more cut sets fail. We can think of this as a series arrangement of k sets of components in parallel. In terms of minimal cut sets, the structure function of the system is

$$\phi(\mathbf{x}) = \prod_{i=1}^{k}[1 - \prod_{i \in b_k}(1 - x_i)]. \tag{5.4}$$

Example 5.1 Using path sets and cut sets to determine a structure function. Consider the system in Fig. 5.5. The minimal path sets are $a_1 = \{1, 2\}$, $a_2 = \{1, 3\}$. Using Eq. 5.3, the structure function for the system is

$$\begin{aligned}\phi(\mathbf{x}) &= 1 - \prod_{j=1}^{2}(1 - \prod_{i \in a_j} x_i) \\ &= 1 - (1 - x_1x_2)(1 - x_1x_3) \\ &= x_1x_2 + x_1x_3 - x_1x_2x_3.\end{aligned}$$

The minimal cut sets for the system are $b_1 = \{1\}$ and $b_2 = \{2, 3\}$. Using Eq. 5.4, the structure function for the system is

$$\begin{aligned}\phi(\mathbf{x}) &= \prod_{k=1}^{2}(1 - \prod_{i \in b_k}(1 - x_i)) \\ &= (1 - (1 - x_1))(1 - (1 - x_2)(1 - x_3)) \\ &= x_1(x_2 + x_3 - x_2x_3) \\ &= x_1x_2 + x_1x_3 - x_1x_2x_3.\end{aligned}$$

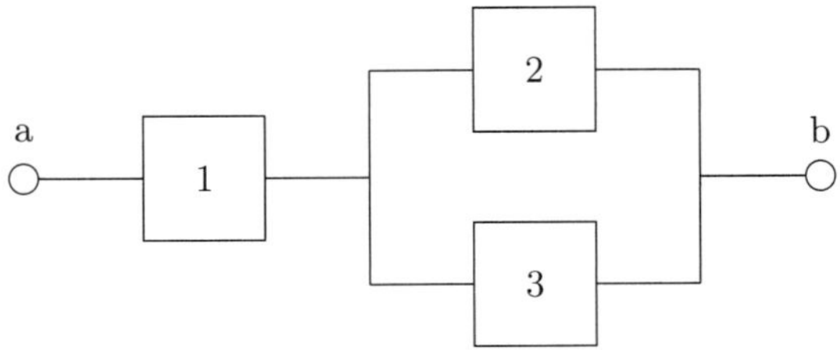

Fig. 5.5. System with minimal path sets $a_1 = \{1, 2\}$ and $a_2 = \{1, 3\}$.

5.1.4 Fault Trees

Another of the most commonly used system representations in risk and reliability analysis is the *fault tree*. A fault tree is a logic diagram that displays the relationships between a critical event, typically a system failure, and the causes of the event, typically component failures. It illustrates how the states of the system's components relate to the state of the system as a whole. *Logic gates* are the graphical symbols used to represent the connections between the components and the system. We discuss here only the most basic logic gates and events; for more details see Vesely et al. (1981).

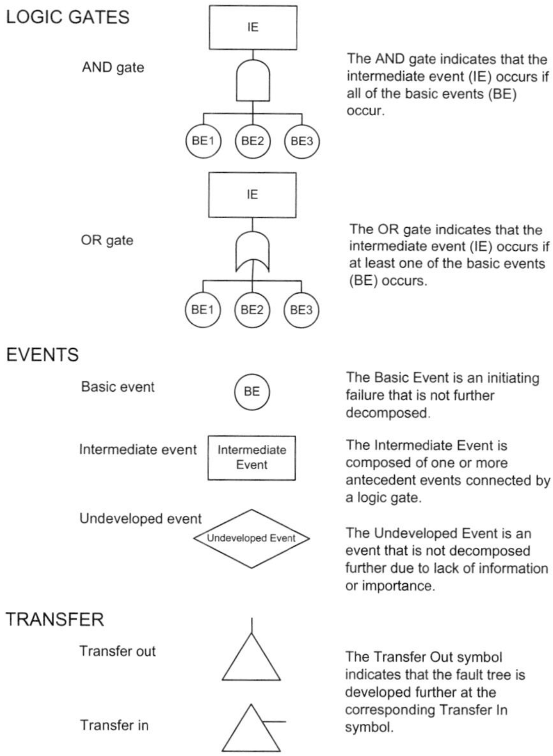

Fig. 5.6. Some common fault tree symbols.

Figure 5.6 summarizes the most common events and logic gates used to construct fault trees. A *basic event* is an initiating fault that requires no further decomposition. Fault trees describe the functioning of a system only to the resolution of its basic events, which are often component failures. An *undeveloped event* is a fault that we choose not to develop, either because we consider it insignificant or because we have insufficient information available to further develop it.

An *intermediate event* is a fault that occurs because one or more antecedent (previous) faults have occurred. Intermediate events often correspond to subsystem faults, where subsystems comprise more than one component. Logic gates connect the antecedent faults to the intermediate event. The two most common logic gates are the AND gate and the OR gate. The intermediate event is the output of the gate; the antecedent events are the inputs. With an AND gate, the output occurs only if all of the inputs occur. With an OR gate, the output occurs if at least one of the inputs occurs.

Example 5.2 Basic fault-tree analysis. Figure 5.7 contains a fault tree with a top event, one intermediate event, and three basic events. The top event is a failure of a fire protection system. This system failure is composed of two other events connected by an AND gate: the basic event that the sprinklers fail, and the intermediate event that the alarm fails. The AND gate means that there is no fire protection only if both the alarm and sprinklers fail. The alarm failure, in turn, is an intermediate event composed of two basic events connected by an OR gate. The alarm fails if either the wiring fails or the power fails.

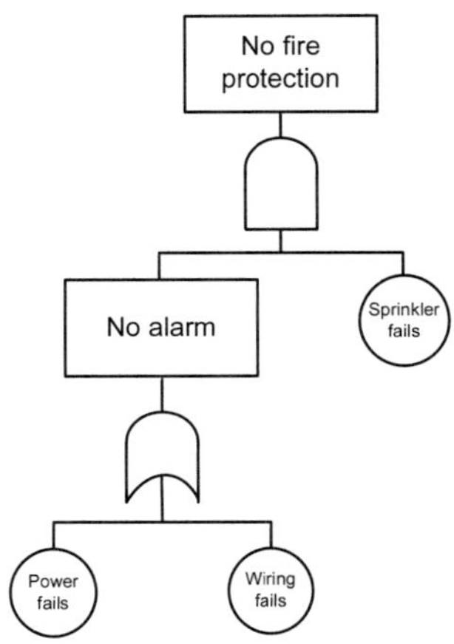

Fig. 5.7. Fault tree for a fire protection system with three basic events, one intermediate event, and one top event.

When a fault tree contains only AND and OR gates, fault trees and reliability block diagrams are equivalent. A reliability block diagram for a series system indicates that all components must work for the system to work; this is equivalent to saying that the system fails if one or more of the components fail. Consequently, we can convert the reliability block diagram of a series system to a fault tree with an OR gate. Figure 5.8 gives some of the simple relationships between reliability block diagrams and fault trees.

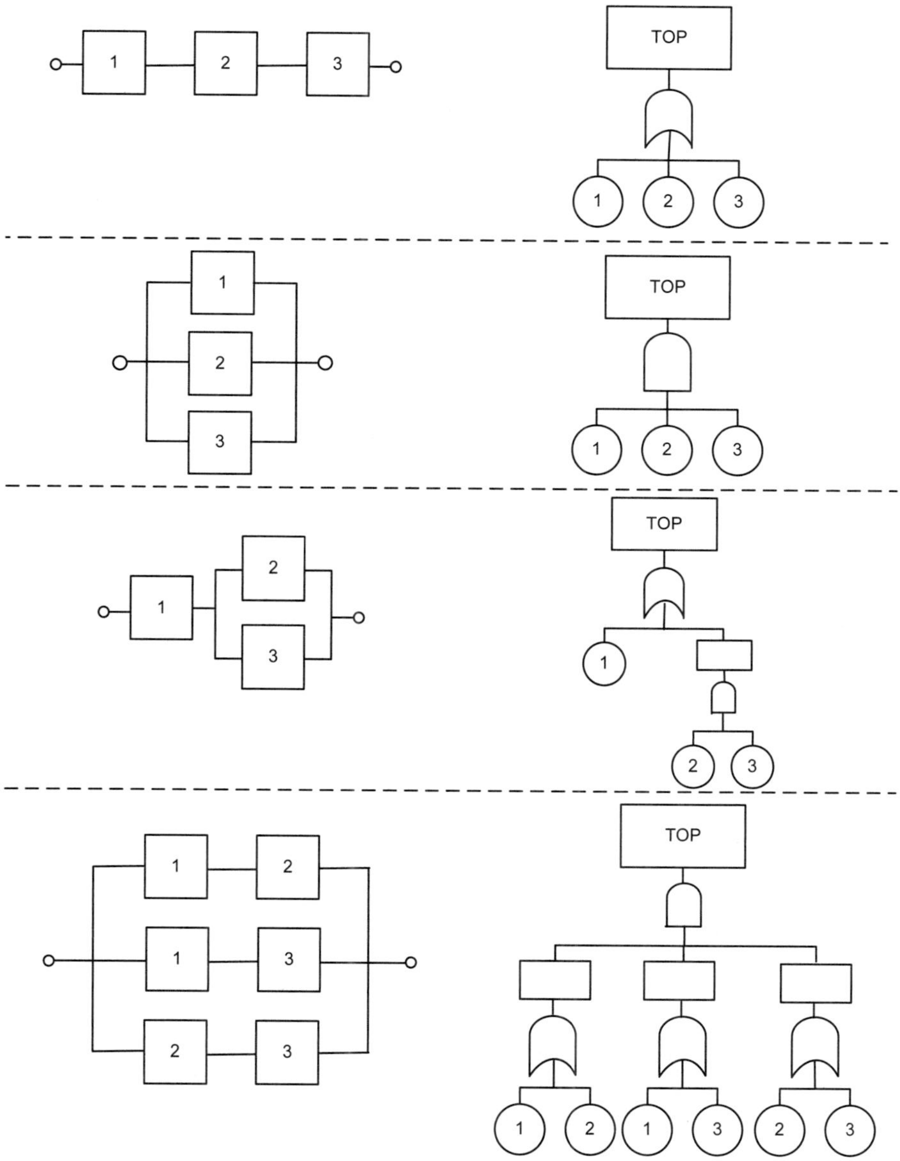

Fig. 5.8. Relationships between reliability block diagrams and fault trees.

Example 5.3 Determining the structure of intermediate and top events for sample fault tree. Consider the fault tree shown in Fig. 5.9 from Hamada et al. (2004). What are the structure functions at the system level (TE) and at intermediate event 1 (IE1)?

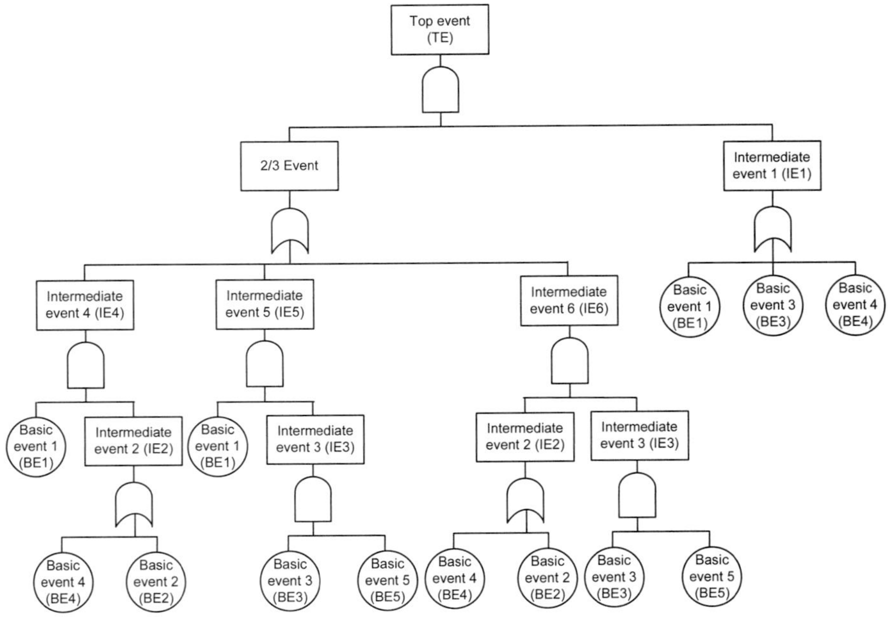

Fig. 5.9. Sample fault tree.

The minimal cut sets for IE1 are {BE1}, {BE3}, {BE4}. Applying Eq. 5.4, we find that the structure function for intermediate event 1 is

$$\phi_{IE1}(\mathbf{x}) = x_1 x_3 x_4.$$

Considered alone, IE1 is a series system with components BE1, BE3, and BE4.

The minimal cut sets for the system are {BE1, BE2}, {BE1, BE4}, {BE1, BE3, BE5}, {BE2, BE3, BE5}, {BE3, BE4, BE5}. Notice that the intermediate event *2/3 Event* in Fig. 5.9 is an example of a 2-of-3 subsystem (see also Fig. 5.4). This event occurs if two of BE1, IE2, or IE3 fail. Applying Eq. 5.4, we find that the structure function for the system is

$$\begin{aligned}\phi_{TE}(\mathbf{x}) = &[1-(1-x_1)(1-x_2)][1-(1-x_1)(1-x_4)]\\ &[1-(1-x_1)(1-x_3)(1-x_5)]\\ &[1-(1-x_2)(1-x_3)(1-x_5)]\end{aligned}$$

$$\begin{aligned}
&[1-(1-x_3)(1-x_4)(1-x_5)]\\
&=(x_1+x_2-x_1x_2)(x_1+x_4-x_1x_4)\\
&\quad(x_1+x_3+x_5-x_1x_3-x_1x_5-x_3x_5+x_1x_3x_5)\\
&\quad(x_2+x_3+x_5-x_2x_3-x_2x_5-x_3x_5+x_2x_3x_5)\\
&\quad(x_3+x_4+x_5-x_3x_4-x_3x_5-x_4x_5+x_3x_4x_5)\\
&=x_1x_3+x_1x_5+x_1x_2x_4+x_2x_3x_4+x_2x_4x_5-x_1x_3x_5\\
&\quad-2x_1x_2x_3x_4-2x_1x_2x_4x_5-x_2x_3x_4x_5+2x_1x_2x_3x_4x_5.
\end{aligned}$$

The final simplification uses $x_i^2 = x_i$ for binary variables.

5.2 System Analysis

5.2.1 Calculating System Reliability

For the ith component and time t, we have defined the random variable $X_i(t)$ so that

$$X_i(t) = \begin{cases} 1 & \text{if the } i\text{th component is functioning at time } t \\ 0 & \text{if the } i\text{th component has failed prior to time } t, \end{cases}$$

and the system structure function

$$\phi(\mathbf{X}(t)) = \begin{cases} 1 & \text{if the system is functioning at time } t \\ 0 & \text{if the system has failed prior to time } t. \end{cases}$$

For nonrepairable components, the *reliability function* (defined in Chap. 1) is $R_i(t) = \mathbf{P}(X_i(t) = 1)$. Similarly, for the system, the reliability function is $R_S(t) = \mathbf{P}(\phi(\mathbf{X}(t)) = 1)$. We often drop the explicit dependence on t in our notation.

Since $\phi(\mathbf{X})$ is a Bernoulli random variable, we have

$$\begin{aligned}
\mathrm{E}[\phi(\mathbf{X})] &= 0 \cdot \mathbf{P}(\phi(\mathbf{X}) = 0) + 1 \cdot \mathbf{P}(\phi(\mathbf{X}) = 1)\\
&= \mathbf{P}(\phi(\mathbf{X} = 1)).
\end{aligned}$$

Suppose that we want to determine the reliability function for a series system of n components, where we assume the functioning of the components is statistically independent. Then we can write

$$\begin{aligned}
R_S(t) &= \mathbf{P}(\phi(\mathbf{X}(t)) = 1) \qquad (5.5)\\
&= \mathbf{P}\Big(\prod_{i=1}^{n} X_i(t) = 1\Big)\\
&= \prod_{i=1}^{n} \mathbf{P}(X_i(t) = 1)\\
&= \prod_{i=1}^{n} R_i(t).
\end{aligned}$$

From Eq. 5.5, we see that the system reliability for a series system is always less than or equal to the reliability of the least reliable component, or

$$R_S(t) \leq \min_i R_i(t).$$

We can use the expression for system reliability to calculate the hazard function for a series system:

$$\begin{aligned} h_S(t) &= \frac{-\mathrm{d}\log[R_S(t)]}{\mathrm{d}t} = \frac{-\mathrm{d}\log[\prod_{i=1}^n R_i(t)]}{\mathrm{d}t} \\ &= \sum_{i=1}^n \frac{-\mathrm{d}\log[R_i(t)]}{\mathrm{d}t} = \sum_{i=1}^n h_i(t). \end{aligned}$$

Consider a parallel system and assume that the functioning of the components is statistically independent. The system fails only if each component fails. Therefore,

$$\begin{aligned} R_S(t) &= \mathbf{P}(\phi(\mathbf{X}(t) = 1)) && (5.6) \\ &= \mathrm{E}[\phi(\mathbf{X}(t))] && (5.7) \\ &= \mathrm{E}[1 - \prod_{i=1}^n (1 - X_i(t))] \\ &= 1 - \mathrm{E}[\prod_{i=1}^n (1 - X_i(t))] \\ &= 1 - \prod_{i=1}^n \mathrm{E}[1 - X_i(t)] \\ &= 1 - \prod_{i=1}^n (1 - R_i(t)). \end{aligned}$$

Next, consider a k-of-n system. In Sect. 5.1.2, we saw that the structure function for a k-of-n system is

$$\phi(\mathbf{X}(t)) = \begin{cases} 1 \text{ if } \sum_{i=1}^n X_i(t) \geq k \\ 0 \text{ if } \sum_{i=1}^n X_i(t) < k. \end{cases}$$

For simplicity, assume that each component in the system has the same reliability function, $R_i(t) = R(t), i = 1, \ldots n$, and that the components fail independently. Define $Y(t) = \sum_{i=1}^n X_i(t)$. At a given time t, $Y(t) \sim$ *Binomial*$(n, R(t))$. We can write

$$\begin{aligned} R_S(t) &= \mathbf{P}(Y(t) \geq k) && (5.8) \\ &= \sum_{y=k}^n \binom{n}{y} [R(t)]^y [1 - R(t)]^{n-y} \\ &= 1 - \sum_{y=0}^{k-1} \binom{n}{y} [R(t)]^y [1 - R(t)]^{n-y}. \end{aligned}$$

Example 5.4 Standby Redundant Systems. A standby redundant system is a special case of a parallel system. In it, only a single component operates at a time, with the remaining components successively brought into operation upon failure of the operating component. A spare tire for a car is a simple example of a standby redundant system. Figure 5.10 shows the block diagram for a standby redundant system. The component ordering $1, \dots, n$ is the order in which the standby components come into operation.

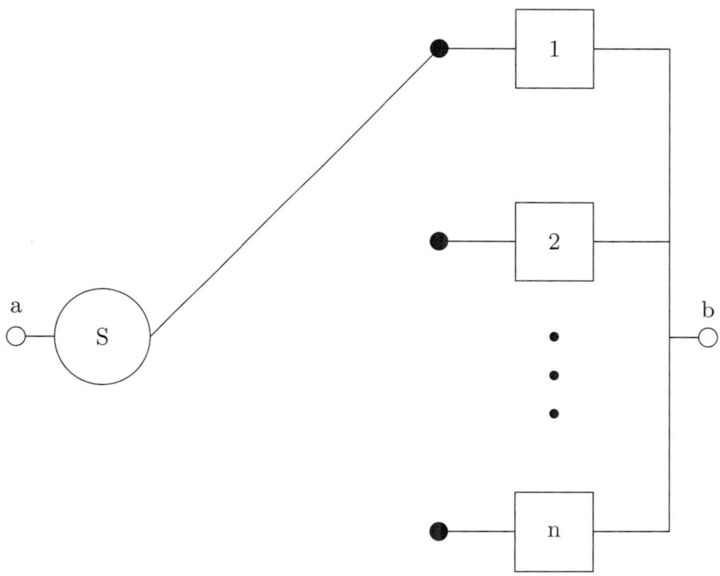

Fig. 5.10. Standby system reliability block diagram.

In a standby redundant system, the reliability of the system depends on the reliability of the *switch*, or the changeover between the failed and new component. Suppose that the switch is perfectly reliable. The lifetime T of the system is $T = \sum_{i=1}^{n} T_i$, where T_i is the lifetime of the ith component.

Let $R_S^n(t)$ denote the reliability of a standby redundant system with n independent components. For a two-component system, we can express the reliability as

$$\begin{aligned} R_S^2(t) &= \mathbf{P}(T_1 > t) + \mathbf{P}(T_1 \leq t \text{ and } T_2 > t - T_1) \\ &= R_1(t) + \int_0^t f_1(t_1) R_2(t - t_1) \mathrm{d}t_1. \end{aligned}$$

In the case of a standby redundant system with imperfect switching, the switch may not activate the standby component when the operating component fails. Let π represent the probability that the switch works. For a two-component system, we have

$$R_S^2(t) = \mathbf{P}(T_1 > t) + \pi \mathbf{P}(T_1 \le t \text{ and } T_2 > t - T_1)$$
$$= R_1(t) + \pi \int_0^t f_1(t_1) R_2(t - t_1) \mathrm{d}t_1.$$

5.2.2 Prior Distributions for Systems

In previous chapters, we have considered how to specify prior distributions for unknown parameters. When analyzing systems, choosing prior distributions becomes more complex, because the prior distributions chosen for the parameters of component models have implications for the system model.

Consider the three-component series system shown in Fig. 5.11. We know from Sect. 5.2.1 that the reliability of the system is $R_S(t) = \prod_{i=1}^{3} R_i(t)$, where $R_i(t)$ is the reliability of the ith component (C_i). Suppose that we describe our prior uncertainty about the component reliabilities using independent uniform distributions, so that at a particular time t, $R_i(t) \sim Uniform(0, 1)$. What does this imply about our prior distribution for $R_S(t)$?

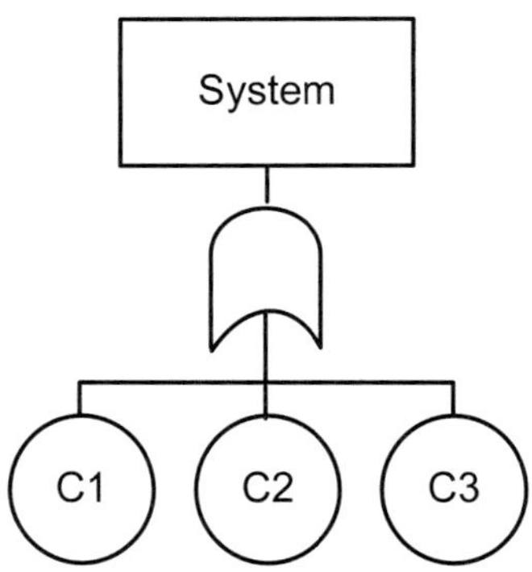

Fig. 5.11. Three-component series system fault tree.

We can address this question by simulation. Suppose that we simulate the reliability of each component by drawing from a $Uniform(0, 1)$ distribution, and that we simulate the system reliability by multiplying the three draws together. If we repeat this procedure 10,000 times, we see a histogram like that in Fig. 5.12.

Notice that describing uncertainty about the reliability of the components of a series system using uniform prior distributions does not imply

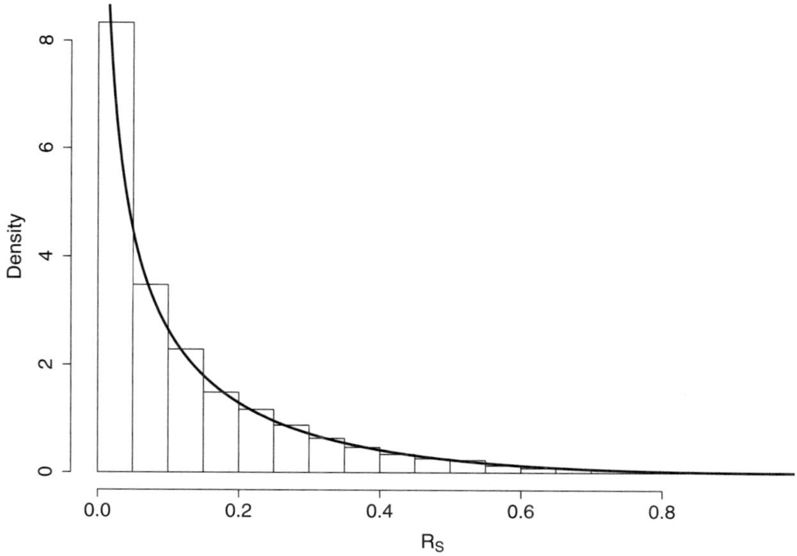

Fig. 5.12. Induced prior on system with three components in series with uniform priors. The histogram comes from simulation and the solid line is actual prior density function.

that there is a uniform prior distribution on the system itself. Given a series system with k components, the prior density induced on the system is $[\Gamma(k)]^{-1}[-\log(R_S)]^{k-1}$, which has mean 2^{-k} (Parker, 1972). This density is drawn as the solid line in Fig. 5.12 for $k = 3$ components.

Now suppose instead that we want to determine what prior distribution we need to place on the component reliabilities of a series system to induce a $Uniform(0,1)$ distribution for the system reliability. If we assume that each of the k components has the same prior distribution, then the prior density of the components is $[\Gamma(1/k)]^{-1}[-\log(\pi_i)]^{\frac{-(k-1)}{k}}$, which has mean $2^{-1/k}$. Figure 5.13 gives a plot of the prior density for one component (of three) that induces a uniform prior distribution on the system.

Suppose that we do not want to assume the same prior distribution for each component reliability in the series system, but that we do want to induce a $Uniform(0,1)$ prior distribution on system reliability. If $R_S \sim Uniform(0,1)$, then $-\log(R_S) \sim Exponential(1)$. Since $R_S = \prod_{i=1}^{k} R_i$, $-\log(R_S) = \sum_{i=1}^{k} -\log(R_i)$. If $-\log(R_i) \sim Gamma(\alpha_i, 1)$, with $\sum_{i=1}^{k} \alpha_i = 1$, then the distribution of $\sum_{i=1}^{k} -\log(R_i) = -\log(R_S) \sim Gamma(1,1) = Exponential(1)$. The distribution plotted in Fig. 5.13 is a special case when $\alpha_i = 1/k$.

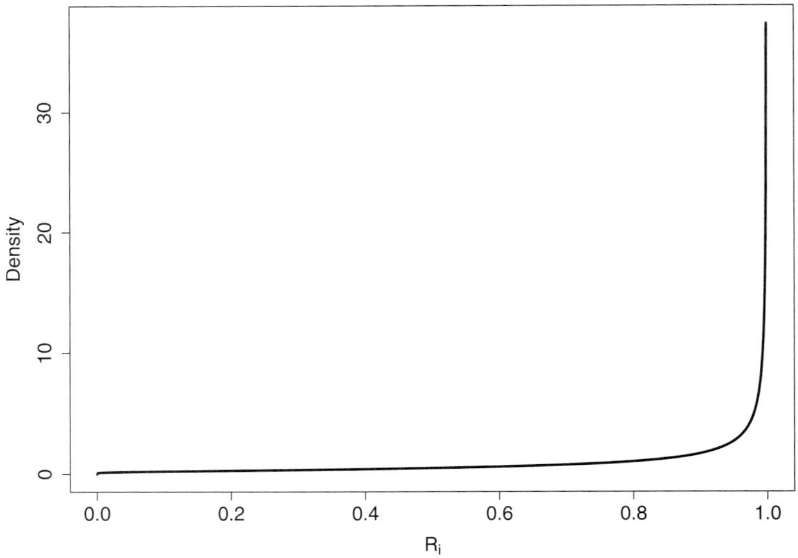

Fig. 5.13. Prior on components with three components in series to induce a uniform prior on the system.

Suppose that instead of a uniform prior distribution for system reliability, we would like a more informative prior distribution (Lawrence and Vander Wiel, 2005). Let $-\log(R_S) \sim Gamma(\alpha, \beta)$. This implies that R_S has a unimodal prior distribution on $[0, 1]$. If $-\log(R_i) \sim Gamma(\alpha_i, \beta)$ with $\sum_{i=1}^{k} \alpha_i = \alpha$, then the $Gamma(\alpha, \beta)$ prior is induced at the system. The distribution induced on R_S is called the *negative log-gamma distribution.*

The analytical results we have discussed are for series systems. However, the simulation approach — make random draws from component reliability prior distributions, use the methods from Sect. 5.2.1 to determine the expression for the intermediate or top event reliability, calculate the expression using the random component draws, draw a histogram of the results — works generally for any system structure.

Example 5.5 Induced prior distribution for sample fault tree. Consider the fault tree shown in Fig. 5.9. In Example 5.3, we determined the structure function for the system. Assuming the components are independent, we can express the system reliability

$$\begin{aligned} R_{TE} = {} & R_1R_3 + R_1R_5 + R_1R_2R_4 + R_2R_3R_4 + R_2R_4R_5 - R_1R_3R_5 \\ & -2R_1R_2R_3R_4 - 2R_1R_2R_4R_5 - R_2R_3R_4R_5 + 2R_1R_2R_3R_4R_5. \end{aligned}$$

Suppose that we specify our prior distributions as

$$R_1 \sim Beta(3,1),$$
$$R_2 \sim Beta(5,1),$$
$$R_3 \sim Beta(7,1),$$
$$R_4 \sim Beta(6,3), \text{ and}$$
$$R_5 \sim Beta(5,1).$$

To examine the induced prior distribution for the top event reliability, we simulate 10,000 random draws from the prior distributions for the component reliabilities and evaluate the expression for R_{TE} for each of the 10,000 sets of draws. Figure 5.14 shows a histogram of the induced prior distribution for R_{TE}.

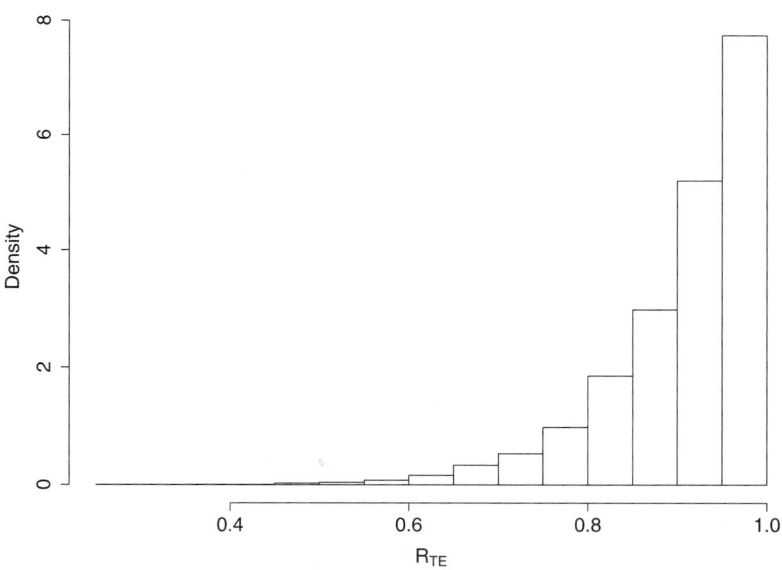

Fig. 5.14. Prior induced on top event of sample fault tree.

5.2.3 Fault Trees with Bernoulli Data

Suppose that a system has been represented as a fault tree with only AND and OR gates, or equivalently as a reliability block diagram. In Sect. 5.1.3 we saw that we can express the reliability of a coherent system (including a fault tree) in terms of its component reliabilities. (See also Barlow and Proschan

(1975).) This expression gives us a way to analyze any system with Bernoulli data at any combination of events in the fault tree.

To analyze a system with Bernoulli data, we follow these steps:

1. Determine the fault tree for the system.
2. Determine which events have data.
3. For each event that has data, determine an expression for its reliability in terms of the reliabilities of the basic events. The formulas in Sect. 5.1.3 are useful for expressing intermediate and top event probabilities in terms of minimal cut or path sets.
4. Determine the likelihood for the data, expressing the likelihood of the intermediate or top events in terms of functions of the basic events.
5. Specify a prior distribution on the reliabilities of the basic events.

As an example, consider the three-component series system pictured in Fig. 5.11. We have collected data on each of the three components, as given in Table 5.1.

Table 5.1. Data for three-component series system with no system data

	Successes	Failures	Units Tested
Component 1	8	2	10
Component 2	7	2	9
Component 3	3	1	4

We model the data for each component as $Binomial(n_i, \pi_i)$, where n_i is the number of tests for each component, π_i is the success probability for each component, and $i = 1, 2, 3$ indexes the component. (Refer to Eqs. 2.1 and 2.2 for an explanation of how a binomial likelihood arises from Bernoulli data.) If we assume that the components fail independently, we know that the reliability of the system, π_S, is $\pi_S = \pi_1\pi_2\pi_3$.

Assume that we specify independent $Uniform(0, 1)$ prior distributions for each π_i. These assumptions imply that the joint posterior distribution for (π_1, π_2, π_3), which is proportional to the prior distribution times the likelihood function, is

$$p(\pi_1, \pi_2, \pi_3 \,|\, \mathbf{x}) \propto \pi_1^8(1-\pi_1)^2\pi_2^7(1-\pi_2)^2\pi_3^3(1-\pi_3),$$

where $\mathbf{x}$ are the data in Table 5.1.

It is straightforward to draw a sample from this posterior distribution by using Metropolis-Hastings sampling or by simulation (see Exercise 5.13). Once we have drawn a sample from the joint posterior distribution, we can get a sample from the posterior distribution of π_S simply by calculating $\pi_1\pi_2\pi_3$ for each sample from the joint posterior distribution.

Figure 5.15 plots the histogram of a sample from the posterior distribution of π_S. The posterior mean is 0.36, and a 95% credible interval is (0.13, 0.64). The posterior mean is low because each of the three components must work for the system to work. Table 5.2 summarizes the posterior distributions for each parameter.

Table 5.2. Posterior distributions for three-component series system given data in Table 5.1

			Quantiles				
Parameter	Mean	Std Dev	0.025	0.050	0.500	0.950	0.975
π_1	0.75	0.12	0.48	0.53	0.76	0.92	0.94
π_2	0.73	0.13	0.44	0.49	0.74	0.91	0.93
π_3	0.67	0.18	0.27	0.34	0.69	0.92	0.95
π_S	0.36	0.13	0.13	0.16	0.36	0.59	0.64

Springer and Thompson (1966) derives an analytic form for the posterior probability density function for π_S:

$$p(\pi_S) = \frac{3960}{7}\pi_S^3 - 1980\pi_S^4 + 99000\pi_S^7 + [374220 + 356400\log(\pi_S)]\pi_S^8$$
$$-[443520 - 237600\log(\pi_S)]\pi_S^9 - \frac{198000}{7}\pi_S^{10}. \tag{5.9}$$

In Fig. 5.15, Eq. 5.9 is overlaid on the histogram. The analytic form of the posterior probability density is not necessarily more useful than the random sample from the posterior we obtain using MCMC. Sample moments and credible intervals can be calculated easily from the sample; we must use integration to calculate them from the analytic form.

Now suppose that instead of the data in Table 5.1, we have the data in Table 5.3, which now include independent observations on the entire system. Once again, the data for each component has a $Binomial(n_i, \pi_i)$ distribution with $i = 1, 2, 3$ indexing the component.

Table 5.3. Data for three-component series system with system data

	Successes	Failures	Units Tested
Component 1	8	2	10
Component 2	7	2	9
Component 3	3	1	4
System	10	2	12

Because we assume independent $Uniform(0, 1)$ prior distributions for each π_i, the joint posterior distribution for (π_1, π_2, π_3) is

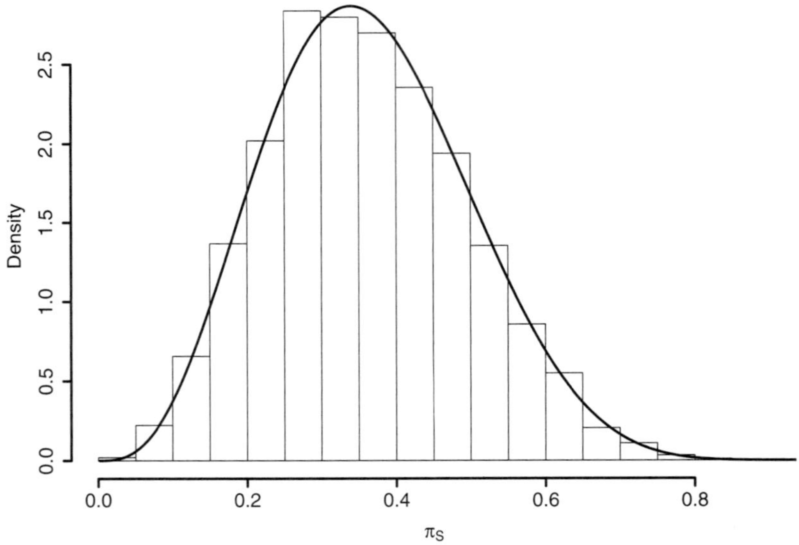

Fig. 5.15. Histogram of sample from posterior distribution of π_S from the three-component series system with analytical density from Eq. 5.9 given data in Table 5.1.

$$p(\pi_1, \pi_2, \pi_3 \mid \mathbf{x}) \propto \pi_1^8(1-\pi_1)^2\pi_2^7(1-\pi_2)^2\pi_3^3(1-\pi_3)(\pi_1\pi_2\pi_3)^{10}(1-\pi_1\pi_2\pi_3)^2.$$

It is straightforward to draw a sample from this posterior distribution by using Metropolis-Hastings sampling. The kernel density estimate of the posterior distribution of π_S is plotted in Fig. 5.16. The posterior mean is 0.60, and a 95% credible interval is (0.40, 0.78). Table 5.4 summarizes the posterior distributions for all of the parameters. Notice that the posterior mean, our point estimate of system reliability, is higher once we add 12 independent observations of the system with 10 successes.

Table 5.4. Posterior distributions for three-component series system given data in Table 5.3

			Quantiles				
Parameter	Mean	Std Dev	0.025	0.050	0.500	0.950	0.975
π_1	0.84	0.079	0.66	0.70	0.85	0.95	0.96
π_2	0.83	0.082	0.65	0.68	0.85	0.95	0.96
π_3	0.85	0.091	0.64	0.68	0.86	0.97	0.98
π_S	0.60	0.097	0.40	0.43	0.60	0.75	0.78

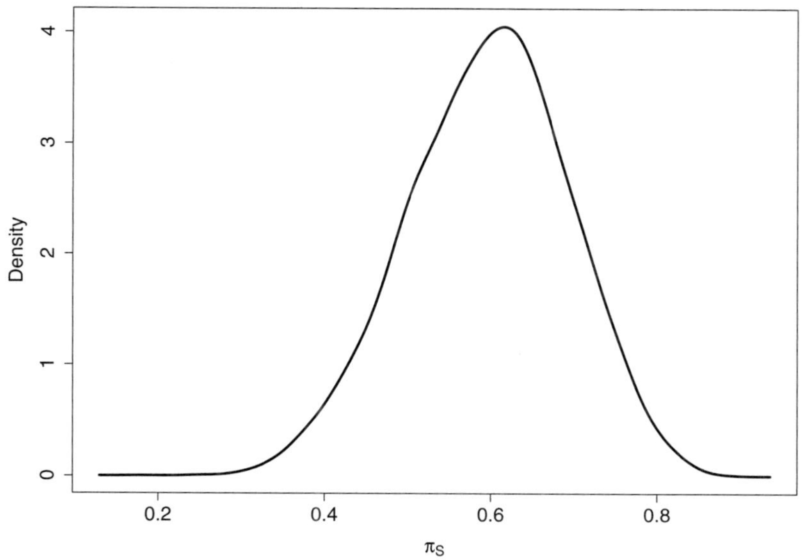

Fig. 5.16. Kernel density estimate of posterior distribution of π_S from the three-component series system given data in Table 5.3.

5.2.4 Fault Trees with Lifetime Data

In Sect. 5.1.3 we saw that we can express the reliability of a coherent system of independent components in terms of its component reliabilities. We can also use this method to model the reliability of systems that have lifetime data at the system or component level.

Recall from Table 1.1 that $R(t) = 1 - F(t)$ and that $f(t) = \frac{\mathrm{d}}{\mathrm{d}t}F(t)$. From Barlow and Proschan (1975), we know that the expression for R_S is multilinear in R_i, which means that it is linear as a function of each R_i. Consequently, we can use the chain rule for differentiation to find an expression for the probability density function for the lifetime of the system.

Suppose that we have a three-component series system as pictured in Fig. 5.11. Let component C_i have reliability R_i. We see that

$$\begin{aligned}
R_S &= R_1 R_2 R_3, \\
1 - F_s(t) &= [1 - F_1(t)][1 - F_2(t)][1 - F_3(t)], \\
\frac{\mathrm{d}}{\mathrm{d}t}(1 - F_s(t)) &= \frac{\mathrm{d}}{\mathrm{d}t}[1 - F_1(t)][1 - F_2(t)][1 - F_3(t)], \\
-f_s(t) &= \frac{\mathrm{d}}{\mathrm{d}t}[1 - F_1(t) - F_2(t) - F_3(t) + \\
&\quad F_1(t)F_2(t) + F_1(t)F_3(t) + F_2(t)F_3(t) - F_1(t)F_2(t)F_3(t)],
\end{aligned}$$

$$
\begin{aligned}
f_s(t) &= f_1(t) + f_2(t) + f_3(t) - F_1(t)f_2(t) - f_1(t)F_2(t) \\
&\quad -F_1(t)f_3(t) - f_1(t)F_3(t) - F_2(t)f_3(t) - f_2(t)F_3(t) \\
&\quad +F_1(t)F_2(t)f_3(t) + F_1(t)f_2(t)F_3(t) + f_1(t)F_2(t)F_3(t), \\
&= f_1(t) - f_1(t)F_2(t) - f_1(t)F_3(t) + f_1(t)F_2(t)F_3(t) + \\
&\quad f_2(t) - F_1(t)f_2(t) - f_2(t)F_3(t) + F_1(t)f_2(t)F_3(t) + \\
&\quad f_3(t) - F_1(t)f_3(t) - F_2(t)f_3(t) + F_1(t)F_2(t)f_3(t), \\
&= \sum_{i=1}^{3} f_i(t) \prod_{j \neq i} R_j(t).
\end{aligned}
$$

Notice that we have been able to express the probability density function of the system lifetime in terms of the probability density functions and reliability functions of the component lifetimes.

Example 5.6 Poly-Weibull distribution. Suppose that component C_i has a $Weibull(\lambda_i, \beta_i)$ distribution. This implies that $f_i(t) = \lambda_i \beta_i t^{\beta_i - 1} \exp(-\lambda_i t^{\beta_i})$, and $R_i(t) = \exp(-\lambda_i t^{\beta_i})$, which means that

$$
\begin{aligned}
f_s(t) &= \sum_{i=1}^{3} \lambda_i \beta_i t^{\beta_i - 1} \exp(-\lambda_i t^{\beta_i}) \prod_{j \neq i} \exp(-\lambda_j t^{\beta_j}) \\
&= (\lambda_1 \beta_1 t^{\beta_1 - 1} + \lambda_2 \beta_2 t^{\beta_2 - 1} + \lambda_3 \beta_3 t^{\beta_3 - 1}) \exp(-\sum_{i=1}^{3} \lambda_i t^{\beta_i}).
\end{aligned}
$$

This distribution for the system is called the *poly-Weibull distribution*. Berger and Sun (1993) discusses its analysis in Bayesian models.

Again consider the system in Fig. 5.11, a three-component series system. Suppose that component C_i has an $Exponential(\lambda_i)$ distribution. This implies that $R_i(t) = e^{-\lambda_i t}$ and that $R_S(t) = \prod_{i=1}^{3} R_i(t) = \exp(\sum_{i=1}^{3} -\lambda_i t)$, which means that the lifetime distribution of the system is $Exponential(\sum_{i=1}^{3} \lambda_i)$. Consider the data in Table 5.5. These data are simulated, with $\lambda_1 = 3$, $\lambda_2 = 1$, $\lambda_3 = 0.5$, and $\lambda_S = 4.5$.

Table 5.5. Exponential data for three-component series system

System	0.0565, 0.259, 0.0934, 0.0323, 0.0618, 0.504, 0.0830, 0.0807, 0.00471, 0.236
Component 1	0.441, 0.0316, 0.533, 0.404, 0.134, 0.00616, 0.444, 0.759, 0.488, 0.490
Component 2	3.040, 0.783, 0.0587, 5.695, 0.317, 0.486, 0.204
Component 3	0.127, 0.292, 2.546, 0.359, 2.741, 1.253, 2.366, 3.953

Suppose that we assign independent $Uniform(0, 10)$ prior distributions to λ_1, λ_2, and λ_3. The likelihood function has the form

$$\exp(-\lambda_1 \sum_{i=1}^{10} x_i) \exp(-\lambda_2 \sum_{i=1}^{7} y_i) \exp(-\lambda_3 \sum_{i=1}^{8} z_i) \exp[-(\lambda_1 + \lambda_2 + \lambda_3) \sum_{i=1}^{10} s_i],$$

where x_i represents the data for component 1, y_i the data for component 2, z_i the data for component 3, and s_i the data for the system. Using MCMC, we calculate the posterior mean for λ_1 as 3.49 with 95% credible interval $(1.95, 5.39)$, the posterior mean for λ_2 as 0.80 with 95% credible interval $(0.35, 1.44)$, the posterior mean for λ_3 as 0.69 with 95% credible interval $(0.32, 1.20)$, and the posterior mean for λ_S as 4.98 with 95% credible interval $(3.39, 6.92)$. Table 5.6 contains summaries of the posterior distributions for all of the parameters.

Table 5.6. Posterior distributions for three-component series system given data in Table 5.5

			Quantiles				
Parameter	Mean	Std Dev	0.025	0.050	0.500	0.950	0.975
λ_1	3.49	0.88	1.95	2.16	3.42	5.04	5.39
λ_2	0.80	0.28	0.35	0.40	0.77	1.31	1.44
λ_3	0.69	0.23	0.32	0.36	0.67	1.11	1.20
λ_S	4.98	0.91	3.39	3.60	4.92	6.57	6.92

Example 5.7 Competing Risks. Often there are k causes of failure in a given situation. For example, a car may fail to start because of a broken starter motor or alternator; a person may die because of heart disease, cancer, accident, suicide, or other causes. In *competing risks* models, an item is subject to k risks or causes of failure. We model this as a series system, where each risk is thought of as a component, and where the item fails when any "component" fails.

Suppose that an item can fail for $k = 2$ reasons, that the time to the first failure mode is $T_1 \sim Exponential(\lambda_1)$, that the time to the second failure mode is $T_2 \sim Exponential(\lambda_2)$, and that T_1 and T_2 are independent. The observed lifetime T is the minimum of T_1 and T_2 and has an $Exponential(\lambda_1 + \lambda_2)$ distribution.

5.2.5 Bayesian Network Models

Fault trees and reliability block diagrams are the most well-known graphical models for system reliability. However, BNs are another way to represent systems. Formally, a BN is a pair $N = \langle (V, E), \mathbf{P} \rangle$, where (V, E) are the nodes and edges of a directed acyclic graph, and $\mathbf{P}$ is a joint probability distribution on V. Each node contains a random variable — in a reliability context, often the reliability of a single component. The directed edges (arrows) between the nodes define conditional dependencies among the random variables.

In a fault tree, the success or failure of the basic events determines the success or failure of the intermediate and top events. In a BN, the success or failure of the components determines the *probability* of the success or failure of the intermediate and top events. For example, a BN can represent that "if component 1 and component 2 are working, there is a 90% chance that subsystem 1 is working."

Figure 5.17 summarizes the three probabilistic relationships that we can specify in a BN. The joint distribution of V, the set of nodes in a BN, is

$$\mathbf{P}(V) = \prod_{v \in V} \mathbf{P}(v \mid \text{parents}[v]), \tag{5.10}$$

where the parents of a node are the set of nodes with an edge pointing to the node. For example, in the serial structure in Fig. 5.17a, the parent of node C is node B, and node A has no parents. Because node A has no parents, we call it a *root node* of the BN.

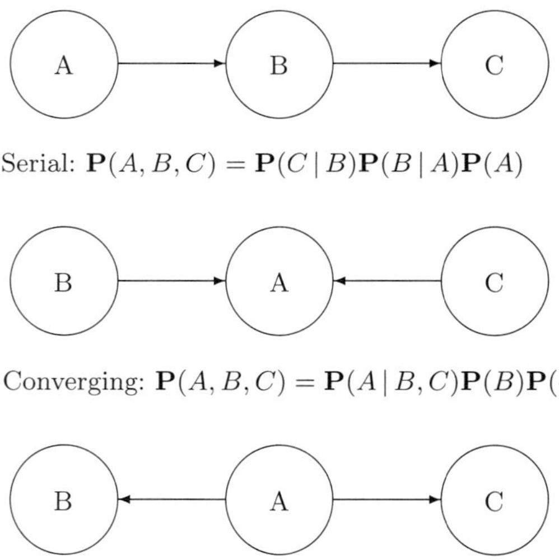

(a) Serial: $\mathbf{P}(A, B, C) = \mathbf{P}(C \mid B)\mathbf{P}(B \mid A)\mathbf{P}(A)$

(b) Converging: $\mathbf{P}(A, B, C) = \mathbf{P}(A \mid B, C)\mathbf{P}(B)\mathbf{P}(C)$

(c) Diverging: $\mathbf{P}(A, B, C) = \mathbf{P}(C \mid A)\mathbf{P}(B \mid A)\mathbf{P}(A)$

Fig. 5.17. Serial, converging, and diverging structures in a BN.

Fault trees are special cases of BNs. Bobbio et al. (2001) gives an algorithm that converts fault trees to BNs:

1. For each basic event, create a root node in the BN. If a basic event occurs more than once in the fault tree, it should appear only once in the BN.
2. Assign to the root node the same probability as its corresponding basic event.
3. Create a node for each intermediate event.
4. Connect each intermediate event as the child of its antecedent events in the fault tree, regardless of the gate connecting them.
5. Assign the conditional probabilities $\mathbf{P}(\text{intermediate event node} \mid \text{basic event antecedents})$ using the logic specified by their connecting gates in the fault tree.

Any inference in the created BN is the same as it would be for the fault tree. Figure 5.18 shows the translation of a two-component parallel system (left) into a BN (right).

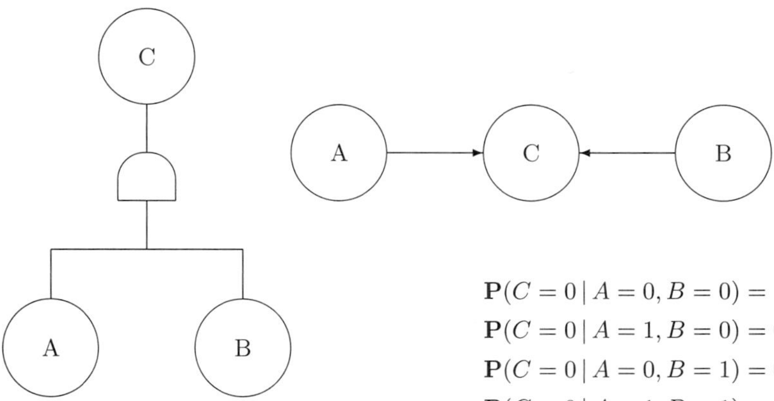

$$\mathbf{P}(C = 0 \mid A = 0, B = 0) = 1$$
$$\mathbf{P}(C = 0 \mid A = 1, B = 0) = 0$$
$$\mathbf{P}(C = 0 \mid A = 0, B = 1) = 0$$
$$\mathbf{P}(C = 0 \mid A = 1, B = 1) = 0$$

Fig. 5.18. Fault tree with AND gate (left) and its conversion to a BN (right).

Because the BN requires us to specify conditional probabilities, we do not need to have nodes with binary states — we can have nodes that are multistate discrete or even continuous random variables. Jensen (2001) and Spiegelhalter et al. (1996) provide algorithms to compute the joint distribution for V and marginal distributions for any subset of variables if all of the random variables V are discrete and all of the conditional probabilities in Eq. 5.10 have point values.

Example 5.8 Developing a Bayesian Network. This example is adapted from Wilson et al. (2007). At 8:40 p.m. on February 25, 1991, parts of an Iraqi Scud missile destroyed the barracks housing members of the U.S. Army's 14th Quartermaster Detachment. This was the single most devastating attack on

U.S. forces during the first Gulf War: 29 soldiers died and 99 were wounded. In the aftermath of this attack, the Army has focused on developing air defense systems capable to defend against ballistic missile attacks. The Critical Measurements and Counter Measures Program (CMCM), run by the U.S. Army Space and Missile Defense Command, conducts exercises to replicate projected ballistic missile threats. These exercises help the U.S. military collect realistic data to evaluate potential defensive measures. The high-fidelity hardware and realistic scenarios created for the exercises provide extensive optical, radar, and telemetry data.

CMCM is organized into campaigns. Each campaign chooses a new ballistic missile threat and develops two to four high-fidelity launch vehicles that emulate the threat as closely as possible, given intelligence information. While CMCM reuses some elements across campaigns, each set of launch vehicles is essentially a complex, one-of-a-kind, one-time-use system built for a specific data collection purpose. Typically, because of cost and schedule constraints, there are no "risk reduction" flights performed, so there are no full-system tests before the actual flights. The systems are designed and built in a distributed fashion, with scientists and engineers from different companies designing, building, and integrating various parts of the vehicle. These campaigns are expensive (millions of dollars) and have a politically high profile. We use one of the CMCM campaigns to illustrate the development of a BN to assess the preflight probability of mission success.

The events that made up the mission fall into three categories: the threat-representative flight, the data collection, and the auxiliary experiments. Failure in any of these categories would cause the mission to be unsuccessful. Figure 5.19 summarizes the events that made up the mission. The threat-representative events are on the left side of the diagram. Notice that there were nine different data collection streams; some started immediately after ignition, and others started after later events.

Defining the mission events allows a specific definition of mission success. We first decided that mission success was a discrete quantity defined as catastrophic failure (RED), degraded (YELLOW), or nominal (GREEN). This language was natural for the CMCM staff and contractors working on the program, as the Department of Defense commonly uses it to describe categories of outcomes in technical and military missions. In addition, we defined RED, YELLOW, and GREEN states for each of the events in Fig. 5.19. Table 5.7 summarizes which event states could cause catastrophic mission failure (RED).

Equation 5.10 shows that a set of conditional distributions determines the joint distribution of the nodes in the BN. For example, in Fig. 5.19, one of the probabilities that we had to assess to determine the joint distribution of all of the events was $\mathbf{P}(\text{Data Collect 1} = \text{RED} \mid \text{Ignition} = \text{GREEN})$. Notice that the conditional dependence structure of the BN greatly decreased the total number of probabilities that we had to specify. If the random variables

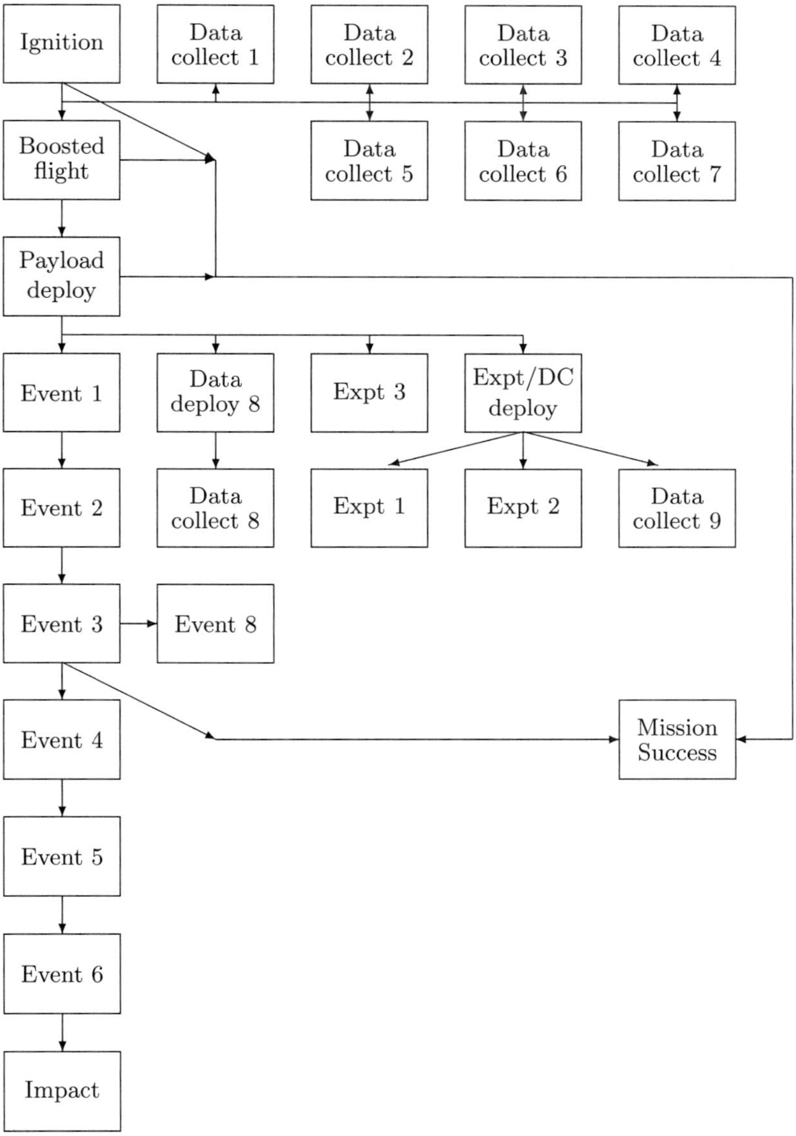

Fig. 5.19. BN showing conditional dependencies between threat-representative events, data collection, and auxiliary experiments.

Table 5.7. Mission success RED

Event	State
Ignition	RED
Boosted Flight	RED
Payload Deploy	RED
Event 3	RED

were discrete and there was no conditional structure, then we would have to assess every possible combination of values of the random variables.

Consider again Fig. 5.19 and Table 5.7. These summarize the events that made up the mission and the event states that define mission success. To quantitatively assess the probability of mission success, all of the conditional probabilities in Fig. 5.19 need to be assessed. We could not elicit these probabilities directly, nor were test data collected that addressed the probabilities directly. Consequently, once we defined mission success, the definition process resumed for each of the mission's component probabilities.

Consider, for example, the event boosted flight. We can decompose boosted flight into the BN given in Fig. 5.20. Still these nodes are not at the right granularity, as there are no data or information about their conditional probabilities. Figure 5.21 is the BN for roll control, which is a further decomposition of part of Fig. 5.20. At this granularity, we can estimate the conditional probabilities. Some of the nodes represent parts that were used in past missions and have existing test data. Other probabilities are elicited using standard expert judgment elicitation techniques, which is what we did for newer parts.

This process was completed for the entire set of events in Fig. 5.19, resulting in a BN with approximately 600 nodes.

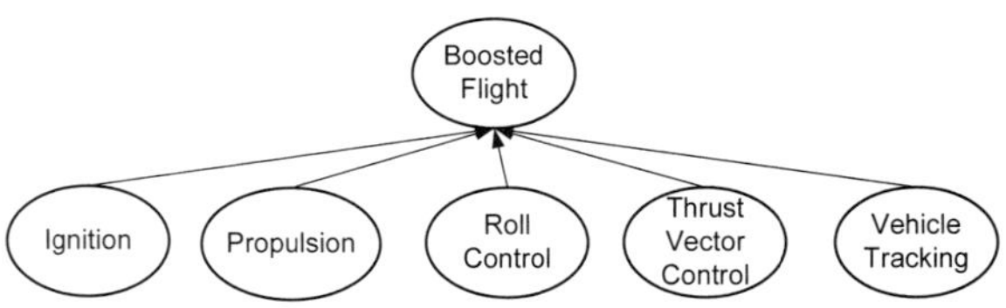

Fig. 5.20. BN decomposing boosted flight.

Consider the three-component system pictured in the converging BN in Fig. 5.22. We have collected data on each of the three components and on the entire system, as given in Table 5.8. We model the data for each component as $Binomial(n_i, \pi_i)$, where n_i is the number of tests for each component,

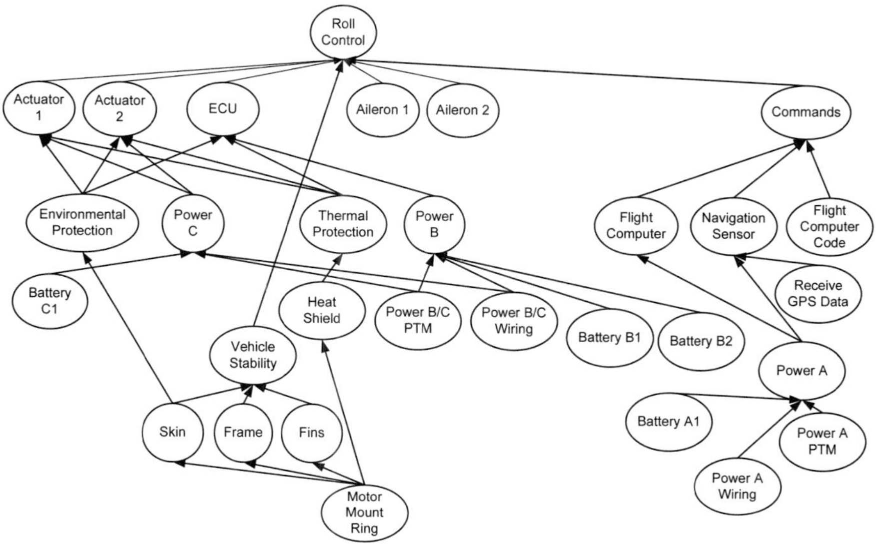

Fig. 5.21. BN decomposing roll control.

π_i is the success probability for each component, and $i = 1, 2, 3$ indexes the component.

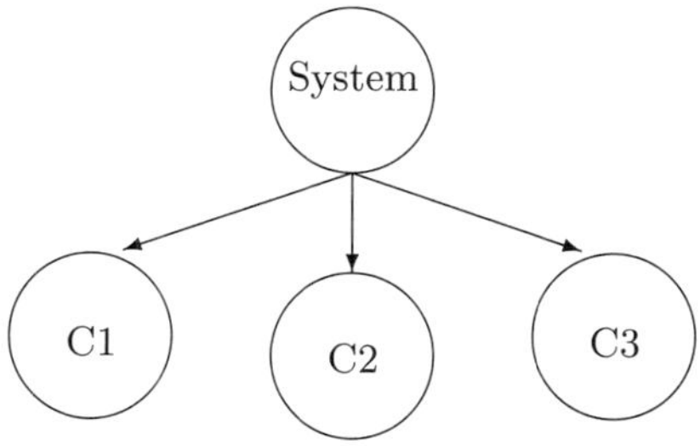

Fig. 5.22. Three-component system BN.

From Eq. 5.10 and Fig. 5.17, to specify the joint distribution of all of the nodes, we need to specify $\mathbf{P}(\text{System} \,|\, C_1, C_2, C_3)$. We specify these conditional distributions in Table 5.9, with Success = 1 and Failure = 0. If all

Table 5.8. Data for three-component series system with system data

	Successes	Failures	Units Tested
Component 1	8	2	10
Component 2	7	2	9
Component 3	3	1	4
System	10	2	12

three components are working, there is a 0.95 probability that the system is working as well. Seven of the eight conditional probabilities are known, but $\mathbf{P}(\text{System} = \text{Success} \mid C_1 = \text{Failure}, C_2 = \text{Success}, C_3 = \text{Success}) = \pi_{FSS}$ is unknown.

Table 5.9. Conditional probabilities for three-component BN

C_1	C_2	C_3	$\mathbf{P}(\text{System} = 1 \mid C_1, C_2, C_3)$
1	1	1	0.95
1	1	0	0.80
1	0	1	0.85
1	0	0	0.50
0	1	1	π_{FSS}
0	1	0	0.40
0	0	1	0.55
0	0	0	0.05

Using Table 5.9, we can write an expression for the reliability of the system as

$$\begin{aligned}
\pi_S &= \mathbf{P}(\text{System} = 1) \\
&= \mathbf{P}(\text{System} = 1 \mid C_1 = 1, C_2 = 1, C_3 = 1)\mathbf{P}(C_1 = 1, C_2 = 1, C_3 = 1) \\
&+ \mathbf{P}(\text{System} = 1 \mid C_1 = 1, C_2 = 1, C_3 = 0)\mathbf{P}(C_1 = 1, C_2 = 1, C_3 = 0) \\
&+ \mathbf{P}(\text{System} = 1 \mid C_1 = 1, C_2 = 0, C_3 = 1)\mathbf{P}(C_1 = 1, C_2 = 0, C_3 = 1) \\
&+ \mathbf{P}(\text{System} = 1 \mid C_1 = 1, C_2 = 0, C_3 = 0)\mathbf{P}(C_1 = 1, C_2 = 0, C_3 = 0) \\
&+ \mathbf{P}(\text{System} = 1 \mid C_1 = 0, C_2 = 1, C_3 = 1)\mathbf{P}(C_1 = 0, C_2 = 1, C_3 = 1) \\
&+ \mathbf{P}(\text{System} = 1 \mid C_1 = 0, C_2 = 1, C_3 = 0)\mathbf{P}(C_1 = 0, C_2 = 1, C_3 = 0) \\
&+ \mathbf{P}(\text{System} = 1 \mid C_1 = 0, C_2 = 0, C_3 = 1)\mathbf{P}(C_1 = 0, C_2 = 0, C_3 = 1) \\
&+ \mathbf{P}(\text{System} = 1 \mid C_1 = 0, C_2 = 0, C_3 = 0)\mathbf{P}(C_1 = 0, C_2 = 0, C_3 = 0) \\
&= 0.95\pi_1\pi_2\pi_3 + 0.80\pi_1\pi_2(1 - \pi_3) + 0.85\pi_1(1 - \pi_2)\pi_3 \\
&+ 0.50\pi_1(1 - \pi_2)(1 - \pi_3) + \pi_{FSS}(1 - \pi_1)\pi_2\pi_3 + 0.40(1 - \pi_1)\pi_2(1 - \pi_3) \\
&+ 0.55(1 - \pi_1)(1 - \pi_2)\pi_3 + 0.05(1 - \pi_1)(1 - \pi_2)(1 - \pi_3).
\end{aligned}$$

Notice that we have written π_S as a function of π_1, π_2, π_3, and π_{FSS}.

Suppose that we specify independent prior distributions for each component reliability $\pi_i, i = 1, \ldots, 3$, with $-\log(\pi_i) \sim \text{Gamma}(1/3, 1)$. If this were a series system, these priors would induce a $Uniform(0, 1)$ distribution on the system reliability. We also assume a $Uniform(0.35, 0.85)$ distribution for π_{FSS}. These assumptions imply that the joint posterior distribution for $(\pi_1, \pi_2, \pi_3, \pi_{FSS})$ is proportional to the prior distributions times the likelihood function:

$$p(\pi_1, \pi_2, \pi_3, \pi_{FSS} \mid \mathbf{x}) \propto \pi_1^8(1-\pi_1)^2\pi_2^7(1-\pi_2)^2\pi_3^3(1-\pi_3)\pi_S^{10}(1-\pi_S)^2$$
$$[-\log(\pi_1)]^{-\frac{2}{3}}[-\log(\pi_2)]^{-\frac{2}{3}}[-\log(\pi_3)]^{-\frac{2}{3}}$$
$$I[\pi_{FSS} \in (0.35, 0.85)],$$

where $\mathbf{x}$ is the data in Table 5.8 and $I(\cdot)$ is the indicator function.

It is straightforward to draw a sample from this posterior distribution by using Metropolis-Hastings sampling. The kernel density estimate of the posterior distribution of π_S is plotted in Fig. 5.23. The posterior mean is 0.82, and the posterior 95% credible interval is (0.70, 0.90). The posterior distribution for π_{FSS} is essentially the same as the prior distribution, which indicates that these data have virtually no information about this conditional probability. To learn about π_{FSS}, we want to collect information about system success or failure given that component 1 has failed and that components 2 and 3 are working.

Table 5.10. Posterior distributions for three-component BN given data in Table 5.8

			Quantiles				
Parameter	Mean	Std Dev	0.025	0.050	0.500	0.950	0.975
π_1	0.81	0.10	0.57	0.61	0.82	0.95	0.96
π_2	0.78	0.12	0.51	0.56	0.80	0.95	0.96
π_3	0.77	0.16	0.41	0.47	0.80	0.97	0.98
π_{FSS}	0.60	0.14	0.36	0.37	0.60	0.83	0.84
π_S	0.82	0.051	0.70	0.72	0.82	0.89	0.90

5.2.6 Models for Dependence

Thus far, we have considered systems with components that fail independently. However, we cannot always assume that failures are independent; the failure of one component may be related to the failure of another. There are two main types of dependence: *positive* and *negative*. If the failure of one component causes the failure of another component to become more likely, their dependence is *positive*. If the failure of one component causes the failure of another component to become less likely, their dependence is *negative*.

Common cause failures are multiple failures that result from a single root cause. Figure 5.24 shows a fault tree with a common cause failure, C_{AB}, for

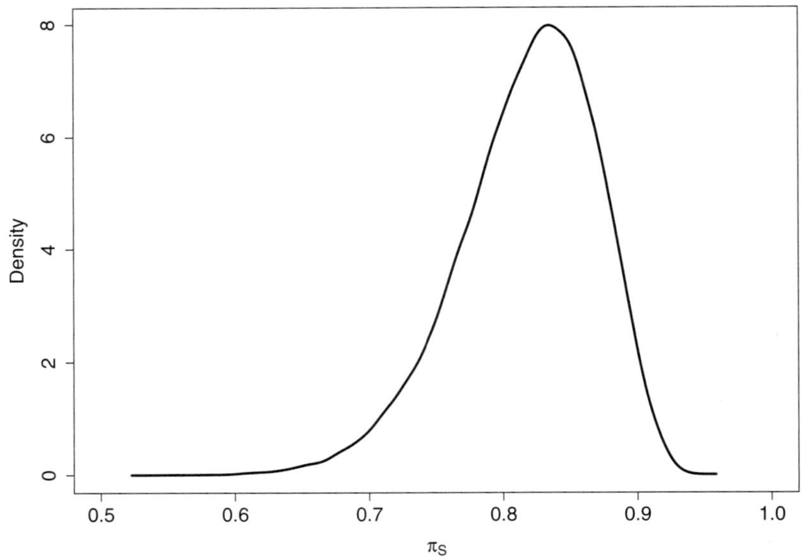

Fig. 5.23. Kernel density estimate of posterior distribution on π_S given data in Table 5.8.

components A and B. If C_{AB} fails, denoted $C_{AB} = 0$, then both components A and B fail. The fault tree explicitly adds the common cause failure using an OR gate.

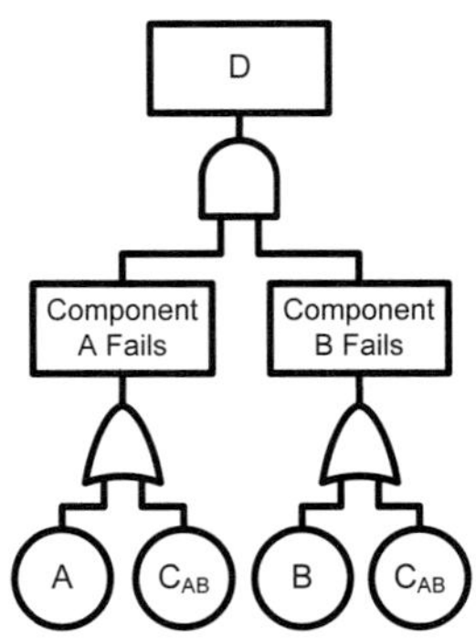

Fig. 5.24. Fault tree with a common cause failure (C_{AB}) for components A and B.

The minimal cut sets for the fault tree are $\{A, B\}$ and $\{C_{AB}\}$. We can calculate the probability of a system failure, $\mathbf{P}(D = 0)$, as

$$\begin{aligned}\mathbf{P}(D = 0) &= \mathbf{P}(A = 0, B = 0) + \mathbf{P}(C_{AB} = 0) - \mathbf{P}(C_{AB} = 0, A = 0, B = 0)\\ &= (1 - \pi_A)(1 - \pi_B) + \mathbf{P}(C_{AB} = 0) - (1 - \pi_A)(1 - \pi_B)\mathbf{P}(C_{AB} = 0)\\ &= \mathbf{P}(C_{AB} = 0) + (1 - \pi_A)(1 - \pi_B)\mathbf{P}(C_{AB} = 1).\end{aligned}$$

The Marshall-Olkin model (Marshall and Olkin, 1967) is one of the most common models for common cause failures. Suppose that we have a system with m components. Different kinds of shocks can occur that cause groups of components to fail. Denote the rates at which these shocks occur as $\lambda_{x_1 x_2 \ldots x_m}$, where $x_i = 1$ if the shock kills the ith component, and 0 otherwise. The shocks are assumed to occur independently. The overall failure rate for the ith component is $\sum_{x_i=1} \lambda_{x_1 x_2 \ldots x_m}$, or the sum over all of the failure rates associated with shocks that cause x_i to fail.

As a specific example, suppose that we have a two-component system with three independent sources of shocks in the environment. A shock from source one causes component 1 to fail, a shock from source two causes component 2 to fail, and a shock from source three causes both components to fail. Suppose that shocks from source one occur with a rate of λ_{10}, shocks from source two occur with a rate of λ_{01}, and shocks from source three occur with a rate of λ_{11}. The joint survival probability, the probability that the lifetimes exceed values (t_1, t_2), is

$$\begin{gathered}F(t_1, t_2) = \mathbf{P}[T_1 > t_1, T_2 > t_2] = \exp[-\lambda_{10} t_1 - \lambda_{01} t_2 - \lambda_{11} \max(t_1, t_2)],\\ t_1 \geq 0, \quad t_2 \geq 0, \quad \lambda_{10} > 0, \quad \lambda_{01} > 0, \quad \lambda_{11} > 0.\end{gathered}$$

We call the joint distribution with this survival function the *bivariate exponential distribution*. The bivariate exponential distribution models T_1 and T_2 as dependent components.

The marginal distributions for T_1 and T_2 are

$$\begin{aligned}F_1(t_1) &= \mathbf{P}[T_1 > t_1] = \exp[(-\lambda_{10} + \lambda_{11}) t_1], \quad t_1 \geq 0,\\ F_2(t_2) &= \mathbf{P}[T_2 > t_2] = \exp[(-\lambda_{01} + \lambda_{11}) t_2], \quad t_2 \geq 0.\end{aligned}$$

Marshall and Olkin (1967) and Barlow and Proschan (1975) give additional properties of the bivariate and multivariate exponential distributions.

The β-factor model, introduced by Fleming (1975), is a special case of the Marshall-Olkin model. Suppose that a system comprises m identical components, each with failure rate λ. Each component may fail for one of two reasons:

1. A cause that affects only the component and is independent of the remaining components. Denote this failure rate as λ_I.
2. A cause that affects all components and causes them all to fail at the same time. Denote this failure rate as λ_C.

Assuming that the two causes of failure are independent, $\lambda = \lambda_I + \lambda_C$. Let $\beta = \frac{\lambda_C}{\lambda}$. Then $\lambda_C = \beta\lambda$, and $\lambda_I = (1 - \beta)\lambda$. The β-factor is the relative fraction of common cause failures among all failures of a component.

Another special case of the Marshall-Olkin model is the binomial failure rate (BFR) model, introduced by Vesely (1977). Suppose that a system comprises m identical components. Each component fails, independently of the other components, with failure rate λ_I. In addition to the individual failures, common shocks hit the system with rate ν. The common shocks cause each component to fail independently of the other components with probability π. In addition, assume that shocks and individual failures occur independently.

These assumptions imply that the time between individual failures has an *Exponential*(λ_I) distribution and the time between common shocks has an *Exponential*(ν) distribution. The component failure rate for one component is $\lambda_I + \pi\nu$.

This is Vesely's original BFR model (Vesely, 1977). Atwood (1986) discusses analysis of these models, and Apostolakis and Moieni (1987) discusses extensions of the model. Hokstad (1988) places a beta prior distribution on π and discusses model estimation from a Bayesian perspective.

Cascading failures are multiple failures initiated by the failure of one component. For example, when several components share a common load, the failure of one leads to an increased load on the others, and consequently to a higher chance of failure among the other components.

Suppose that we have a two-component series system and that each component has an *Exponential*(λ_0) lifetime distribution. When the first component fails, the failure rate for the second component increases to $\lambda_1 > \lambda_0$. The time to first failure is the minimum time to failure for the components and has an *Exponential*$(2\lambda_0)$ distribution. The time to system failure is the sum of the time to first failure and the time to second failure, which has an *Exponential*(λ_1) distribution.

The probability density function for the lifetime of the system is

$$f(t) = \frac{2\lambda_0\lambda_1}{2\lambda_0 - \lambda_1}\exp(-\lambda_1 t) - \exp(-2\lambda_0 t),$$
$$t > 0, \quad \lambda_0 > 0, \quad \lambda_1 > 0, \quad 2\lambda_0 \neq \lambda_1.$$

If $2\lambda_0 = \lambda_1$, the system lifetime has a *Gamma*$(2, \lambda_1)$ distribution.

For a more complete mathematical development of probability models for dependent failures, see Singpurwalla (2006).

5.3 Related Reading

Mastran (1976) and Mastran and Singpurwalla (1978) are early references for the development of Bayesian systems assessment. References on system assessment with Bernoulli data include Martz et al. (1988), Martz and Waller

(1990), and Martz and Almond (1997). Graves et al. (2007) addresses multistate fault trees and partial information in reliability block diagrams. Johnson et al. (2003) and Hamada et al. (2004) develop this further using MCMC techniques. Hulting and Robinson (1990) and Reese et al. (2005) develop system assessment with lifetime distributions. Lee and Gross (1991) proposes a class of models where lifetime distributions are conditionally independent given the distribution of a common environmental factor. This model uses the generalized gamma distribution, which has the exponential, Weibull, gamma, and lognormal distributions as special cases. Crowder (2001) presents a detailed treatment of competing risk models.

There is a small literature on the use of BNs in failure modes and effects analysis (Lee, 2001) and reliability (for example, Sigurdsson et al. (2001), Bobbio et al. (2001), Portinale et al. (2005)), although there is quite a broad literature on using BNs for probabilistic modeling (e.g., Spiegelhalter (1998), Neil et al. (2000), Laskey and Mahoney (2000), Jensen (2001)). Neil et al. (2000) identifies five *idioms* or patterns that appear frequently in BNs and that can be used to develop representations of complex systems. See Wilson et al. (2007) for a fuller treatment of Example 5.8. Huzurbazar (2005) develops *flowgraph models*, which are another useful class of system reliability models.

5.4 Exercises for Chapter 5

5.1 Draw a reliability block diagram describing how to successfully perform an everyday task.

5.2 Draw the reliability block diagram and fault tree corresponding to a 3-of-5 system.

5.3 Determine the structure function for a 3-of-5 system.

5.4 Draw the reliability block diagram corresponding to Fig. 5.9.

5.5 Determine the minimal path sets and minimal cut sets for IE6 in Fig. 5.9. Calculate the structure function for IE6.

5.6 Define the *structural importance* of component i in a coherent system of n components as

$$I_\phi(i) = \frac{1}{2^{n-1}} \sum_{\mathbf{x}\,|\,x_i=1} [\phi(1_i, \mathbf{x}) - \phi(0_i, \mathbf{x})].$$

The sum is over the 2^{n-1} vectors for which $x_i = 1$. Calculate the structural importance of each component in Fig. 5.5.

5.7 Derive Eq. 5.8 from Eq. 5.1 by assuming that each component has reliability $R_i(t) = R(t)$.

5.8 Calculate the hazard function for a series system with n components when each component lifetime has a Weibull distribution.

5.9 Show that the mean time to failure (MTTF) for a standby system with perfect switching is equal to the sum of the MTTFs for each component:

$$MTTF_S = \sum_{i=1}^{n} MTTF_i.$$

5.10 Suppose that each of the n components of a standby system with perfect switching has an *Exponential*(λ) distribution. Show that the lifetime of the system has a *Gamma*(n, λ) distribution.

5.11 Reanalyze the data from Table 5.3 assuming that the prior distribution for the reliability of each component is $[\Gamma(1/3)]^{-1}(-\log(\pi_i))^{-\frac{2}{3}}$.

5.12 There are a variety of different measures of the reliability importance of a component (Rausand and Høyland, 2003). Birnbaum's measure of importance of the ith component at time t is

$$I_B(i \mid t) = \frac{\mathrm{d}R_S(t)}{\mathrm{d}\pi_i(t)}.$$

Birnbaum's measure is the partial derivative of the system reliability with respect to each component reliability $\pi_i(t)$. A larger value of $I_B(i \mid t)$ means that a small change in the reliability of the ith component results in a comparatively large change in the system reliability. Show that in a series system, Birnbaum's measure selects the component with the lowest reliability as the most important one.

5.13 Show how to calculate the posterior distribution for π_1, π_2, and π_3 using the data in Table 5.1 using simulation and the Metropolis-Hastings algorithm.

5.14 Assume a two-component series system. One component has an *Exponential*(3) prior distribution; the other has a *Weibull*$(5, 2)$ prior distribution. Using simulation, determine the probability density function of the prior distribution for the system.

5.15 Translate the fault tree in Fig. 5.9 into a BN.

5.16 Translate the fault tree in Fig. 5.24 into a BN. Write down the conditional probabilities specified by the fault tree.

5.17 Suppose that the data in Table 5.3 come from a three-component parallel system. Using independent *Uniform*$(0, 1)$ prior distributions for the reliability of each component, calculate the posterior distributions for the reliability of each component and the system.

5.18 Suppose that we have a three-component system like that in Example 5.1, and suppose that each component has an *Exponential*(λ) lifetime. Write an expression for the probability density function of the lifetime of the system.

5.19 Reanalyze the BN in Fig. 5.22 with data from Tables 5.8 and 5.9 assuming that we have also observed 20 observations with $C_1 = 0, C_2 = 1, C_3 = 1$ that resulted in 6 system successes and 14 system failures.

5.20 In Example 5.7, determine the probability that the item fails because of risk 1.

6

Repairable System Reliability

This chapter considers the reliability of multiple-time-use systems that are repaired when they fail. The effectiveness of repairs varies from restoring a system to a brand new state to restoring it to the reliability just before the system last failed. Several models for failure count and failure time data collected on repairable systems allow for different degrees of repair effectiveness. The models considered include renewal processes, homogeneous and nonhomogeneous Poisson processes, modulated power law processes, and a piecewise exponential model. This chapter also addresses how well these models fit the data and evaluates current reliability and other performance criteria, which characterize the reliability of repairable systems.

6.1 Introduction

When modeling failure time data, there is a distinction that needs to be made between one-time-use and multiple-time-use (or repairable) systems. When a one-time-use system fails, we simply replace it with a new system of the same type. A light bulb is an example of a one-time-use system. To assess the reliability of a one-time-use system, testing a sample of these systems and treating their failure times as a random sample from an appropriate distribution usually suffices.

A distinguishing feature of repairable system models is that they allow for the reliability growth or decay of a system. For example, consider a complex computer system made up of many subsystems. When the computer fails because a cooling fan fails, the fan can be replaced and the computer system restored to full operation. The fan's failure may perhaps have affected the reliability of other components, however, and so the computer system may have experienced reliability decay. If so, system failures should occur with increasing frequency. Unlike those for one-time-use systems, failure times for repairable systems are necessarily dependent.

We covered system reliability in detail in Chap. 5. The system reliability models presented in Chap. 5 implicitly assumed that when a system fails, it is not repairable. In this chapter, we present methods for assessing system reliability when repairs restore a system to an operable condition after a component or subsystem failure. For example, an automobile (system) fuel pump (component) could fail, then be repaired, restoring the vehicle to an operable condition. These types of systems are aptly called *repairable systems* and this chapter focuses on various probability models appropriate for such systems.

Next we consider different data types that arise in assessing repairable systems and introduce some needed notation.

6.1.1 Types of Data

We begin with notation for failure time data from a single repairable system. First, let T_i be the time the ith failure occurs. The failure times of a single repairable system satisfy $0 < T_1 < T_2 < \ldots$. Collecting such failure time data can employ Type I- and Type II-censoring schemes. The Type I-censoring scheme stops collection at a specified time t_c, resulting in $0 < T_1 < T_2 < \ldots < T_n < t_c$. For example, we collect failure times for the first 12 months of a system's operation. Note that the number of failures n is a random variable and the system has been operating for $t_c - T_n$ since the last failure when data collection stops. The Type II-censoring scheme stops collection at the nth failure time for a specified n. That is, the failure times satisfy $0 < T_1 < T_2 < \ldots < T_n$ under Type II censoring. Note that the repairable systems literature refers to *Type I* and *Type II censoring* as "time truncation" and "failure truncation," respectively. In the remainder of this chapter, however, we use the standard Type I- and Type II-censoring terminology. Also, note that *interfailure times* $X_i = T_i - T_{i-1}$ (with $T_0 = 0$) are equivalent to the failure times.

Sometimes, the failure times may be available as failure count data, where $N(a, b)$ is the number of failures in an interval $(a, b]$. For example, we may have the number of failures per month for the first 12 months of a system's operation, resulting in $N(0, 1], \ldots, N(11, 12]$. Finally, let $N(t)$ denote the number of failures in the interval $(0, t]$.

These failure time and failure count data provide information about the effectiveness of system repairs, which we characterize next.

6.1.2 Characteristics of System Repairs

When a system fails and a repair restores it to full operation, questions arise regarding the repair's effect on the reliability of a system, especially as the total operating time of the system increases. Two extreme descriptions of effectiveness are "*good-as-new*" and "*bad-as-old*." A "good-as-new" repair means that the repair has returned the system to a brand-new state, so that the time to next failure has the same distribution as the first system failure time. That is, the time to next failure does not depend on the age of the system at the last failure time.

In contrast, a "bad-as-old" repair means that the repair has brought the system back to the state it was at just before the last failure. That is, the time to next failure depends on the last failure time. Under this scenario, some systems exhibit reliability growth while others exhibit reliability decay. For those systems exhibiting reliability growth, failures occur less often than before over time. When work continues on a system after being brought online, the system can exhibit reliability growth. For reliability decay, failures occur more often than just before the last failure over time. Reliability decay is what we typically think of for systems that are aging and failing more often over time.

Finally, the effectiveness of repairs for some systems falls between these two extremes, which are better than bad-as-old but not as good as good-as-new, i.e., "*better-than-old*" but "*worse-than-new.*"

This chapter considers various repairable system models, which have some or all of these characteristics, depending on the values that their model parameters take on. In turn, we consider renewal processes, Poisson processes, modulated power law processes, and piecewise exponential models.

6.2 Renewal Processes

A *renewal process* is a simple model for failure times characterized by the interfailure times $X_i = T_i - T_{i-1}$ (with $T_0 = 0$) being a random sample from a specified distribution, that is, the X_i are independent and identically distributed (i.i.d.). For example, interfailure times of the *exponential renewal process* and the *gamma renewal process* have exponential and gamma distributions, respectively. Under a renewal process, the time to the next failure has the same distribution whether the system is brand new or has just been repaired for the 100th time. That is, the renewal process describes the effectiveness achieved by "good-as-new" repairs. Note that a renewal process is unable to model the reliability growth or reliability decay often observed in repairable systems, however. A figure of merit that characterizes a renewal process is the *mean time between failure* (MTBF), which is the mean of the interfailure time distribution, denoted by $\mathrm{E}(X)$.

Because interfailure times characterize renewal processes, the most natural data to model are the interfailure times $X_i = T_i - T_{i-1}$ (with $T_0 = 0$); denote their probability density and cumulative distribution functions by $f(x|\boldsymbol{\theta})$ and $F(x|\boldsymbol{\theta})$, respectively, where $\boldsymbol{\theta}$ are the interfailure time distribution parameters.

For Type I censoring with data collection stopping at time t_c, $0 < T_1 < \ldots T_n < t_c$ and $t_c - T_n$ is a Type I-censored observation. The equivalent interfailure times $X_1, \ldots, X_n$ are conditionally independent with probability density function $f(x|\boldsymbol{\theta})$ and their likelihood function takes the form

$$\left(\prod_{i=1}^{n} f(x_i|\boldsymbol{\theta})\right) [1 - F(t_c - t_n|\boldsymbol{\theta})] ,$$

where $x_1, \ldots, x_n$ are the observed interfailure times and t_n is the last observed failure time. For Type II censoring with data collection stopping at the nth failure, the interfailure times $X_1, \ldots, X_n$ are conditionally independent with density $f(x|\boldsymbol{\theta})$ and their likelihood function takes the form

$$\prod_{i=1}^{n} f(x_i|\boldsymbol{\theta}) .$$

Generally, we collect interfailure times for a renewal process. With failure count data, reliability assessments are much harder to do and determining the likelihood function is generally a challenging problem for such data in arbitrary intervals. For exponential interfailure times, however, the model is a homogeneous Poisson process (as discussed in Sect. 6.3.1).

Example 6.1 Renewal process analysis of failure time data. Consider the simulated failure time data from a supercomputer. The Blue Mountain supercomputer at Los Alamos National Laboratory consists of 48 SGI Origin 2000 shared memory processors (SMPs). When an SMP fails, it is restarted. Consequently, we can view each of the SMPs as a repairable subsystem. When the scientists first brought Blue Mountain online, they were interested in whether Blue Mountain was experiencing reliability growth. To illustrate the analysis of failure time data, we simulated failure times for 15 months of operation, as presented in Table 6.1. Note that after the last failure, the SMP was still operating when data collection was stopped at the 15th month or 457 days, and had been operating for $457 - 456.5085 = 0.4915$ day.

Table 6.1. Blue Mountain supercomputer simulated failure times for one SMP

1.06, 5.63, 16.16, 35.65, 56.05, 59.39, 64.31, 64.84, 77.90, 97.29,
98.44, 110.38, 112.07, 137.01, 137.12, 170.84, 179.54, 237.06,
247.52, 272.82, 337.23, 348.07, 353.88, 365.64, 456.51

One possible model for the monthly number of failures is an exponential renewal process. To compare with fitting other models later, let us focus on $\eta = 1/\lambda$. The Blue Mountain supercomputer engineers expect that an SMP will fail approximately twice per month, which is about 2/30 failure per day or once every 15 days.

$$\eta \sim Gamma(15, 1) ,$$

which has a mean of 15 and a standard deviation of $\sqrt{15}$.

Recall that a Type I-censoring scheme at 15 months (i.e., 457 days) collected these failure time data, so that the likelihood function for the observed failure times $0 < t_1 < \ldots < t_{25} < t_c = 457$ is

$$\begin{aligned} f(t_1, \ldots, t_{25}|\lambda) &= \left(\prod_{i=1}^{25} \lambda \exp[-\lambda(t_i - t_{i-1})]\right) \exp[-\lambda(t_c - t_n)] \\ &= \lambda^n \exp(-\lambda t_c). \end{aligned}$$

We use MCMC to obtain draws from the posterior distribution of η. See Fig. 6.1, which presents the prior and posterior distributions for η as dashed and solid lines, respectively; its posterior median is 16.19 days and a 95% credible interval is $(12.19, 21.63)$. Using the reciprocals of the η posterior draws, the posterior median for the constant intensity λ is 0.0618 failure per day and a 95% credible interval is $(0.0462, 0.0820)$.

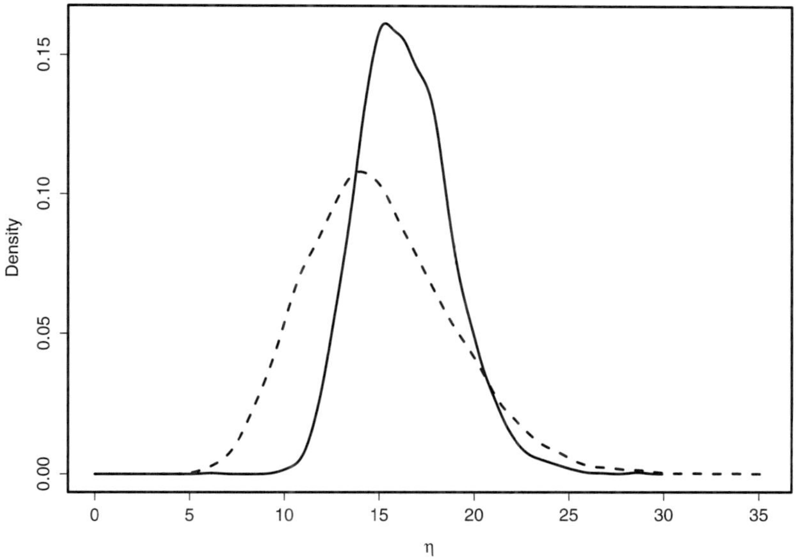

Fig. 6.1. Prior (dashed line) and posterior (solid line) distributions for $\eta = 1/\lambda$ for the exponential renewal process analysis of the failure time data for one Blue Mountain supercomputer SMP.

6.3 Poisson Processes

While renewal processes are an important class of models, they are inappropriate to use in situations where reliability growth (or decay) may occur.

Consequently, we turn our attention to a broad class of models, which allows the possibility of reliability growth or decay. To begin, let us define an important type of counting process called a *Poisson process*.

Definition 6.1 A counting process $N(t)$ is a Poisson process if

1. $N(0) = 0$.
2. For any $a < b \leq c < d$, the random variables $N(a,b]$ and $N(c,d]$ are independent. That is, counts in nonoverlapping intervals are independent.
3. A function $\lambda(\cdot)$, called the *intensity function*, exists as defined by

$$\lambda(t) = \lim_{\Delta t \to 0} \frac{P(N(t, t+\Delta t] = 1)}{\Delta t} \,. \tag{6.1}$$

4. There are no simultaneous failures, expressed as

$$\lim_{\Delta t \to 0} \frac{P(N(t, t+\Delta t] \geq 2)}{\Delta t} = 0\,.$$

A consequence of these four conditions presented in the Poisson process definition is that

$$P[N(t) = x] = \frac{\Lambda(t)^x \exp[-\Lambda(t)]}{x!}\,, \tag{6.2}$$

where

$$\Lambda(t) = \int_0^t \lambda(z) dz\,. \tag{6.3}$$

The probability statement in Eq. 6.2 implies that $N(a,b]$ has a Poisson distribution with parameter $\int_a^b \lambda(z)dz$. In other words,

$$\mathrm{E}(N(a,b]) = \mathrm{Var}(N(a,b]) = \int_a^b \lambda(z)dz = \Lambda(b) - \Lambda(a)\,.$$

One performance measure of Poisson processes is the *rate of occurrence of failures* (ROCOF), defined as

$$\frac{\mathrm{d}}{\mathrm{d}t}\mathrm{E}[N(t)],$$

for differentiable $\mathrm{E}[N(t)]$. It turns out that the ROCOF and intensity function in Eq. 6.1 are equal when the probability of simultaneous failures is zero (i.e., Definition 6.1, point 4). (See Rigdon and Basu (2000), Theorem 13, p. 28.) Consequently, when the intensity function $\lambda(t)$ given in Eq. 6.1 is large, we expect many failures in a time interval, and if $\lambda(t)$ is small, we expect few failures in a time interval.

A useful characterization of the failure times from a Poisson process is as follows:

$$\Lambda(T_i) - \Lambda(T_{i-1}) \sim \mathit{Exponential}(1)\,. \tag{6.4}$$

That is, $\Lambda(T_i) - \Lambda(T_{i-1})$, $i = 1, \ldots, n$, (with $T_0 = 0$) are i.i.d. *Exponential*(1). Using Eq. 6.4, we can express T_i in terms of T_{i-1} as

$$T_i \sim \Lambda^{-1}[\Lambda(T_{i-1}) + Exponential(1)]\,, \tag{6.5}$$

which suggests how to simulate failure times from a Poisson process, where $\Lambda^{-1}(\cdot)$ denotes the inverse of $\Lambda(\cdot)$ for $\Lambda(\cdot)$ defined in Eq. 6.3. Note that for specific models, $\Lambda(\cdot)$, which involves an integral, simplifies to a function that is easily invertible.

Within this class of models are *homogeneous Poisson processes* (HPPs) and *nonhomogeneous Poisson processes* (NHPPs). Next, we consider each type of Poisson process in turn.

6.3.1 Homogeneous Poisson Processes (HPPs)

An HPP is a Poisson process that has a constant intensity function. That is, $\lambda(t) = \lambda$, which implies that the mean number of failures in the interval $(a, b]$ is

$$\mathrm{E}(N(a,b]) = \mathrm{Var}(N(a,b]) = \int_a^b \lambda(z)dz = \int_a^b \lambda dz = \lambda(b-a)\,,$$

a linear function of λ. The mean number of failures depends only on the interval length, which is the most restrictive feature of HPPs. See Fig. 6.2, which plots the intensity function $\lambda(t)$ and mean $\Lambda(t)$ function for $\lambda(t) = \lambda = 0.8$ over the interval (0, 100). From Eq. 6.4, it follows that the interfailure times have an exponential distribution with rate parameter λ. That is, a renewal process (with exponentially distributed interfailure times) generates the failure times, so that HPPs describe "good-as-new" repairs. Consequently, the analyst should take care when modeling repairable systems with HPPs, because the constant intensity function assumption does not hold for many systems.

For HPPs, the interfailure times are i.i.d. *Exponential*(λ), so that for Type I censoring with data collection stopping at time t_c, the likelihood function for the observed failure times $0 < t_1 < \ldots < t_n < t_c$ is

$$\begin{aligned} f(t_1, \ldots, t_n|\lambda) &= \left(\textstyle\prod_{i=1}^n \lambda \exp[-\lambda(t_i - t_{i-1})]\right) \exp[-\lambda(t_c - t_n)] \\ &= \lambda^n \exp(-\lambda t_c) \end{aligned} \tag{6.6}$$

for $n > 0$, where $t_0 = 0$, and

$$\exp(-\lambda t_c)$$

for $n = 0$. For Type II censoring with data collection stopping at the nth failure, the likelihood function for failure times $0 < T_1 < \ldots < T_n$ is

$$f(t_1, \ldots, t_n|\lambda) = \lambda^n \exp(-\lambda t_n)\,.$$

For count data, the number of failures N_i for the ith interval $(a_i, b_i]$, $i = 1, \ldots, m$, has a *Poisson*$[\lambda(b_i - a_i)]$ distribution, so that the likelihood function for counts N_i, for $(a_i, b_i]$, $i = 1, \ldots, m$, is

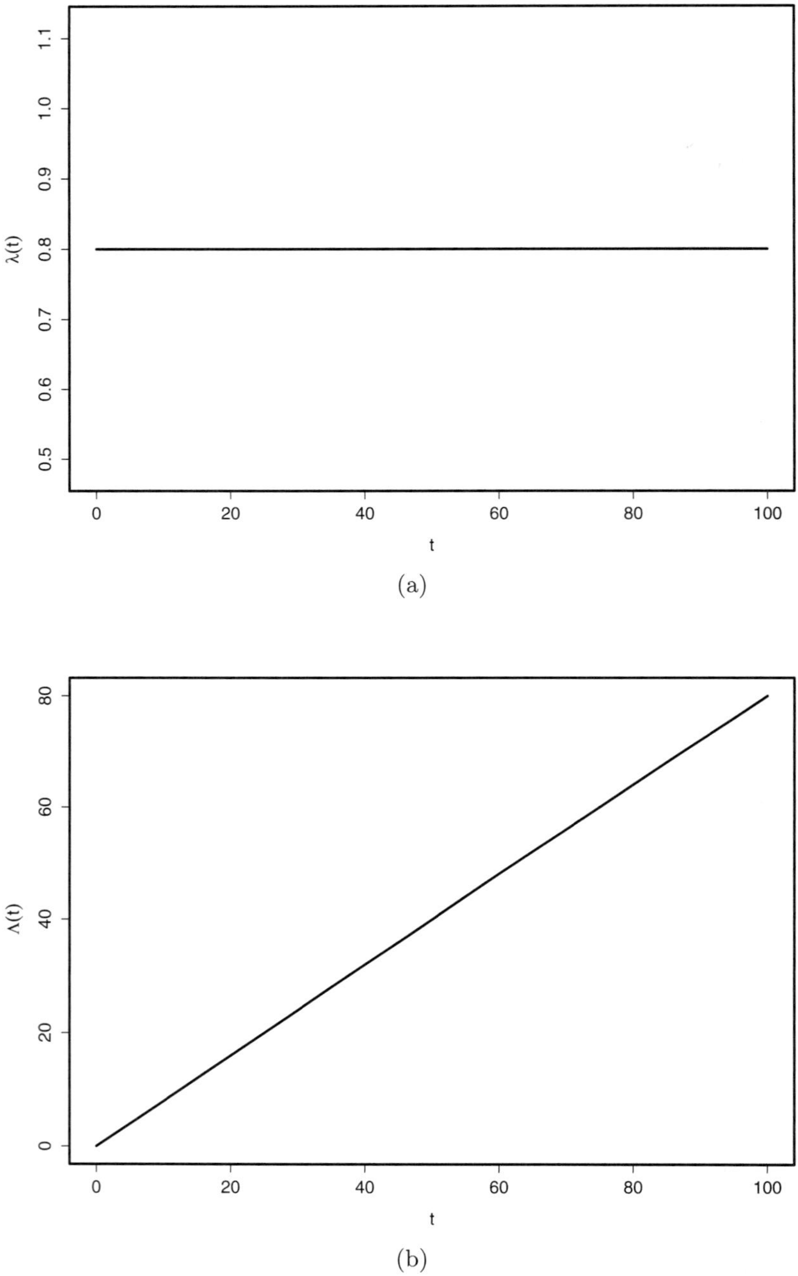

Fig. 6.2. Plots of the (a) intensity function $\lambda(t)$ and (b) mean function $\Lambda(t)$ of an HPP with $\lambda = 0.8$.

$$\prod_{i=1}^{m}[\lambda(b_i - a_i)]^{n_i} \exp[-\lambda(b_i - a_i)]/n_i! \,. \tag{6.7}$$

Regarding the choice of prior distributions for HPPs, λ is positive and real valued, so that one appropriate choice is the gamma distribution, which is conjugate for failure count data, Type I-censored and Type II-censored failure time data, and uncensored failure time data.

Example 6.2 HPP analysis of failure count data. Consider an analysis of the failure count data in Table 6.2 for one Blue Mountain supercomputer SMP. (See also the discussion in Example 6.1.) Table 6.2 displays failure counts (i.e., number of failures) in its first 15 months of operation, where Eq. 6.7 gives the likelihood function for these failure count data.

Table 6.2. Blue Mountain supercomputer monthly failure counts for one SMP (with $b_0 = 0$ and $a_i = b_{i-1}$) (Ryan and Reese, 2001)

Month (b_i in cumulative number of days)	Number of Failures ($N(a_i, b_i]$)	Cumulative Number of Failures ($N(b_i)$)
Jul ($b_1 = 31$)	5	5
Aug ($b_2 = 62$)	4	9
Sep ($b_3 = 92$)	6	15
Nov ($b_4 = 123$)	1	16
Dec ($b_5 = 153$)	2	18
Jan ($b_6 = 184$)	1	19
Feb ($b_7 = 215$)	4	23
Mar ($b_8 = 243$)	4	27
Apr ($b_9 = 274$)	2	29
May ($b_{10} = 304$)	1	30
Jun ($b_{11} = 335$)	1	31
Jul ($b_{12} = 365$)	1	32
Aug ($b_{13} = 396$)	1	33
Sep ($b_{14} = 427$)	1	34
Nov ($b_{15} = 457$)	1	35

As in Example 6.1, we employ the same $Gamma(15, 1)$ prior distribution for η. To analyze the supercomputer failure count data, we make draws from the posterior distribution using Markov chain Monte Carlo (MCMC). Figure 6.3 presents the prior and posterior distributions for η as dashed and solid lines, respectively. With η being the expected number of days before the SMP will next fail, note that its posterior distribution is centered at approximately 14 days. The posterior median for η is 13.64 days and a 95% credible interval is $(10.35, 18.24)$. We can easily obtain draws from the posterior distribution for the constant intensity λ by taking reciprocals of the η posterior draws; its median is 0.0733 failure per day and 95% credible interval is $(0.0558, 0.0966)$.

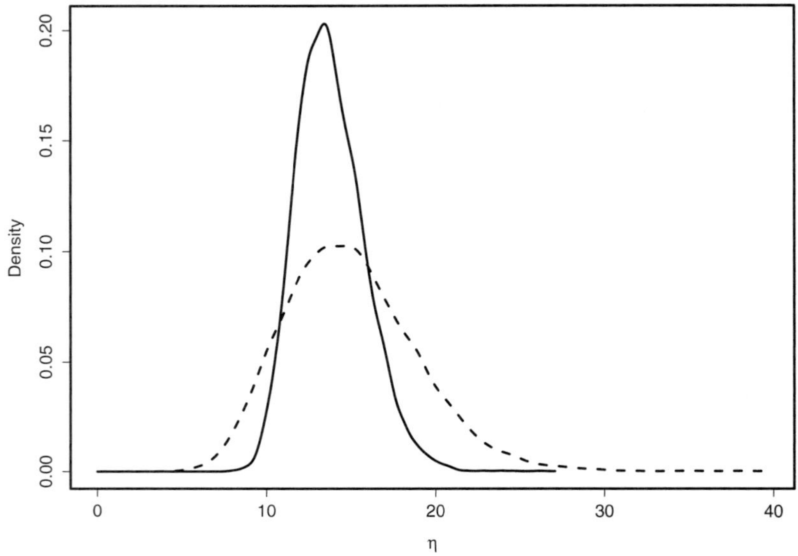

Fig. 6.3. Prior (dashed line) and posterior (solid line) distributions for $\eta = 1/\lambda$ for the exponential renewal process analysis of the failure count data for one Blue Mountain supercomputer SMP.

6.4 Nonhomogeneous Poisson Processes (NHPPs)

NHPPs are Poisson processes for which the intensity function $\lambda(t)$ is nonconstant, i.e., a function over time. For an NHPP, $N(a, b]$ has a Poisson distribution with mean

$$\mathrm{E}(N(a,b]) = \int_a^b \lambda(t)dt = \Lambda(b) - \Lambda(a)\,.$$

In the next two sections, we present two major classes of NHPPs, power law processes and log-linear processes.

6.4.1 Power Law Processes (PLPs)

Historically, the observation that plots of "cumulative failure rates" versus cumulative operating hours for some repairable systems were approximately linear on log-log paper suggested the NHPP model with a *power law intensity function*. The power law intensity function takes the form

$$\lambda(t) = \frac{\phi}{\eta}\left(\frac{t}{\eta}\right)^{\phi-1},$$

where both the scale parameter η and the shape parameter ϕ are positive. Consequently, we refer to an NHPP with power law intensity function as a *power law process* (PLP). The repairable systems literature has also referred to this model as the Weibull process model, but several authors have noted that this nomenclature is confusing; the failure times do not have a Weibull distribution (except for the first failure), and neither do the interfailure times. Note that for a PLP, the mean number of failures up to time t is

$$\mathrm{E}[N(t)] = \Lambda(t) = \left(\frac{t}{\eta}\right)^{\phi}.$$

Also, the PLP is an HPP when ϕ is equal to 1 (and $\eta = 1/\lambda$), and for values of $\phi > 1$, the intensity function is increasing, which implies reliability decay. Similarly, values of $\phi < 1$ imply reliability growth, because the intensity function is decreasing. For example, see Fig. 6.4, which displays the intensity and mean functions for $\phi = 1$ as well as for $\phi = 0.5$ and 1.2 when $\eta = 10$.

For NHPPs under Type I censoring with data collection stopping at time t_c, the likelihood function for observed failure times $0 < t_1 < \ldots < t_n < t_c$ is

$$\begin{aligned} f(t_1, \ldots, t_n|\boldsymbol{\theta}) &= \left(\textstyle\prod_{i=1}^n \lambda(t_i) \exp[-\{\Lambda(t_i) - \Lambda(t_{i-1})\}]\right) \\ &\quad \times \exp[-\{\Lambda(t_c) - \Lambda(t_n)\}] \\ &= \left(\textstyle\prod_{i=1}^n \lambda(t_i)\right) \exp[-\Lambda(t_c)], \end{aligned} \tag{6.8}$$

for $n > 0$, where $t_0 = 0$, and

$$\exp[-\Lambda(t_c)] \text{ for } n = 0,$$

where $\lambda(t)$ is the intensity function and $\Lambda(t) = \int_0^t \lambda(x)dx$.

For Type II censoring with data collection stopping at the nth failure, the likelihood function for observed failure times $0 < t_1 < \ldots < t_n$ is

$$\begin{aligned} f(t_1, \ldots, t_n|\boldsymbol{\theta}) &= \left(\textstyle\prod_{i=1}^n \lambda(t_i) \exp[-\{\Lambda(t_i) - \Lambda(t_{i-1})\}]\right) \\ &= \left(\textstyle\prod_{i=1}^n \lambda(t_i)\right) \exp[-\Lambda(t_n)], \end{aligned}$$

where $t_0 = 0$.

For count data, the number of failures N_i for the ith interval $(a_i, b_i]$, $i = 1, \ldots, m$, has a $Poisson[\Lambda(b_i) - \Lambda(a_i)]$ distribution, so that the likelihood function for observed counts n_i, for $(a_i, b_i]$, $i = 1, \ldots, m$, is

$$\prod_{i=1}^m [\Lambda(b_i) - \Lambda(a_i)]^{n_i} \exp[-\{\Lambda(b_i) - \Lambda(a_i)\}]/n_i!\,. \tag{6.9}$$

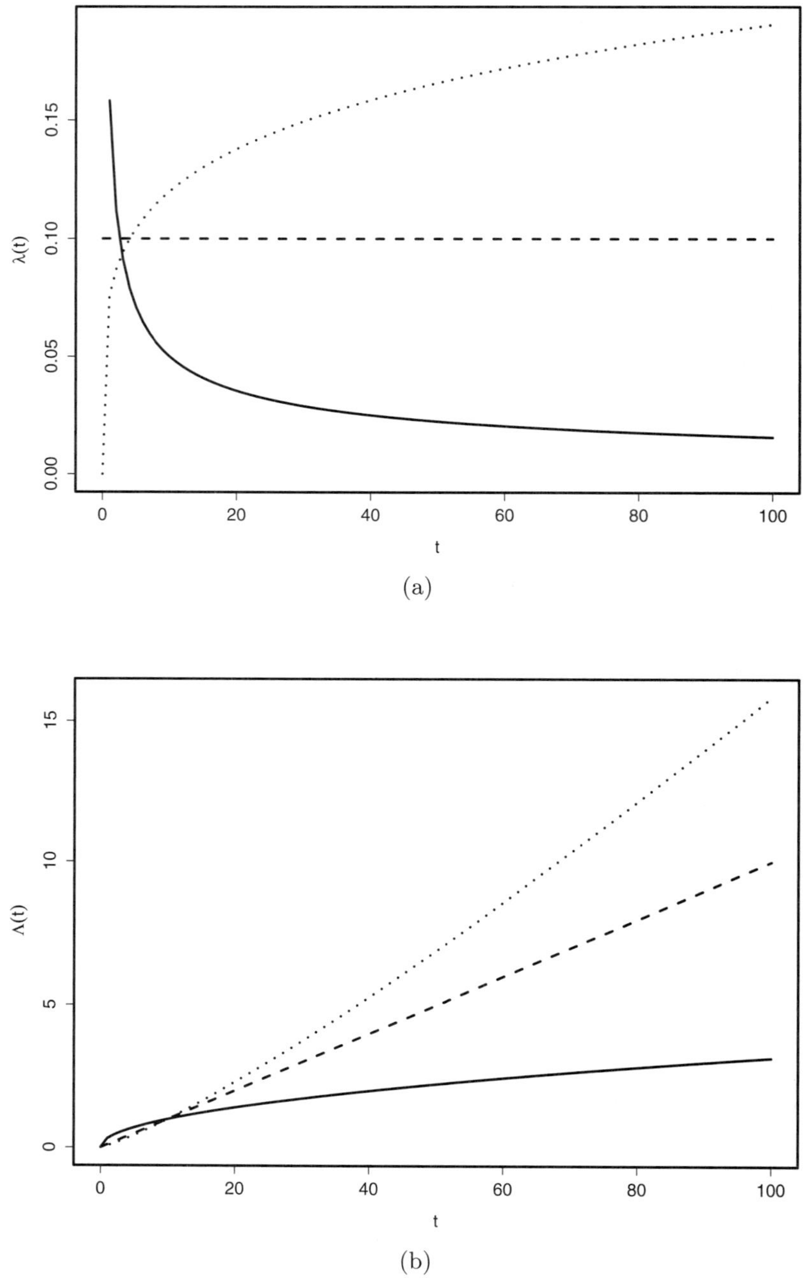

Fig. 6.4. Plots of the (a) intensity function $\lambda(t)$ and (b) mean function $\Lambda(t)$ of an NHPP with a power law intensity function or PLP (with $\eta = 10$ and solid line for $\phi = 0.5$, dashed line for $\phi = 1$, and dotted line for $\phi = 1.2$).

When choosing prior distributions for PLPs, remember that $\phi < 1$ implies reliability growth, $\phi = 1$ implies "good-as-new" repairs, and $\phi > 1$ implies reliability decay. Some of the repairable systems literature has developed prior distributions with elicited information. Guida et al. (1989) presents an easy-to-elicit informative prior on (η, ϕ), when an expert provides information about the expected number of failures up to some specified time $\tilde{t}$, denoted by $\Lambda(\tilde{t})$. The expert only needs to provide a mean μ and standard deviation σ summarizing his/her best guess and uncertainty for $\Lambda(\tilde{t})$. (With a normal distribution in mind, $\mu \pm 2\sigma$ accounts for approximately 95% of the expert's uncertainty about $\Lambda(\tilde{t})$.) After a change of variables, the expression for the conditional prior density function for η given ϕ is

$$p(\eta|\phi) = \phi\left(\frac{\mu}{\sigma^2}\right)^{\left(\frac{\mu}{\sigma}\right)^2} \tilde{t}^{\phi\left(\frac{\mu}{\sigma}\right)^2} \eta^{-\phi\left(\frac{\mu}{\sigma}\right)^2-1} \exp\left[-\frac{\mu}{\sigma^2}(\tilde{t}/\eta)^{\phi}\right] / \Gamma\left[(\mu/\sigma)^2\right] .$$

With the conditional prior distribution for η given ϕ specified, all that remains is to specify a prior distribution for ϕ. Guida et al. (1989) suggests a uniform distribution on the following intervals, which we slightly modify:

- (0.3, 1.1) when there is a strong belief of reliability growth, but there is no information about what the value of ϕ (< 1) is,
- (0.3, 5.0) when there is weak information about ϕ, and
- (0.9, 5.0) when there is a strong belief of reliability decay, but weak information about what the value of ϕ (> 1) is.

Note that when $\tilde{t} = \eta$, $\Lambda(\tilde{t}) = 1$. Therefore, we can interpret η as the time to expect the first failure. If $\phi = 3$, then one expects one failure in the first η time units, and $2^3 - 1 = 7$ failures in the second η time units. In many applications, a $\phi > 3$ indicates more decay than an analyst would expect and an expert would be able to rule out such values. Consequently, the Guida et al. (1989) choice of prior distributions for ϕ seems to make sense.

Kyparisis and Singpurwalla (1985) proposes a more flexible prior distribution for ϕ, a scaled beta distribution, which has the following probability density function:

$$p(\phi) = \frac{\Gamma(k_1 + k_2)(\phi - L)^{k_1-1}(U - \phi)^{k_2-1}}{\Gamma(k_1)\Gamma(k_2)(U - L)^{k_1+k_2-1}} ,$$

where $0 \leq L < \phi < U$ and $\kappa_1, \kappa_2 > 0$. For a scaled beta distribution with mean $\mu \in (L, U)$, the variance σ^2 must be less than $(U-L)^2\mu(1-\mu)$. Beginning with a valid mean μ and variance σ^2 for a scaled beta distribution, the following expressions provide values for the parameters k_1 and k_2:

$$k_1 = \frac{\mu^2(1-\mu)}{\sigma^2} - \mu$$

and

$$k_2 = \frac{\mu(1-\mu)^2}{\sigma^2} - (1-\mu) .$$

Example 6.3 NHPP with power law intensity analyses of failure count and failure time data. Consider an analysis of the data in Tables 6.2 and 6.1 using an NHPP with power law intensity function, i.e., a PLP.

First, consider an analysis of the failure count data (monthly number of failures) in Table 6.2 for one Blue Mountain supercomputer SMP. Under a PLP, the likelihood function is given in Eq. 6.9, where $\Lambda(t) = (\frac{t}{\eta})^{\phi}$. For the failure count data in Table 6.2, $m = 15$. For the first failure count $n_1 = N(a_1, b_1] = 5$, $a_1 = 0$ and $b_1 = 31$, so that its likelihood contribution is

$$\left[\left(\frac{31}{\eta}\right)^{\phi} - \left(\frac{0}{\eta}\right)^{\phi}\right]^5 \exp\left(-\left[\left(\frac{31}{\eta}\right)^{\phi} - \left(\frac{0}{\eta}\right)^{\phi}\right]\right) /5!\,.$$

Regarding the choice of prior distributions, the same distribution for η used in Example 6.1 is

$$\eta \sim Gamma(15, 1)\,,$$

based on the Blue Mountain supercomputer engineers' expectation that an SMP will fail approximately twice per month or about 2/30 failures per day. (Informally, we think the shape $\phi \approx 1$, but more formally a conditional prior distribution for η given ϕ might be warranted.) The choice for ϕ is the following independent prior distribution:

$$\phi \sim Gamma(2, 2)\,,$$

which has a mean of 1 to allow for $\phi = 1$, the special case of an HPP.

To analyze these failure count data, we make draws from the posterior distribution of (η, ϕ) using MCMC. See Fig. 6.5, which presents the prior and posterior distributions for (η, ϕ) as dashed and solid lines, respectively. The posterior median for η is 11.14 days and a 95% credible interval is $(6.45, 17.86)$. The posterior median for ϕ is 0.92 and a 95% credible interval for ϕ is $(0.78, 1.07)$. The posterior probability that ϕ exceeds 1 is 0.14. Because the credible interval for ϕ contains 1, there is no strong evidence against an HPP (i.e., $\phi = 1$) providing an adequate fit.

Next, consider an analysis of the simulated failure time data in Table 6.1 for one SMP using a PLP. Recall that the failure times are Type I censored at 15 months (i.e., 457 days). Consequently, the appropriate likelihood function for the failure time data has the form given in Eq. 6.8, whose expression for these failure time data, t_i, $i = 1, \ldots, 25$, is

$$\prod_{i=1}^{25} \phi \left(\frac{t_i}{\eta}\right)^{\phi-1} \exp\left(-\left[\left(\frac{457}{\eta}\right)^{\phi}\right]\right)\,.$$

As before, we use the same independent $Gamma(15, 1)$ prior distribution for η and independent $Gamma(2, 2)$ prior distribution for ϕ and employ MCMC to obtain draws from the joint posterior distribution of $(\eta,\ \phi)$. See Fig. 6.6,

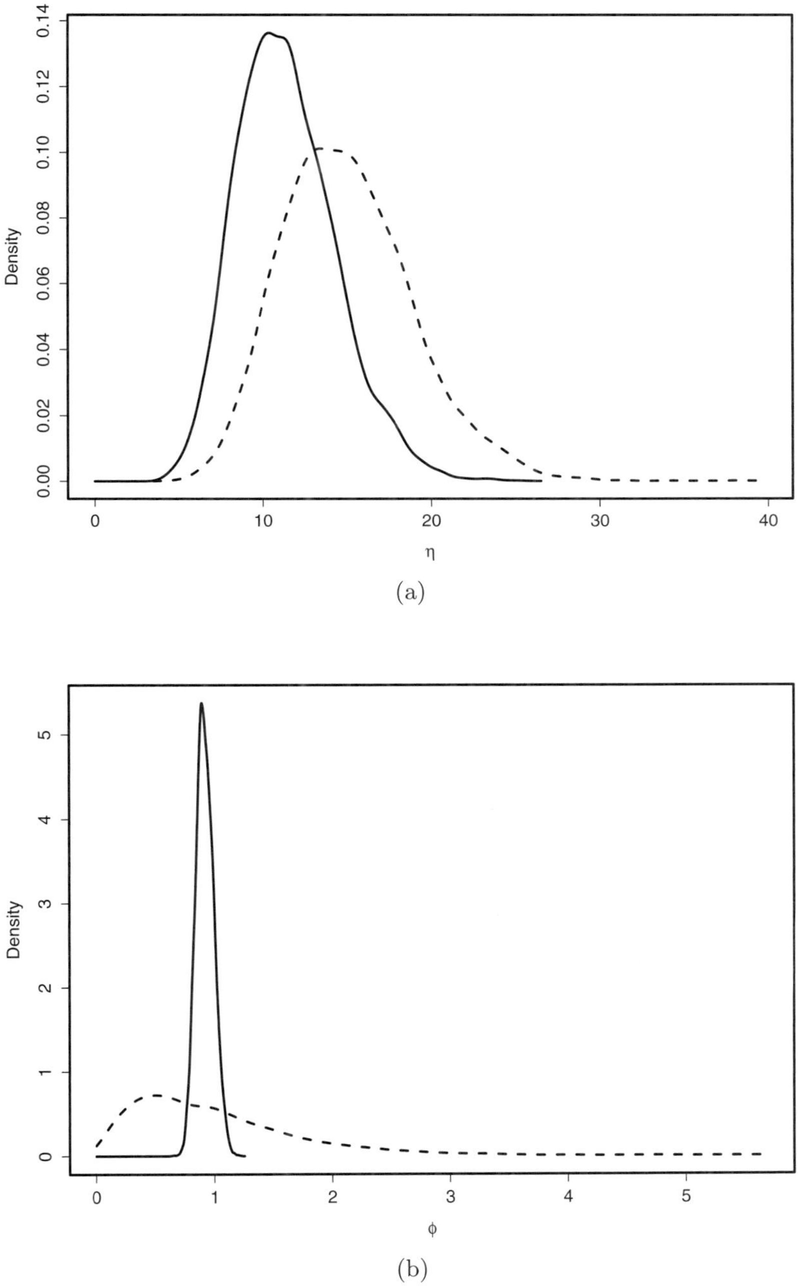

Fig. 6.5. Prior (dashed line) and posterior (solid line) distributions for (a) η and (b) ϕ for an NHPP (with a power law intensity function) analysis of failure count data for one Blue Mountain supercomputer SMP.

which presents the prior and posterior distributions for η as dashed and solid lines, respectively. For η, its posterior median is 12.50 days and a 95% credible interval is $(7.51, 19.54)$. The posterior median for ϕ is 0.88 and a 95% credible interval is $(0.75, 1.02)$. Moreover, the posterior probability that ϕ exceeds 1 is 0.057. For these data, the credible interval for ϕ barely contains 1, which is evidence for reliability growth (i.e., $\phi < 1$).

6.4.2 Log-Linear Processes

An alternative to a PLP is a *log-linear process*, which is an NHPP with intensity function

$$\lambda(t) = \exp(\gamma_0 + \gamma_1 t),$$

where the parameters γ_0 and γ_1 are both defined on the real line. For a log-linear process, the mean number of failures up to time t is

$$\Lambda(t) = \exp(\gamma_0)[\exp(\gamma_1 t) - 1]/\gamma_1 .$$

Similar to the PLP, a log-linear process with $\gamma_1 = 0$ is an HPP, where $\lambda = \exp(\gamma_0)$. Note that $\gamma_1 > 0$ implies reliability decay, whereas $\gamma_1 < 0$ implies reliability growth.

We leave the fitting of log-linear process to the data in Tables 6.2 and 6.1 as Exercises 6.5 and 6.6.

6.5 Alternatives to NHPPs

Next, we consider two alternatives to NHPP models, modulated PLPs and a piecewise exponential (PEXP) model.

6.5.1 Modulated Power Law Processes (MPLPs)

A criticism of the NHPP and the PLP in particular is that after a repair, the system is "bad-as-old"; the intensity function before the failure is the same as that after the repair. Some of the literature has argued that a compromise is warranted, because there are cases when a repair does not make a system brand new but does improve it relative to "bad-as-old." Consequently, researchers developed the *modulated power law process* (MPLP), which allows for such improvements. A characterization of the MPLP failure times is

$$\Lambda(T_i) - \Lambda(T_{i-1}) \sim Gamma(\kappa, 1), \tag{6.10}$$

where $\Lambda(t) = \left(\frac{t}{\eta}\right)^{\phi}$ and $\kappa > 0$.

Compare this characterization with that for the PLP failure times in Eq. 6.4, where the difference has an *Exponential*(1) distribution. Recall that

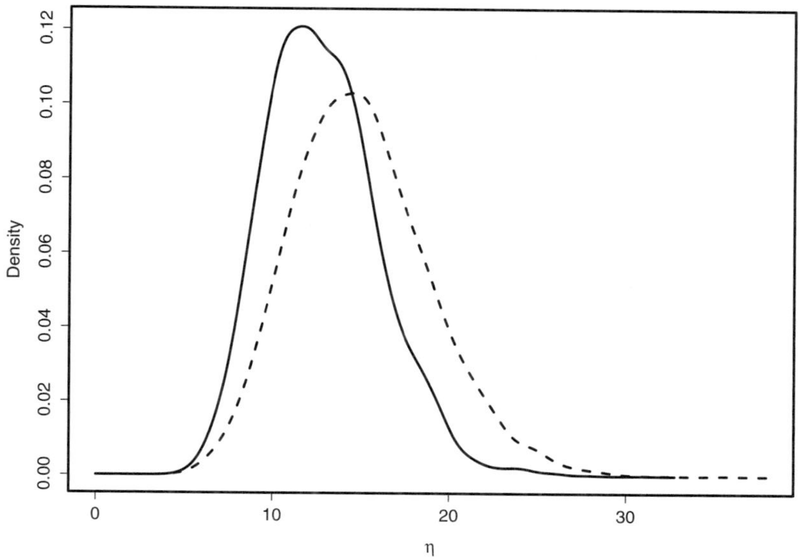

(a)

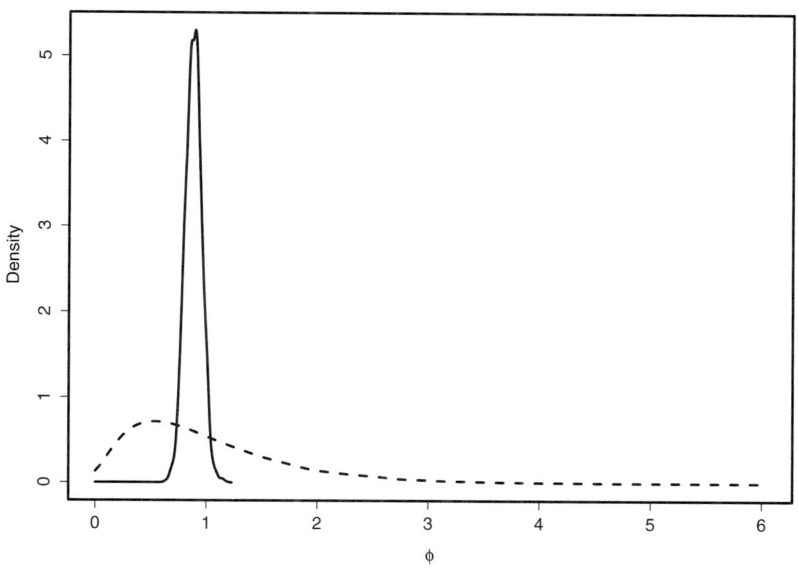

(b)

Fig. 6.6. Prior (dashed line) and posterior (solid line) distributions for (a) η and (b) ϕ for an NHPP (with a power law intensity function) analysis of failure time data for one Blue Mountain supercomputer SMP.

an *Exponential*(1) distribution is a *Gamma*(1, 1) distribution and that for integer κ, the sum of κ independent *Exponential*(1) random variables is distributed *Gamma*(κ, 1). Consequently, an interpretation of the MPLP is that a failure happens after κ exponentially distributed shocks have occurred, instead of just one for the PLP. Consequently, κ is called a shock parameter. If the intensity of an MPLP is increasing (i.e., $\phi > 1$) and $\kappa > 1$, then the probability of a failure in a small interval just after a failure is smaller than the probability of a failure in an interval of the same length just before the failure; and it is larger than the probability of a failure in an interval of the same length when the system was brand new. Note that an MPLP is (a) a gamma renewal process when $\phi = 1$, (b) a PLP when $\kappa = 1$, and (c) an HPP when $\phi = \kappa = 1$.

From Eq. 6.10, we can express T_i in terms of T_{i-1} as

$$T_i \sim \Lambda^{-1}[\Lambda(T_{i-1}) + Gamma(\kappa, 1)] \,, \tag{6.11}$$

where $\Lambda^{-1}(\cdot) = \eta(\cdot)^{1/\phi}$, which suggests how to simulate failure times from an MPLP.

For MPLPs under Type I censoring with data collection stopping at time t_c, the likelihood function for observed failure times $0 < t_1 < \ldots < t_n < t_c$ is

$$\begin{aligned}
&f(t_1, \ldots, t_n | \eta, \phi, \kappa) \\
&= \left(\prod_{i=1}^{n} \frac{\lambda(t_i)}{\Gamma(\kappa)} [\Lambda(t_i) - \Lambda(t_{i-1})]^{\kappa-1} \exp[-\{\Lambda(t_i) - \Lambda(t_{i-1})\}] \right) \\
&\quad \times \exp[-\{\Lambda(t_c) - \Lambda(t_n)\}] \\
&= \left(\prod_{i=1}^{n} \lambda(t_i)[\Lambda(t_i) - \Lambda(t_{i-1})]^{\kappa-1} \right) \frac{\exp[-\Lambda(t_c)]}{[\Gamma(\kappa)]^n} \,,
\end{aligned}$$

for $n > 0$, where $t_0 = 0$, and

$$1 - F_{T_1}(t_c) \,, \tag{6.12}$$

for $n = 0$, where $\lambda(t) = (\phi/\eta)(t/\eta)^{\phi-1}$ and $\Lambda(t) = (t/\eta)^{\phi}$. Recall that for MPLPs, $\Lambda(T_1) \sim Gamma(\kappa, 1)$, so that $F_{T_1}(t_c)$ is the cumulative distribution function of a *Gamma*(κ, 1) random variable evaluated at t_c.

For Type II censoring with data collection stopping at the nth failure, the likelihood function for failure times $0 < t_1 < \ldots < t_n$ is

$$\begin{aligned}
f(t_1, \ldots, t_n | \eta, \phi, \kappa) &= \textstyle\prod_{i=1}^{n} \frac{\lambda(t_i)}{\Gamma(\kappa)} [\Lambda(t_i) - \Lambda(t_{i-1})]^{\kappa-1} \exp[-\{\Lambda(t_i) - \Lambda(t_{i-1})\}] \\
&= \left(\textstyle\prod_{i=1}^{n} \lambda(t_i)[\Lambda(t_i) - \Lambda(t_{i-1})]^{\kappa-1}\right) \frac{\exp[-\Lambda(t_n)]}{[\Gamma(\kappa)]^n} \,,
\end{aligned}$$

where $t_0 = 0$.

We leave the fitting of an MPLP to the data in Tables 6.2 and 6.1 as Exercises 6.3 and 6.4.

6.5.2 Piecewise Exponential Model (PEXP)

One way to generalize an exponential renewal process (or HPP) is to allow for interfailure times X_i that are independent but not identically distributed. For example, the interfailure times X_i are independent and exponentially distributed with means μ_i described by

$$\mu_i = \left(\frac{\delta}{\lambda}\right) i^{\delta-1}, \tag{6.13}$$

where $\lambda > 0$. This model is the *piecewise exponential* or *PEXP* model. For $\delta = 1$, the model is an HPP with intensity function $\lambda(t) = \lambda$. For $\delta > 1$, μ_i is strictly increasing in i, so that the system after a repair is better than after the last repair or is "*better-than-old.*" A natural application for "better-than-old" repairs arises when making significant improvements to the system after each failure as happens in prototyping systems. The expectation is that the next prototype is better than the previous one and the literature refers to this scenario as "*test, analyze, and fix*" or *TAAF*. For $\delta < 1$, μ_i is strictly decreasing in i, so that the system is worse than after the last repair, i.e., "*worse-than-old.*" Note that the PEXP intensity function is equal to $1/\mu_i$ between the ith and $(i+1)$st failures and makes jumps after each failure. Consequently, the PEXP intensity is not directly comparable to the NHPP, because the NHPP intensity function is continuous.

Recall that we defined the PEXP model in terms of interfailure times X_i, which have independent exponential distributions with means μ_i as described by Eq. 6.13.

For Type I censoring with data collection stopping at time t_c, the observed failure times, $0 < t_1 < \ldots < t_n < t_c$, yield the observed interfailure times, $x_i = t_i - t_{i-1}$, $i = 1, \ldots, n$, where $t_0 = 0$. Also, the system had been working for $t_c - t_n$ when the data collection stops. The likelihood function for these data is

$$\begin{aligned} f(x_1, \ldots, x_n, t_n | \delta, \lambda) = &\left(\textstyle\prod_{i=1}^{n} (1/\mu_i) \exp[-(1/\mu_i) x_i]\right) \\ &\times \exp[-(1/\mu_{n+1})(t_c - t_n)] \end{aligned}$$

for $n > 0$, and

$$\exp[-(1/\mu_1) t_c] \text{ for } n = 0.$$

For Type II censoring with data collection stopping at the nth failure, the likelihood function for observed interfailure times x_i, $i = 1, \ldots, n$, is

$$f(x_1, \ldots, x_n | \delta, \lambda) = \prod_{i=1}^{n} (1/\mu_i) \exp[-(1/\mu_i) x_i] \, .$$

Regarding the choice of prior distributions for the PEXP model, δ and λ are both positive and real valued, so that one appropriate choice for prior distributions is to use independent gamma distributions. Choosing a ϕ prior distribution centered at 1 allows for HPPs as a special case.

We leave the fitting of a PEXP model to the data in Table 6.1 as Exercise 6.7.

6.6 Goodness of Fit and Model Selection

To assess how well the model fits the data, the analyst can use a Bayesian χ^2 goodness-of-fit test of Sect. 3.4. Recall that the method involves calculating $F_i(y_i \mid \tilde{\boldsymbol{\theta}})$ for the ith observation y_i, where $F_i(\cdot, \cdot)$ is the cumulative distribution function for the ith observation y_i and $\tilde{\boldsymbol{\theta}}$ is a posterior draw of the model parameters $\boldsymbol{\theta}$.

For failure count data (with $Y_i = N_i$), the method depends on uniform draws from the interval $[F_i(y_i - 1 \mid \tilde{\boldsymbol{\theta}}), F(y_i \mid \tilde{\boldsymbol{\theta}})]$, i.e., $Uniform[F_i(y_i - 1 \mid \tilde{\boldsymbol{\theta}}), F(y_i \mid \tilde{\boldsymbol{\theta}})]$. For HPPs and NHPPs, where y_i is the observed failure count for the ith interval (a_i, b_i), Y_i has a $Poisson[\Lambda(b_i) - \Lambda(a_i)]$ distribution. Regarding the other models, we do not expect to analyze count data with the PEXP model because of the PEXP models' motivation based on interfailure times. For gamma renewal processes as well as MPLPs, the count distributions for arbitrary intervals (a_i, b_i) are complicated and beyond the scope of this book.

For failure time data, the analyst can employ a Bayesian χ^2 goodness-of-fit test with the interfailure times using Eqs. 6.4 and 6.10 for Poisson processes (HPPs and NHPPs) and MPLPs, respectively. That is, knowing that the $\Lambda(T_i) - \Lambda(T_{i-1})$ are i.i.d. $Exponential(1)$ for HPPs and NHPPs, calculate $\Delta_i = \Lambda(t_i) - \Lambda(t_{i-1})$ and evaluate $F_i(\Delta_i) = 1 - \exp(-\Delta_i)$. For MPLPs, knowing that the $\Lambda(T_i) - \Lambda(T_{i-1})$ are i.i.d. $Gamma(\kappa, 1)$, calculate $\Delta_i = \Lambda(t_i) - \Lambda(t_{i-1})$ and evaluate $F_i(\Delta_i)$, where $F_i(\cdot)$ is the $Gamma(\kappa, 1)$ cumulative distribution function. For renewal processes, the interfailure times are i.i.d. following the distribution associated with the process. For the PEXP model, $F_i(\cdot)$ associated with the ith interfailure time X_i is the appropriate exponential cumulative distribution function.

Note that for failure time data collected under a Type I-censoring scheme with collection stopping at time t_c, $\Lambda(t_c) - \Lambda(T_n)$ is a Type I-censored observation from $Exponential(1)$ for HPPs and NHPPs and from $Gamma(\kappa, 1)$ for MPLPs, respectively. That is, make a uniform draw from the interval $(1 - F[\Lambda(t_c) - \Lambda(t_n)], 1)$, where $F(\cdot)$ is the $Exponential(1)$ cumulative distribution function for HPPs and NHPPs and is the $Gamma(\kappa, 1)$ cumulative distribution function for MPLPs. For renewal processes or PEXP models, draw a uniform from the interval $(1 - F(t_c - T_n), 1)$, where $F(\cdot)$ is the cumulative distribution function of the appropriate interfailure time distribution (i.e., $Exponential(1/\mu_{n+1})$). One thing to be careful of is that if there are n failures before stopping time t_c, then for a Bayesian χ^2 goodness-of-fit test, there are $n + 1$ observations.

Example 6.4 Goodness-of-fit assessment of PLP analyses. Consider the use of a Bayesian χ^2 goodness-of-fit test for the fits to the data in Tables 6.2 and 6.1 using a PLP from Example 6.3.

For the PLP fit to the failure count data in Table 6.2, there are $n = 15$ intervals (months), so that we use $K = 3$ ($\approx n^{0.4}$) bins as suggested for a Bayesian χ^2 goodness-of-fit test. With three equal probability bins, 10.7% of the test statistic R^B values exceed the 0.95 quantile of the $ChiSquared(2)$ reference distribution, which suggests no lack-of-fit.

For the PLP fit to the failure time data in Table 6.1, $n+1 = 28$, so that we use the suggested $(n+1)^{0.4} \approx 4$ bins. With four equal probability bins, 1.4% of the test statistic R^B values exceed the 0.95 quantile of the $ChiSquared(3)$ reference distribution, which suggests no lack-of-fit.

We leave the application of a Bayesian χ^2 goodness-of-fit test to fits of these data using log-linear process, MPLP, and PEXP models as Exercises 6.3, 6.4, 6.5, 6.6, and 6.7.

After assessing goodness of fit, there may remain several competing models. For example, if an HPP fits well, an NHPP should also fit well; consequently, the simpler HPP is preferable unless the HPP does not fit that well. Also inspect the posterior distribution of the parameters. For example, the ϕ posterior distribution for an NHPP may include 1, which does not rule out the simpler HPP. More formally, we can use the Bayesian information criterion (BIC) in Chap. 4 to select a model. See Sect. 4.6 for more details.

6.7 Current Reliability and Other Performance Criteria

Current reliability and other performance criteria characterize repairable system reliability. Next, we consider each in turn.

6.7.1 Current Reliability

One criterion for characterizing repairable system reliability is *current reliability*. Given the last failure at time t^*, current reliability is the reliability at $t^* + t$, which is the probability that the next failure occurs after $t^* + t$, or equivalently, the probability of no failures in the interval $(t^*, t^* + t)$. For example, for the Blue Mountain supercomputer example based on the failure time data in Table 6.1, $t^* = 456.5085$ corresponds to almost 15 months of operation and $t^* + t$ might be the time after a month or even a year of additional operation.

Denote the current reliability function by $R_{t^*}(t)$. Then, for given model parameter values $\boldsymbol{\theta}$, an analyst can evaluate $R_{t^*}(t)$ under the appropriate model. For a renewal process, the reliability function depends on the interfailure time distribution; for example, $R_{t^*}(t) = 1 - \exp(-\lambda t)$ for the exponential renewal process. Similarly, for PEXP models the next interfailure time follows an appropriate exponential distribution; the next interfailure time follows an *Exponential*(μ_{n+1}) distribution if the last failure was the nth failure with μ_{n+1} defined in Eq. 6.13.

For HPPs and NHPPs, we use the fact that the next failure time is related to the previous failure time by $\Lambda(T_{*+1}) - \Lambda(t^*) \sim Exponential(1)$, which yields

$$R_{t^*}(t) = 1 - \exp\{-[\Lambda(t^* + t) - \Lambda(t^*)]\}\,.$$

Similarly, for MPLPs $\Lambda(T_{*+1}) - \Lambda(t^*) \sim Gamma(\kappa, 1)$, so that current reliability function $R_{t^*}(t)$ depends on a $Gamma(\kappa, 1)$ cumulative distribution function evaluated at $\Lambda(t^*+t)-\Lambda(t^*)$. We leave the development of the MPLP current reliability function $R_{t^*}(t)$ as Exercise 6.8.

6.7.2 Other Performance Criteria

Depending on the situation, other performance criteria of a repairable system are relevant. For example, for a new system that is identical to an existing system, its current reliability is calculated using $t^* = 0$.

Other criteria include:

- Given the last failure at time t^*, what is the distribution of the number of failures in the interval $(t^*, t^* + t)$? Summarize this distribution by a mean or specified quantile. For HPPs and NHPPs, use the fact that the failure counts have a $Poisson[\Lambda(t^*+t)-\Lambda(t^*)]$ distribution. For the other models, the failure count distribution is complicated and beyond the scope of this book.
- Given the last failure at time t^*, what is the distribution of the time until the mth additional failure? For example, a major overhaul may be done after m additional failures so that quantifying the amount of time until the major overhaul may be of interest. For $m = 1$, the interest is in the time until the next failure. For $m = 1$, employ the same relationships used in the preceding section for current reliability. For $m > 1$, a simple method simulates additional interfailure times with the model under study, e.g., for HPPs and NHPPs, use $\Lambda(T_i) - \Lambda(T_{i+1}) \sim Exponential(1)$, and for MPLPs, use $\Lambda(T_i) - \Lambda(T_{i+1}) \sim Gamma(\kappa, 1)$.
- For availability, see Sect. 6.9.

Example 6.5 Evaluation of various performance criteria for PLPs. As an illustration of evaluating performance criteria, consider the PLP from Example 6.3 based on the analysis of the failure time data in Table 6.1.

For $t^* = 456.5085$ and $t = 31$ for an additional month of operation, we obtain the posterior distribution for current reliability $R_{t^*}(t) = 1-\exp[-\{\Lambda(t^*+t) - \Lambda(t^*)\}]$, where $\Lambda(t) = (\frac{t}{\eta})^\phi$, by evaluating the current reliability function with draws from the joint posterior distribution of (η, ϕ). The posterior median of current reliability is 0.757 with a 95% credible interval of (0.555, 0.895).

To evaluate the predictive failure count distribution for an additional month of operation, we make a draw from a $Poisson[\Lambda(t^* + t) - \Lambda(t^*)]$ distribution for each draw of the joint posterior distribution of (η, ϕ), where $\Lambda(\cdot)$ is a function of (η, ϕ). The median of the predictive failure count distribution is 1 with a 95% credible interval of (0, 4). The user expects one failure and no more than four failures in the next month of operation.

Next, consider the predictive distribution for the next failure time T from $\Lambda(T) - \Lambda(t^*) \sim Exponential(1)$; because $\Lambda(\cdot)$ is a function of (η, ϕ), we can make a draw from the next failure time predictive distribution by using $\Lambda(T) - \Lambda(t^*) \sim Exponential(1)$ for each draw of the joint posterior distribution of (η, ϕ). The median of the next failure time predictive distribution is 471.81 days with a 95% credible interval of (457.05, 547.11). In terms of the next interfailure time $T - t^*$, the median is 15.30 days with a 95% credible interval of (0.54, 90.60).

We leave the evaluation of these performance criteria for other models as Exercises 6.9 and 6.10.

6.8 Multiple-Unit Systems and Hierarchical Modeling

A natural model for failure count or failure time data from multiple-unit systems is a hierarchical model. Take, for example, the Blue Mountain supercomputer with 48 SMPs. (Actually, SMP 21 is different from the rest, so that we focus on the remaining 47 SMPs and index them by 1 to 47.) Table 6.3 displays actual failure count data for the first 15 months of operation, where the months listed correspond to the same ones listed in Table 6.2 for the single SMP example. While the 47 SMPs are identical, they perform similarly but not exactly the same.

A hierarchical model easily handles the similarity as follows: model each SMP's data by a PLP with a common shape parameter ϕ, but with individual scale parameters, i.e., η_i is the scale parameter for the ith SMP. Describe the similarity of the η_i by a distribution defined on the positive real line, such as

$$\eta_i | \alpha, \beta \sim Gamma(\alpha, \beta)\,, \tag{6.14}$$

where the η_i are conditionally independent. Consequently, the description of the PLP model for the Blue Mountain supercomputer failure count data is

$$N_{ij} \sim Poisson[\Lambda_i(b_i) - \Lambda_i(a_i)], \quad i = 1, \ldots, 47, \quad j = 1, \ldots, 15\,, \tag{6.15}$$

where N_{ij} is the number of failures for the ith SMP in the jth interval (a_j, b_j) (in days) and $\Lambda_i(t) = (\frac{t}{\eta_i})^{\phi}$.

On further inspection of Eq. 6.14, note that the η_i follow a distribution defined by parameters α and β. These parameters are likely not known in practice, and therefore, need their prior distributions specified. To complete the model, one appropriate choice of the prior distributions for these positively real valued hyperparameters is

$$\begin{aligned} \alpha &\sim Gamma(a_\alpha, b_\alpha) \quad \text{and} \\ \beta &\sim Gamma(a_\beta, b_\beta)\,, \end{aligned} \tag{6.16}$$

where the means of the α and β prior distributions are a_α/b_α and a_β/b_β, respectively.

Consider the simulated failure time data for the 47 SMPs in Table 6.4, where data collection stopped at time $t_c = 457$ days. For failure time data, the distributions for η_i, α, and β defined above in Eqs. 6.14 and 6.16 are the same. The only difference from the previous analysis is that we need to use the appropriate likelihood function for the failure time data. If $t_{i1}, \ldots, t_{im_i}$ denote the m_i observed failure times for the ith SMP, then the contribution of the ith SMP's data to the likelihood function is

$$f(t_{i1}, \ldots, t_{im_i}|\boldsymbol{\theta}_i) = \left(\prod_{i=1}^{m_i} \lambda_i(t_i)\right) \exp[-\Lambda_i(t_c)] \text{ for } m_i > 0, \tag{6.17}$$

and

$$\exp[-\Lambda_i(t_c)] \text{ for } m_i = 0\,,$$

where $\lambda_i(t) = \frac{\phi}{\eta_i}(\frac{t}{\eta_i})^{\phi-1}$ is the intensity function and $\Lambda_i(t) = (\frac{t}{\eta_i})^{\phi}$.

Example 6.6 Blue Mountain supercomputer failure count data analysis. Consider the analysis of the failure count data in Table 6.3 excluding the SMP 21 data and use the hierarchical PLP model presented above in Eqs. 6.14 and 6.15. The prior distributions we employ for α, β, and ϕ are

$$\begin{aligned} \alpha &\sim Gamma(15, 1)\,, \\ \beta &\sim Gamma(2, 2)\,, \text{ and} \\ \phi &\sim Gamma(2, 2)\,. \end{aligned}$$

This hierarchical PLP model has 50 parameters: η_i ($i = 1, \ldots, 47$), α, β, and ϕ. The likelihood function for these data is the product of probabilities of the observed counts n_{ij} based on Eq. 6.15, $i = 1, \ldots, 47$, $j = 1, \ldots, 15$. The joint prior distribution is the product of η_i probability density functions and the independent α, β, and ϕ prior density functions and has the form

$$\begin{aligned} &\left(\prod_{i=1}^{47} \tfrac{\beta^\alpha}{\Gamma(\alpha)} \eta_i^{\alpha-1} \exp(-\beta\eta_i)\right) \times \tfrac{1^{15}}{\Gamma(15)} \alpha^{15-1} \exp(-1 \times \alpha) \\ &\times \tfrac{2^2}{\Gamma(2)} \beta^{2-1} \exp(-2 \times \beta) \times \tfrac{2^2}{\Gamma(2)} \phi^{2-1} \exp(-2 \times \phi)\,. \end{aligned}$$

To analyze the supercomputer failure count data, we then use MCMC to obtain draws from the joint posterior distribution of the model parameters (η_i, $i = 1, \ldots, 47$, α, β, ϕ). Figure 6.7 presents the prior and posterior distributions for α and β, and ϕ as dashed and solid lines, respectively. The posterior

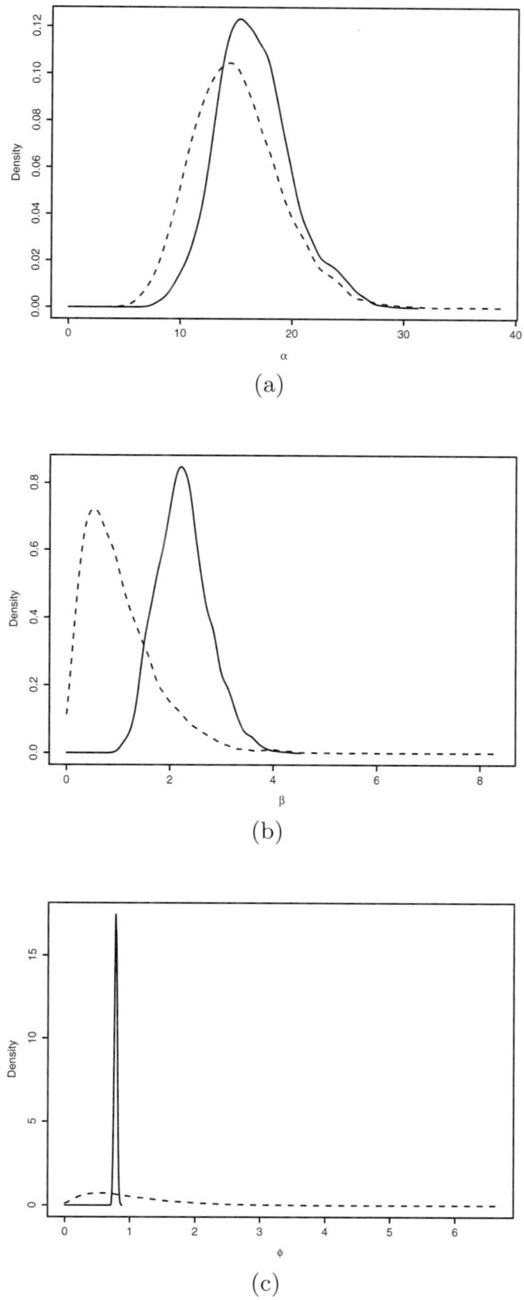

Fig. 6.7. Prior (dashed line) and posterior (solid line) distributions for (a) α, (b) β, and (c) ϕ for an NHPP (with a power law intensity function) analysis of count data for all SMPs of the Blue Mountain supercomputer.

Table 6.3. Blue Mountain supercomputer monthly failure counts

	Month														
SMP	1	2	3	4	5	6	7	8	9	10	11	12	13	14	15
1	1	5	2	1	0	1	1	2	1	0	1	0	1	2	5
2	5	4	6	1	2	1	4	4	2	1	1	1	1	1	1
3	1	5	3	0	0	0	0	2	3	1	1	1	0	1	3
4	4	3	2	1	2	1	1	3	1	0	3	0	0	2	3
5	2	2	2	0	1	1	1	4	2	0	0	1	0	0	1
6	3	4	3	1	1	4	1	4	2	1	1	3	1	1	4
7	1	7	3	6	1	2	3	3	4	0	0	0	0	6	4
8	3	3	3	1	0	3	2	4	3	1	2	3	0	0	2
9	6	2	3	1	0	1	0	5	1	0	2	0	1	0	2
10	4	4	5	1	1	1	2	7	4	0	0	1	1	1	1
11	4	7	3	2	0	0	0	4	2	0	0	0	1	3	1
12	4	4	3	3	0	0	1	4	2	2	1	3	2	2	2
13	2	4	3	1	0	0	0	3	1	0	1	0	0	2	1
14	3	4	3	2	0	3	1	2	1	0	0	0	2	2	1
15	2	5	3	3	1	0	1	5	1	0	0	1	1	2	2
16	2	2	3	1	2	1	1	4	2	0	1	3	0	1	1
17	4	3	3	2	0	1	2	3	1	0	1	1	1	2	3
18	5	4	3	0	1	4	1	2	1	0	1	1	1	3	2
19	5	2	3	3	2	4	2	5	1	0	1	0	2	2	2
20	2	3	1	1	0	3	0	1	1	2	2	0	1	2	2
21	10	5	3	6	6	4	10	5	3	8	3	2	8	2	5
22	5	2	5	1	1	1	1	2	2	4	2	0	3	3	8
23	3	3	2	1	1	1	2	5	1	0	2	0	2	1	2
24	2	3	2	1	2	3	0	3	2	0	0	1	4	1	4

distribution for ϕ is particularly interesting because it strongly suggests that $\phi < 1$. (Its posterior median is 0.782 with a 95% credible interval of (0.740, 0.828).) That is, the Blue Mountain supercomputer SMPs appear to be experiencing reliability growth over the first 15 months of operation and a closer examination provides a sensible explanation. Namely, the engineers collected the initial data during the early part of the learning curve for the team that built this supercomputer, where the early failures resulted from them learning how to build this supercomputer. The predictive distribution of η shown in Fig. 6.8 displays the variation of the η across the SMPs for the Blue Mountain supercomputer. The η predictive median is 7.19 with a 95% credible interval of $(3.93, 12.14)$. Consequently, we would expect that another identical SMP would have an η drawn from this distribution.

We leave the analysis of the failure time data in Table 6.4 as Exercise 6.11.

After fitting hierarchical models for multiple unit repairable systems, the analyst needs to assess their goodness of fit and choose among the best fitting models. Next, we consider goodness of fit and model selection for these

Table 6.3 (cont.)

	Month														
SMP	1	2	3	4	5	6	7	8	9	10	11	12	13	14	15
25	2	3	2	0	1	1	1	3	6	1	0	0	1	1	1
26	3	2	3	0	2	5	4	1	1	1	0	2	4	2	3
27	1	2	1	2	2	4	3	1	2	0	5	0	3	1	5
28	1	2	2	1	2	2	0	3	2	1	1	2	2	3	2
29	2	3	1	2	2	3	0	1	3	0	0	0	1	1	1
30	5	4	1	1	1	4	0	1	4	0	1	0	1	1	2
31	1	5	1	3	1	4	3	3	1	1	0	1	1	1	2
32	1	3	4	1	3	4	0	2	2	0	0	0	2	1	2
33	4	3	4	0	1	1	1	3	7	1	0	0	0	0	2
34	1	2	1	2	0	1	0	2	1	1	1	1	1	2	3
35	1	3	2	1	2	1	0	2	4	0	0	0	2	2	3
36	2	6	3	1	0	2	0	2	2	0	0	1	2	3	3
37	1	3	1	2	0	0	1	2	1	0	0	1	1	1	2
38	3	2	1	1	2	0	1	2	2	0	0	0	2	1	3
39	2	2	1	0	0	0	2	3	2	0	0	1	1	2	1
40	5	3	2	0	2	2	2	2	1	0	1	4	2	3	2
41	3	3	3	3	4	2	0	4	1	1	0	1	2	0	1
42	5	4	2	2	5	0	0	3	1	0	0	1	3	1	4
43	2	4	3	0	4	2	0	3	4	0	3	0	1	1	3
44	5	2	3	2	3	0	1	3	1	0	0	0	2	3	2
45	1	3	1	1	5	1	0	3	1	0	0	0	2	3	2
46	1	2	2	0	2	1	0	3	1	0	1	1	4	1	4
47	5	2	2	1	1	1	0	2	1	2	0	0	2	1	2
48	2	2	1	0	2	0	1	2	1	0	0	0	2	1	2

hierarchical models. Then, with the selected model, the analyst can assess repairable system reliability and other performance criteria for multiple-unit systems.

Goodness of Fit, Model Selection, and Performance Criteria

For hierarchical models, we can use a Bayesian χ^2 goodness-of-fit test as demonstrated in Chap. 3 for such models. For the hierarchical PLP model, the distribution of the data depend only on the η_i and ϕ. Consequently, the Bayesian χ^2 goodness-of-fit test uses the joint posterior distribution of the η_i and ϕ.

For model selection, because the model is hierarchical (i.e., the η_i follow a distribution), we can use the deviance information criterion (DIC) in Chap. 4 to compare different models. See Sect. 4.6 for more details.

For current reliability and other performance criteria, we can adapt them appropriately for a multiple-unit system. With a hierarchical PLP for the Blue Mountain supercomputer in mind, they include:

Table 6.4. Blue Mountain supercomputer simulated failure times

SMP	Failure Time (days)
1	4.74 20.23 22.21 26.02 35.09 37.34 58.04 59.71 67.67 72.89 74.16 85.683 94.59 112.38 124.13 124.84 129.14 135.98 141.93 146.52 163.07 201.48 215.10 239.87 244.75 244.97 260.63 268.35 292.90 300.76 380.19
2	5.21 14.49 42.31 54.52 59.09 66.09 74.99 119.27 127.47 146.48 172.77 190.21 203.52 223.37 252.64 265.52 284.06 312.68 322.67 333.70 348.48 411.39 434.76
3	2.99 36.72 47.54 87.11 94.82 116.97 128.36 129.40 139.33 145.92 146.83 172.27 198.52 233.19 240.46 249.28 258.04 268.39 286.56 365.35 373.33 380.91 381.11
4	42.26 45.46 55.27 80.06 87.56 111.32 144.79 164.18 214.58 243.81 260.25 283.78 317.26 322.27 338.86 346.07 356.86 397.68 422.27 425.41 440.49
5	9.48 67.27 76.12 90.91 105.43 147.54 173.73 175.84 179.46 185.45 195.94 200.14 207.40 208.35 208.54 241.41 340.28 359.49 360.95 383.89 390.13 396.31 423.98
6	2.92 15.12 21.37 37.946 50.10 50.82 57.63 58.12 87.22 90.70 100.21 104.66 127.41 156.67 157.65 169.65 211.26 213.83 239.47 244.72 253.85 256.67 261.12 273.67 281.09 285.91 288.59 302.28 306.79 307.83 326.10 373.31 435.77
7	17.71 37.58 41.11 102.74 107.18 107.95 109.81 110.64 116.22 163.64 193.91 231.12 234.50 256.82 274.48 276.68 281.12 287.11 297.60 298.25 299.85 306.84 314.70 317.10 326.65 356.16 376.71 407.81 414.18 450.93
8	2.31 5.33 5.55 23.25 24.81 24.99 35.43 71.04 86.81 89.35 102.83 103.07 119.92 139.11 156.55 157.97 187.32 270.21 296.37 316.14 348.45 355.55 360.17 374.29 391.85 408.31 413.17 445.63
9	14.04 16.76 18.92 25.00 27.91 54.41 63.05 101.81 151.78 160.52 165.67 176.95 192.48 230.24 231.88 248.37 254.00 255.23 257.66 275.18 275.24 303.78 434.53 442.76
10	6.33 7.51 23.24 26.07 32.82 35.14 39.69 57.55 69.03 72.90 84.50 87.87 96.38 101 213.18 213.32 251.67 287.73 376.82 423.34
11	5.07 22.55 63.93 92.42 98.19 123.51 132.63 133.14 139.87 171.51 190.23 205.25 245.36 280.36 285.87 311.83 315.17 325.40 362.05 365.85 381.39 386.24 389.76 393.30 417.56 428.36 430.87 440.46
12	3.47 14.16 15.91 31.63 55.77 62.60 83.49 101.45 122.86 158.42 166.82 172.60 219.55 231.88 249.27 255.38 264.73 271.39 282.75 310.56 311.98 313.81 328.48 331.01 340.30 418.31 427.03 432.79 451.31
13	12.46 14.59 14.89 23.88 28.15 37.28 44.56 102.92 152.57 163.76 171.65 250.90 280.22 296.61 319.93 324.62 326.32 366.95 372.64 390.26 394.24 394.64 445.71
14	0.39 4.96 12.92 22.91 25.58 31.11 41.56 48.02 69.30 74.77 87.00 119.82 123.04 127.75 135.46 141.75 145.28 163.28 163.97 172.67 194.20 211.92 231.63 234.31 283.38 310.90 326.992 367.59 416.13

Table 6.4 (cont.)

SMP	Failure Time (days)
15	16.08 24.63 25.87 36.24 52.79 65.81 78.15 81.19 82.27 95.89 120.91 136.70 140.67 173.57 185.39 203.04 208.16 238.60 264.60 297.60 302.32 320.78 341.65 418.53
16	3.15 4.78 7.45 7.74 25.81 32.40 83.79 129.46 140.23 166.38 211.81 214.92 221.02 221.61 222.67 225.72 246.59 281.50 295.78 298.47 302.15 309.20 319.26 333.64 342.92 346.70 380.06 388.85 399.20
17	0.70 6.03 7.24 13.54 18.00 36.71 43.81 55.61 57.56 77.43 88.48 91.01 91.15 100.47 116.43 126.77 130.034 132.454 135.24 139.03 164.19 170.65 183.29 186.68 203.86 206.31 206.84 212.804 267.21 280.24 295.80 302.04 305.36 324.70 330.95 343.83 389.55 400.68 410.89 433.36 442.24
18	5.87 17.83 18.15 30.07 58.60 110.28 126.17 143.21 154.50 160.66 170.05 176.12 178.27 208.00 210.87 214.22 228.17 247.60 258.09 264.14 264.54 284.73 307.00 332.03 335.80 360.85 365.85 384.79 392.50 403.93 404.38 431.86 445.38
19	7.88 32.46 37.01 72.83 109.83 112.61 119.80 174.97 180.84 181.22 186.42 211.01 213.84 254.90 258.63 288.40 295.22 307.53 311.70 312.37 325.70 334.29 337.09 366.18 377.39 405.69 445.68
20	3.43 21.43 31.06 46.66 62.18 70.53 74.61 82.96 83.62 137.70 172.44 213.21 269.85 334.47 368.24 400.74 448.52
21	4.80 7.45 9.82 10.63 12.12 24.23 28.04 36.00 38.45 43.85 89.26 114.62 130.88 132.69 137.06 146.30 153.91 167.51 199.37 199.49 200.90 214.02 214.60 224.19 244.96 250.35 272.87 284.55 324.05 378.26 397.10 402.24
22	24.22 25.49 31.24 61.98 84.32 87.34 122.86 140.59 185.39 189.66 224.16 246.84 338.90 344.35 369.57 381.26 394.70 403.89 424.43
23	14.61 24.4 45.00 58.31 73.57 106.29 194.06 215.43 221.15 230.38 263.93 265.72 298.25 303.26 344.48 365.01 376.42 380.24 434.89
24	0.10 3.72 27.91 30.03 34.59 37.76 48.28 75.48 83.02 83.64 89.80 111.75 112.89 122.19 133.97 145.40 207.39 263.17 276.59 307.37 333.20 336.08 369.71 390.81
25	6.68 14.12 19.22 20.93 26.03 32.60 35.41 37.80 59.37 60.69 89.44 91.75 95.43 104.09 108.96 112.88 118.78 119.29 167.34 184.54 198.70 200.95 235.06 249.52 254.19 277.33 310.21 348.91 349.79 351.21 399.84 431.59 446.07 448.45
26	1.01 2.41 15.58 16.27 24.68 41.27 43.96 46.69 54.43 71.69 71.98 82.80 98.32 119.42 135.33 163.09 166.02 178.62 185.04 203.25 205.56 237.75 243.09 305.73 341.17 346.88 352.93 358.92 378.80 390.15 404.60 411.75 419.16 422.22 437.56 444.36 445.20
27	4.08 18.52 23.22 36.52 54.49 84.40 90.44 105.16 154.22 173.99 178.13 207.80 210.30 211.41 244.14 255.32 257.74 265.89 268.74 271.35 280.20 302.76 355.92 359.47 370.24 371.32 391.32 419.83 424.60
28	1.09 23.28 29.31 34.06 70.20 81.51 84.48 103.78 114.67 161.57 201.24 244.70 252.16 315.38 349.36 358.17 364.85 374.97 430.41 432.36 449.44
29	9.47 10.99 77.32 83.53 96.42 97.49 142.79 165.45 179.13 195.31 196.67 205.24 222.18 243.44 247.05 252.24 281.95 293.60 310.40 335.20 430.02 453.96

Table 6.4 (cont.)

SMP	Failure Time (in days)
30	19.56 27.34 33.87 60.21 60.66 71.31 123.22 138.68 145.64 152.48 163.88 208.29 244.21 316.24 327.91 361.54 392.95 409.86 414.46 416.83 422.10 435.33 436.37 447.43
31	3.86 3.95 8.38 10.01 15.83 21.05 38.44 56.93 60.93 68.70 92.21 93.23 143.64 156.27 189.08 211.22 214.72 232.72 234.53 247.44 281.64 302.46 368.82 373.13 389.31 433.16
32	10.18 14.98 43.82 58.23 65.66 68.26 72.06 91.74 102.63 126.19 143.71 169.95 180.48 234.46 242.24 267.69 286.16 299.25 301.67 316.69 335.71 379.36 384.61 428.16 432.78
33	0.83 23.45 30.19 35.96 54.60 63.37 80.97 114.49 131.13 207.48 245.51 289.81 339.13 357.80 364.81 419.67
34	1.99 4.52 13.66 16.61 25.15 26.4565716 29.01 40.52 40.93 47.37 66.49 81.76 108.95 142.63 187.67 197.68 204.68 251.43 287.77 290.99 333.57 399.20 402.03 412.32 430.59 436.84 442.42
35	2.91 15.07 17.04 24.81 38.56 67.47 72.17 100.91 155.78 156.71 238.59 244.13 267.37 277.54 294.70 300.23 305.40 312.98 323.06 375.52 388.411 396.05
36	1.87 3.96 7.81 12.78 19.28 27.08 32.64 41.23 67.60 71.06 71.42 82.88 96.46 152.69 199.12 243.67 248.64 253.94 263.35 264.97 278.58 279.46 335.86 382.17 414.88 421.70 436.04 438.70
37	7.63 8.35 29.85 97.91 223.65 259.71 300.88 307.31 353.61 407.32 444.93
38	0.16 5.26 11.01 25.961 33.88 46.37 76.20 79.80 88.94 96.32 114.08 184.72 206.83 222.50 247.79
39	4 0.43 1.34 9.39 23.38 32.36 38.44 53.15 74.301 93.48 101.95 116.27 118.86 126.89 133.28 135.64 139.69 166.89 197.56 207.67 215.89 217.44 237.65 276.59 338.95 363.96 401.752 405.97 452.58
40	2.25 5.06 21.65 26.52 44.52 79.00 97.32 111.98 131.35 188.10 209.57 217.30 254.68 310.35 320.90 326.93 332.91 374.47 396.85 401.51 415.36 418.95 420.13 447.88
41	4.61 13.53 15.43 19.28 20.09 30.42 41.18 51.01 52.93 82.89 98.87 106.41 154.46 157.11 163.23 166.06 233.07 242.28 303.52 340.23 380.67 382.28 394.17 408.60 442.35 456.10
42	7.88 14.63 29.11 31.13 47.18 50.77 57.64 90.29 101.85 121.98 123.02 169.67 192.55 220.77 232.55 287.42 300.40 306.34 306.72 355.14 403.39 449.62
43	4.78 10.71 28.92 35.77 39.97 41.47 59.71 64.34 67.01 74.47 107.91 174.62 182.04 189.98 223.53 224.57 270.58 280.91 316.45 321.99 361.63 429.61 441.27
44	11.53 34.18 48.81 66.12 92.63 106.22 135.86 144.09 173.37 196.60 199.46 205.31 221.45 228.07 233.43 251.18 271.68 274.24 284.63 320.86 353.80 367.24 386.66 397.75 404.77 422.52 425.94 444.80
45	0.01 9.62 4.6.58 59.46 64.51 81.31 109.50 145.062 172.69 200.09 282.49 327.92 352.72 357.66 396.49 421.80 446.03

Table 6.4 (cont.)

SMP	Failure Time (in days)
46	60.69 83.15 102.36 142.44 148.28 153.07 251.35 253.70 255.91 264.82 279.42 302.01 324.50 333.98 354.27 356.36 382.19 414.86 451.51
47	0.41 2.9 14.57 21.76 36.44 64.96 74.75 86.59 158.88 178.28 180.82 194.04 198.24 255.30 290.43 361.85 362.61

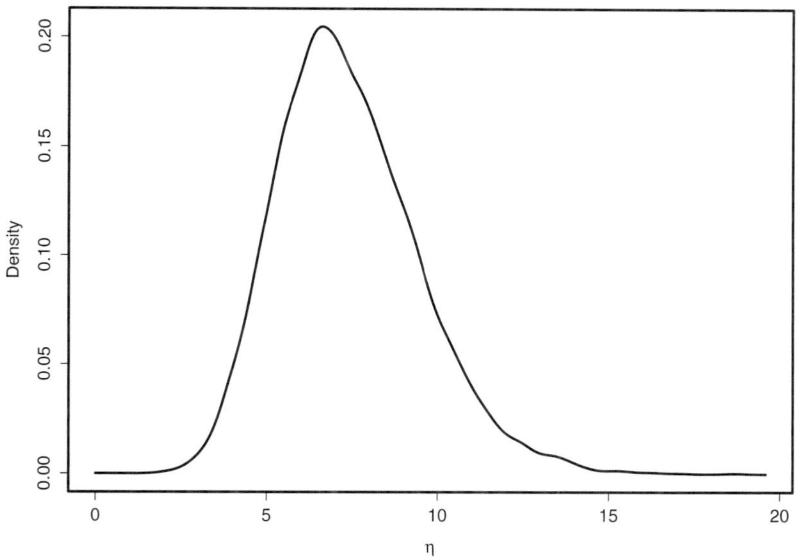

Fig. 6.8. Predictive distribution for η for an NHPP (with a power law intensity function) analysis of count data for all SMPs of the Blue Mountain supercomputer.

- For current system reliability, compute the current reliability of each SMP and multiply them together assuming that the system fails if one of its SMPs fails. That is, assume that the Blue Mountain supercomputer is a series system. For each draw of the joint posterior distribution of the η_i and ϕ, compute the current system reliability to obtain a draw from its posterior distribution. For a system, t^* is the last failure time of the system. We can also evaluate the current system reliability when the system was brand new by letting $t^* = 0$.
- Given time t^*, what is the distribution of the total number of failures in the interval $(t^*, t^* + t)$? This distribution may be summarized by a mean or specified quantile. For HPPs and NHPPs (which includes PLPs), employ the $Poisson[\Lambda_i(t^* + t) - \Lambda_i(t^*)]$ distribution for the ith SMP failure count.
- Given time t^*, what is the distribution of the time until the mth additional failure? For example, a major overhaul may be done after m more

failures so that the amount of time until the major overhaul may be of interest. Note that the m failures need not be on the same SMP. For $m = 1$, the interest is in the time until the next failure and we can employ the same relationships discussed previously for current reliability. For $m > 1$, a simple method simulates additional interfailure times appropriately using the model under study. For the ith SMP under HPPs and NHPPs (which includes PLPs), use $\Lambda_i(T_j) - \Lambda_i(T_{j-1}) \sim \textit{Exponential}(1)$, and under MPLPs, use $\Lambda_i(T_j) - \Lambda_i(T_{j-1}) \sim \textit{Gamma}(\kappa, 1)$; then simulate failure times for all the SMPs.

- We can evaluate the criteria discussed above from a new supercomputer consisting of k SMPs, which have η_i that is drawn from the same distribution as that for the Blue Mountain supercomputer. That is, for each draw of the joint posterior distribution of (α, β), make k draws from $\textit{Gamma}(\alpha, \beta)$ to obtain values for η_i, $i = 1, \ldots, k$.

Example 6.7 Goodness-of-fit assessment for Blue Mountain supercomputer failure count data analysis and evaluation of performance criteria. To assess the goodness of fit for the hierarchical PLP model in Example 6.6, we use a Bayesian χ^2 goodness-of-fit test. Based on $47 \times 15 = 705$ failure counts, the recommended number of bins is $K = 14$ ($\approx 705^{0.4}$) bins. Note that 8.6% of the test statistic R^B values exceed the 0.95 quantile of the $\textit{ChiSquared}(13)$ reference distribution, which suggests no lack of fit.

Now consider the distribution of the total number of failures in the next month ($t = 30$ days) as the performance criterion; start after the 15th month ($t^* = 457$ days) and evaluate it each month for the next two years (assuming 30-day months). See Fig. 6.9, which plots the median and 0.025 and 0.975 quantiles of the predictive distribution of the monthly total number of failures for the Blue Mountain supercomputer after time t^*. Note that the plot is not strictly decreasing because of simulation error; for a particular current month, draw a single total failure count for each draw of the joint posterior distribution of the model parameters. We can decrease the simulation error by simulating more total failure counts for each draw of the joint posterior distribution of the model parameters.

6.9 Availability

In this chapter, we have considered the reliability of a repairable system. So far, the assumption has been that the repairs are instantaneous. But repairs take time and now let us consider their impact. When a manufacturing system is up (operating), it is making products, but when it is down, it is not. Consequently, a manufacturer is interested in how often the manufacturing system is up. This is the idea behind availability.

Assume that the repair times or downtimes D follow a given distribution. Also the failure times or uptimes U follow a different distribution. Then,

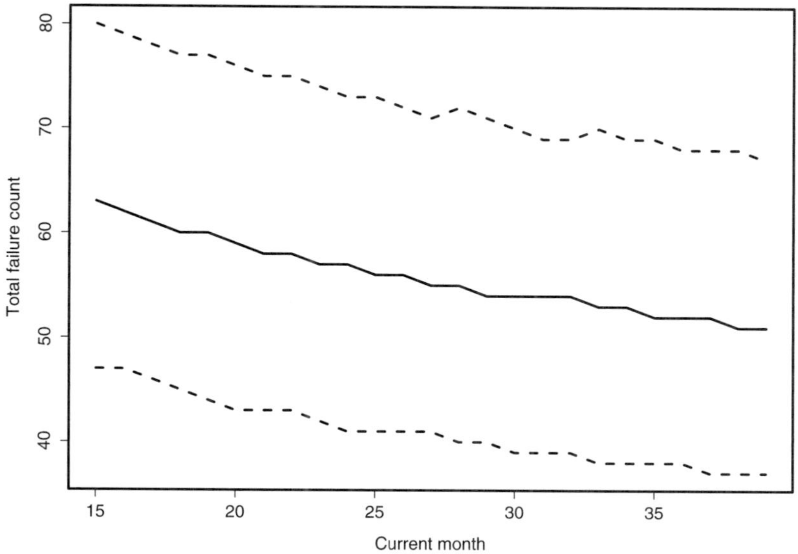

Fig. 6.9. Medians (solid line) and 0.025 and 0.975 quantiles (dashed lines) of the predictive distribution of the total number of failures for the Blue Mountain supercomputer for the next month after the current month from 15 months to 39 months.

whether the system is operating or not is what is of most interest. Let state $S(t) = 1$ if the system is operating at time t, and $S(t) = 0$, otherwise. Then, define *availability at time t* as $A(t) = \mathbf{P}(S(t) = 1)$. A related quantity is *average availability*, defined as $(1/T)\int_0^T A(t)dt$ over a time period of length T. Finally, define *long-run* or *steady-state availability* as

$$A = \lim_{t\to\infty}\mathbf{P}[S(t) = 1]\,.$$

In the remainder of this chapter, we concentrate on long-run availability. The successive uptimes (U_is) and downtimes (D_is) characterize a two-state (up and down) renewal process. Renewal theory (Barlow and Proschan, 1975) provides an expression for long-run availability:

$$A = \frac{\mathrm{E}(U)}{\mathrm{E}(U) + \mathrm{E}(D)}\,,$$

where $\mathrm{E}(U)$ is the mean failure time (uptime) and $\mathrm{E}(D)$ is the mean repair time (downtime).

Consider the case of failure times independently distributed as *Exponential*(λ) and repair times independently distributed as *Exponential*(μ). A well-known result is that

$$\mathbf{P}[S(t) = 1] = \tfrac{\mu}{\mu+\lambda} + \tfrac{\lambda}{\lambda+\mu} \exp[-(\lambda + \mu)t] \ ,$$

from which

$$\begin{aligned} A &= \lim_{t\to\infty} \mathbf{P}[S(t) = 1] = \lim_{t\to\infty} \tfrac{\mu}{\mu+\lambda} + \tfrac{\lambda}{\lambda+\mu} \exp[-(\lambda + \mu)t] \\ &= \tfrac{\mu}{\mu+\lambda} = \tfrac{1/\lambda}{1/\lambda+1/\mu} = \tfrac{\mathrm{E}(U)}{\mathrm{E}(U)+\mathrm{E}(D)} \end{aligned} \qquad (6.18)$$

follows.

We can assess long-run availability in Eq. 6.18 by evaluating its posterior distribution. In simple situations, evaluate the posterior distribution of long-run availability using the joint posterior distributions of the parameters, λ and μ. Even in Eq. 6.18, note that the long-run availability is a ratio of two dependent random variables (i.e., the posterior distributions of μ and $\mu + \lambda$), which has a distribution that is generally nontrivial to evaluate. Consequently, in complex situations, make a draw from the posterior distributions of λ and μ and evaluate Eq. 6.18 to obtain a draw from the posterior distribution of long-run availability, as illustrated in the next example.

Example 6.8 Illustration of availability evaluation. To illustrate evaluating availability in Eq. 6.18, suppose that 25 successive uptimes and downtimes are recorded in hours for a system. The interfailure times or uptimes are (9.17, 5.22, 15.93, 13.99, 13.70, 7.22, 13.10, 2.43, 0.75, 54.52, 1.58, 4.94, 5.80, 5.12, 11.91, 0.29, 10.54, 0.14, 12.83, 7.65, 4.04, 2.31, 44.01, 4.16, 6.22). The repair times or downtimes are (1.35, 1.62, 2.01, 0.39, 1.90, 1.12, 0.42, 1.33, 3.14, 3.44, 1.72, 1.09, 1.26, 1.93, 2.16, 0.63, 0.82, 3.46, 3.34, 0.15, 2.46, 0.71, 0.24, 1.05, 4.36). Using diffuse prior distributions for λ and μ, i.e., $Gamma(0.00001, 0.00001)$, Fig. 6.10 displays their posterior distributions in (a) and (b), respectively. Moreover, Fig. 6.10(c) displays the posterior distribution for long-run availability A by making draws from the (λ, μ) posterior distribution and evaluating Eq. 6.18 to obtain draws from the log-run availability posterior distribution. The posterior median of long-run availability is 0.859 with a 95% credible interval of (0.777, 0.914).

6.9.1 Other Data Types for Availability

As in Example 6.8, we consider only uptime and downtime data for assessing availability in the remainder of this chapter. However, there are other types of data that are discussed more fully in the literature (Martz and Waller, 1982). In this chapter, we focus on analyzing failure count and failure time data associated with the failure time or uptime distribution, which is exponential or gamma that underlie the HPPs or NHPPs and MPLPs, respectively. Other types of data considered by the literature include:

- "Snapshots" — sample the system at different times and record whether the system is working or not.

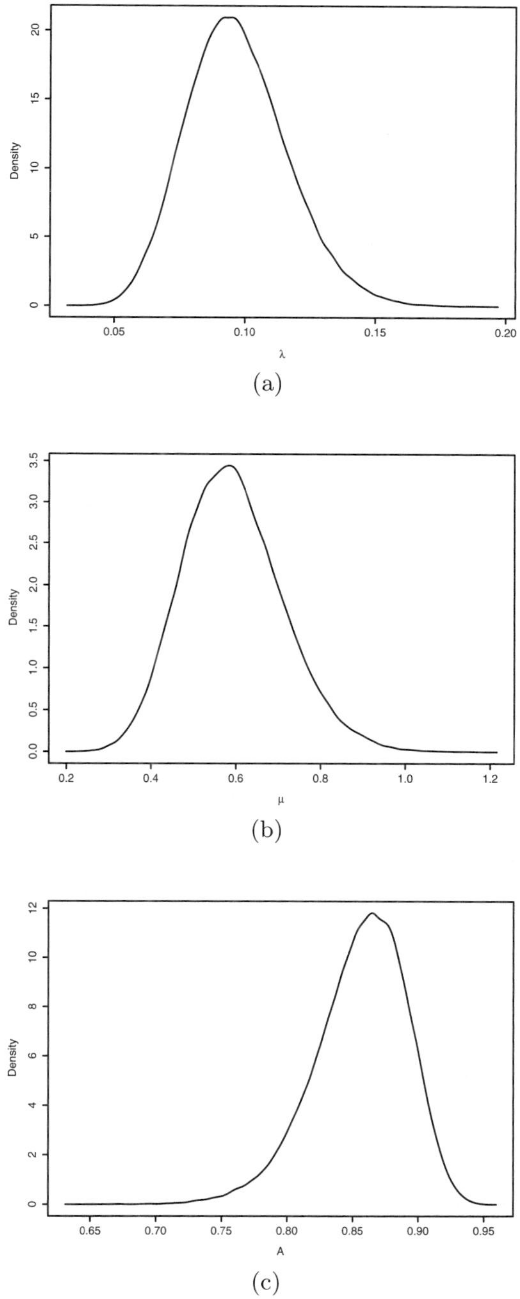

Fig. 6.10. Posterior distributions for (a) λ, (b) μ, and (c) long-run availability A based on Eq. 6.18. Failure times have an *Exponential*(λ) distribution and repair times have an independent *Exponential*(μ) distribution.

- Collect initial k cycles of uptimes and downtimes, followed by n "snapshots."
- Observe a system at n random times and continue to observe the system until its state changes.

6.9.2 Complex System Availability

In this section, we consider the availability of more complex systems. For purely series and purely parallel systems, which have M independently operating and repairable components and M independently operating repair facilities (one for each component), the long-run availability of the purely series system is

$$A = \prod_{i=1}^{M} A_i \,, \tag{6.19}$$

where A_i is the long-run availability of the ith component, and the long-run availability of the purely parallel system is

$$A = 1 - \prod_{i=1}^{M} (1 - A_i) \,. \tag{6.20}$$

To assess long-run availability, we can make a draw from the posterior distribution of $\boldsymbol{\theta}$, the vector of parameters for the failure time and repair time distributions. For example, if the ith component's failure and repair times have *Exponential*(λ_i) and *Exponential*(μ_i) distributions, respectively, then $\boldsymbol{\theta} = \{\lambda_i, \mu_i, i = 1, \ldots, M\}$.

The assumptions made in Eqs. 6.19 and 6.20 ignore the complexity of actual systems. Continuing to operate components that have not failed when the system is down, independent repair facilities for each component, and independent failure and repair times do not usually reflect reality. We can relax these assumptions in many directions, such as:

- There are fewer repair facilities than components.
- In a series system, do not operate the components that have not failed when repairing the system.
- The parallel system is set up as a standby system, in which the nonoperating components are not aging.
- The component failure times (across different components and within components) are dependent.
- The components are worse after each failure/repair, i.e., components are not as good-as-new after repairs.
- There is periodic maintenance performed on the components.

While the literature addresses special cases of these relaxed assumptions, a closed form solution becomes harder to obtain when relaxing these assumptions. Instead, we evaluate long-run availability by simulation. As illustrated by Hamada et al. (2006), simulation can handle very complex systems. The next example illustrates the use of simulation to evaluate long-run availability for a two-component parallel system with one repair station.

Example 6.9 Evaluation of system availability by simulation. Let us evaluate the long-run availability of a two-component parallel system with one repair station. Assume that the two components have conditionally independent failure times, distributed as $Exponential(0.1)$. There is a single repair station, which has repair times that follow an $Exponential(0.5)$ distribution. Further, assume that both components are operating when neither component has failed, i.e., there is no standby.

We evaluate the long-run availability using discrete event simulation by simulating failure and repair times up to time $limitTime$. Let $upTime$ and $downTime$ denote the current cumulative uptime and downtime, respectively, and $totalTime$ denote the current cumulative time. Also, let T_1 and T_2 denote the two component failure times, ordered so that $T_1 \leq T_2$. Finally, let $rTime$ denote the current repair time. The following algorithm describes the discrete event simulation for the two-component parallel system with one repair station, $Exponential(\lambda)$ failure times and $Exponential(\mu)$ repair times.

1. Assign values to λ, μ, and $limitTime$.
2. Initialize $totalTime$, $upTime$, and $downTime$ to 0.
3. Draw initial component failure times, T_i, $i = 1, 2$, from an $Exponential(\lambda)$ distribution, and order so that $T_1 \leq T_2$.
4. If $totalTime$ is less than $limitTime$, go to Step 5, else go to Step 14.
5. If both components are functioning ($T_i > 0$, $i = 1, 2$), go to Step 6, else go to Step 9.
6. Advance forward in time to next failure. Add T_1 to $upTime$ and $totalTime$.
7. Update failure times by subtracting T_1 from T_2, and setting $T_1 = 0$.
8. Draw a repair time $rTime$ from an $Exponential(\mu)$ distribution. Go to Step 4.
9. If one of the components is still functioning ($T_2 > 0$), go to Step 10, else go to Step 13.
10. If the functioning component fails before repairing the other component ($T_2 < rTime$), go to Step 11, else go to Step 12.
11. Add T_2 to $upTime$ and $totalTime$. Subtract T_2 from $rTime$ and set $T_2 = 0$. Go to Step 4.
12. Add $rTime$ to $upTime$ and $totalTime$. Subtract $rTime$ from T_2. Draw new failure time T_1 from an $Exponential(\lambda)$ distribution. Order T_1 and T_2 so that $T_1 \leq T_2$. Go to Step 4.
13. Since both components have failed, the system is not functioning. Advance forward in time to next repair. Add $rTime$ to $downTime$ and $totalTime$. Draw a failure time T_2 from an $Exponential(\lambda)$ distribution. Draw a repair time $rTime$ from an $Exponential(\mu)$ distribution. Go to Step 4.
14. Estimate long-run availability by $upTime/totalTime$.

We can use this algorithm to simulate the system for a long time (say, 10,000,000 time units) and estimate the long-run availability as the proportion of total uptime to total simulated time. Two code runs yielded simulated availabilities of 0.94567 and 0.94564. The actual availability is 0.94595, but by simulating the system for a longer time, we can obtain even closer approximations.

In Example 6.9, we use simulation to evaluate long-run availability for specific values of the failure time and repair time distribution parameters. For making inferences about long-run availability, take a draw from both the posterior distribution of the failure time and repair time distribution parameters $\boldsymbol{\theta}$ and the vector of failure time and repair time distribution parameters associated with the system's components, and evaluate the long-run availability $A(\boldsymbol{\theta})$ by simulation to obtain a draw from the long-run availability posterior distribution.

6.10 Related Reading

There is a large literature on repairable system reliability. For example, Rigdon and Basu (2000) provides a book-length treatment on this subject. In the remainder of this section, we mostly point to topics not covered in this chapter. Englehardt (1995) discusses various incomplete data types including truncation, left-time censoring, left-failure censoring, gaps, and grouping. Guida and Pulcini (2006) and Ryan (2003) propose alternative intensity functions. Guida and Pulcini (2006) considers a PLP with bounded intensity to account for applications in which the intensity function does not increase indefinitely. Also, Ryan (2003) proposes several flexible families of intensity functions. Only a few papers discuss inference for hierarchical NHPPs using MCMC. Ryan and Reese (2001) introduces the Blue Mountain supercomputer example and considers more complicated hierarchical models and extensions than we presented here. This chapter only introduced the PEXP model for modeling reliability growth under TAAF for failure time data. There is also substantial literature on discrete reliability growth, for which the data are the number of successes for a given number of tests after each improvement. Fries and Sen (1996) provides a comprehensive review of the discrete reliability growth literature.

Regarding availability, Barlow and Proschan (1975) presents renewal theory to obtain expressions for long-run availability for specific systems. Martz and Waller (1982) summarizes the Bayesian literature before 1982 for long-run availability for specific system structures and some of the specialized data types mentioned in Sect. 6.9.1. See also Brender (1968a), Brender (1968b), Gaver and Mazumdar (1969), Thompson and Springer (1972), Thompson and Palicio (1975), and Lie et al. (1977). Since 1982, subsequent articles by Tillman et al. (1982), Bacon-Shone (1983), Kuo (1984), Kuo (1985), Kuo (1986), Sharma and Krishna (1995), Pham-Gia and Turkkan (1999), and Cha and Kim (2001)] consider additional system structures and data types. However, we can obtain all these results in a practical setting by the simulation approach discussed in Sect. 6.9.2. See also Hamada et al. (2006), which uses this simulation approach for a very complex manufacturing system.

Finally, there is a large literature on recurrent events. See Cook and Lawless (2007) for a recent treatment, which provides other models that we might apply to repairable systems reliability data.

6.11 Exercises for Chapter 6

6.1 Using Eq. 6.4, show that for the exponential renewal process, the inter-failure times are exponentially distributed with parameter λ.

6.2 Develop the likelihood function for failure count data for a gamma renewal process. Also, develop the likelihood function for failure time data under Type I- and Type II-censoring schemes.

6.3 Fit an MPLP to the data in Table 6.2. Does the analysis support the need for an MPLP over a PLP? Assess goodness of fit using a Bayesian χ^2 goodness-of-fit test.

6.4 Fit an MPLP to the data in Table 6.1. Does the analysis support the need for an MPLP over a PLP? Assess goodness of fit using a Bayesian χ^2 goodness-of-fit test.

6.5 Fit a log-linear process to the data in Table 6.2. How does the fit compare with that of a PLP in terms of a Bayesian χ^2 goodness-of-fit test? Also compare the log-linear process and PLP fits using BIC discussed in Sect. 4.6.

6.6 Fit a log-linear process to the data in Table 6.1. How does the fit compare with that of a PLP in terms of a Bayesian χ^2 goodness-of-fit test? Also compare the log-linear process and PLP fits using BIC.

6.7 Fit a PEXP model to the data in Table 6.1. Assess goodness of fit using a Bayesian χ^2 goodness-of-fit test.

6.8 Develop an expression for the MPLP current reliability $R_{t^*}(t)$.

6.9 Evaluate current reliability and other performance criteria as in Example 6.5 using the data in Table 6.2 for the PLP, NHPP, and MPLP models. How do these results from these different models compare?

6.10 Evaluate current reliability and other performance criteria as in Example 6.5 using the data in Table 6.1 for the log-linear process, MPLP, and PEXP models. How do these results compare with those for the PLP evaluated in Example 6.5?

6.11 Analyze the failure time data in Table 6.4 using a hierarchical PLP model. Assess goodness of fit using a Bayesian χ^2 goodness-of-fit test.

6.12 Continuing with the preceding exercise, propose hierarchical log-linear process, MPLP, and PEXP models and fit them. How do the fits compare with using a Bayesian χ^2 goodness-of-fit test? Also compare the fits using DIC presented in Sect. 4.6.

6.13 As in Example 6.6, propose hierarchical log-linear process and MPLP models and fit them to the failure count data in Table 6.3. How do the fits compare with that of a hierarchical PLP in terms of a Bayesian χ^2 goodness-of-fit test? Also compare the hierarchical PLP, log-linear process, and MPLP fits using DIC.

6.14 Continuing with Example 6.7, evaluate other repairable system performance criteria.

6.15 As with Example 6.7, evaluate current reliability and other performance criteria using the hierarchical log-linear process and MPLP models and fit them to the failure count data in Table 6.3. How do these results compare with those from the hierarchical PLP model?

6.16 Evaluate current reliability and other performance criteria for the hierarchical PLP, log-linear process, MPLP, and PEXP models fit to the failure time data in Table 6.4. How do these results compare?

6.17 Proschan (1963) analyzes interfailure times in hours of air conditioning systems for a fleet of Boeing 720 jet planes as displayed in Table 6.5. An asterisk indicates a major overhaul and the first failure time afterwards is not reported.
   a) Fit hierarchical HPP, PLP, log-linear process, and MPLP models.
   b) Choose the model that fits these data the best.
   c) Evaluate appropriate performance criteria for this fleet of jet planes.

Table 6.5. Interfailure times for various Boeing 720 jet plane air conditioning systems. (An asterisk indicates a major overhaul and we do not report the first failure time afterwards)

Plane	Interfailure Time (in hours)
7907	194 15 41 29 33 181
7908	413 14 58 37 100 65 9 169 447 184 36 201 118 * 34 31 18 18 67 57 62 7 22 34
7909	90 10 60 186 61 49 14 24 56 20 79 84 44 59 29 118 25 156 310 76 26 44 23 62 * 130 208 70 101 208
7910	74 57 48 29 502 12 70 21 29 386 59 27 * 153 26 326
7911	55 320 56 104 220 239 47 246 176 182 33 * 15 104 35
7912	23 261 87 7 120 14 62 47 225 71 246 21 42 20 5 12 120 11 3 14 71 11 14 11 16 90 1 16 52 95
7913	97 51 11 4 141 18 142 68 77 80 1 16 106 206 82 54 31 216 46 111 39 63 18 191 18 163 24
7914	50 44 102 72 22 39 3 15 197 188 79 88 46 5 5 36 22 139 210 97 30 23 13 14
7915	359 9 12 270 603 3 104 2 438
1916	50 254 5 283 35 12
1917	130 493
8044	487 18 100 7 98 5 85 91 43 230 3 130
8045	102 209 14 57 54 32 67 59 134 152 27 14 230 66 61 34

6.18 Convert Table 6.5 to failure count data and analyze as in the preceding exercise. Compare these results with those from the preceding exercise.

6.19 Kumar and Klefsjö (1992) analyzes interfailure times in hours of load-haul-dump (LHD) machine hydraulic systems as displayed in Table 6.6. LHD machines are the primary machinery for loading rock in Swedish underground mines.
  a) Fit hierarchical HPP, PLP, log-linear process, and MPLP models.
  b) Choose the model that fits these data the best.
  c) Evaluate appropriate performance criteria for a population of 25 LHD machines.

Table 6.6. Interfailure times (in hours) for several LHD machine hydraulic systems (Kumar and Klefsjö, 1992)

LHD1	327, 125, 7, 6, 107, 277, 54, 332, 510, 110, 10, 9, 85, 27, 59, 16, 8, 34, 21 152, 158, 44, 18
LHD3	637, 40, 197, 36, 54, 53, 97, 63, 216, 118, 125, 25, 4, 101, 184, 167, 81, 46 18, 32, 219, 405, 20, 248, 140
LHD9	278, 261, 990, 191, 107, 32, 51, 10, 132, 176, 247, 165, 454, 142, 39, 249 212, 204, 182, 116, 30, 24, 32, 38, 10, 311, 61
LHD11	353, 96, 49, 211, 82, 175, 79, 117, 26, 4, 5, 60, 39, 35, 258, 97, 59, 3, 37 8, 245, 79, 49, 31, 259, 283, 150, 24
LHD17	401, 36, 18, 159, 341, 171, 24, 350, 72, 303, 34, 45, 324, 2, 70, 57, 103 11, 5, 3, 144, 80, 53, 84, 218, 122
LHD20	231, 20, 361, 260, 176, 16, 101, 293, 5, 119, 9, 80, 112, 10, 162, 90, 176 360, 90, 15, 315, 32, 266

6.20 Evaluate the long-run availability for a system with gamma distributed failure and repair times.

6.21 Specify a complex system (e.g., number of components, structure such as a parallel-series system, component failure and repair time distributions) and evaluate its long-run availability using the simulation approach.

6.22 Assess the availability of the system in Exercise 6.20. Simulate failure and repair times and obtain the posterior distribution for the failure and repair time distribution parameters using diffuse prior distributions.

6.23 Evaluate the availability of a PLP by the simulation approach using the relationship in Eq. 6.4 for successive failure times.

6.24 Evaluate the availability of a log-linear process by the simulation approach using the relationship in Eq. 6.4 for successive failure times.

6.25 Evaluate the availability of an MPLP by the simulation approach using the relationship in Eq. 6.10 for successive failure times.

7

Regression Models in Reliability

The distribution of reliability data may depend on covariates, also known as explanatory variables, independent variables, predictors, or regressors. This chapter shows how to incorporate covariates in the analysis of binomial success/failure data, Poisson count data, and lifetime data. Covariates allow us to compare the reliability between two or more different situations. We also discuss how covariates arise in accelerated life testing and in experiments to improve reliability.

7.1 Introduction

This chapter considers situations in which the reliability data distribution depends on covariates. That is, the data distribution changes when changing the values of the *covariates*. The literature also refers to covariates as explanatory variables, independent variables, predictors, or regressors. In general, *regression models* are models involving covariates. In regression models, we can express the relationship between the data distribution and the covariates by a distribution parameter (possibly transformed) as a function of the covariates. For example, in the well-known multiple regression model, the response Y has a $Normal[\mu(\mathbf{x}), \sigma^2]$ distribution with mean μ that is related to k covariates $x_1, \ldots, x_k$ through $\mu = \beta_0 + \beta_1 x_1 + \cdots + \beta_k x_k = \mathbf{x}^T\boldsymbol{\beta}$, where β_0, $\beta_1, \ldots, \beta_k$ are the model parameters known as regression coefficients.

The data distributions that commonly arise in reliability applications, such as the binomial, Poisson, lognormal, and Weibull distributions, often depend on covariates. For success/failure data having a $Binomial[n, \pi(\mathbf{x})]$ distribution, the probability of success/failure π may depend on the covariates through $\text{logit}(\pi) = \log[\pi/(1-\pi)] = \mathbf{x}^T\boldsymbol{\beta}$. Similarly, for failure count data following a $Poisson[\lambda(\mathbf{x})]$ distribution, $\log(\lambda) = \mathbf{x}^T\boldsymbol{\beta}$ for the mean count λ, and for lifetimes having a $LogNormal[\mu(\mathbf{x}), \sigma^2]$ distribution, $\mu = \mathbf{x}^T\boldsymbol{\beta}$ for the mean logged lifetime μ. Finally, for lifetimes following a $Weibull[\lambda(\mathbf{x}), \gamma]$ distribution, the scale parameter λ may be related to the covariates through

$\log(\lambda) = \mathbf{x}^T\boldsymbol{\beta}$. We provide more details including alternative relationships in subsequent sections that separately deal with each of these distributions.

Before considering regression models for these distributions, we discuss different types of covariates and covariate relationships.

7.1.1 Covariate Types

Covariates can either be discrete or continuous. A commonly used type of continuous covariate is a polynomial of a variable, e.g., linear or quadratic such as T or T^2, where T denotes temperature. Consequently, $\mu = \mathbf{x}^T\boldsymbol{\beta}$ takes the form

$$\mu = \beta_0 + \beta_1 T \text{ or}$$
$$\mu = \beta_0 + \beta_1 T + \beta_2 T^2 .$$

For discrete covariates with values that are nominal (i.e., names), such as supplier or plant location, use *dummy variables* as covariates to make comparisons between the different values. In the simplest case, where there are two suppliers, say A and B, one covariate x is required. Its values are 0 and 1, which correspond to A and B, respectively. For the multiple regression model mentioned in Sect. 7.1,

$$\mu = \begin{cases} \beta_0 + \beta_1 x = \beta_0 + \beta_1 \times 0 = \beta_0 \text{ for } A \\ \beta_0 + \beta_1 x = \beta_0 + \beta_1 \times 1 = \beta_0 + \beta_1 \text{ for } B\,. \end{cases} \tag{7.1}$$

Therefore, β_0 is the supplier A effect, and β_1 is the difference $\mu_B - \mu_A$. More generally, for m values of the discrete variable, there are $m-1$ dummy variables; the ith dummy variable's value is 1 for the $(i+1)$st value of the discrete covariate, and 0 otherwise. Also, there is an alternate set of dummy variables in which β_i can be interpreted as the difference $\mu_i - \bar{\mu}.$, where $\bar{\mu}. = \sum_{i=1}^{m} \mu_i/m$. See Wu and Hamada (2000), Sect. 1.7, for more details.

Chapters 3 and 4 introduced hierarchical models to capture more complex situations. Likewise, certain situations with covariates may require hierarchical regression models. Suppose that there are data from m nuclear power plants observed at various times. Begin modeling the data by

$$\mu_{ij} = \beta_0 + \beta_1 t_{ij}\,, \tag{7.2}$$

where $i = 1, \ldots, m$, and $j = 1, \ldots, n_i$. That is, inspect the ith plant n_i times, denoted by $t_{i1}, \ldots, t_{in_i}$. Note that Eq. 7.2 indicates a trend over time if $\beta_1 \neq 0$. Now, suppose that the m nuclear power plants are similar but not identical, where their similarity arises from the strict enforcement of standards mandated by the U.S. Nuclear Regulatory Commission. Now describe the differences between the plants by adding an ith plant effect η_i to Eq. 7.2, yielding

$$\mu_{ij} = \beta_0 + \beta_1 t_{ij} + \eta_i\,. \tag{7.3}$$

As done previously, we reflect the similarity of the plants by assuming that the η_i are conditionally independent with a common distribution, such as

$Normal(0, \sigma_\eta^2)$. Consequently, the analyst's interest is in estimating σ_η^2, because it describes the population variation of the plant effects. Sometimes a particular η_i is of interest, however. For example, we may want to know the safety of the nuclear plant next door rather than the distribution of safeties of all the nuclear plants scattered across the United States. Note that in the classical literature, the η_i are referred to as *random effects*, and Eq. 7.3, as a random effects model.

7.1.2 Covariate Relationships

In the previous section, we introduced different types of covariates and briefly discussed the relationship between covariates and the reliability data distributions. This section explores these relationships more fully.

The reliability data distributions depend on covariates through relationships with the parameters of the distribution. For example, for μ (perhaps, the mean of the distribution) and the covariates $\mathbf{x}$, $\mu = \mathbf{x}^T\boldsymbol{\beta}$ describes a linear relationship. This is the so-called linear model because of the linearity in the parameters $\boldsymbol{\beta}$.

We may express the relationship by first transforming the parameter. Letting $g(\cdot)$ be some monotonic function, then $g(\mu)$ may have a linear relationship defined by $g(\mu) = \mathbf{x}^T\boldsymbol{\beta}$. For example, for $\mu = \gamma_0 \exp(-\gamma_1 x)$, $g(\mu) = \log(\mu) = \log(\gamma_0) - \gamma_1 x = \beta_0 + \beta_1 x$. The binomial, Poisson, and Weibull regression models in subsequent sections of this chapter use such transformations.

Finally, we may specify relationships that are intrinsically nonlinear in the parameters, $\mu = h(\mathbf{x}, \boldsymbol{\beta})$, where no transformation leads to a linear relationship. Take, for example, $\mu = \beta_0 + \beta_1 \cos(x - \beta_2)$, which cannot be transformed into a linear function of the parameters.

The next three sections consider regression models for binomial success/failure data, Poisson count data, and lifetime data, respectively.

7.2 Logistic Regression Models for Binomial Data

In this section, we focus on the binomial regression model in more detail. For binomial success/failure data, the success or failure probability π may depend on covariates. If $Y \sim Binomial(n, \pi)$, then the *logistic regression model* relates π to the covariates through the logit *link function*

$$\text{logit}(\pi) = \log[\pi/(1-\pi)] = \mathbf{x}^T\boldsymbol{\beta}\,. \tag{7.4}$$

Note that a link function connects or links a distribution parameter to the covariates. A desirable feature of the logit transformation of π is that it is defined on $(-\infty, \infty)$ so that there are no restrictions on $\boldsymbol{\beta}$. Without restrictions, there is more flexibility in specifying prior distributions for $\boldsymbol{\beta}$. By inverting Eq. 7.4, an expression for the probability π is

$$\pi = \exp(\mathbf{x}^T\boldsymbol{\beta})/[1 + \exp(\mathbf{x}^T\boldsymbol{\beta})]\,. \tag{7.5}$$

Equation 7.5 has the form of the logistic cumulative distribution function, which means that there is symmetry about zero, i.e., $F(-w) = 1 - F(w)$ for all w. Link functions other than the logit function have been used primarily in medical applications. These include the probit function $\Phi^{-1}(\cdot)$, where $\Phi(\cdot)$ is the standard normal cumulative distribution function. The probit function is also symmetric about zero. Another common link function is the complementary log-log function, $\log[-\log(1-\pi)]$, which is not symmetric about zero. By inverting the complementary log-log function, we can express the success/failure probability π as the extreme value cumulative distribution function. Because the exponential and Weibull lifetime distributions are special cases of the extreme value distribution, in reliability applications, this warrants serious consideration of the complementary log-log function if we suspect an underlying exponential or Weibull distribution.

In certain situations, there may be random effects ω_i associated with the binomial data (y_i, n_i), $i = 1, \ldots, m$. For example, the ith dataset, consisting of y_i successes/failures out of n_i tests, may be collected from the ith situation. We may express the success/failure probability π using the logit link function as

$$\text{logit}(\pi) = \log[\pi/(1-\pi)] = \mathbf{x}^T\boldsymbol{\beta} + \omega_i\,, \tag{7.6}$$

where the ω_i are conditionally independent and follow some distribution, such as $Normal(0, \sigma_\omega^2)$.

In the model for binomial data, the likelihood contribution for Y_i distributed $Binomial[n_i, \pi(\mathbf{x}_i)]$ is

$$\pi(\mathbf{x}_i)^{y_i}[1 - \pi(\mathbf{x}_i)]^{n_i - y_i}\,,$$

where $\mathbf{x}_i$ is the vector of covariate values associated with (y_i, n_i).

Regarding the choice of prior distributions for the regression coefficients $\boldsymbol{\beta}$, if little is known about each of the regression coefficients, one choice is to use independent $\beta_i \sim Normal(0, 10^k)$ that are suitably diffuse prior distributions for sufficiently large values of k. If more is known about a regression coefficient, we use a normal distribution with mean possibly different from zero and a much smaller variance. When there are random effects as in Eq. 7.6 that follow a $Normal(0, \sigma_\omega^2)$ distribution, then if little is known about σ_ω^2, one choice is to use a suitably diffuse prior distribution for σ_ω^2, such as an $InverseGamma(0.001, 0.001)$ distribution. If the number of random effects is small, also consider using a $Uniform(0, U)$ (large U) distribution as a diffuse prior for σ_ω (Gelman, 2006).

Example 7.1 Logistic regression model for high-pressure coolant injection (HPCI) system demand data. The reliability of U.S. commercial nuclear power plants is an extremely important consideration in managing public health risk. The high-pressure coolant injection (HPCI) system is a frontline safety system in a boiling water reactor (BWR) that injects water into a pressurized reactor core when a small break loss-of-coolant accident occurs. Grant et al. (1999) lists 63 unplanned demands to start for the HPCI system at 23 U.S. commercial BWRs during 1987–1993. See Table 7.1, which presents these data. For these demands, all failures are counted together, including failure to start, failure to run, failure of the injection valve to reopen after operating successfully earlier in the mission, and unavailability because of maintenance. In Table 7.1, asterisks identify the 12 demands for which the HPCI system failed.

Table 7.1. Dates of unplanned HPCI system demands and failures during 1987–1993 (Grant et al., 1999). An asterisk indicates a failure

01/05/87*	08/03/87*	03/05/89	08/16/90*	08/25/91
01/07/87	08/16/87	03/25/89	08/19/90	09/11/91
01/26/87	08/29/87	08/26/89	09/02/90	12/17/91
02/18/87	01/10/88	09/03/89	09/27/90	02/02/92
02/24/87	04/30/88	11/05/89*	10/12/90	06/25/92
03/11/87*	05/27/88	11/25/89	10/17/90	08/27/92
04/03/87	08/05/88	12/20/89	11/26/90	09/30/92
04/16/87	08/25/88	01/12/90*	01/18/91*	10/15/92
04/22/87	08/26/88	01/28/90	01/25/91	11/18/92
07/23/87	09/04/88*	03/19/90*	02/27/91	04/20/93
07/26/87	11/01/88	03/19/90	04/23/91	07/30/93
07/30/87	11/16/88*	06/20/90	07/18/91*	
08/03/87*	12/17/88	07/27/90	07/31/91	

We are interested in whether there is a trend in the HPCI failure on demand probability π over time. To informally look for a trend, first use a *cumulative plot*, which graphs the cumulative number of demands versus the cumulative number of failures. See Fig. 7.1, which presents a cumulative plot of these data.

In the cumulative plot, the horizontal axis gives the number of demands that have occurred, whereas the vertical axis gives the corresponding number of failures. Consequently, the slope provides an estimate of π. A constant slope (i.e., a straight line) suggests a constant π, whereas a changing slope indicates changes in π over time. To help detect any curvature, graph a corresponding straight line on the cumulative plot with a slope equal to the average number of failures ($= 23/63$). Note in Fig. 7.1 that π appears to be relatively constant over this time period. However, the slight departure from the diagonal line in the right half of the cumulative plot suggests that π depends somewhat on time. Formally, we investigate this possible dependence by fitting a logistic regression model to the HPCI system demand data. Here, assume that for the ith demand, $\text{logit}(\pi_i) = \log[\pi_i/(1-\pi_i)] = \beta_0 + \beta_1 t_i$, where t_i denotes the

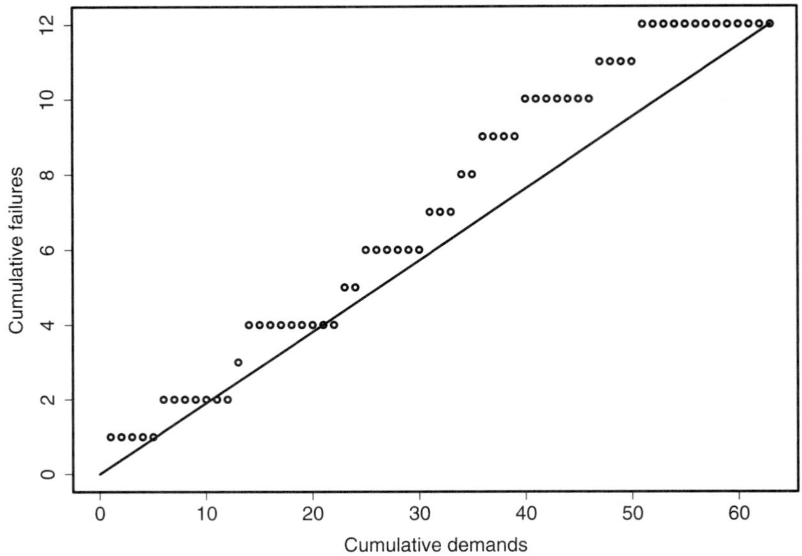

Fig. 7.1. Cumulative number of HPCI system demands versus cumulative number of failures.

number of elapsed days from a chosen reference date, such as 01/01/87, for $i = 1, \ldots, 63$.

Because each demand results in either an HPCI failure or success, we assume $Y_i \sim Bernoulli(\pi_i)$, where $y_i = 1(0)$ denotes an HPCI failure (success). Therefore, the likelihood contribution of y_i is $\pi_i^{y_i}(1-\pi_i)^{1-y_i}$, and $\text{logit}(\pi_i) = \log[\pi_i/(1-\pi_i)] = \beta_0 + \beta_1 t_i$ implies that $\pi_i = \exp(\beta_0+\beta_1 t_i)/[1+\exp(\beta_0+\beta_1 t_i)]$. We use independent and diffuse $Normal(0, 10^6)$ distributions for β_0 and β_1. We use MCMC to obtain draws from the joint posterior distribution of β_0 and β_1, as summarized in Table 7.2. The Bayesian χ^2 goodness-of-fit test suggests that the logistic regression model fits the HPCI system safety data well and is left as Exercise 7.3.

Table 7.2. Posterior distribution summaries of the HPCI system demand data model parameters

			Quantiles				
Parameter	Mean	Std Dev	0.025	0.050	0.500	0.950	0.975
β_0	−0.9713	0.5456	−2.0800	−1.8800	−0.9542	−0.1051	0.0584
β_1	−5.96E-4	5.10E-4	−16.34E-4	−14.44E-4	−5.89E-4	2.24E-4	3.81E-4

Figure 7.2 plots the posterior medians of π_i along with the corresponding 0.05 and 0.95 quantiles as a function of time t_i for $i = 1, \ldots, 63$, where solid and dashed lines connect the plotted quantities. Note the decreasing trend in the posterior median of π over the seven-year period. However, the evidence is weak that β_1 is actually nonzero, where a zero β_1 means that π does not depend on time. As seen from Fig. 7.3, which presents the marginal posterior distribution of β_1, the posterior probability that β_1 is less than zero is only 0.885.

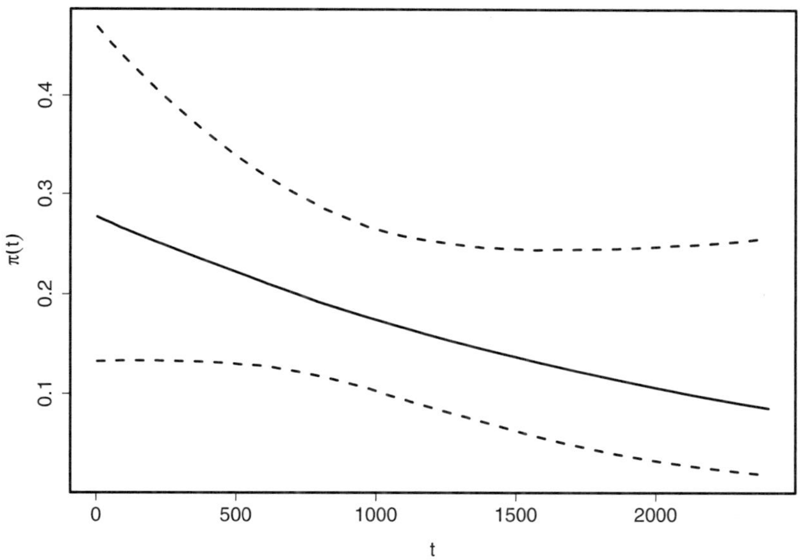

Fig. 7.2. Posterior medians (solid line) and 0.05 and 0.95 quantiles (dashed lines) of the HPCI system failure upon demand probability π over time t (in days).

In the next example with multiple units, we consider the use of a hierarchical model.

Example 7.2 Hierarchical logistic regression for emergency diesel generators (EDGs) demand data. EDGs provide backup power during external power outages at commercial nuclear power plants. To ensure safety and to control the risk of severe core damage during station blackouts, EDGs must be sufficiently reliable. Poloski and Sullivan (1980) presents EDG failure to start on demand data at U.S. commercial nuclear power plants. The weekly test data are derived from Licensee Event Reports, mandated by the U.S. Nuclear Regulatory Commission, from January 1, 1976, to December 31, 1978.

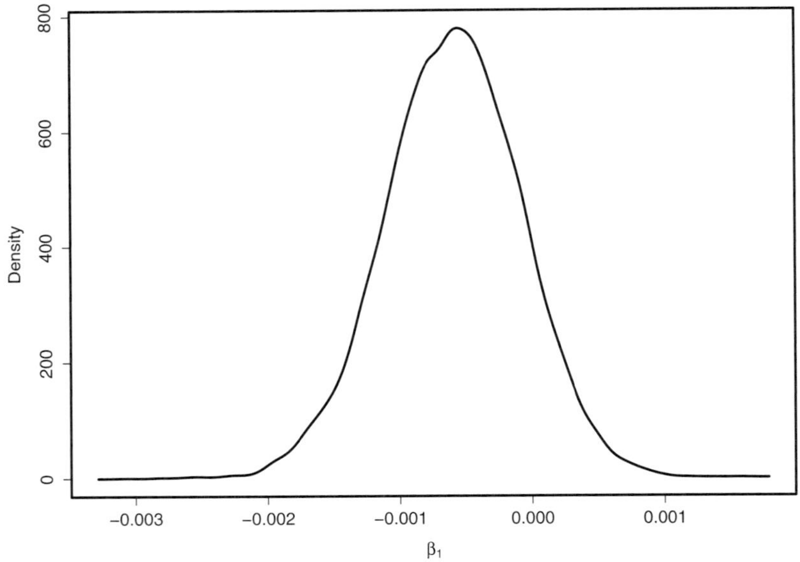

Fig. 7.3. Posterior distribution for β_1 of the HPCI system demand data model.

Table 7.3 presents the combined annual number of demands and failures for 1976–1978 by plant and (coded) nuclear steam supply system (NSSS) vendor for 58 nuclear power plants. The table also shows the date that each plant first attained criticality.

What is of interest is to determine whether the EDG probability of failure to start on demand, known as a *demand failure rate*, exhibits a time trend. If there is a time trend, we want to know if these demand failure rates differ by NSSS vendor and to quantify any differences in the EDG rates between the different U.S. commercial nuclear power plants.

We address the above questions formally by using a hierarchical logistic regression model. For the 163 demand/failure datasets in Table 7.3, assume that the number of failures $Y_i \sim Binomial(n_i, \pi_i)$, $i = 1, \ldots, 163$, where n_i denotes the number of demands for the ith dataset. Further, use the logit link function to relate the EDG demand failure rate π_i with time by

$$\text{logit}(\pi_i) = \log[\pi_i/(1-\pi_i)] = \mu + \alpha_{\text{ind}_i} + \beta t_i + \gamma_1 z_{1i} + \gamma_2 z_{2i} + \gamma_3 z_{3i}, \quad (7.7)$$

where, on the logit(π_i) scale, μ is the overall (average) effect, and βt_i represents the (linear) effect of time t_i (measured in days since the criticality date). Because the plants have all been built and operated to the same Nuclear Regulatory Commission safety standards, assume that the plant effects follow some distribution. That is, the plant effects are conditionally independent and have an assumed normal distribution. Notationally, the plant effects $\alpha_j \sim$

$Normal(0, \sigma_\alpha^2)$, $j = 1, \ldots, 58$. Because we do not know the actual dates when failures occurred, simply assume that all failures occurred on the last day of the respective year in which they were reported.

A model for the NSSS categorical vendor effects on the logit(π_i) scale in Eq. 7.7 uses three dummy variables z_1, z_2, and z_3, which have the associated regression coefficients γ_1, γ_2, and γ_3, respectively; ($z_{1i} = 0$, $z_{2i} = 0$, $z_{3i} = 0$) represents vendor A, ($z_{1i} = 1$, $z_{2i} = 0$, $z_{3i} = 0$) denotes vendor B, ($z_{1i} = 0$, $z_{2i} = 1$, $z_{3i} = 0$) represents vendor C, and ($z_{1i} = 0$, $z_{2i} = 0$, $z_{3i} = 1$) denotes vendor D. Regarding the regression coefficients, γ_1 quantifies the comparative effect of vendor B relative to vendor A (or effect B−A), γ_2 quantifies the comparative effect of vendor C relative to vendor A (or effect C−A), and γ_3 quantifies the comparative effect of vendor D relative to vendor A (or effect D−A) on the logit(π_i) scale. Consequently, $\gamma_1 - \gamma_2$ represents the comparative effect of vendor B versus vendor C (or effect B−C), $\gamma_1 - \gamma_3$ represents the comparative effect of vendor B versus vendor D (or effect B−D), and $\gamma_2 - \gamma_3$ represents the comparative effect of vendor C versus vendor D (or effect C−D) on the logit(π_i) scale. Also, in Eq. 7.7, the vector **ind** $= (1, 1, 1, 2, 2, \ldots, 58, 58, 58)$ of length 163, indicates the plant corresponding to the ith dataset, where ind_i is the ith entry of **ind**. The use of this vector correctly associates the same plant effect with all the datasets from that plant.

In the model for the EDG demand data, the likelihood contribution for y_i is $\pi_i^{y_i}(1 - \pi_i)^{n_i - y_i}$, which by simplifying Eq. 7.5, yields

$$\pi_i = \frac{\exp(\mu + \alpha_{\text{ind}_i} + \beta t_i + \gamma_1 z_{1i} + \gamma_2 z_{2i} + \gamma_3 z_{3i})}{1 + \exp(\mu + \alpha_{\text{ind}_i} + \beta t_i + \gamma_1 z_{1i} + \gamma_2 z_{2i} + \gamma_3 z_{3i})} .$$

We complete the model by choosing the following independent and diffuse prior distributions:

$$\begin{aligned}
&\sigma_\alpha^2 \sim InverseGamma(0.001, 0.001),\\
&\beta \sim Normal(0, 10^6),\\
&\mu \sim Normal(0, 10^6),\\
&\gamma_1 \sim Normal(0, 10^6),\\
&\gamma_2 \sim Normal(0, 10^6), \text{ and}\\
&\gamma_3 \sim Normal(0, 10^6).
\end{aligned}$$

Then, we analyze the EDG demand data by using MCMC to obtain draws from the joint posterior distribution of μ, β, γ_1, γ_2, γ_3, σ_α, and $\boldsymbol{\pi}$, given $\mathbf{y}$. Table 7.4 summarizes the marginal posterior distributions of all of these parameters (but only for selected π_i), as well as $\gamma_1 - \gamma_2$, $\gamma_1 - \gamma_3$, and $\gamma_2 - \gamma_3$.

The results from Table 7.4 suggest several conclusions. Because the posterior distribution of β is centered close to 0, the time since criticality has almost no effect on the demand failure rates. There also appears to be little, if any, difference in the demand failure rates between NSSS vendors, except for the vendor D plants. The fact that the posterior distribution of γ_3 is concentrated below zero (i.e., all its quantiles listed in Table 7.4 are negative) suggests that

Table 7.3. EDG failure to start and demand data during 1976–1978 (Poloski and Sullivan, 1980)

		Criticality	1976		1977		1978	
Plant	NSSS	Date	Failures	Demands	Failures	Demands	Failures	Demands
Arkansas Nuclear One 1	A	08/06/74	1	104	1	104	0	104
Crystal River 3	A	01/14/77			4	100	2	104
Davis-Besse 1	A	09/10/77			1	32	2	104
Rancho Seco	A	09/16/74	2	104	1	104	0	104
Three Mile Island 1	A	06/05/74	1	104	0	104	2	104
Three Mile Island 2	A	03/28/78					2	80
Arkansas Nuclear One 2	B	12/05/78					0	8
Calvert Cliffs 1	B	10/07/74	3	104	2	104	2	104
Calvert Cliffs 2	B	11/30/76	1	8	3	104	0	104
Fort Calhoun	B	08/06/73	3	104	2	104	1	104
Millstone 2	B	10/17/75	4	104	2	104	0	104
Maine Yankee	B	10/23/72	0	104	0	104	2	104
Palisades	B	05/24/71	0	104	0	104	0	104
St. Lucie	B	04/22/76	1	72	3	104	1	104
Browns Ferry 1	C	08/17/73	1	208	1	208	0	208
Big Rock Point	C	09/27/62	11	52	7	52	1	52
Brunswick 2	C	03/20/75	3	208	3	208	1	208
Cooper Station	C	02/21/74	1	104	0	104	1	104
Duane Arnold	C	03/23/74	1	104	1	104	1	104
Dresden 1	C	10/15/59	0	12	0	52	5	52
Dresden 2	C	01/07/70	2	104	7	104	7	104
Dresden 3	C	01/31/71	2	104	2	104	0	104
Edwin I. Hatch 1	C	09/12/74	6	156	2	156	0	156
James A. Fitzpatrick	C	11/17/74	2	208	2	208	1	208
Millstone 1	C	10/26/70	0	52	0	52	0	52
Monticello	C	12/10/70	1	104	0	104	0	104
Nine Mile Point 1	C	09/05/69	0	104	0	104	0	104
Oyster Creek 1	C	05/03/69	2	104	0	104	1	104
Peach Bottom 2	C	09/16/73	0	208	4	208	3	208
Pilgram 1	C	06/16/72	0	104	0	104	0	104

Table 7.3. (cont.)

Plant	NSSS	Criticality Date	1976 Failures	1976 Demands	1977 Failures	1977 Demands	1978 Failures	1978 Demands
Quad-Cities 1	C	10/18/71	1	104	3	104	0	104
Quad-Cities 2	C	04/26/72	0	104	0	104	0	104
Vermont Yankee	C	03/24/72	1	104	1	104	0	104
Beaver Valley 1	D	05/10/76	2	66	7	104	4	104
Donald C. Cook 1	D	01/18/75	1	104	0	104	0	104
Donald C. Cook 2	D	03/10/78					2	84
Haddam Neck	D	07/24/67	1	104	0	104	0	104
Indian Point 2	D	05/22/73	0	156	0	156	0	156
Indian Point 3	D	04/06/76	0	114	0	156	0	156
Joseph M. Farley 1	D	08/09/77			3	100	8	260
Kewaunee	D	03/07/74	0	104	1	104	0	104
North Anna 1	D	04/05/78					0	76
Prairie Island 1	D	12/01/73	0	104	0	104	0	104
Prairie Island 2	D	12/17/74	0	104	0	104	0	104
Point Beach 1	D	11/02/70	0	104	2	104	1	104
Point Beach 2	D	05/30/72	0	104	0	104	0	104
Robert E. Ginna	D	11/08/69	0	104	0	104	1	104
H. B. Robinson 2	D	09/20/70	0	104	0	104	0	104
Salem 1	D	12/11/76	0	9	2	156	0	156
San Onofre 1	D	06/14/67	0	104	0	104	2	104
Surry 1	D	07/01/72	1	104	0	104	0	104
Surry 2	D	03/07/73	0	104	0	104	0	104
Trojan	D	12/15/75	0	104	1	104	0	104
Turkey Point 3	D	10/20/72	0	104	2	104	0	104
Turkey Point 4	D	06/11/73	0	104	0	104	0	104
Yankee Rowe	D	08/19/60	0	156	1	156	0	156
Zion 1	D	06/19/73	1	156	0	156	4	156
Zion 2	D	12/24/73	0	156	0	156	0	156

Table 7.4. Posterior distribution summaries of EDG demand data model parameters

Parameter	Mean	Std Dev	Quantiles 0.025	0.050	0.500	0.950	0.975
μ	−4.6460	0.4699	−5.4640	−5.3390	−4.6820	−0.8240	−3.6550
β	−1.072E-4	2.132E-4	−3.655E-4	−3.169E-4	−8.872E-5	1.002E-4	1.352E-4
γ_1	0.0697	0.4552	−0.9390	−0.7062	0.0586	0.8216	1.0090
γ_2	0.0099	0.4444	−1.0410	−0.7985	0.0373	0.6795	0.8362
γ_3	−0.7429	0.6000	−2.0810	−1.8600	−0.6555	0.0177	0.0662
$\gamma_1 - \gamma_2$	0.0598	0.4595	−0.8725	−0.6836	0.0427	0.8417	1.0420
$\gamma_1 - \gamma_3$	0.8126	0.6016	−0.1097	−0.0371	0.7785	1.8760	2.0740
$\gamma_2 - \gamma_3$	0.7528	0.5033	−0.0587	−0.0078	0.7465	1.6150	1.7930
σ_α	1.2040	0.3157	0.8546	0.8996	1.1670	1.5510	1.6500
π_1	0.0075	0.0045	0.0016	0.0021	0.0066	0.0160	0.0186
π_2	0.0072	0.0043	0.0015	0.0020	0.0063	0.0154	0.0179
π_3	0.0070	0.0041	0.0014	0.0019	0.0061	0.0149	0.0174
π_4	0.0263	0.0108	0.0099	0.0115	0.0247	0.0460	0.0515
π_5	0.0253	0.0103	0.0095	0.0111	0.0239	0.0443	0.0492
$\vdots$	$\vdots$	$\vdots$	$\vdots$	$\vdots$	$\vdots$	$\vdots$	$\vdots$
π_{161}	0.0018	0.0015	0.0002	0.0002	0.0014	0.0049	0.0060
π_{162}	0.0018	0.0015	0.0002	0.0002	0.0013	0.0047	0.0058
π_{163}	0.0017	0.0015	0.0001	0.0002	0.0013	0.0046	0.0056
π_{1avg}	0.0072	0.0043	0.0015	0.0020	0.0063	0.0154	0.0180
π_{2avg}	0.0258	0.0105	0.0097	0.0113	0.0243	0.0450	0.0502
$\vdots$	$\vdots$	$\vdots$	$\vdots$	$\vdots$	$\vdots$	$\vdots$	$\vdots$
π_{58avg}	0.0018	0.0015	0.0002	0.0002	0.0013	0.0047	0.0058

vendor D plants have smaller EDG demand failure rates than those of vendor A plants.

Similarly, both $\gamma_1 - \gamma_3$ and $\gamma_2 - \gamma_3$ are concentrated above zero and suggest that vendor D plants have smaller EDG demand failure rates than those of either vendor B or vendor C plants. The data also directly support these results, as shown in Table 7.5, which presents the average failure rate estimates aggregated by NSSS vendor. Note that vendor D plants have EDG demand failure rates that are roughly one-half as large as those of the other vendors.

Note that the concentration of the σ_α posterior distribution is away from zero, which clearly indicates a plant effect. In other words, different plants have different EDG demand failure rates.

Finally, while there is no apparent time effect, but a significant plant effect on the demand failure rate, we obtain an average demand failure rate for each plant in Table 7.3 by averaging the individual demand failure rates associated with each plant. For example, π_1, π_2, and π_3 all correspond to Arkansas Nuclear One 1 (plant number 1); obtain draws from the posterior distribution

Table 7.5. Average EDG demand failure rate estimates by NSSS vendor for EDG reliability example

NSSS Vendor	Total Failures	Total Demands	Average Failure Rate
A	19	1356	0.0140
B	30	2064	0.0145
C	88	6824	0.0129
D	47	8093	0.0058

of the average demand failure rate for this plant by averaging draws from the joint posterior distribution of π_1, π_2, and π_3, i.e., $\pi_{1avg} = (\pi_1 + \pi_2 + \pi_3)/3$. See Table 7.4, which presents a summary of the π_{1avg} posterior distribution. Similarly, $\pi_{2avg} = (\pi_4 + \pi_5)/2$, and $\pi_{58avg} = (\pi_{161} + \pi_{162} + \pi_{163})/3$, which have posterior distributions as summarized in Table 7.4. We can now report the corresponding means of these "average" posterior distributions as estimates of the EDG demand failure rates for Arkansas Nuclear One 1, Crystal River 3, and Zion 2 plants, respectively.

7.3 Poisson Regression Models for Count Data

In this section, we focus on the *Poisson regression model* in more detail. For Poisson distributed counts, the mean number of counts λ may depend on covariates. The loglinear model is a regression model that incorporates covariates as follows. For Poisson counts, where $Y \sim Poisson(\lambda)$, the *loglinear model* connects λ and the covariates by

$$\log(\lambda) = \mathbf{x}^T \boldsymbol{\beta}\,, \tag{7.8}$$

where $\boldsymbol{\beta}$ is the vector of regression coefficients. Because $\log(\lambda)$ is defined on $(-\infty, \infty)$, $\boldsymbol{\beta}$ has no restrictions, which allows more flexibility in specifying prior distributions for $\boldsymbol{\beta}$. By inverting Eq. 7.8, an expression for the mean number of counts is

$$\lambda = \exp(\mathbf{x}^T \boldsymbol{\beta}). \tag{7.9}$$

The likelihood contribution for y_i is

$$\lambda_i^{y_i} \exp(-\lambda_i)/y_i!\,,$$

where λ_i is obtained by evaluating Eq. 7.9 with $\mathbf{x}_i$, the values of the covariates $\mathbf{x}$ associated with y_i, i.e., $\lambda_i = \exp(\mathbf{x}_i^T \boldsymbol{\beta})$.

As discussed in Sect. 7.1.1, we can also incorporate random effects in a regression model. Similar to the logistic regression model for binomial data (as in Example 7.2), the Poisson regression model with random effects takes the form $\log(\lambda_i) = \mathbf{x_i}^T \boldsymbol{\beta} + \omega_i$, where the random effects ω_i are conditionally independent and follow a $Normal(0, \sigma_\omega^2)$ distribution.

Example 7.3 Hierarchical Poisson regression for nuclear power plant scram rate data. The reactor protection system is an important frontline safety system in a nuclear power plant. When a transient event occurs, such as a loss of off-site power, the reactor protection system, also called the scram system, rapidly changes the reactor from a critical to a noncritical status. The rate at which unplanned scrams occur is an important consideration in assessing overall plant reliability. Martz et al. (1999) presents unplanned scram rate data for 66 U.S. commercial nuclear power plants during 1984–1993, which Table 7.6 displays. The data consist of the annual number of unplanned scrams y_{ij} in T_{ij} total critical operating hours for the ith plant ($i = 1, \ldots, 66$) and jth coded year ($j = 1, \ldots, 10$). The 66 nuclear plants are believed similar, but not identical, and we incorporate their similarity by a hierarchical model. Using these data, estimates of trends in the scram rate at each plant over this 10-year period and comparisons to the overall population trend are of interest.

In modeling the scram rate data, we assume that given the true unknown scram rate λ_{ij}, $y_{ij} \sim Poisson(\lambda_{ij}T_{ij}/1000)$. Note that λ_{ij} has the interpretation as the scram rate (or mean number of scrams) per 1,000 critical operating hours. To graphically assess a trend, calculate an estimate (maximum likelihood estimate (MLE)) for each of the plants by year using $\hat{\lambda}_{ij} = y_{ij}/(T_{ij}/1000)$ and plot them. Figure 7.4 graphs the logged estimates as well as the yearly average, which appears as a solid line. Note the decreasing trend in the logged average scram rate over time, which is approximately linear over time. This pattern in Fig. 7.4 suggests the following loglinear model for λ_{ij}:

$$\log(\lambda_{ij}) = \beta_0 + \beta_1 t_j + \omega_i \,, \tag{7.10}$$

where coded year $t_j = \text{year} - 1983$ and ω_i is the ith plant effect. On the $\log(\lambda_{ij})$ scale, β_0 denotes the overall effect, β_1 represents the decrease each year, and ω_i denotes a plant effect, which are assumed conditionally independent and follow a $Normal(0, \sigma_\omega^2)$ distribution.

In the model for the scram rate data, the contribution of y_{ij} to the likelihood function is

$$\lambda^{y_{ij}} \exp(-\lambda_i T_{ij}/1000)/y_{ij}! \,.$$

To complete the model, we use the following independent and diffuse prior distributions:

$$\begin{aligned} \sigma_\omega^2 &\sim InverseGamma(0.001, 0.001), \\ \beta_0 &\sim Normal(0, 10^6), \text{ and} \\ \beta_1 &\sim Normal(0, 10^6). \end{aligned}$$

To analyze the scram rate data, we use MCMC to obtain draws from the joint posterior distribution of β_0, β_1, and σ_ω (as well as for the ω_i) given all the data denoted by $\mathbf{y}$. Table 7.7 summarizes the marginal posterior distributions of these parameters. By inverting Eq. 7.10, an expression for λ_{ij} is

$$\lambda_{ij} = \exp(\beta_0 + \beta_1 j + \omega_i) \,, \tag{7.11}$$

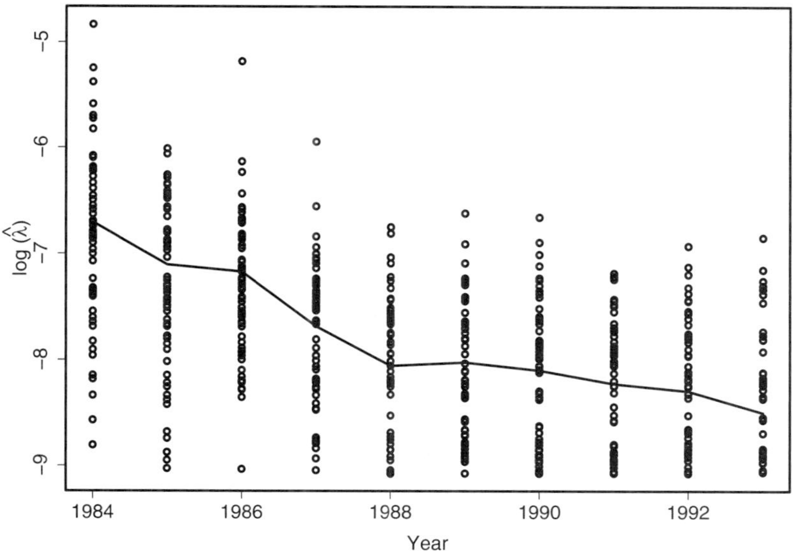

Fig. 7.4. Logged estimates (MLEs) of the scram rate per 1,000 critical operating hours λ over time (year). The solid line is the yearly averages of the scram rate per 1,000 critical operating hours.

used to evaluate the posterior distribution of scram rates for a specified plant and year. That is, for the ith plant and jth year, we obtain draws from the posterior distribution of λ_{ij} by evaluating Eq. 7.11 with the $(\beta_0, \beta_1, \omega_i)$ joint posterior draws. Table 7.7 summarizes the posterior distributions of the Arkansas 1 and Arkansas 2 scram rates for each of the 10 years ($i = 1, 2$ correspond to these two plants, and $j = 1, \ldots, 10$ correspond to these years).

In Table 7.7, note that the β_1 posterior draws are mostly negative (i.e., all the listed posterior quantiles are negative). Consequently, there is a decreasing trend in the scram rate over time. Because the σ_ω posterior is concentrated away from zero, there is a significant plant effect on the scram rate. In other words, different plants have different scram rates. Figure 7.5 plots the posterior means of λ_{1j}, $j = 1, \ldots, 10$, from Table 7.7, along with the posterior 0.05 and 0.95 quantiles for Arkansas 1. We can produce plots for the other plants as we did for Arkansas 1 using the joint posterior distribution of β_0, β_1, and $\boldsymbol{\omega}$, the vector of plant effects.

Now consider the population of plants and suppose that including the variability in the scram rates over this population is of interest. In other words, we are interested in estimating the population scram rate over time. To do this, use the joint posterior distribution draws on β_0, β_1, and σ_ω as follows. For each joint posterior draw of β_0, β_1, and σ_ω, first draw a corresponding

Table 7.6. U.S. commercial nuclear power plant scram rate data (number of scrams y in T total critical operating hours) from 1984–1993 (Martz et al., 1999)

Plant	1984		1985		1986		1987		1988		1989		1990		1991		1992		1993	
	y	T	y	T	y	T	y	T	y	T	y	T	y	T	y	T	y	T	y	T
Arkansas 1	3	6250.0	8	7017.5	2	5536.7	2	7855.7	1	6156.6	5	5999.1	0	6500.2	2	8149.8	0	7137.8	1	7599.4
Arkansas 2	12	7643.3	9	6383.0	5	6370.0	2	7715.4	2	6032.0	2	6610.1	4	8246.6	1	7341.1	0	6454.2	0	8390.4
Beaver Valley 1	4	6451.6	8	8247.4	3	6243.8	4	7339.4	3	7066.7	4	5887.6	1	8155.9	2	5029.2	1	8226.7	1	5980.6
Big Rock Point	2	6896.6	3	6521.7	1	8387.3	1	6215.5	3	6394.2	1	6920.8	0	6759.0	0	7460.5	3	4790.5	0	6958.8
Brunswick 2	3	2654.9	3	7142.9	2	4232.4	2	8328.5	2	5645.8	1	5779.9	6	5926.6	2	5236.2	1	2378.3	0	5915.3
Callaway	12	1503.8	19	8154.5	7	7307.6	1	6227.7	6	8202.1	2	7481.6	4	7365.0	1	8734.1	3	7289.2	0	7569.0
Calvert Cliffs 1	5	7575.8	6	5357.1	4	6906.2	6	6615.6	3	6398.5	0	1806.6	0	1924.5	1	6687.0	1	5050.2	2	8619.0
Cook 1	3	8108.1	1	2564.1	5	7536.4	2	6012.2	3	8433.8	2	6169.8	0	6944.8	1	7754.3	1	5752.1	0	8760.0
Cook 2	7	5303.0	4	5970.1	2	5560.5	5	6290.3	0	2715.5	1	6580.9	3	4958.9	3	8053.2	1	3169.4	2	8491.5
Cooper Station	3	6000.0	1	2040.8	2	6570.1	6	8424.2	3	5967.9	3	6672.9	1	6953.3	0	6898.8	0	8466.7	1	5146.8
Crystal River 3	2	8333.3	7	4375.0	1	3691.1	2	5333.6	2	7457.3	1	4274.4	0	5591.1	4	7187.2	2	6684.2	1	7445.8
Davis-Besse	4	5555.6	5	2840.9	1	178.0	5	7425.7	1	2126.7	2	8547.1	2	4966.6	0	7054.6	1	8759.2	2	7305.4
Diablo Canyon 1	5	1084.6	9	6521.7	3	5967.4	5	8475.7	5	5682.3	1	7189.1	4	8504.3	4	7197.4	2	7297.6	1	8631.1
Dresden 2	3	6521.7	7	4964.5	6	7110.1	4	5763.7	0	6974.7	2	7252.5	3	5958.8	4	5279.9	0	7553.4	0	4886.7
Dresden 3	8	3883.5	4	6666.7	6	2766.4	6	7208.7	1	6346.3	3	7311.6	1	7453.4	1	5356.0	2	5689.3	3	7116.7
Duane Arnold	6	6593.4	0	4733.2	2	7348.2	0	5668.3	1	6609.9	5	6921.1	6	6641.2	3	8277.5	2	7192.9	1	6963.4
Farley 1	2	6896.6	4	7547.2	4	7276.4	4	8307.2	1	7428.3	1	7613.4	1	8695.9	5	6987.0	1	7210.4	0	8542.6
Farley 2	6	8333.3	5	6849.3	4	7549.7	2	6537.7	0	8784.0	6	7205.2	1	6501.1	4	8480.1	7	7157.6	1	6931.8
Fort Calhoun	1	5263.2	0	6466.1	2	8485.2	0	6608.3	0	6510.0	1	7816.5	1	5622.4	0	8030.0	3	5791.6	2	7081.4
Ginna	1	6666.7	8	7843.1	4	7716.3	0	8014.5	2	7679.2	1	6648.5	6	7393.2	0	7591.6	2	7633.7	2	7561.8
Grand Gulf	7	2089.6	14	5714.3	6	5624.6	2	7203.3	6	8498.1	5	7005.5	5	6911.1	6	8230.3	3	7349.0	1	7140.5
Haddam Neck	3	6521.7	5	8620.7	7	5060.9	1	4728.9	3	6177.0	0	5883.3	2	2824.5	0	6693.2	1	7039.6	2	7145.9
Hatch 1	7	5645.2	7	6930.7	4	5521.2	4	7191.7	5	6008.8	0	8760.0	4	5939.6	5	6790.3	4	8566.3	5	7099.4
Hatch 2	7	3111.1	5	7352.9	7	6451.9	4	8519.6	7	6359.2	1	6495.8	2	8684.7	2	6778.8	3	7004.9	1	7873.9

Table 7.6. (cont.)

Plant	1984		1985		1986		1987		1988		1989		1990		1991		1992		1993	
	y	T	y	T	y	T	y	T	y	T	y	T	y	T	y	T	y	T	y	T
Indian Point 2	4	4705.9	11	8527.1	10	5101.9	2	6347.3	4	7491.8	2	5644.2	0	5837.0	2	4762.7	3	8625.4	0	6630.7
Indian Point 3	7	6930.7	9	5882.4	8	6581.6	5	5496.5	4	7312.7	1	5352.0	2	5511.3	2	7668.5	2	5397.0	0	1303.5
Kewaunee	4	7547.2	4	7272.7	3	7584.3	2	7860.9	3	7755.6	1	7436.2	0	7700.5	1	7306.0	2	7726.0	2	7607.6
LaSalle 1	9	6293.7	9	5769.2	1	2395.7	6	5609.1	0	5931.1	1	6114.8	2	8475.3	1	6747.1	1	6568.3	3	7402.3
LaSalle 2	11	5472.6	1	3846.2	4	6614.0	1	4781.4	2	6648.2	1	6693.0	2	6343.1	3	8445.6	3	6077.7	0	5912.2
Maine Yankee	7	6666.7	8	7017.5	6	7791.0	2	5724.4	3	6949.7	2	8210.0	0	6215.9	4	7585.4	1	6950.9	0	6991.8
McGuire 1	4	6060.6	5	6849.3	3	5022.2	4	6835.7	4	6783.8	2	7210.8	3	4807.9	2	6327.6	2	6862.8	1	5164.3
McGuire 2	16	6986.9	9	5487.8	6	5770.4	5	7046.9	2	7313.5	3	6943.4	1	5937.3	3	8561.3	5	6214.9	4	6425.7
Millstone 1	0	6990.2	3	7317.1	3	8276.5	4	6970.7	1	8661.6	3	7377.3	2	8021.0	1	3099.9	1	5983.6	1	8481.2
Millstone 2	3	8571.4	1	4545.5	4	6599.6	5	8242.0	1	6953.1	0	6027.7	2	6551.5	3	5141.0	0	3204.0	5	7689.9
Monticello	0	810.6	3	8108.1	2	6984.9	4	7173.6	1	8768.7	2	6679.1	1	8487.3	4	7075.6	0	8566.3	3	7391.0
North Anna 1	8	4761.9	2	6896.6	6	7560.0	4	4585.4	4	8019.5	3	5023.1	1	8748.4	1	6697.6	0	7242.3	0	6474.9
North Anna 2	4	6153.8	2	8695.7	4	7301.3	0	6842.2	0	8734.9	0	6918.9	1	7012.2	1	8601.6	2	7308.2	2	7329.4
Oconee 1	3	7500.0	4	8510.6	2	5948.7	0	6913.9	1	8769.0	3	7371.0	1	7774.7	2	7287.5	3	7586.1	2	7928.0
Oconee 2	0	8784.0	4	6779.7	5	7253.7	3	8604.9	1	6989.2	3	7385.8	0	7505.7	0	8760.0	1	7229.3	2	7422.5
Oconee 3	4	6557.4	2	6060.6	2	7835.4	0	6142.2	2	7229.7	2	7682.9	3	8730.6	4	6740.6	4	6803.2	1	8655.4
Oyster Creek	2	1694.9	6	6818.2	3	2389.1	3	5620.0	0	5789.0	5	5015.2	3	7804.6	1	5297.6	4	7545.8	0	7690.6
Palisades	1	1562.5	2	7407.4	2	1490.5	6	4226.6	0	4990.4	1	6050.6	2	5143.1	3	6845.5	5	6686.0	0	4707.4
Point Beach 1	0	6420.1	1	7142.9	2	7905.4	2	7389.4	0	7847.7	0	7728.3	0	7423.8	2	7622.9	1	7492.8	0	7835.6
Point Beach 2	0	7544.2	1	7692.3	2	7262.7	1	7583.1	1	7707.8	2	7243.6	0	7738.8	1	7645.2	0	7546.1	1	7924.7
Prairie Island 1	4	8333.3	3	7317.1	2	7898.4	2	7287.6	0	7835.6	1	8740.7	1	7840.4	1	7988.3	0	6850.8	1	8507.9
Prairie Island 2	0	7844.0	0	7408.6	3	7972.1	0	8760.0	0	7813.9	3	7852.4	5	7785.7	0	8760.0	0	6538.2	0	7381.4
Quad Cities 1	3	4761.9	2	8333.3	4	6151.3	1	6251.6	1	8477.9	3	6621.4	1	7318.1	1	5032.2	1	6249.8	1	7020.4
Quad Cities 2	2	6896.6	4	6349.2	2	5728.0	5	6941.4	3	6292.8	2	8434.7	2	6304.6	1	7794.5	0	5692.6	5	4725.8
Robinson 2	0	616.1	12	7843.1	10	7118.8	4	6354.3	3	5791.5	3	4262.0	2	5674.7	1	7131.0	1	5867.4	0	6191.2

Table 7.6. (cont.)

Plant	1984		1985		1986		1987		1988		1989		1990		1991		1992		1993	
	y	T	y	T	y	T	y	T	y	T	y	T	y	T	y	T	y	T	y	T
Salem 1	10	2673.8	1	8333.3	9	7097.2	1	6412.5	3	6937.1	3	6276.4	3	6055.0	1	6636.8	0	5581.8	4	5949.9
Salem 2	10	3389.8	10	5235.6	9	5629.3	4	6423.0	7	5992.9	4	7650.0	2	5350.5	1	7259.9	3	5149.4	2	5513.9
San Onofre 2	5	5263.2	10	5235.6	8	6480.0	3	6192.6	0	8286.3	1	5227.0	1	7692.8	2	5732.7	2	8242.0	0	7280.2
San Onofre 3	7	5072.5	5	4807.7	6	7422.3	2	7135.2	1	5930.8	2	8251.6	1	6297.7	1	8270.3	1	6701.5	2	6726.6
St. Lucie 1	6	5555.6	1	7142.9	4	8424.0	6	6971.6	3	7554.3	2	8290.1	1	5569.7	3	7151.0	1	8561.0	3	6859.5
St. Lucie 2	9	7377.0	7	7446.8	5	7326.7	5	7382.3	0	8784.0	2	6626.9	1	6691.4	0	8760.0	4	6039.9	2	6759.3
Summer	11	5555.6	12	6451.6	6	8453.2	4	6222.4	4	6067.7	5	7276.2	0	7346.3	0	7265.5	2	8553.1	1	7357.9
Surry 1	8	5298.0	7	7954.5	5	6233.2	3	6178.3	2	3755.2	2	4272.2	2	6723.4	0	8760.0	2	7140.8	2	8432.2
Surry 2	14	7446.8	1	5882.4	4	6171.1	1	6555.2	3	5028.3	2	1504.3	3	7973.7	2	6035.8	0	8478.8	5	6389.4
Susquehanna 1	7	6542.1	4	5633.8	0	6196.3	1	6464.6	2	8289.7	4	6592.5	0	6769.1	1	8622.5	1	6747.2	1	5275.4
Susquehanna 2	7	2147.2	5	7692.3	2	5946.6	1	8484.0	0	6156.9	0	6916.4	2	8197.5	1	7119.1	1	7255.8	0	8275.5
Turkey Point 3	8	7339.4	6	5405.4	6	6988.4	5	1909.7	0	5408.1	1	5806.6	2	5283.7	1	2252.1	0	6034.2	0	8501.0
Turkey Point 4	9	5084.7	9	7894.7	3	3048.1	1	4503.2	1	5050.1	2	4147.1	3	6802.7	0	1426.3	1	7226.1	2	7441.7
Vermont Yankee	2	7142.9	1	6250.0	2	4359.6	3	7374.6	3	8404.4	0	7416.2	3	7522.8	3	8265.0	1	7742.8	0	7021.0
Washington Nuclear 2	23	4364.3	12	6896.6	7	6391.5	6	6199.4	2	6310.9	4	6857.8	2	5908.9	2	4406.5	3	5758.0	4	6961.5
Zion 1	6	6315.8	3	5357.1	2	5491.0	1	6877.3	4	6747.9	1	5268.3	2	5097.0	1	4652.6	0	4605.3	1	6987.6
Zion 2	7	6306.3	1	5882.4	4	7783.5	0	5569.7	2	7004.6	0	8333.9	4	3122.7	2	5544.4	0	5758.7	0	5427.4

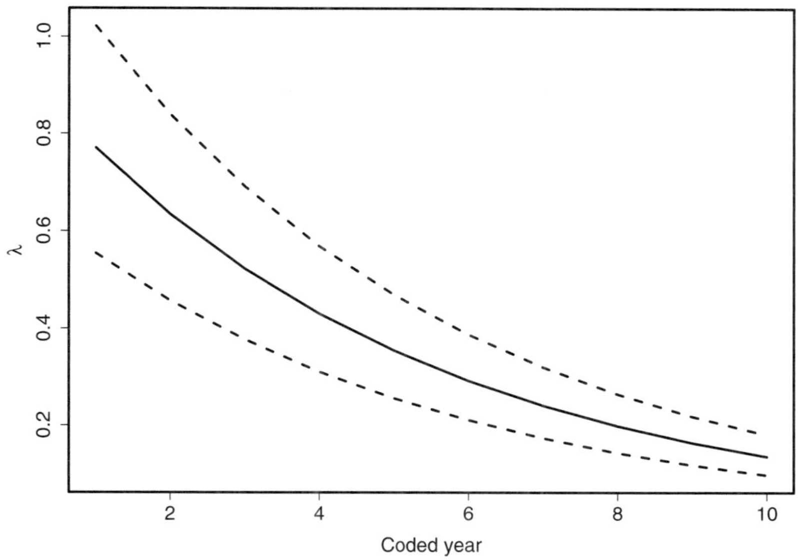

Fig. 7.5. Posterior means (solid line) and 0.05 and 0.95 quantiles (dashed lines) on the scram rate per 1,000 critical operating hours λ over time (coded year) for Arkansas 1.

plant effect using $\omega \sim Normal(0, \sigma_\omega^2)$, and then evaluate $\exp(\beta_0 + \beta_1 j + \omega)$ to obtain a posterior draw from the population scram rate λ. See Fig. 7.6, which plots the posterior means along with the 0.05 and 0.95 quantiles of the population scram rate versus time. The 0.05 and 0.95 quantiles are clearly wider apart than those for an individual plant, which reflects the population variability in the scram rate over time. Another interpretation of Fig. 7.6 is that it displays the predictive mean and 90% credible interval for a randomly chosen plant, which has a plant effect $\omega \sim Normal(0, \sigma_\omega^2)$. For example, we can use Fig. 7.6 to predict the performance over time of a newly built plant, believed a member of the same population of 66 plants listed in Table 7.6.

7.4 Regression Models for Lifetime Data

The two preceding sections presented regression models for count data. This section focuses on regression models for lifetime data.

It is always advantageous to fit the lifetime data using a model dictated by the science or engineering of the problem (e.g., the accelerated life testing models used in Sect. 7.7). When such theoretical models are not available, there are several models to choose from that have proven useful in practice.

Table 7.7. Posterior distribution summaries for scram rate data model parameters

			Quantiles				
Parameter	Mean	Std Dev	0.025	0.050	0.500	0.950	0.975
β_0	0.0486	0.0759	−0.0832	−0.0616	0.0472	0.1574	0.1775
β_1	−0.1943	0.0105	−0.2116	−0.2088	−0.1942	−0.1799	−0.1766
σ_ω	0.4202	0.0492	0.3362	0.3481	0.4168	0.5017	0.5219
$\lambda_{1,1}$	0.7707	0.1428	0.5213	0.5543	0.7599	1.0220	1.0820
$\lambda_{1,2}$	0.6344	0.1165	0.4301	0.4571	0.6259	0.8393	0.8887
$\lambda_{1,3}$	0.5224	0.0955	0.3549	0.3771	0.5156	0.6911	0.7316
$\lambda_{1,4}$	0.4301	0.0786	0.2921	0.3104	0.4247	0.5686	0.6023
$\lambda_{1,5}$	0.3542	0.0648	0.2400	0.2554	0.3495	0.4688	0.4961
$\lambda_{1,6}$	0.2917	0.0536	0.1973	0.2104	0.2878	0.3870	0.4091
$\lambda_{1,7}$	0.2403	0.0445	0.1624	0.1730	0.2370	0.3197	0.3368
$\lambda_{1,8}$	0.1979	0.0370	0.1334	0.1419	0.1952	0.2638	0.2781
$\lambda_{1,9}$	0.1630	0.0308	0.1095	0.1166	0.1608	0.2176	0.2297
$\lambda_{1,10}$	0.1343	0.0257	0.0898	0.0958	0.1324	0.1796	0.1897
$\lambda_{2,1}$	1.0530	0.1656	0.7582	0.8007	1.0440	1.3400	1.3990
$\lambda_{2,2}$	0.8667	0.1348	0.6274	0.6605	0.8597	1.1020	1.1500
$\lambda_{2,3}$	0.7136	0.1104	0.5177	0.5442	0.7076	0.9060	0.9455
$\lambda_{2,4}$	0.5876	0.0908	0.4262	0.4481	0.5834	0.7462	0.7778
$\lambda_{2,5}$	0.4839	0.0749	0.3509	0.3689	0.4804	0.6139	0.6409
$\lambda_{2,6}$	0.3985	0.0621	0.2882	0.3035	0.3954	0.5060	0.5287
$\lambda_{2,7}$	0.3282	0.0516	0.2368	0.2494	0.3254	0.4179	0.4381
$\lambda_{2,8}$	0.2704	0.0430	0.1940	0.2047	0.2677	0.3454	0.3612
$\lambda_{2,9}$	0.2227	0.0359	0.1589	0.1679	0.2203	0.2858	0.2994
$\lambda_{2,10}$	0.1835	0.0301	0.1301	0.1376	0.1814	0.2359	0.2482

Consider the following regression model often used for lognormal lifetimes:

$$Y \sim LogNormal[\mu(\mathbf{x}), \sigma^2], \text{ and } \mu = \mathbf{x}^T \boldsymbol{\beta}.$$

From $\log(Y) \sim Normal(\mu, \sigma^2)$, another way to express the *lognormal regression model* is

$$\log(Y) = \mu + \sigma\varepsilon \quad \text{and } \varepsilon \sim Normal(0, 1)\,. \tag{7.12}$$

As seen in Eq. 7.12, this model is a location-scale model (for the logged lifetimes); that is, to the location μ, there is a scaled random variable added, where the random variable ε is scaled by σ.

If, instead in Eq. 7.12, $\varepsilon \sim ExtremeValue(0, 1)$, the standard extreme distribution, then $Y \sim Weibull(\lambda, \gamma)$, where $\mu = -\log(\lambda)$, and $\sigma = 1/\gamma$; here, the Weibull probability density function has the form

$$f(t|\lambda, \gamma) = \lambda\gamma(\lambda t)^{\gamma-1} \exp[-(\lambda t)^{\gamma}]\,.$$

A more direct way to state the *Weibull regression model* is to let

$$Y \sim Weibull(\lambda, \gamma) \quad \text{and} \log(\lambda) = \mathbf{x}^T \boldsymbol{\beta}\,.$$

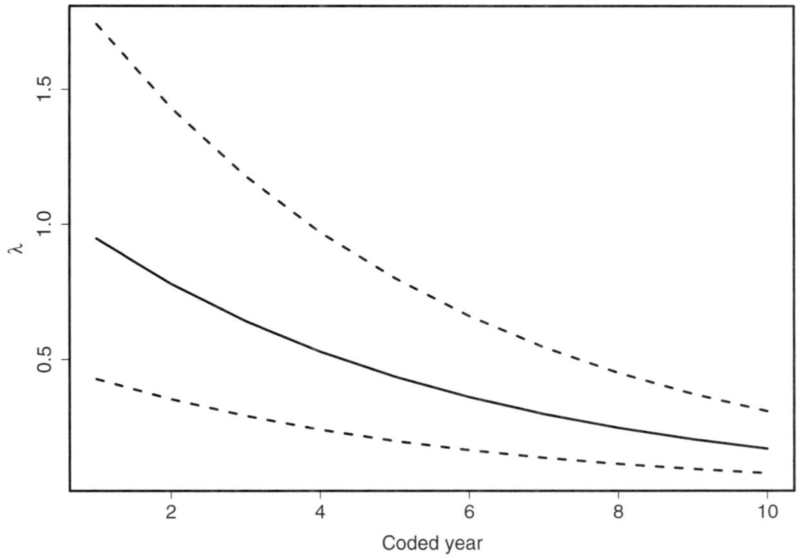

Fig. 7.6. Posterior means (solid line) and 0.05 and 0.95 quantiles (dashed lines) of the population scram rate per 1,000 critical operating hours λ over time (coded year) for scram system example.

Typically, the Weibull shape is constant because a change in shape suggests a switch to a different regime, which indicates a different failure mechanism. Example 7.4 illustrates how both the scale and shape parameters can depend on the covariates, however. Recall that the exponential distribution is a special case of the Weibull distribution with shape parameter $\gamma = 1$, so that the *exponential regression model* is a special case of the Weibull regression model.

The Weibull regression model has two interesting properties. First, it exhibits the proportional hazards property, where the ratio of the hazard associated with covariate values $\mathbf{x}$ to the hazard associated with covariate values $\mathbf{x}_0$ has the following form:

$$h(t|\mathbf{x})/h(t|\mathbf{x}_0) = g(\mathbf{x}), \tag{7.13}$$

for some positive function $g(\cdot)$, and $g(\mathbf{x}_0) = 1$. In terms of the reliability (or survival) function, $R(t|\mathbf{x}) = R(t|\mathbf{x}_0)^{g(\mathbf{x})}$, which for $g(\mathbf{x}) > 1$, $R(t|\mathbf{x}) < R(t|\mathbf{x}_0)^{g(\mathbf{x})}$. Second, the model is *time-scale accelerated*, that is, $F(t|\mathbf{x}) = F(a(\mathbf{x})t)$, where $a(\cdot)$ is a positive acceleration factor and $a(\mathbf{x}_0) = 1$. Time is accelerated for $a(\mathbf{x}) > 1$, so that lifetimes are shorter than those at $\mathbf{x}_0$. Also, time is decelerated for $a(\mathbf{x}) < 1$, where lifetimes are longer than those at $\mathbf{x}_0$. Because of this time-scale acceleration property, the Weibull regression model arises naturally in accelerated life testing in which higher than usual conditions such as higher temperature or pressure lead to earlier failures (see Sect. 7.7 for more details).

We can view the lognormal and Weibull regression models as *generalized linear models* (GLMs) (McCullagh and Nelder, 1989), a family of models that also includes the logistic and Poisson regression models. Consider the gamma regression model, in which $Y \sim Gamma(\alpha, \lambda)$, with mean $\mu = \alpha/\lambda$ and variance α/λ^2. Because μ is positive, it is natural to express the relationship between μ and the covariates by $\log(\mu) = \mathbf{x}^T\boldsymbol{\beta}$, which puts no restrictions on $\boldsymbol{\beta}$. The GLM literature recommends, however,

$$\mu^{-1} = \mathbf{x}^T\boldsymbol{\beta}, \tag{7.14}$$

using the so-called *canonical link function*, which here is the reciprocal function. Equation 7.14 offers an alternate relationship that may fit the data well in a particular situation and should be considered. See McCullagh and Nelder (1989) for more details. Equation 7.14 does restrict $\boldsymbol{\beta}$, however, because $\mathbf{x}^T\boldsymbol{\beta}$ must be positive.

In an analysis of lifetime data, when there are no restrictions on the regression coefficients $\boldsymbol{\beta}$, one choice is to use an independent $Normal(0, 10^k)$ distribution as a prior distribution for each β_i, with large enough k if little is known. If more is known about a particular regression coefficient, we can use a normal distribution with mean possibly different than zero and a much smaller variance.

Example 7.4 Weibull (both scale and shape) regression model for fiber strength data. Zok et al. (1995) presents data on the tensile strength of silicon carbide fibers. Tensile strength is measured as the stress applied in megapascals (MPa) until fracture failure of a fiber occurs. Table 7.8 displays the results of the strength tests for gauge lengths of 265, 25.4, 12.7, and 5.0 mm with test sizes of 50, 64, 50, and 50 fibers, respectively. From these data, a determination of the strength distribution of the fibers as a function of gauge length is of interest.

Weibull probability plots, one for the data at each of the four gauge lengths, indicate that a Weibull strength distribution is a reasonable assumption. A *Weibull probability plot* graphs the ordered observations against quantiles of the Weibull distribution so that the points will plot as a straight line if the observations follow a Weibull distribution. See Meeker and Escobar (1998) for more details. Consequently, assume now that, for a given fiber length x, fiber strength $S \sim Weibull[\lambda(x), \beta(x)]$, where $\lambda(x)$ and $\beta(x)$ are the Weibull scale and shape parameters that both depend on x.

Before performing a formal analysis using a regression model, we consider the following preliminary analysis. For each of the four datasets in Table 7.8 (i.e., one for each fiber length), compute the MLEs of the Weibull scale and shape parameters λ and β for the first parameterization of the Weibull distribution given in Appendix B. Figure 7.7 plots the MLEs of the Weibull scale

Table 7.8. Silicon carbide fiber tensile strengths (in MPa) for four gauge lengths for fiber strength example (Zok et al., 1995)

265 mm	0.36	0.50	0.57	0.95	0.99	1.09	1.09	1.33	1.33	1.37	1.38	1.38	1.39
	1.41	1.42	1.42	1.45	1.49	1.50	1.56	1.57	1.57	1.75	1.78	1.79	1.79
	1.82	1.83	1.86	1.89	1.90	1.92	1.93	1.96	1.97	1.99	2.04	2.06	2.06
	2.08	2.11	2.26	2.27	2.27	2.38	2.39	2.47	2.48	2.73	2.74		
25.4 mm	1.25	1.50	1.57	1.85	1.92	1.94	2.00	2.02	2.13	2.17	2.17	2.20	2.23
	2.24	2.30	2.33	2.42	2.43	2.45	2.49	2.51	2.54	2.57	2.62	2.66	2.68
	2.71	2.72	2.76	2.79	2.79	2.80	2.81	2.82	2.90	2.92	2.93	3.02	3.11
	3.11	3.14	3.20	3.20	3.22	3.26	3.29	3.30	3.34	3.35	3.37	3.43	3.43
	3.47	3.61	3.61	3.62	3.64	3.72	3.79	3.84	3.93	4.03	4.07	4.13	
12.7 mm	1.96	1.98	2.06	2.07	2.07	2.11	2.22	2.25	2.39	2.42	2.63	2.67	2.75
	2.75	2.75	2.89	2.93	2.95	2.96	2.97	3.00	3.03	3.04	3.05	3.07	3.08
	3.13	3.20	3.22	3.23	3.26	3.27	3.29	3.30	3.36	3.39	3.39	3.41	3.41
	3.43	3.52	3.72	3.96	4.07	4.09	4.13	4.13	4.14	4.15	4.29		
5.0 mm	2.36	2.40	2.54	2.67	2.68	2.69	2.70	2.77	2.77	2.79	2.83	2.91	3.04
	3.05	3.06	3.24	3.27	3.28	3.34	3.36	3.39	3.51	3.53	3.59	3.63	3.64
	3.64	3.66	3.71	3.73	3.75	3.78	3.81	3.88	3.93	3.94	3.94	3.94	3.70
	4.04	4.07	4.08	4.08	4.16	4.18	4.22	4.24	4.35	4.37	4.50		

parameter λ as a function of fiber length x on the log-log scale. The straight line relationship in Fig. 7.7 suggests the following model for $\lambda(x)$:

$$\log[\lambda(x)] = \gamma_1 + \gamma_2 \log(x) ,$$

where γ_1 and γ_2 are two unknown regression parameters. An expression for the model in terms of $\lambda(x)$ is

$$\lambda(x) = \exp(\gamma_1) x^{\gamma_2} , \tag{7.15}$$

referred to as a power law model for $\lambda(x)$.

Similarly, Fig. 7.8 plots the MLEs of the Weibull shape parameter $\beta(x)$ as a function of fiber length x on the log-log scale. The apparent linearity displayed in Fig. 7.8 also suggests a power law model for β. That is, the plot suggests

$$\log[\beta(x)] = \gamma_3 + \gamma_4 \log(x) ,$$

which in terms of $\beta(x)$, is

$$\beta(x) = \exp(\gamma_3) x^{\gamma_4}, \tag{7.16}$$

where γ_3 and γ_4 are two unknown regression parameters. Note that, for the above regression models, $\lambda(x) > 0$, and $\beta(x) > 0$, regardless of the values of γ_1, γ_2, γ_3, and γ_4, so that there are no restrictions on these regression coefficients.

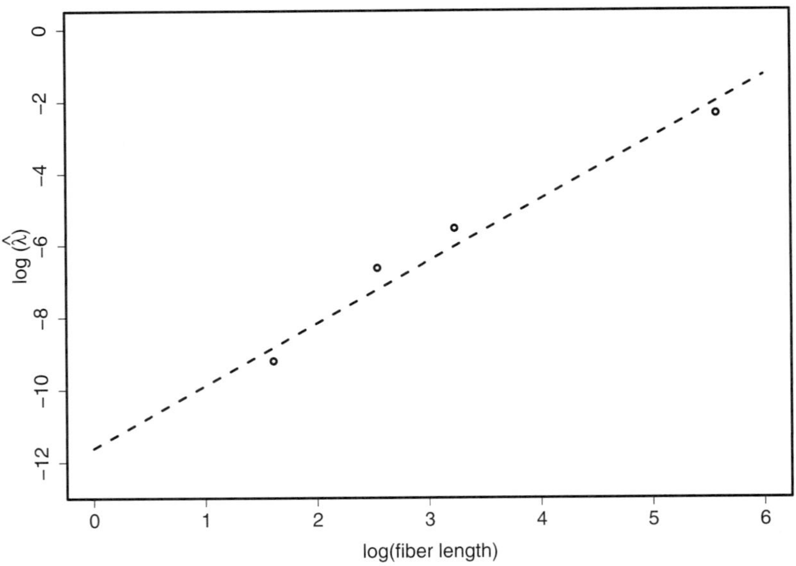

Fig. 7.7. Logged fiber lengths versus logged estimates (MLEs) of Weibull scale λ for fiber strength example.

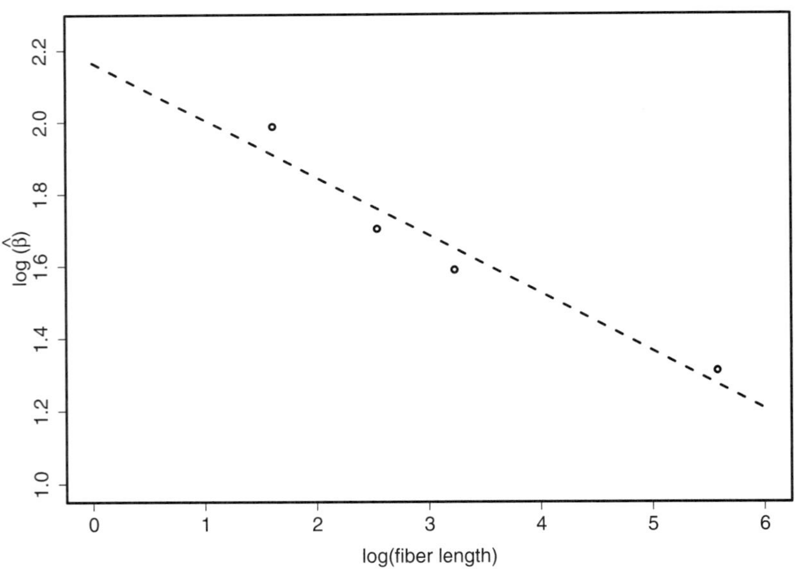

Fig. 7.8. Logged fiber lengths versus logged estimates (MLEs) of Weibull shape β for fiber strength example.

In a model for the fiber strength data, $S_{ij} \sim Weibull[\lambda(x_i), \beta(x_i)]$, $i = 1, \ldots, 4$, $j = 1, \ldots, n_i$, where $\mathbf{n} = (50, 64, 50, 50)$, $\mathbf{x} = (265, 25.4, 12.7, 5.0)$, and Eqs. 7.15 and 7.16 provide expressions for $\lambda(x_i)$ and $\beta(x_i)$, respectively. Consequently, the observed strength s_{ij} contributes a Weibull likelihood function to the model likelihood function. To complete the model, we use independent and diffuse $Normal(0, 10^6)$ prior distributions for the regression coefficients γ_1, γ_2, γ_3, and γ_4 and analyze the fiber strength data by employing MCMC to obtain draws from the joint posterior distribution of the four regression coefficients. Table 7.9 summarizes the marginal posterior distributions of these four parameters.

Table 7.9. Posterior distribution summaries for the fiber strength data model parameters

			Quantiles				
Parameter	Mean	Std Dev	0.025	0.050	0.500	0.950	0.975
γ_1	−10.650	0.597	−11.800	−11.610	−10.640	−9.638	−9.412
γ_2	1.471	0.115	1.225	1.276	1.473	1.662	1.711
γ_3	2.055	0.072	1.900	1.932	2.057	2.174	2.205
γ_4	−0.1285	0.0192	−0.1664	−0.1593	−0.1288	−0.0958	−0.0889

Consider the posterior distributions of several quantities of interest. Using the fact that the Weibull mean is $\alpha^{-1/\beta}\Gamma(1 + 1/\beta)$ and substituting α and β with Eqs. 7.15 and 7.16, the mean strength to failure (MSTF) for a given fiber length x becomes

$$\mathrm{E}(S \mid x, \gamma_1, \gamma_2, \gamma_3, \gamma_4) = [\exp(\gamma_1) x^{\gamma_2}]^{-\exp(-\gamma_3) x^{-\gamma_4}} \Gamma[1 + \exp(-\gamma_3) x^{-\gamma_4}] . \tag{7.17}$$

We now evaluate Eq. 7.17 for each of the posterior distribution draws of γ_1, γ_2, γ_3, and γ_4 to obtain corresponding draws from the posterior distribution of the MSTF given in Eq. 7.17. See Fig. 7.9, which plots the posterior means and 0.05 and 0.95 quantiles of the MSTF as a function of the fiber length x.

Similarly, again upon substituting Eqs. 7.15 and 7.16, the probability that fiber strength exceeds a particular strength s (i.e., the reliability) for a fiber length x becomes

$$R(s|x, \gamma_1, \gamma_2, \gamma_3, \gamma_4) = \exp[-\exp(\gamma_1) x^{\gamma_2} s^{\exp(\gamma_3) x^{\gamma_4}}] . \tag{7.18}$$

For example, Fig. 7.10 plots the posterior means and 0.05 and 0.95 quantiles of fiber reliability for strength $s = 1.5$ MPa as a function of fiber length x. In both Figs. 7.9 and 7.10, note the significant decrease in both MSTF and reliability as fiber length increases.

Next, we present some model selection tools for determining which covariates to include in a regression model.

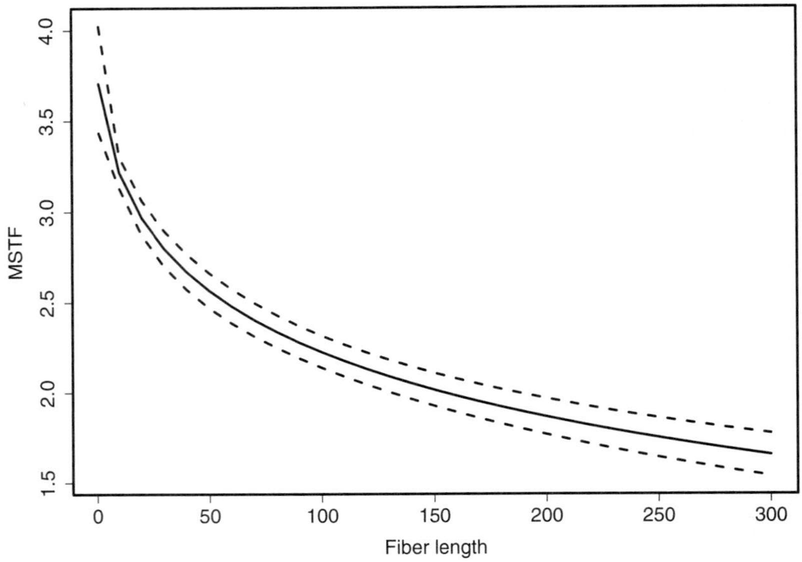

Fig. 7.9. Fiber length (in mm) versus posterior mean (solid line) and 0.05 and 0.95 quantiles (dashed lines) for the MSTF in fiber strength example.

7.5 Model Selection

Model selection means different things to different people, because it is a broad concept. It includes determining the appropriate distribution for the data, as discussed in Chap. 4. In specifying regression models, the relationship between the distribution parameters and the covariates also needs specification, e.g., should the model first transform the distribution parameter? The goodness-of-fit methods introduced in Sect. 3.4 can address such decisions. In this chapter on regression models, what we mean by model selection is the decision of which covariates from a specified list of covariates to include in the model. For example, does the mean logged lifetime for lognormally distributed lifetimes depend on temperature and if so, is it related by a linear, quadratic, or higher order polynomial? For regression models without a hierarchical structure (i.e., without random effects), use the Bayesian information criterion (BIC). For hierarchical regression models, use the deviance information criterion (DIC). See Sect. 4.6 for more details on the BIC and DIC model selection methods.

Example 7.5 Model selection for a logistic regression model. To illustrate the use of model selection, let us return to the EDG demand data models in Example 7.2. The DIC for the full model in Eq. 7.7 is 418.16; the DIC for the model without the time since criticality and NSSS vendor

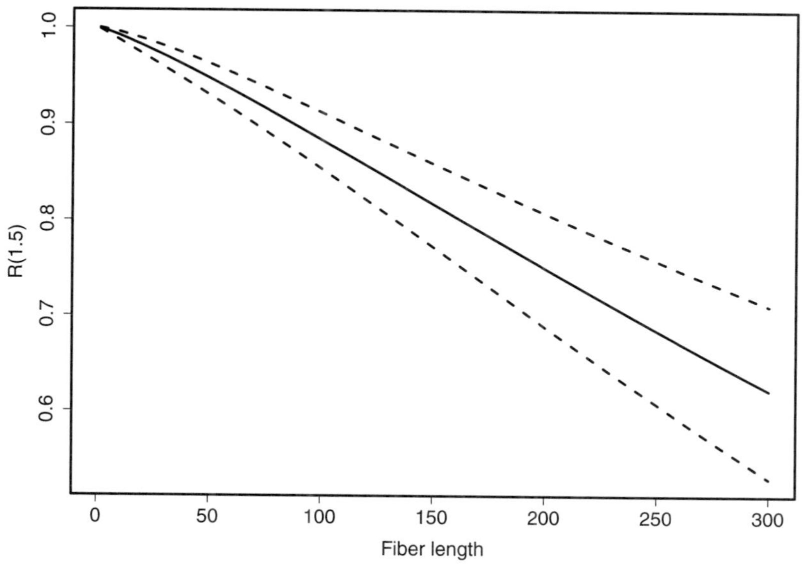

Fig. 7.10. Fiber length (in mm) versus posterior mean (solid line) and 0.05 and 0.95 quantiles (dashed lines) of the reliability for strength $s = 1.5$ MPa for fiber strength example.

covariates is 422.17. Consequently, DIC favors the full model, which has the lower DIC.

Next, we consider residual analysis, which is a graphical tool to assess the regression model fit in terms of the covariates included in the model, as well as those that are not.

7.6 Residual Analysis

Residual analysis graphically assesses how well the assumed model structure fits the data. *Residuals* are what's left in the data after removing the structure of the fitted model and are the basis of various plots. Lack of patterns in these residual plots suggests that the assumed model is consistent with the data. These plots can also identify outliers, i.e., a few data points that the assumed model does not explain well.

Let us begin by focusing on residuals for the lognormal, exponential, and Weibull distributions. For these distributions, we can write $Y = g(\boldsymbol{\theta}, \varepsilon)$ for a random variable Y, a vector of parameters $\boldsymbol{\theta}$, a standardized random variable ε, and a function $g(\cdot)$; a standardized random variable has a distribution with known parameters such as a standard normal distribution (with mean 0 and variance 1). For this situation, find a function $k(\cdot)$ such that

$$\varepsilon = k(Y, \boldsymbol{\theta}). \tag{7.19}$$

Then for an observation y, define the observed residual as $\varepsilon = k(y, \boldsymbol{\theta})$, called a *Cox-Snell residual* (Cox and Snell, 1968). In a Bayesian approach, the residual has a posterior distribution. We obtain the residual posterior distribution by simply propagating the posterior distribution of $\boldsymbol{\theta}$ through $k(\cdot)$; obtain draws from the posterior distribution of $\boldsymbol{\theta}$ by MCMC and obtain draws from the residual posterior distribution by evaluating $k(y, \boldsymbol{\theta})$ for each of the $\boldsymbol{\theta}$ draws. Finally, summarize the residual posterior distribution by several quantiles (such as (0.05, 0.25, 0.5, 0.75, 0.95)) and use the median (i.e., the 0.5 quantile) as an estimate of the residual.

A more satisfying, but time-consuming procedure calculates a residual posterior distribution for the ith observation y_i by using the posterior distribution of $\boldsymbol{\theta}$ obtained by excluding y_i. That is, base the posterior distribution of $\boldsymbol{\theta}$ on $\mathbf{y}_{(-i)} = (y_1, \ldots, y_{i-1}, y_{i+1}, \ldots, y_n)$, so that y_i does not influence the resulting posterior distribution.

Next, consider the lognormal regression model in which lifetimes $Y_i \sim LogNormal[\mu(\mathbf{x}_i), \sigma^2]$, where $\mu = \mathbf{x}_i^T \boldsymbol{\beta}$. Now, taking logarithms ($Z_i = \log(Y_i)$) leads to the well-known normal regression model in which the $Z_i \sim Normal[\mu(\mathbf{x}_i), \sigma^2]$. Another way to write this model is

$$Z_i = \mu(\mathbf{x}_i) + \sigma\varepsilon_i,$$

where $\varepsilon \sim Normal(0, 1)$. Define the residual for the ith observation z_i (= $\log(y_i)$) by

$$\varepsilon_i = [z_i - \mu(\mathbf{x}_i)]/\sigma\,. \tag{7.20}$$

As we did previously, obtain the posterior distribution of the residual by propagating the posterior distribution of $\boldsymbol{\theta} = (\boldsymbol{\beta}, \sigma^2)$ through Eq. 7.20.

The residuals in Eq. 7.20 follow a $Normal(0, 1)$ distribution if the lognormal model for Y_i is correct. Consequently, we can use a *normal probability plot* to assess the normality assumption. To construct a normal probability plot, order the observations by their residual posterior medians, and plot their summaries against the expectations of the standard normal order statistics. (The standard normal order statistics are the ordered n independent standard normal random variables, which have expectations that are tabled and used in commercial implementations of normal probability plots. We can also obtain their expectations by simulation: draw n independent standard normal random variables and order them for many sets of draws, take the average of all the smallest order statistics, and plot against the smallest residual summary, and so on.) The ordered medians of the residual posterior distributions should plot approximately as a straight line if normality holds. Such graphical methods help in assessing the distributional assumptions made in the data model and complement the analytical Bayesian χ^2 goodness-of-fit test of Sect. 3.4.

In this chapter, however, we are concerned with assessing the structural assumptions made in the regression model involving the covariates. Questions arise such as: Is a linear polynomial enough? or Is a quadratic or higher-order polynomial needed? A plot of these summaries against their corresponding linear covariate values should not show a pattern to suggest that a quadratic covariate is needed; that is, in a *residual plot*, the residuals plotted against a linear covariate should not display a quadratic shape. The residuals should plot as an equal-width band with half the residuals on either side of zero (or median if the standardized random variable does not have a symmetric distribution). If the variability of the residuals depends on the covariate, then the constant variance assumption (σ^2) is suspect. We can also look for patterns in plots of residual summaries against the medians of the $\mu(\mathbf{x}_i)$ posterior distributions, i.e., the posterior distributions of the logged lifetime data means for the lognormal regression model. Figure 7.11 displays typical residual plots for (a) no pattern indicating no missing term in variable x, (b) a missing quadratic term in variable x, (c) a variable z that a model should include as a covariate, and (d) increasing error variance in variable x. In Fig. 7.11(b)-(c), the missing term or variable refers to covariates that are not in the current model that produced these residuals.

Type I-censored (or time-censored or right-censored) lifetimes often arise in reliability analyses so that residuals for censored data need to be addressed. For the lognormal regression model, where $Z_i \sim Normal(\mu[\mathbf{x}), \sigma^2]$ and lifetime $Y_i = \exp(Z_i)$, suppose that z_i is Type I censored at c_i. Then the distribution for the residual

$$\varepsilon_i = [z_i - \mu(\mathbf{x}_i)]/\sigma = k(z_i, \boldsymbol{\theta}),$$

conditioned on $\boldsymbol{\theta} = (\boldsymbol{\beta}, \sigma)$ is a truncated standard normal distribution with $\varepsilon_i \geq k(c_i, \boldsymbol{\theta})$. Consequently, the posterior distribution is the mixture of these truncated distributions over the posterior distribution of $\boldsymbol{\theta}$. More specifically, the probability density function of the ith residual conditioned on $\boldsymbol{\theta}$ is

$$g(\varepsilon_i|\boldsymbol{\theta}) = \phi(\varepsilon_i)/\{1 - \Phi[k(c_i, \boldsymbol{\theta})]\},$$

where $\phi(\cdot)$ and $\Phi(\cdot)$ are the probability density and cumulative distribution functions of the standard normal distribution, respectively. Then, the posterior density function of the ith residual is

$$g(\varepsilon_i|\mathbf{z}) = \int \phi(\varepsilon_i) I\{\varepsilon_i \geq k(c_i, \boldsymbol{\theta})\}/\{1 - \Phi[k(c_i, \boldsymbol{\theta})]\} p(\boldsymbol{\theta}|\mathbf{z}) \mathrm{d}\boldsymbol{\theta},$$

where $I\{\cdot\}$ is the indicator function. We can evaluate the residual posterior distribution for a censored observation by sampling from its distribution as follows. First, draw a $\boldsymbol{\theta}$ from its posterior distribution and calculate $k(c_i, \boldsymbol{\theta})$. Then make a draw from $\phi(\varepsilon_i)/\{1 - \Phi[k(c_i, \boldsymbol{\theta})]\}$ for $\varepsilon_i > k(c_i, \boldsymbol{\theta})$. Recall that drawing $u \sim Uniform(0, 1)$ and evaluating $F^{-1}(u)$ yields a draw from any random variable with cumulative distribution function $F(\cdot)$. In this case, draw a uniform u and evaluate

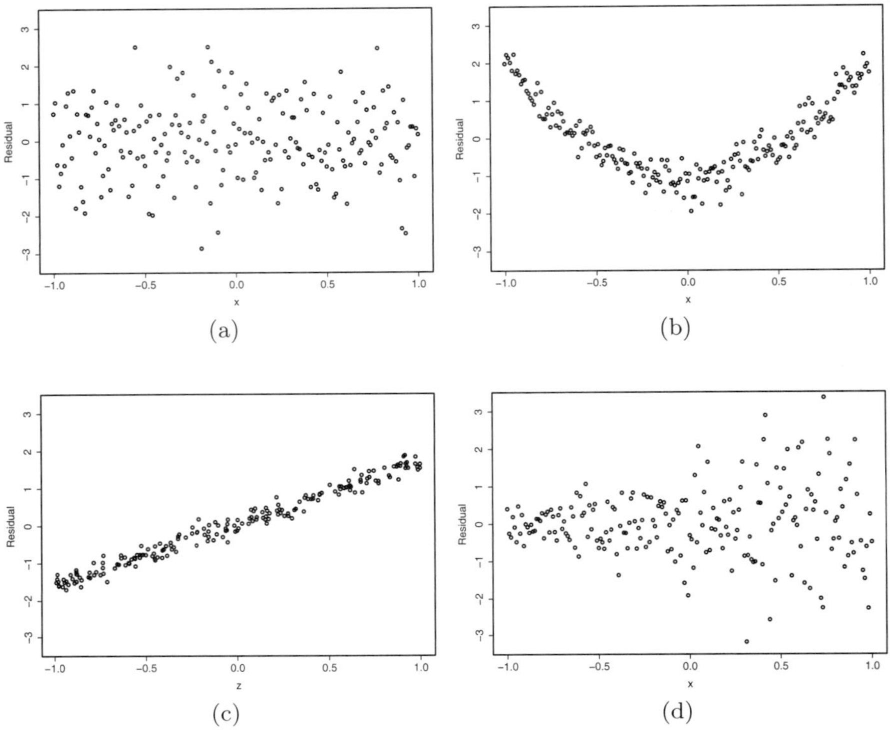

Fig. 7.11. Typical residual plots: (a) no pattern, (b) missing quadratic term in variable x, (c) missing variable z, and (d) increasing error variance in variable x.

$$\varepsilon_i = \Phi^{-1}(\Phi[k(c_i, \boldsymbol{\theta})] + u\{1 - \Phi[k(c_i, \boldsymbol{\theta})]\}) .$$

As with the lognormal regression model, we use the Cox-Snell residuals (Cox and Snell, 1968) for the exponential and Weibull regression models. In the same way that the residuals for the normal regression model conditioned on the model parameters are standard normal, define residuals for these regression models conditioned on the model parameters that are standard exponential (i.e., with rate parameter $\lambda = 1$). Writing the exponential regression model as

$$y_i = \varepsilon_i / \exp(\mathbf{x}^T \boldsymbol{\beta}),$$

where ε_i has a standard exponential distribution, an expression for the exponential regression residual is

$$\varepsilon_i = y_i \exp(\mathbf{x}^T \boldsymbol{\beta}) = k(y_i, \boldsymbol{\beta}).$$

Similarly, for the Weibull regression model, we can write the model as

$$y_i = \varepsilon_i^{\gamma} / \exp(\mathbf{x}^T \boldsymbol{\beta}),$$

where ε_i has a standard exponential distribution and γ is the Weibull shape parameter; that is, $Y \sim Weibull[\lambda = \exp(\mathbf{x}^T \boldsymbol{\beta}), \gamma]$ for the first parameterization listed in Appendix B. Then, define the Weibull regression residual by

$$\varepsilon_i = y_i^\gamma \exp(\mathbf{x}^T \boldsymbol{\beta}) = k(y_i, \boldsymbol{\theta}),$$

where $\boldsymbol{\theta} = (\boldsymbol{\beta}, \gamma)$. An assessment of these residual plots should account for the asymmetry of the standard exponential distribution. For example, 0.25, 0.50, and 0.75 quantiles of the standard exponential distribution might be plotted to see if 25% of the residuals are below the 0.25 quantile, and so on.

Example 7.6 Residual analysis for a Weibull regression model. As an illustration of residual analysis for the Weibull regression model, let us return to the fiber strength data model in Example 7.4. A plot in Fig. 7.12 of logged median posterior residual against fiber length shows no discernible pattern. The solid line in the figure is the logged median of a standard exponential distribution. Consequently, there is no evidence to suggest that there are missing covariates that need to be added to the model.

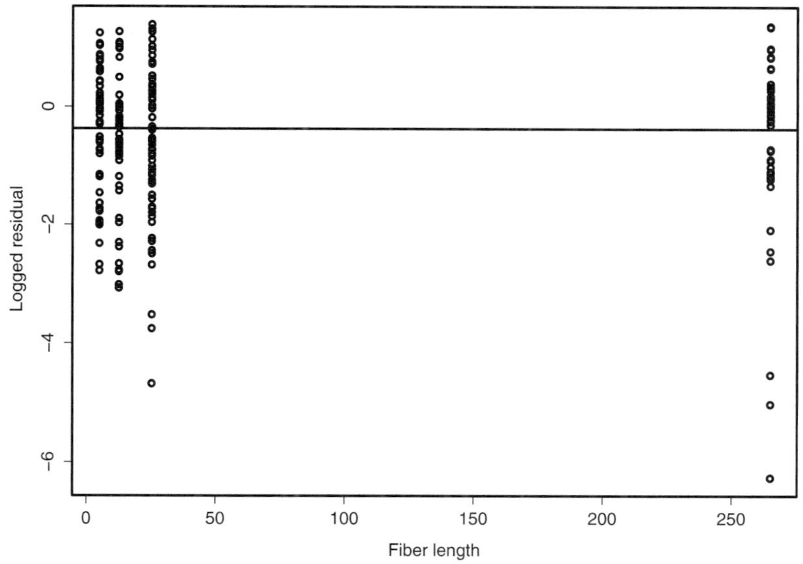

Fig. 7.12. Fiber length versus logged median posterior residuals for fiber strength example. Solid line is the logged median of a standard exponential distribution.

For Type I-censored data, we can obtain the posterior distributions of the residuals using the same procedure described for the lognormal regression model. Note that for the exponential and Weibull regression models, where the standardized random variable has the standard exponential distribution, $\Phi(w) = 1 - \exp(-w)$ and $\Phi^{-1}(w) = -\log(1 - w)$.

Finally, we need to consider residuals for the discrete response regression models for the binomial and Poisson distribution cases. For these cases, use *deviance residuals*, which McCullagh and Nelder (1989) and Pierce and Schafer (1986) define as

$$\text{sign}(y_i - \mu_i(\hat{\boldsymbol{\theta}}))\{2[l(y_i, y_i) - l(\hat{\boldsymbol{\theta}}, y_i)]\}^{0.5},$$

where $\hat{\boldsymbol{\theta}}$ is the MLE of $\boldsymbol{\theta}$, $l(\boldsymbol{\theta}, y_i) = \log f(y|\boldsymbol{\theta})$ is the log-likelihood function for the ith observation, $\mu_i(\hat{\boldsymbol{\theta}})$ is the MLE of the mean of Y_i, and sign$(\cdot)$ is the sign $(+/-)$ of its argument. One motivation for deviance residuals is that they approximately follow a standard normal distribution. In the Bayesian approach, instead of the MLE $\hat{\boldsymbol{\theta}}$, propagate the posterior distribution of $\boldsymbol{\theta}$ through Eq. 7.6 to obtain the posterior distribution of the residuals.

For the binomial regression model, the deviance residual is

$$\text{sign}(y_i - n_i\pi_i)\{2[y_i \log(y_i/n_i\pi_i) + (n_i - y_i)\log((n_i - y_i)/(n_i - n_i\pi_i))]\}^{0.5}.$$

For the Poisson regression model, the deviance residual is

$$\text{sign}(y_i - \lambda_i)\{2[y_i \log(y_i/\lambda_i) - y_i + \lambda_i]\}^{0.5}.$$

Note that for the binomial data regression model in which $\text{logit}(\pi_i) = \mathbf{x}_i^T\boldsymbol{\beta}$ and for Poisson regression model in which $\log(\lambda_i) = \mathbf{x}_i^T\boldsymbol{\beta}$,

$$\pi_i = \exp(\mathbf{x}_i^T\boldsymbol{\beta})/[1 + \exp(\mathbf{x}_i^T\boldsymbol{\beta})] \text{ and } \lambda_i = \exp(\mathbf{x}_i^T\boldsymbol{\beta}),$$

respectively.

The Bernoulli regression model for binary data is a special case of the binomial regression model with $n_i = 1$. Consequently, we can use the binomial data deviance residuals above. Exercise 7.16 suggests a study to understand what needs to be looked at when using these deviance residuals.

Example 7.7 Residual analysis for a logistic regression model. As an example of residual analysis for the logistic regression model, let us return to the EDG demand data model in Example 7.2. Figure 7.13 plots the median posterior residuals against time since criticality (in days); the graph shows no discernible pattern. Consequently, there is no evidence to suggest that there are missing covariates that need to be added to the model.

Next, let us consider an example that shows the use of residuals in a Poisson regression model.

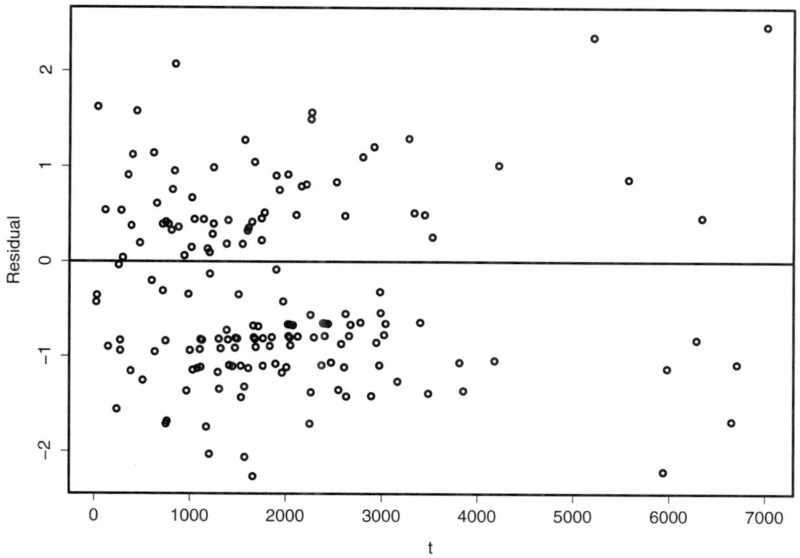

Fig. 7.13. Median posterior residuals over time since criticality t (in days) for the EDG reliability example.

Example 7.8 Residual analysis for a Poisson regression model. As an illustration of residual analysis for the Poisson regression model, consider the scram rate data model in Example 7.3. Figure 7.14 plots the median posterior residuals against the coded year; the graph shows no discernible pattern. Consequently, there is no evidence to suggest that there are missing covariates that need to be added to the model.

Finally, residuals for the gamma regression model have not yet been discussed. Exercise 7.17 suggests that Cox-Snell residuals cannot be found for this model. However, deviance residuals can be developed for this model (McCullagh and Nelder, 1989).

Next, we consider accelerated life testing, which employs accelerating variables to shorten the lifetimes that appear as covariates in regression models used to analyze the accelerated life test data.

7.7 Accelerated Life Testing

Covariates arise naturally in predicting the reliability of a highly reliable component during the component's design phase. Unfortunately, prediction of a

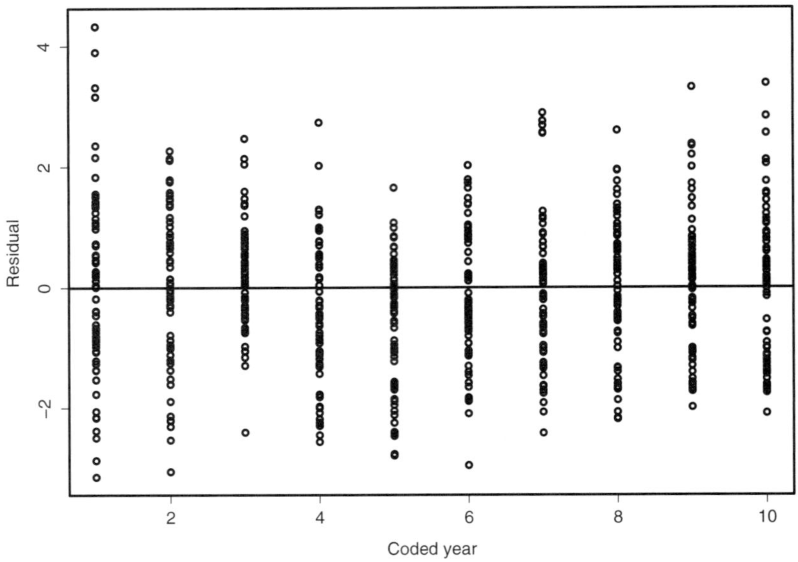

Fig. 7.14. Median posterior residuals over time (coded year) for scram system example.

prototype's reliability in the design phase usually requires extrapolating the regression relationship outside the range of covariate values that we are able to collect informative failure data on. The practical difficulty is that few, if any, highly reliable components fail during testing conducted under normal operating conditions. For example, consider testing a new filament in a long-life light bulb, which has a filament design life of five years. After two years of testing a batch of 100 light bulbs, no light bulbs may have failed. With such test results, the analyst could conclude that the failure rate is probably less than 1 per 100 per two years of use and that the expected lifetime exceeds two years; but, it would be difficult to obtain a more precise estimate of the failure rate or light bulb mean lifetime based only on these results. Few manufacturers are willing to wait two years (much less five) to bring a new product to market.

The standard method to assess highly reliable components is to test them under extreme operating conditions, referred to as *accelerated testing* (ALT). In the light bulb example, increasing the voltage applied to the filaments to induce more failures or turning the light bulbs on and off at an abnormally high rate may induce shock-related failures. Under sufficiently extreme environmental or operating conditions, most components will fail in an acceptably short period of time.

The statistical challenge in analyzing ALT data is modeling the relationship between either the failure rate (or the component lifetime) at an extreme condition to that under normal use conditions. Modeling becomes even more challenging when there is more than one accelerating variable (e.g., both temperature and voltage). Choosing the model is important because the predictions made by extrapolating to normal use conditions critically depend on it. That is, extrapolation is prediction under conditions that differ substantially from the experimental or observational conditions under which we collected the ALT data. In general, we try to avoid extrapolating too far outside the range of observed values of a covariate, because the validity of extrapolation relies heavily on the validity of the model. Moreover, to ensure quality predictions under ALT, it helps to choose a model that is based on an experimentally confirmed physical relationship or well-established empirical relationship.

Before planning for ALT, investigators should examine the nature of such relationships to both identify those variables that are eligible for acceleration and determine the magnitudes of the accelerating constants that may be required. It is also important to consider variables that cannot be accelerated and to assess the likely impact that these variables may have on failure rates.

While it is beyond this chapter's scope to provide a comprehensive coverage of the many physical models for ALT, we illustrate the general statistical principles involved by considering two of the more common models for temperature acceleration below, the Arrhenius and Eyring models. Moreover, the analyst can extend these statistical methods in a relatively simple way to a host of related accelerated testing problems. Readers interested in a more detailed exploration of the physical models underlying temperature and other accelerating variables should consult Meeker and Escobar (1998), Nelson (1990), and Mann et al. (1974).

7.7.1 Common Accelerating Variables and Relationships

We can categorize accelerating variables according to whether the variables speed the aging process of the device, the amount of stress applied to the device, or the number of times the device is used. Examples of aging variables include temperature, humidity, or exposure to environmental chemicals, and electromagnetic radiation. Stress variables include voltage, pressure, vibration, and temperature cycling. Finally, in some cases, accelerated life testing simply involves increasing the use rate of the device. For example, continuously activating a switch or valve over several days yields the equivalent of several years of normal use.

In this section, let us focus on temperature acceleration. See Exercise 7.22, which presents an ALT employing voltage. We can often accelerate aging by operating the device at an elevated temperature. Because temperature often plays an important role in aging, researchers have developed several models that describe its effect on the chemical reactions that underly material degradation. The *Arrhenius model* is one of the most popular of these models.

The Arrhenius relationship describes the effect that temperature has on chemical reaction rates. Denote the chemical reaction rate for a particular material at temperature T by $r(T)$. Then, the Arrhenius relationship is

$$r(T) = \gamma \exp\left(-\frac{E_a}{k_B T}\right). \tag{7.21}$$

In this expression, γ and E_a are constants specific to the material being tested, k_B is Boltzmann's constant (equal to 8.62×10^{-5} electron-volts per degree Celsius), and T is absolute temperature measured in kelvins (°C +273.15).

If we assume that the chemical reaction rate is proportional to the amount of material degradation, then the lifetime is inversely proportional to the rate. The *acceleration factor* at temperature T_1 relative to temperature T_0 — the factor by which the lifetime at temperature T_1 is decreased from its baseline value at T_0 — is

$$\frac{r(T_1)}{r(T_0)} = \exp\left[-\frac{E_a}{k_B}\left(\frac{1}{T_1} - \frac{1}{T_0}\right)\right].$$

The Arrhenius relation predicts that differences in logged lifetimes is a constant $(-E_a/k_B)$ times the difference in inverse temperatures. That is,

$$\Delta \log(\text{lifetime}) = \frac{-E_a}{k_B}\left(\frac{1}{T_1} - \frac{1}{T_0}\right).$$

In general, the value of $-E_a/k_B$, while not known, requires an estimate for planning purposes; the analyst may base the estimate on previous experimental results conducted on similar devices. Following the ALT, we estimate its value for the particular device using this prior information along with the observed ALT data.

In analyzing accelerated lifetime data, many investigators have noted that the Arrhenius model is inadequate for extrapolating between three or more values of the temperature acceleration factors. This led Eyring et al. (1941) to propose a generalization of the Arrhenius model, known as the expanded *Eyring model*. The Eyring model expresses chemical reaction rates as

$$r(t) = \gamma A(T) \exp\left(-\frac{E_a}{k_B T}\right),$$

where $A(\cdot)$ is a function that provides additional flexibility in describing the change in the underlying reaction rate as temperature varies. Based on an extensive literature review, Meeker and Escobar (1998) (p. 473) reports that most investigators assume that the function $A(T)$ is a simple power function of T. That is, $A(T) = T^m$, where m is typically assigned a value between 0 and 1. From a Bayesian perspective, however, we can obtain the posterior distribution of m for a given device using the observed ALT data. Consequently, its value need not be fixed, although the analyst should consider the effectiveness of an estimate from the available data. In the following example, we explore this point further.

The acceleration factor at temperature T_1 relative to temperature T_0 for the Eyring model is

$$\frac{r(T_1)}{r(T_0)} = \frac{A(T_1)}{A(T_0)} \exp\left[-\frac{E_a}{k_B}\left(\frac{1}{T_1} - \frac{1}{T_0}\right)\right].$$

When $A(T) = T^m$, the Eyring accelerating factor becomes

$$\frac{r(T_1)}{r(T_0)} = \left(\frac{T_1}{T_0}\right)^m \exp\left[-\frac{E_a}{k_B}\left(\frac{1}{T_1} - \frac{1}{T_0}\right)\right].$$

Example 7.9 A mechanical component with temperature as an accelerating factor. Table 7.10 displays lifetimes in hours for a hypothetical mechanical component tested at four different temperatures. In this example, we fit both the Arrenhius and Eyring models to these data, assess the adequacy of each of these models, and compare predictions based upon these models.

Table 7.10. Lifetimes in hours for a mechanical component during temperature-accelerated life testing. The experiment ended after 100 hours, and an asterisk indicates that the component was still operating at the end of the experiment

300 K	350 K	400 K	500 K
100*	47.5	29.5	80.9
100*	73.7	100*	76.6
100*	100*	52.0	53.4
80.7	100*	63.5	100*
100*	86.2	100*	47.5
29.1	100*	99.5	26.1
100*	100*	56.3	77.6
100*	100*	92.5	100*
100*	100*	100*	61.8
100*	71.8	100*	56.1

Arrhenius Model

For purposes of illustration, we assume a Weibull lifetime model for these data and that the log characteristic lifetime (ψ in the third parameterization of the Weibull distribution in Appendix B) is inversely proportional to the reaction rate. Consequently, under the Arrhenius model for reaction rate,

$$\log(\psi_i) = \alpha_0 + \alpha_1/T_i, \qquad (7\ 22)$$

where T_i denotes the absolute temperature in kelvins applied to the ith experimental unit, and α_0 and α_1 correspond to $-\log(\gamma)$ and E_a/k_B in Eq. 7.21, respectively. Moreover, we parameterize the Weibull probability density function so that the expected lifetime of the ith unit, say $\mu_i = \mathrm{E}(Y_i)$, equals $\psi_i \Gamma(1+1/\beta)$, and its variance is

$$\mathrm{Var}(Y_i) = \psi_i^2 \left[\Gamma\left(1+\frac{2}{\beta}\right) - \Gamma\left(1+\frac{1}{\beta}\right)^2 \right] .$$

For an analysis of the ALT data, we choose independent $Normal(0, 10^2)$ and $Normal(0, 10^6)$ prior distributions for α_0 and α_1, respectively, and a $Gamma(0.1, 0.1)$ prior distribution for β. We use MCMC to obtain draws from the joint posterior distribution of $(\alpha_0, \alpha_1, \beta)$. Table 7.11 presents summaries of the posterior distributions of the model parameters. (Note that the values $\alpha_0 = 2.5$, $\alpha_1 = 800$, and $\beta = 2$ generated the data in Table 7.10.)

Table 7.11. Posterior distribution summaries for mechanical component Arrhenius model parameters

			Quantiles				
Parameter	Mean	Std Dev	0.025	0.050	0.500	0.950	0.975
β	2.318	0.479	1.475	1.591	2.287	3.140	3.329
α_0	3.157	0.580	1.938	2.172	3.189	4.060	4.222
α_1	621.9	237.5	207.8	262.0	604.9	1042.0	1146.0

In an actual testing application, an analyst would want the posterior distribution of quantities like the mean lifetime or the predictive distribution for the lifetime of a single item at normal use conditions. A primary advantage of the Bayesian approach in ALT is that approximating these posterior and predictive distributions is straightforward.

To obtain draws from the posterior distribution for the mean lifetime at a given operating temperature T_{pred}, use MCMC to obtain draws from the joint posterior distribution of $(\alpha_0, \alpha_1, \beta)$, and for each $(\alpha_0, \alpha_1, \beta)$ draw, evaluate the mean lifetime using

$$\mu = \exp(\alpha_0 + \alpha_1/T_{pred})\Gamma(1+1/\beta) .$$

For example, for a normal operating temperature of $T_{pred} = 293$ K, Table 7.12 summarizes the resulting posterior distribution for mean lifetime.

Similarly, to obtain draws from the predictive distribution for a single component at T_{pred}, for each posterior draw of $(\alpha_0, \alpha_1, \beta)$, we can randomly draw a Weibull lifetime with parameters (ψ, β) using

$$\psi = \exp\left(\alpha_0 + \alpha_1/T_{pred}\right) .$$

Figure 7.15 displays the resulting predictive distribution for $T_{pred} = 293$ K as a solid line.

Table 7.12. Posterior distribution summaries of mean lifetime for the mechanical component Arrhenius and Eyring models

Model	Quantiles						
	0.025	0.050	0.250	0.500	0.750	0.950	0.975
Arrhenius	112.7920	119.5196	144.7638	169.0767	286.1581	286.1581	332.2496
Eyring	112.7602	119.0651	144.4720	169.6633	286.6463	286.6463	325.8400

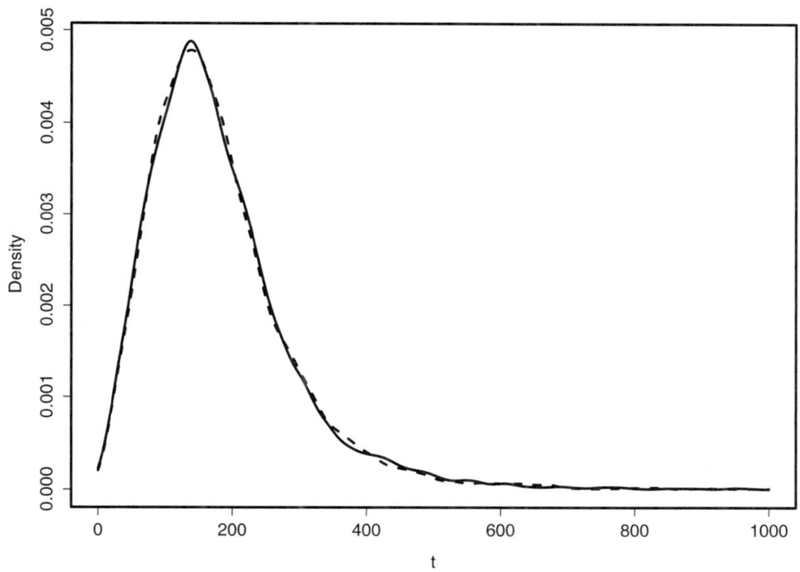

Fig. 7.15. Predictive distributions of lifetime t (in hours) for mechanical component Arrhenius (solid line) and Eyring (dotted line) models.

Eyring Model

Although we generated the data in Table 7.10 using an Arrhenius model, let us consider fitting them to the more general Eyring model. As with the Arrhenius model, assume that the lifetimes have a Weibull distribution and that the log characteristic life of a unit is inversely proportional to the reaction rate. That is, assume that the characteristic life of the ith unit equals

$$\log(\psi_i) = \alpha_0 + \alpha_1/T_i - m\log(T_i), \tag{7.23}$$

where, as before, T_i denotes the absolute temperature in kelvins, and α_0, α_1, β, and m are the model parameters.

In an analysis of the ALT data, besides employing the prior distributions used previously for the Arrhenius model, we use a $Normal(0, 10^2)$ prior distribution for m. Table 7.13 summarizes the posterior distributions of the parameters for the Eyring model. Recall that the true value of m is 0 for the data generated using the Arrhenius model.

Table 7.13. Posterior distribution summaries for mechanical component Eyring model parameters

			Quantiles				
Parameter	Mean	Std Dev	0.025	0.050	0.500	0.950	0.975
β	2.317	0.486	1.457	1.571	2.282	3.167	3.353
α_0	2.93	8.29	−13.51	−10.87	2.94	16.62	19.39
α_1	633.7	492.7	−340.3	−174.5	628.0	1442.0	1610.0
m	−0.033	1.196	−2.403	−2.014	−0.031	1.942	2.370

It is worth noting that the posterior distributions on the model parameters do not assign high probability to the true values of the model parameters, which in this case are $\alpha_0 = 2.5$, $\alpha_1 = 800$, $\beta = 2$, and $m = 0$. We can explain this apparent failure of the posterior distributions to capture the true parameter values by examining Eq. 7.23 more carefully. By expanding $\log(T_i)$ in a one-term Taylor series around the mean of the temperatures used in this ALT experiment,

$$\log(\psi_i) \approx \alpha_0 - m + \frac{\alpha_1}{T_i} - \frac{mT_i}{\bar{T}}, \qquad \text{where} \qquad \bar{T} = \frac{1}{40}\sum_{i=1}^{40} T_i.$$

If $1/T_i$ is approximately linear over the range of observed temperatures, then the intercept parameter α_0 is collinear with m, i.e., if α_0 and m increase by the same amount, the value of $\alpha_0 - m$ does not change, and the estimated value of α_1 depends on the estimated value of m. Figure 7.16 further illustrates this collinearity between the parameters α_0 and m by showing trace plots of 10,000 draws of the posterior distribution of α_0 and m (after an initial burn-in period of 4,000 draws and recording every 1,000th draw thereafter), which still have a sample correlation of 0.9976.

Fortunately, the collinearity of the parameter estimates does not cause difficulties in approximating either the posterior distribution of the population mean lifetime or of the predictive distribution of the lifetimes of individual components. See how the summaries of the posterior mean lifetime at $T_{pred} =$ 293 K in Table 7.12 for the Eyring model are very similar to those for the Arrhenius model. Also, see how the Eyring model predictive distribution for $T_{pred} = 293$ K in Fig. 7.15 compares favorably with that obtained under the simpler Arrhenius model.

Finally, as an example of using model selection tools, the BIC for the Eyring model is 234.5865 and that for the Arrhenius model is 234.5818, which favors the Arrhenius model ever so slightly; in practice, the Arrhenius model would be chosen because of its simplicity with one less parameter.

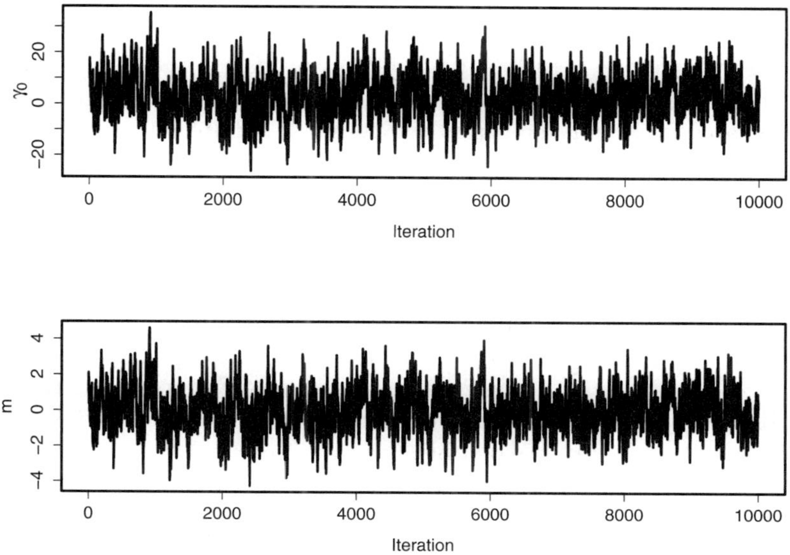

Fig. 7.16. Trace plots of posterior distribution draws for the mechanical component Eyring model parameters γ_0 and m.

7.8 Reliability Improvement Experiments

In this section, we consider regression analysis of reliability data from statistically designed experiments. Statistically designed experiments provide information to improve the reliability of a product (or process) by identifying variables referred to as *factors* that most impact its reliability. We can empirically identify such factors by deliberately changing the factor values (referred to as levels) and observing the resulting lifetimes. An analysis of these lifetimes not only identifies important factors, but suggests recommended factor levels that yield improved reliability.

Statistically designed experiments provide a systematic and efficient experimental plan to study several factors simultaneously. We study each factor at a few values referred to as *factor levels*. Such experimental plans include full factorial designs, which consist of all combinations of the factor levels; for example, k factors each at 2 levels has 2^k combinations of factor levels. Let us also refer to the combinations as *runs*. The notation 2^k and 3^k denotes a full factorial design with all k factors at two and three levels, respectively. Other experimental plans include fractional factorial designs, which consist of a fraction of a full factorial design, and mixed-level designs, which have at least two factors with different number of levels.

We can also apply Taguchi's *robust parameter design* paradigm (Taguchi, 1986) to reliability. That is, the analyst seeks *robust reliability* situations, where reliability is high and insensitive to noise variables that are difficult or impossible to control. Noise variables include manufacturing variables that cannot easily be controlled and the environmental conditions under which the product is used. Such experiments, however, must control the noise variables, called *noise factors*. The other experimental factors from which we seek robust reliability are *control factors*. Consequently, experimental plans for finding robust reliability need to involve both control and noise factors. *Product array designs* are such plans that consist of one plan for the control factors, called the control array, and one for the noise factors, called the noise array. Recall that a row of the array specifies the level of each of the factors associated with the array. We obtain the entire experimental plan by running each of the control array rows with all of the noise array rows. The entire experimental plan is typically a fraction of a full factorial design, because each of the arrays is usually a fractional factorial design. Product array designs allow estimates of all *interactions* between the control and noise factors, i.e., control-noise interactions; if there is an interaction effect between two factors, then a covariate in the regression model involves both factors, e.g., x_1x_2 for factors x_1 and x_2. Other so-called *combined array designs* use a different fraction of a full factorial design, which is chosen to allow estimates of particular control-control, control-noise, and noise-noise interactions. See Wu and Hamada (2000), Chap. 12, for more details.

We can analyze lifetime data from such experimental plans, using the lifetime regression models presented in Sect. 7.4. In fact, reliability improvement experiments also collect binomial successes/failures or Poisson counts, so that the analyst can use the logistic and Poisson regression models presented in Sects. 7.2 and 7.3, respectively. See Exercises 7.13 and 7.14 for examples of such experiments.

Next, we consider the covariates associated with the experimental factors used in the associated regression models, where we assume that the reliability data distribution is related to k covariates $x_1, \ldots, x_k$ through $\beta_0 + \beta_1 x_1 + \cdots + \beta_k x_k = \mathbf{x}^T\boldsymbol{\beta}$, where β_0, $\beta_1, \ldots, \beta_k$ are the regression coefficients. When an experiment studies a factor at evenly spaced levels or values, use covariates that correspond to polynomials known as *orthogonal polynomials*. When a factor has two levels with labels $(0, 1)$, there is no notion of equally spaced levels. For factors with two levels, assess linearity by using the linear orthogonal polynomial, which has values $(-1, +1)$. That is, for data collected at factor A set at level 0, the covariate x_A takes the value -1, and for data taken at factor A set at level 1, this covariate x_A takes the value $+1$. When a factor has three evenly spaced levels labeled $(0, 1, 2)$, assess linearity and curvature using linear and quadratic orthogonal polynomials, which have covariate values of $(-1, 0, 1)$ and $(-1, 2, -1)$, respectively. That is, for data collected at factor A set at level 0, the linear covariate $x_{Al} = -1$ and the quadratic covariate $x_{Aq} = -1$; similarly, for data taken at factor A set at level 1, $x_{Al} = 0$ and $x_{Aq} = 2$, and for data taken at factor A set at level 2, $x_{Al} = 1$ and $x_{Aq} = -1$. See Sect. 1.8 of Wu and Hamada (2000) for more discussion of orthogonal polynomials. For example, orthogonal polynomials allow classical estimates of the associated regression coefficients that are statistically independent. When the factor levels are not evenly spaced, however, the analyst can use the factor levels directly as covariate values; often using centered values produces estimated regression coefficients with less statistical dependence between them. For example, a temperature factor, which has levels of 10, 35, and 45°C, has centered covariate values of −20, 5, and 15 by subtracting the average $30 = (10 + 35 + 45)/3$. Consequently, use the centered covariate and the square of the centered covariate to capture linearity and curvature for factor levels that are not evenly spaced.

Next, we consider the case when the experiment has two or more factors. Besides the covariates for each factor, consider covariates that capture the joint impact of multiple factors on the reliability data distribution; we refer to the regression coefficients associated with such covariates as interaction effects. First, consider two factors A and B, both studied at two levels (coded as 0 and 1) in a full factorial design known as a 2^2 design. For example, factor A could be temperature, with its two levels at 170°C and 180°C, and factor B could be pressure, with its two levels at 50 Torr and 80 Torr. See Table 7.14, which lists the four combinations of the factor levels. Three covariates that explain the data correspond to the A and B *main effects* (i.e., the factor effects) and their *interaction*; denote these covariates by x_A, x_B, and x_{AB}, respectively, and also, refer to the main effects and interactions collectively as *factorial effects*. See Table 7.14 for their corresponding covariates. The four runs (i.e., rows) allow estimates of four parameters, with the fourth parameter being the intercept, denoted by β_0 in previously presented regression models. Note that the A and B main effects are also linear orthogonal polynomial effects. The interaction effect compares the difference between the data distribution at the two levels of B (1 vs. 0) when A=1 with the difference between the data distribution at the two levels of B (1 vs. 0) when A=0. If there is no interaction, then the interaction effect is zero. That is, the regression coefficient β_{AB} corresponding to this covariate is zero.

For the two factors each at two levels case, let us be more explicit how to evaluate $\mu = \mathbf{x}^T\boldsymbol{\beta}$. As seen from Table 7.14, we have data of four different combinations of factor levels and there are three associated covariates x_A, x_B, and x_{AB}. The vector of regression coefficients $\boldsymbol{\beta}$ includes β_A, β_B, β_{AB}, as well as the β_0; i.e., $\boldsymbol{\beta} = (\beta_0, \beta_A, \beta_B, \beta_{AB})$. Then, for data at the first factor level combination or run, where $A = 0$ and $B = 0$, $\mathbf{x} = (1, x_A, x_B, x_{AB}) = (1, -1, -1, +1)$. Consequently,

$$\begin{aligned}\mu = \mathbf{x}^T\boldsymbol{\beta} &= \beta_0(1) + \beta_A(-1) + \beta_B(-1) + \beta_{AB}(+1)\\ &= \beta_0 - \beta_A - \beta_B + \beta_{AB}\,.\end{aligned}$$

Similarly, for data for the fourth run, where $A = 1$ and $B = 1$,

$$\mu = \mathbf{x}^T\boldsymbol{\beta} = \beta_0(1) + \beta_A(+1) + \beta_B(+1) + \beta_{AB}(+1)$$
$$= \beta_0 + \beta_A + \beta_B + \beta_{AB} \, .$$

Next, consider two factors A and B both at three levels (coded as 0, 1, and 2). We refer to the full factorial design consisting of nine factor level combinations as a 3^2 design. For example, A could be temperature, which has three levels at 170°C, 180°C, and 190°C, and B could be pressure at 50 Torr, 80 Torr, and 110 Torr. Again, the levels for each factor need to be evenly spaced. See Table 7.15, which lists the nine factor level combinations. For the 3^2 design, there are eight covariates, which explain the observed data, corresponding to linear and quadratic A and B main effects and linear-linear, linear-quadratic, quadratic-linear, and quadratic-quadratic interaction effects, and are denoted by x_{Al}, x_{Aq}, x_{Bl}, x_{Bq}, x_{AlBl}, x_{AlBq}, x_{AqBl}, and x_{AqBq}, where the subscripts l and q refer to linear and quadratic effects, respectively. See Table 7.15, which also lists their corresponding covariates. Note that the main effect covariates are orthogonal polynomials. The interaction effects have interpretations similar to those given above for the two-level factors. For example, the A_lB_l interaction effect compares the difference between the experimental responses at the $(0, 2)$ levels of B (2 vs. 0) when $A = 2$ with the difference between the responses at the $(0, 2)$ levels of B (2 vs. 0) when $A = 0$. In other words, the A_lB_l interaction effect compares the linear effect of factor B at the $(0, 2)$ levels of factor A. When the factor levels are not evenly spaced, the analyst can employ the usual polynomials x_1, x_2, x_1x_2, x_1^2, and x_2^2 as covariates using the factor levels directly; as an alternative, use the centered factor levels because the resulting regression coefficient estimates tend to be less statistically dependent.

For the two factors each at three levels case, let us be more explicit how to evaluate $\mu = \mathbf{x}^T\boldsymbol{\beta}$. As seen from Table 7.15, we have data at eight different combinations of factor levels or runs and there are eight associated covariates x_{Al}, x_{Aq}, x_{Bl}, x_{Bq}, x_{AlBl}, x_{AlBq}, x_{AqBl}, and x_{AqBq}. The vector of regression coefficients $\boldsymbol{\beta}$ includes β_{Al}, β_{Aq}, β_{Bl}, β_{Bq}, β_{AlBl}, β_{AlBq}, β_{AqBl}, and β_{AqBq}, as well as the intercept β_0; i.e., $\boldsymbol{\beta} = (\beta_0, \beta_{Al}, \beta_{Aq}, \beta_{Bl}, \beta_{Bq}, \beta_{AlBl}, \beta_{AlBq}, \beta_{AqBl}, \beta_{AqBq})$. Then, for data at the first factor level combination or run, where $A = 0$ and $B = 0$,

$$\mathbf{x} = (1, x_{Al}, x_{Aq}, x_{Bl}, x_{Bq}, x_{AlBl}, x_{AlBq}, x_{AqBl}, x_{AqBq})$$
$$= (1, -1, 1, -1, 1, 1, -1, -1, 1) \, .$$

Consequently,

$$\mu = \mathbf{x}^T\boldsymbol{\beta} = \beta_0(1) + \beta_{Al}(-1) + \beta_{Aq}(1) + \beta_{Bl}(-1) + \beta_{Bq}(1)$$
$$+\beta_{AlBl}(1) + \beta_{AlBq}(-1) + \beta_{AqBl}(-1) + \beta_{AqBq}(1)$$
$$= \beta_0 - \beta_{Al} + \beta_{Aq} - \beta_{Bl} + \beta_{Bq}$$
$$+\beta_{AlBl} - \beta_{AlBq} - \beta_{AqBl} + \beta_{AqBq} \, .$$

Table 7.14. 2^2 design and covariates

Factor		Covariate		
A	B	x_A	x_B	x_{AB}
0	0	−1	−1	+1
0	1	−1	+1	−1
1	0	+1	−1	−1
1	1	+1	+1	+1

Table 7.15. 3^2 design and covariates

Factor		Covariate							
A	B	x_{Al}	x_{Aq}	x_{Bl}	x_{Bq}	x_{AlBl}	x_{AlBq}	x_{AqBl}	x_{AqBq}
0	0	−1	1	−1	1	1	−1	−1	1
0	1	−1	1	0	−2	0	2	0	−2
0	2	−1	1	1	1	−1	−1	1	1
1	0	0	−2	−1	1	0	0	2	−2
1	1	0	−2	0	−2	0	0	0	4
1	2	0	−2	1	1	0	0	−2	−2
2	0	1	1	−1	1	−1	1	−1	1
2	1	1	1	0	−2	0	−2	0	−2
2	2	1	1	1	1	1	1	1	1

In the next four examples, we consider statistically designed experiments, which have various types of factors and data.

Example 7.10 Lognormal regression model for spring experiment lifetime data. Taguchi (1986) presents a well-known spring reliability experiment, which studies seven factors: shape (A), hole ratio (B), coining (C), stress σ_t (D), stress σ_c (E), shot peening (F), and outer perimeter planing (G). Table 7.16 presents the 27-run experimental plan, which studies all the factors, except B and C, at three levels. The three levels (0, 1, 2) of the combined BC factor correspond to levels (0,0), (1,0), (0,1) of the B and C factors, respectively. This experimental plan is a 3^{6-3} fractional factorial design because 27 runs is 1/27th (or 3^{-3}) of a three-level full factorial design (or 3^6 design) in six factors (with factors B and C combined into one factor). Denote this fractional factorial design by $3^{6-3} = 3^6 3^{-3}$. The experimenter tested three springs at each of the 27 runs, by inspecting each spring every 100,000 cycles for failure up to 1.1 million cycles. Table 7.16 presents the experimental design and the lifetime data. For springs still working at the 11th inspection, their lifetimes are Type I censored (i.e., their lifetime exceeds 1.1 million cycles), denoted by $(11, \infty)$ in Table 7.16. The remaining lifetimes are interval censored; for example, $(1, 2)$ means that a spring failed between the 1st and 2nd inspections or between 100,000 and 200,000 cycles. Assuming that the lifetimes follow a $LogNormal(\mu, \sigma^2)$ distribution, then the likelihood contribution for the interval-censored observation $(1, 2)$ is the probability of the lifetime failing between 100,000 and 200,000 cycles, whose expression is

$$\Phi[(\log(2) - \mu)/\sigma] - \Phi[(\log(1) - \mu)/\sigma],$$

where Φ is the standard normal cumulative distribution function and $\mu = \mathbf{x}^T \boldsymbol{\beta}$. The covariates $\mathbf{x}$ are those associated with the seven experimental factors. Table 7.17 displays the covariates for the main effects D_l, D_q, E_l, E_q, A_l, A_q, B_l, C_l, F_l, F_q, G_l, and G_q. We obtain the covariates for the interactions D_lE_l, D_lE_q, D_qE_l, D_qE_q, D_lF_l, D_lF_q, D_qF_l, and D_qF_q, by multiplying the associated main effect covariates together, e.g., D_lE_l is the product of D_l and E_l in Table 7.17.

Table 7.16. 3^{6-3} design and lifetime data for spring experiment (Taguchi, 1986)

Factor						Lifetime		
D	E	A	BC	F	G	(in 100,000 cycles)		
0	0	0	0	0	0	(1,2)	(1,2)	(1,2)
0	0	1	1	1	1	(4,5)	(5,6)	(11,∞)
0	0	2	2	2	2	(2,3)	(2,3)	(11,∞)
0	1	0	0	1	2	(2,3)	(3,4)	(3,4)
0	1	1	1	2	0	(5,6)	(11,∞)	(11,∞)
0	1	2	2	0	1	(1,2)	(1,2)	(1,2)
0	2	0	0	2	1	(1,2)	(1,2)	(3,4)
0	2	1	1	0	2	(1,2)	(1,2)	(2,3)
0	2	2	2	1	0	(3,4)	(3,4)	(4,5)
1	0	0	1	0	0	(1,2)	(1,2)	(2,3)
1	0	1	2	1	1	(11,∞)	(11,∞)	(11,∞)
1	0	2	0	2	2	(6,7)	(11,∞)	(11,∞)
1	1	0	1	1	2	(11,∞)	(11,∞)	(11,∞)
1	1	1	2	2	0	(2,3)	(2,3)	(2,3)
1	1	2	0	0	1	(1,2)	(2,3)	(2,3)
1	2	0	1	2	1	(2,3)	(3,4)	(4,5)
1	2	1	2	0	2	(2,3)	(2,3)	(2,3)
1	2	2	0	1	0	(11,∞)	(11,∞)	(11,∞)
2	0	0	2	0	0	(3,4)	(4,5)	(4,5)
2	0	1	0	1	1	(11,∞)	(11,∞)	(11,∞)
2	0	2	1	2	2	(11,∞)	(11,∞)	(11,∞)
2	1	0	2	1	2	(11,∞)	(11,∞)	(11,∞)
2	1	1	0	2	0	(11,∞)	(11,∞)	(11,∞)
2	1	2	1	0	1	(5,6)	(11,∞)	(11,∞)
2	2	0	2	2	1	(4,5)	(4,5)	(6,7)
2	2	1	0	0	2	(2,3)	(2,3)	(3,4)
2	2	2	1	1	0	(11,∞)	(11,∞)	(11,∞)

For an analysis of the spring experiment lifetime data, we use independent $Normal(0, 10)$ prior distributions for the regression coefficients β_i and

Table 7.17. Spring experiment covariates

D_l	D_q	E_l	E_q	A_l	A_q	C_l	$-B_l$	F_l	F_q	G_l	G_q
−1	−1	−1	−1	−1	−1	−1	−1	−1	−1	−1	−1
−1	−1	−1	−1	0	2	0	2	0	2	0	2
−1	−1	−1	−1	2	−1	1	−1	1	−1	1	−1
−1	−1	0	2	−1	−1	−1	−1	0	2	1	−1
−1	−1	0	2	0	2	0	2	1	−1	−1	−1
−1	−1	0	2	2	−1	1	−1	−1	−1	0	2
−1	−1	1	−1	1	−1	−1	−1	1	−1	0	2
−1	−1	1	−1	1	2	0	2	−1	−1	1	−1
−1	−1	1	−1	1	−1	1	−1	0	2	−1	−1
0	2	−1	−1	−1	−1	0	2	−1	−1	−1	−1
0	2	−1	−1	0	2	1	−1	0	2	0	2
0	2	−1	−1	2	−1	−1	−1	1	−1	1	−1
0	2	0	2	−1	−1	0	2	0	2	1	−1
0	2	0	2	0	2	1	−1	1	−1	−1	−1
0	2	0	2	2	−1	−1	−1	−1	−1	0	2
0	2	1	−1	1	−1	0	2	1	−1	0	2
0	2	1	−1	1	2	1	−1	−1	−1	1	−1
0	2	1	−1	1	−1	−1	−1	0	2	−1	−1
1	−1	−1	−1	−1	−1	1	−1	−1	−1	−1	−1
1	−1	−1	−1	0	2	−1	−1	0	2	0	2
1	−1	−1	−1	2	−1	0	2	1	−1	1	−1
1	−1	0	2	−1	−1	1	−1	0	2	1	−1
1	−1	0	2	0	2	−1	−1	1	−1	−1	−1
1	−1	0	2	2	−1	0	2	−1	−1	0	2
1	−1	1	−1	1	−1	1	−1	1	−1	0	2
1	−1	1	−1	1	2	−1	−1	−1	−1	1	−1
1	−1	1	−1	1	−1	0	2	0	2	−1	−1

an independent $InverseGamma(0.5, 0.1)$ prior distribution for σ^2. Note that β_0 is the intercept, and β_i, $i = 1, \ldots, 20$, correspond to the factorial effects D_l, D_q, E_l, E_q, D_lE_l, D_lE_q, D_qE_l, D_qE_q, A_l, A_q, $-B_l$, C_l, F_l, F_q, D_lF_l, D_lF_q, D_qF_l, D_qF_q, G_l, and G_q, respectively. The subscripts l and q refer to linear and quadratic effects, respectively, corresponding to appropriate orthogonal polynomials. Note that C_l and $-B_l$ correspond to the BC linear and quadratic orthogonal polynomials, respectively. Table 7.18 presents posterior distribution summaries of the model parameters. From these results, note that factors D, E, A, B, and F are important, as are the interactions between D and E and D and F; that is, these factors have regression coefficients (or factorial effects) with posterior distributions that are concentrated away from zero.

Next, we consider the choice of optimal factor levels (i.e., the best factor-level combination) that provides the best reliability. By varying the seven factor levels, there are 972 factor-level combinations. For each combination, obtain the posterior distribution of the probability of exceeding a warranty period, say 1,000 (× 100,000) cycles, and use the 0.1 quantile (i.e., the 90% lower credible bound on the probability of exceeding the warranty period) as a figure of merit. Recall that for any factor other than B and C, the covariate values for the linear and quadratic covariates are $(-1, 1)$, $(0, -2)$, and $(1, 1)$, which correspond to the factor levels 0, 1, and 2, respectively. For the B and C factors, the covariates values for C_l and $-B_l$ are $(-1, 1)$, $(0, -2)$, and $(1, 1)$, which correspond to the B and C factor-level combinations $(0, 0)$, $(1, 0)$, and $(0, 1)$, respectively. We find that $(D, E, A, B, C, F, G) = (1, 2, 2, 0, 0, 1, 0)$ is the best factor-level combination, which has a 90% lower credible bound on the probability of exceeding the warranty period equal to 0.850.

Table 7.18. Posterior distribution summaries for spring lifetime data model parameters

Parameter	Effect	Mean	Std Dev	Quantiles 0.025	0.050	0.500	0.950	0.975
β_1	D_l	2.492	1.103	1.073	1.140	2.259	4.615	5.332
β_2	D_q	−0.6357	0.9862	−3.2920	−2.5210	−0.3713	0.6111	0.7982
β_3	E_l	−0.3873	0.1079	−0.6020	−0.5637	−0.3866	−0.2072	−0.1723
β_4	E_q	0.0263	0.0636	−0.1073	−0.0783	0.0268	0.1291	0.1477
β_5	D_lE_l	−0.4235	0.1422	−0.7151	−0.6503	−0.4253	−0.1899	−0.1433
β_6	D_lE_q	−0.0456	0.0877	−0.2242	−0.1895	−0.0451	0.0932	0.1254
β_7	D_qE_l	−0.2421	0.0839	−0.4064	−0.3839	−0.2420	−0.1065	−0.0815
β_8	D_qE_q	−0.1547	0.0498	−0.2566	−0.2387	−0.1538	−0.0748	−0.0589
β_9	A_l	0.6097	0.1269	0.3582	0.3961	0.6083	0.8185	0.8557
β_{10}	A_q	−0.1730	0.0716	−0.3128	−0.2942	−0.1726	−0.0514	−0.0343
β_{11}	B_l	−0.4420	0.1290	−0.7054	−0.6552	−0.4404	−0.2234	−0.1884
β_{12}	C_l	−0.1055	0.1260	−0.3689	−0.3170	−0.1009	0.0985	0.1386
β_{13}	F_l	0.7648	0.1112	0.5475	0.5786	0.7589	0.9470	0.9944
β_{14}	F_q	−2.5830	1.1710	−5.3760	−4.8950	−2.4120	−1.0670	−0.8990
β_{15}	D_lF_l	0.3861	0.1539	0.0865	0.1289	0.3802	0.6519	0.7084
β_{16}	D_lF_q	−1.5050	1.1000	0.1614	−3.6540	−1.2500	−0.1694	−0.1227
β_{17}	D_qF_l	0.1790	0.0649	0.0472	0.0726	0.1786	0.2842	0.3040
β_{18}	D_qF_q	0.9128	0.9856	−0.5113	−0.3367	0.6466	2.7910	3.5850
β_{19}	G_l	−0.1517	0.1091	−0.3675	−0.3304	−0.1513	0.0333	0.0721
β_{20}	G_q	0.0677	0.0642	−0.0522	−0.0401	0.0668	0.1732	0.1935
β_0		4.080	1.166	2.413	2.564	3.922	6.348	6.887
σ		0.4311	0.0616	0.3248	0.3367	0.4265	0.5437	0.5639

As an illustration of residual analysis for the lognormal regression model, Fig. 7.17 displays a plot of median posterior residuals against the six experimental factor levels, which shows no discernible pattern. Recall that the residuals are based on the logged lifetimes, which have a normal distribution.

Finally, this example demonstrates how we analyze interval and Type I-censored data as discussed in Sect. 7.6.

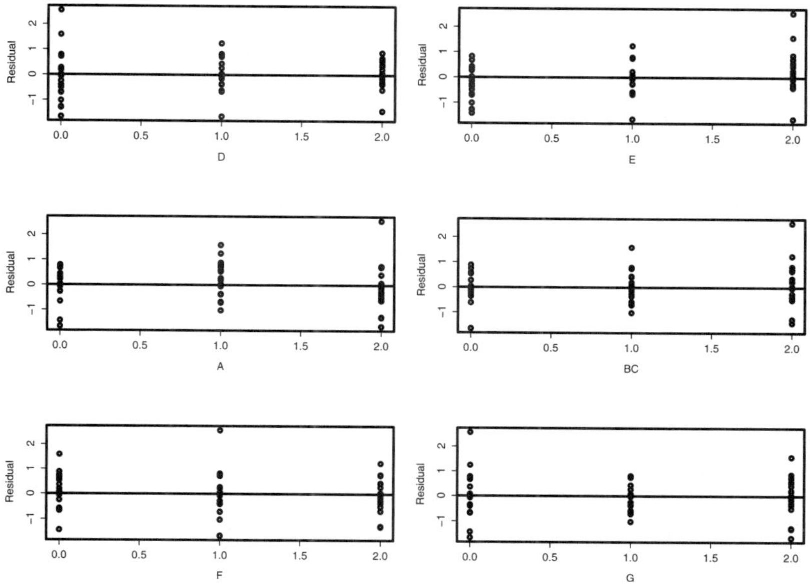

Fig. 7.17. Experimental factor levels versus median posterior residuals for spring experiment.

In the next example, we consider an experiment with qualitative and quantitative factors.

Example 7.11 Weibull regression (with qualitative and quantitative factors) for solder joint lifetime data. Lau et al. (1988) reports on an experiment studying the reliability of solder joints that attach surface mount components to printed circuit boards (PCBs); knowing that solder properties depend on temperature, the experiment tested three types of PCBs under three different temperatures. The experiment subjected a solder joint to a mechanical testing fatigue test under a particular temperature and recorded its lifetime as the number of cycles at which the solder joint fails. The experiment tested 10 solder joints for each combination of the three PCB types (copper-nickel-tin, copper-nickel-gold, and copper-tin-lead) and three temperatures (20, 60, and 100°C). Table 7.19 displays the lifetime data for the nine PCB type-temperature combinations. Like the experimenters in Lau et al. (1988), we assume that the lifetimes follow a Weibull distribution.

In this experiment, PCB type is a qualitative factor, whereas temperature is a quantitative factor. To handle the qualitative factor, use two dummy variables, one which compares copper-nickel-tin with copper-nickel-gold, and the other, which compares copper-nickel-tin with copper-tin-lead. Without available subject-matter knowledge suggesting a particular transformation of temperature, here we only consider linear and quadratic effects for the temperature factor. To handle the interaction between PCB type and temperature, include a linear and quadratic temperature effect for each of the three PCB types. If all three linear effects are the same, and if all three quadratic effects are the same, then there is no interaction between PCB type and temperature. Using these covariates, as presented in Table 7.20 (where the qualitative factor B has three levels $B0$-$B2$, and the quantitative factor A also has three levels $A0$-$A2$), we employ a Weibull regression model in which the logged scale parameter depends on the covariates through $\log(\lambda) = \mathbf{x}^T\boldsymbol{\gamma}$. Regression coefficient γ_0 is the intercept, and γ_i, $i = 1, \ldots, 8$, correspond to the two dummy variables for PCB type and the linear and quadratic temperature effects for each PCB type, respectively. An additional assumption is a common Weibull shape parameter β.

In the analysis of these data, the Weibull probability density function evaluated for each of the data make up the contributions to the likelihood; as a choice for prior distributions, use an independent $Normal(0, 100)$ distribution for each of the regression coefficients γ_i, and an independent $Gamma(1.5, 0.5)$ distribution for the shape parameter β. We employ MCMC to obtain draws from the joint distribution of these model parameters; Table 7.21 summarizes their marginal posterior distributions. (In Table 7.21, the codes for the three PCB types are 0, 1, 2. Also, D with appropriate subscripts denotes the dummy variables, and PCB stands for PCB type, and T for temperature.) These results indicate that reliability for copper-nickel-tin is better than that for the other PCB types. The linear and quadratic temperature effects for each PCB type are similar, which may suggest that there is no interaction between PCB type and temperature, i.e., the impact of temperature on reliability does not depend on the PCB type. We can use BIC of Sect. 4.6 to see whether the simpler model without interaction provides a better fit, but leave this as Exercise 7.1. To obtain the predictive distributions at each of the nine PCB type and temperature factor-level combinations, use the posterior draws for the model parameters as summarized in Table 7.21. We also leave this as Exercise 7.24 to determine what is the recommended factor-level combination.

In the next example, we consider an experiment that collected censored data.

Example 7.12 Weibull regression with Type II-censored data for capacitor experiment. Zelen (1959) presents an early reliability experiment that studied capacitor reliability by varying temperature and voltage. The experiment used two temperatures (170 and 180°C) and four voltages (200, 250, 300, and 350 volts). At each of the 8 factor-level combinations, the experimenter put 10 capacitors on test until the fourth capacitor failed. Table 7.22 presents the recorded failure times in hours.

We analyze these data using the same model proposed by Zelen (1959). That is, assume that the lifetimes have a Weibull distribution, in which the

Table 7.19. 3^2 design and lifetimes for solder joint experiment (Lau et al., 1988)

Factor											
PCB Type	Temp (°C)	Lifetime (cycles)									
copper-nickel-tin	20	218	265	279	282	336	469	496	507	685	685
	60	185	242	254	280	305	353	381	504	556	697
	100	7	46	52	82	90	100	101	105	112	151
copper-nickel-gold	20	685	899	1020	1082	1207	1396	1411	1417	1470	1999
	60	593	722	859	863	956	1017	1038	1107	1264	1362
	100	188	248	266	269	291	345	352	381	385	445
copper-tin-lead	20	791	1140	1169	1217	1267	1409	1447	1476	1488	1545
	60	704	827	925	930	984	984	1006	1166	1258	1362
	100	98	154	193	230	239	270	295	332	491	532

Table 7.20. Covariates for solder joint experiment (with quantitative factor A (temperature) and qualitative factor B (PCB type)

Factor		Covariate							
A	B	$B1$ vs. $B0$	$B2$ vs. $B0$	$A_l\|A0$	$A_q\|B0$	$A_l\|B1$	$A_q\|A1$	$A_l\|B3$	$A_q\|B3$
0	0	0	0	−1	1	0	0	0	0
1	0	0	0	0	−2	0	0	0	0
2	0	0	0	1	1	0	0	0	1
0	1	1	0	0	0	−1	1	0	0
1	1	1	0	0	0	0	−2	0	0
2	1	1	0	0	0	1	1	0	0
0	2	0	1	0	0	0	0	−1	1
1	2	0	1	0	0	0	0	0	−2
2	2	0	1	0	0	0	0	1	1

scale parameter depends on temperature and voltage, i.e., $\log(\lambda) = \mathbf{x}^T\boldsymbol{\gamma}$. Zelen (1959) uses a regression model with reciprocal temperature on the Kelvin scale [1/(temperature in °C +273.15)] and log voltage. Because the levels of these transformed factors are not equally spaced, we employ the usual polynomials directly as covariates. That is, use covariates corresponding to a linear effect for transformed temperature; linear, quadratic, and cubic effects for transformed voltage; and linear, quadratic, and cubic by linear interaction effects for transformed voltage and transformed temperature, as displayed in Table 7.23. These effects (i.e., regression coefficients) correspond to γ_i, $i = 1$, …, 7, respectively; γ_0 is the intercept.

Our analysis must account for the collected data. Under a Type II-censoring scheme, where the data consist of y_j, $j = 1, \ldots, r$, the first r failure times out of n (here $r = 4$ and $n = 10$) and $n - r$ right-censored failure times

Table 7.21. Posterior distribution summaries for solder joint lifetime data model parameters

Parameter	Effect	Mean	Std Dev	Quantiles 0.025	0.050	0.500	0.950	0.975
β		3.28	0.27	2.77	2.85	3.28	3.74	3.83
γ_0		−18.47	1.59	−21.68	−1.15	−18.44	−15.90	−15.42
γ_1	$D_{l \text{ vs. } 0}$	−3.38	0.37	−4.11	−4.00	−3.38	−2.78	−2.66
γ_2	$D_{2 \text{ vs. } 0}$	−3.21	0.36	−3.93	−3.81	−3.21	−2.62	−2.51
γ_3	$T_l\|PCB = 0$	2.60	0.32	1.98	2.09	2.60	3.12	3.22
γ_4	$T_q\|PCB = 0$	−0.77	0.15	−1.06	−1.02	−0.76	−0.52	−0.48
γ_5	$T_l\|PCB = 1$	2.30	0.30	1.72	1.81	2.30	2.79	2.90
γ_6	$T_q\|PCB = 1$	−0.46	0.14	−0.74	−0.69	−0.46	−0.24	−0.20
γ_7	$T_l\|PCB = 2$	2.50	0.30	1.90	2.01	2.50	2.98	3.07
γ_8	$T_q\|PCB = 2$	−0.58	0.16	−0.90	−0.85	−0.57	−0.32	−0.27

at y_r, the joint probability density function of these data takes the form

$$\frac{n!}{(n-r)!}\left(\prod_{j=1}^{r} f(y_j)\right) R(y_r)^{n-r}, \tag{7.24}$$

where $f(\cdot)$ is the lifetime probability density function, and $R(\cdot)$ is the corresponding reliability or survival function. As discussed in Chaps. 1 and 4, only the pattern of the data determines the likelihood function, which is

$$\left(\prod_{j=1}^{r} f(y_j)\right) R(y_r)^{n-r}.$$

For $Weibull(\lambda, \beta)$ lifetimes, the likelihood function is

$$\left(\prod_{j=1}^{r} \lambda\beta y_j^{\beta-1}\exp(-\lambda y_j^{\beta})\right)[\exp(-\lambda y_j^{\beta})]^{n-r}. \tag{7.25}$$

In the analysis of these data, we use an independent $Normal(0, 10)$ prior distribution for each of the $\boldsymbol{\gamma}$ components and an independent $Gamma(1.5, 0.5)$ prior distribution restricted to the interval (0,10) for β (because of rarely seen values of the shape parameter exceeding 10 in practice). We employ MCMC to obtain draws from the joint distribution of the model parameters and summarize their marginal posterior distributions in Table 7.24. These results show that the quadratic and cubic effects (γ_3 and γ_4, respectively) in log voltage impact reliability. Use the posterior draws for the model parameters as summarized in Table 7.24 to obtain predictive distributions at each of the eight temperature and voltage factor-level combinations. We leave this as Exercise 7.25 to determine what is the recommended factor-level combination.

Table 7.22. Capacitor experiment factors and failure time data (Zelen, 1959). The first 4 failures out of 10 are listed

Factor		Lifetime			
Temperature (°C)	Voltage (volts)	(hours)			
170	200	439	904	1092	1105
	250	572	690	904	1090
	300	315	315	439	628
	350	258	258	347	588
180	200	959	1065	1065	1087
	250	216	315	455	473
	300	241	315	332	380
	350	241	241	435	435

Table 7.23. Capacitor experiment covariates

Factor		Covariate						
Temperature	Voltage	T_l	V_l	V_q	V_c	T_lV_l	T_lV_q	T_lV_c
170	200	−1	−3	1	−1	3	−1	1
	250	−1	−1	−1	3	1	1	−3
	300	−1	1	−1	−3	−1	1	3
	350	−1	3	1	1	−3	−1	−1
180	200	1	−3	1	−1	−3	1	−1
	250	1	−1	−1	3	−1	−1	3
	300	1	1	−1	−3	1	−1	−3
	350	1	3	1	1	3	1	1

Table 7.24. Posterior distribution summaries for capacitor lifetime data model parameters

				Quantiles				
Parameter	Effect	Mean	Std Dev	0.025	0.050	0.500	0.950	0.975
γ_0		−0.302	3.215	−6.930	−5.802	−0.244	4.845	5.727
γ_1	T_l	0.007	3.162	−6.208	−5.216	0.028	5.197	6.177
γ_2	V_l	−1.790	2.541	−6.126	−5.704	−2.072	2.941	3.761
γ_3	V_q	−1.621	0.892	−3.445	−3.322	−1.492	−0.374	−0.039
γ_4	V_c	0.2534	0.0864	0.1117	0.1329	0.2437	0.4072	0.4374
γ_5	T_lV_l	−0.029	3.138	−6.149	−5.153	−0.032	5.056	6.114
γ_6	T_lV_q	−0.099	3.156	−6.320	−5.247	−0.097	5.084	6.078
γ_7	T_lV_c	−0.405	3.145	−6.579	−5.646	−0.387	4.834	5.706
β		2.429	0.352	1.763	1.869	2.418	3.024	3.147

In the next example, we consider an experiment that used a split-plot design with correlated lifetime data.

Example 7.13 Lognormal regression for battery split-plot experiment. Ostle (1963), Sect. 13.2, reports on a battery experiment that studied the effect of temperature and electrolyte on reliability. The experiment involved three temperatures (low, medium, and high) and four electrolytes (A, B, C, and D). Table 7.25 presents the experimental data, which are battery activated lifetimes in hours. The experimental plan, known as a split-plot design (Cochran and Cox, 1957), tested all four electrolytes at a given temperature for each replicate. Namely, temperature is the whole-plot factor, and electrolyte is the sub-plot factor. The battery experiment replicated the split-plot design (i.e., the 12 temperature by electrolyte combinations) six times. The special feature of data from a split-plot design is the correlated lifetimes corresponding to the electrolytes with the same temperature within a replicate. For example, we have correlated lifetimes for the electrolytes A-D at low temperature for replicate 1, and so on. Note that the observations across different temperatures or different replicates are independent, however.

To analyze the data, we use a standard *split-plot model* on the logged lifetimes, i.e., the logged lifetimes have a normal distribution, which implies that the lifetimes have a lognormal distribution. A description of the model for the logged lifetimes from one replicate is

$$\begin{aligned} \log(Y_{ij}) &\sim Normal(\mu_{ij}, \sigma_s^2), \\ \mu_{ij} &= \mathbf{x}_{ij}^T \boldsymbol{\beta} + w_i, \text{ and} \\ w_i &\sim Normal(0, \sigma_w^2), \end{aligned} \tag{7.26}$$

where σ_s^2 is the sub-plot error variance and σ_w^2 is the whole-plot error variance. Note that the whole-plot effect w_i is a random effect, which makes the observations within the same whole-plot correlated, i.e., the whole-plot error is common to all of the observations within the whole plot. The regression coefficients $\boldsymbol{\beta}$ correspond to the intercept, two temperature comparisons (low vs. medium and low vs. high), three electrolyte comparisons for the low temperature (A vs. B, A vs. C, and A vs. D), and three electrolyte comparisons for both the medium and high temperatures. Table 7.26 shows the covariates including the column of ones associated with the intercept.

In the analysis of these data, the lognormal probability density function evaluated at each of the lifetimes makes up the contributions to the likelihood; we use an independent $Normal(0, 10)$ prior distribution for each of the regression coefficients β_i, as well as independent $InverseGamma(0.001, 0.001)$ prior distributions for σ_w^2 and σ_s^2. We employ MCMC to obtain draws from the joint posterior distribution of the model parameters; Table 7.27 summarizes their marginal posterior distributions. These results suggest that electrolytes A and D are similar to each other, but are different from electrolytes B and C. The results also suggest that the difference between electrolytes A and D and electrolyte B lessens as temperature increases.

Table 7.25. Battery experiment design lifetime (in hours) data (Ostle, 1963)

Factor		Replicate					
Temperature	Electrolyte	1	2	3	4	5	6
low	A	2.17	1.88	1.62	2.34	1.58	1.66
	B	1.58	1.26	1.22	1.59	1.25	0.94
	C	2.29	1.60	1.67	1.91	1.39	1.12
	D	2.23	2.01	1.82	2.10	1.66	1.10
med	A	2.33	2.01	1.70	1.78	1.42	1.35
	B	1.38	1.30	1.85	1.09	1.13	1.06
	C	1.86	1.70	1.81	1.54	1.67	0.88
	D	2.27	1.81	2.01	1.40	1.31	1.06
high	A	1.75	1.95	2.13	1.78	1.31	1.30
	B	1.52	1.47	1.80	1.37	1.01	1.31
	C	1.55	1.61	1.82	1.56	1.23	1.13
	D	1.56	1.72	1.99	1.55	1.51	1.33

Table 7.26. Battery experiment covariates. (Columns correspond to intercept; low vs. medium temperature; low vs. high temperature; A vs. B electrolytes, A vs. C electrolytes, and A vs. D electrolytes for low temperature; A vs. B electrolytes, A vs. C electrolytes, and A vs. D electrolytes for medium temperature; and A vs. B electrolytes, A vs. C electrolytes, and A vs. D electrolytes for high temperature)

1	0	0	0	0	0	0	0	0	0	0	0
1	0	0	1	0	0	0	0	0	0	0	0
1	0	0	0	1	0	0	0	0	0	0	0
1	0	0	0	0	1	0	0	0	0	0	0
1	1	0	0	0	0	0	0	0	0	0	0
1	1	0	0	0	0	1	0	0	0	0	0
1	1	0	0	0	0	0	1	0	0	0	0
1	1	0	0	0	0	0	0	1	0	0	0
1	0	1	0	0	0	0	0	0	0	0	0
1	0	1	0	0	0	0	0	0	1	0	0
1	0	1	0	0	0	0	0	0	0	1	0
1	0	1	0	0	0	0	0	0	0	0	1

To compare the 12 temperature-electrolyte combinations in Table 7.25, we focus on the 0.1 quantile of the lifetime distribution at each of these combinations, expressed by

$$\exp(\mu - 1.28\sqrt{\sigma_s^2 + \sigma_w^2}) \,, \tag{7.27}$$

where $\mu = \mathbf{x}^T\boldsymbol{\beta}$. We obtain draws from its posterior distribution by taking the posterior draws of $(\boldsymbol{\beta}, \sigma_s^2, \sigma_w^2)$ and evaluating the 0.1 quantile of the lifetime distribution in Eq. 7.27. As a figure of merit to compare the 12 factor-level combinations, use the 0.1 quantiles of these posterior distributions, which are 90% lower credible bound. The figures of merit corresponding to the rows of

Table 7.25 are

$$1.209, 0.840, 1.056, 1.154, 1.132, 0.831, 0.996, 1.037, 1.088, 0.904, 0.950, 1.037,$$

which show that the low temperature and electrolyte A is the best factor-level combination.

Table 7.27. Posterior distribution summaries for battery experiment data model parameters

			Quantiles				
Parameter	Mean	Std Dev	0.025	0.050	0.500	0.950	0.975
β_0	0.6168	0.0943	0.4305	0.4622	0.6167	0.7696	0.8016
β_1	−0.0666	0.1329	−0.3310	−0.2838	−0.0678	0.1500	0.1971
β_2	−0.1023	0.1325	−0.3605	−0.3181	−0.1028	0.1185	0.1637
β_3	−0.3652	0.0630	−0.4917	−0.4702	−0.3653	−0.2611	−0.2419
β_4	−0.1339	0.0640	−0.2599	−0.2383	−0.1339	−0.0281	-0.0070
β_5	−0.0433	0.0631	−0.1658	−0.1462	−0.0427	0.0590	0.0797
β_6	−0.3058	0.0641	−0.4308	−0.4111	−0.3057	−0.1997	−0.1775
β_7	−0.1234	0.0642	−0.2494	−0.2273	−0.1235	−0.0181	0.0017
β_8	−0.0869	0.0641	−0.2112	−0.1919	−0.0870	0.0187	0.0392
β_9	−0.1848	0.0635	−0.3109	−0.2899	−0.1852	−0.0784	−0.0584
β_{10}	−0.1341	0.0633	−0.2598	−0.2389	−0.1332	−0.0301	−0.0116
β_{11}	−0.0476	0.0633	−0.1744	−0.1526	−0.0478	0.0558	0.0759
σ_s	0.1094	0.0118	0.0893	0.0918	0.1083	0.1303	0.1359
σ_w	0.1989	0.0417	0.1342	0.1423	0.1931	0.2750	0.2956

7.9 Other Regression Situations

Regression models arise in other situations considered in previous chapters. For example, in assessing system reliability as presented in Chap. 5, we may describe component data by appropriate regression models. For example, see Exercise 7.15, which considers such a problem for a three-component system. Regression models are likely to arise in repairable system reliability, although Chap. 6 does not specifically address this topic. For example, the reliability of a repairable system may depend on who makes it, who operates it, and in what conditions it is used. Finally, we discuss regression models for degradation data in Chap 8.

7.10 Related Reading

See McCullagh and Nelder (1989) for more details on GLMs. For GLM residuals, see McCullagh and Nelder (1989) and Pierce and Schafer (1986). For censored residuals, see Chaloner (1991) and Collett (1994).

For accelerated life testing, see Nelson (1990). Dorp and Mazzuchi (2005) presents a general Weibull accelerated testing model, which does not require strict adherence to a parametric time-transformation function. Rather, Dorp and Mazzuchi (2005) uses prior information indirectly to define a multivariate prior distribution for the Weibull scale parameters at various stress levels and the common Weibull shape parameter.

For reliability improvement experiments, see Wu and Hamada (2000) and Condra (1993).

Gelman (2006) suggests using a $Uniform(0, U)$ distribution (large U) as a diffuse prior distribution for the standard deviation of the random effects normal distribution if the number of random effects is small. This choice of prior distribution has little impact on the results for Examples 7.2, 7.3, and 7.13, although these datasets involve a large number of random effects.

7.11 Exercises for Chapter 7

7.1 The analysis presented in Example 7.11 suggests that the simpler model without the PCB type by temperature interaction may fit well. Use BIC of Sect. 4.6 to choose between the simpler and full models.

7.2 Cox (1970) presents data from an experiment involving five levels of soaking time and four levels of heating time, which Table 7.28 displays. Assume that the times are in minutes. This experiment tested n ingots, of which r were the number of ingots not ready for rolling. Incorporate the soaking time and heating time factors in an appropriate model for these data. What do you conclude about the impact of these factors on the probability of an ingot not being ready for rolling?

Table 7.28. Ingot experiment data (r/n, where r is the number of ingots not ready for rolling and n is the number of tested ingots) (Cox, 1970)

Soaking	Heating Time			
Time	7	14	27	51
1.0	0/10	0/31	1/56	3/13
1.7	0/17	0/43	4/44	0/1
2.2	0/7	2/33	0/21	0/1
2.8	0/12	0/31	1/22	0/0
4.0	0/9	0/19	1/16	0/1

7.3 Assess the fit of the logistic regression model for the HPCI system safety data in Example 7.1 using a Bayesian χ^2 goodness-of-fit test.

7.4 Suppose that we bin the data in Table 7.1. In other words, suppose that we ignore the actual dates at which the HPCI demands occurred, simply using the fact that, for example, in 1987 there were a total of 4 binomial failures in 16 demands. Using such binned binomial data for each of the seven years, develop and use a logistic regression model to determine the marginal posterior distributions on the probability of HPCI failure π_i in year i =1, ..., 7. How do these results compare to those in Example 7.1?

7.5 Table 7.29 gives the number of demands for the HPCI system during 1987–1993 and the corresponding exposure times (in reactor-critical-years) for 23 commercial U.S. boiling water nuclear power plants. Reactor-critical-years is a relevant variable because the HPCI system uses a turbine-driven pump upon which we can only make demands when the reactor is producing steam. Use a loglinear analysis to estimate and identify any trend in the HPCI demand rate over time.

Table 7.29. HPCI demand data

Calendar Year	HPCI Demands	Reactor-Critical-Years
1987	16	14.63
1988	10	14.15
1989	7	15.75
1990	13	17.77
1991	9	17.11
1992	6	17.19
1993	2	17.34

7.6 Table 7.30 presents data from Shaw et al. (1998) on the number of leaks in pressurized water reactor (PWR) stainless-steel primary-coolant-system piping in 217 PWRs from initial criticality through May 31, 1998. Note that plant age is summarized in five-year bins.

Table 7.30. Piping leak data (Shaw et al., 1998)

Age	Number of Leaks	Reactor Years
0.0–5.0	2	1052.0
5.0–10.0	1	982.5
10.0–15.0	4	756.9
15.0–20.0	4	442.4
20.0–25.0	2	230.9
25.0–30.0	0	43.9

Taking the midpoint of each age bin as the "age" of the plants in that bin, is the leak rate changing as the plants age? Support the claim with a loglinear analysis of these data.

7.7 Perform an analysis of the data in Table 7.8, assuming a lognormal distribution of tensile strength. Compare the results to those in Example 7.4.

7.8 Take one of the lognormal examples in this chapter and try a Weibull or gamma regression model. Do these models fit the data better?

7.9 Similar to the experiment in Example 4.8, Ku et al. (1972) reports on an experiment with a different lubricant called O-67-22. Table 7.31 presents the resulting lifetimes. Is σ^2 for O-67-22 similar to that found in Example 4.8? Is μ for this lubricant different from that for the other lubricant? Also fit the data from both experiments in one model using an appropriate covariate and assuming a common σ^2 for the two lubricants. Based on the combined analysis, is one lubricant better than the other?

Table 7.31. Bearing fatigue failure times (in hours) for lubricant O-67-22 (Ku et al., 1972)

Tester	Failure Time									
1	140.3	158.0	183.9	132.7	117.8	98.7	164.8	136.6	93.4	116.6
2	193.0	172.5	173.3	204.7	172.0	152.7	234.9	216.5	422.6	262.6
3	73.5	263.7	192.3	37.1	160.3	159.2	133.5	200.7	189.6	157.1
4	196.5	218.9	196.9	253.3	212.5	239.6	181.3	193.0	178.3	262.8
5	145.7	116.5	150.5	141.6	129.0	178.4	133.4	120.2	192.6	179.0
6	171.9	188.1	191.6	154.3	171.3	157.4	132.2	156.7	194.8	173.3
7	183.2	222.4	197.5	211.0	178.0	130.5	160.9	197.6	90.0	213.0
8	244.0	179.2	176.2	207.7	148.2	121.6	195.0	133.6	167.2	98.5
9	187.4	202.0	175.0	171.7	230.5	174.2	220.2	166.3	239.8	223.7
10	186.0	202.0	200.9	137.1	195.8	162.4	134.6	174.5	272.9	173.8

7.10 In Example 7.13, assume a Weibull distribution for the lifetimes. Propose a data model for this split-plot experiment and analyze the data from Example 7.13 using this model. How do the results obtained with this model compare with those found for Example 7.13?

7.11 Nelson (1984) reports on strain-controlled, low-cycle fatigue testing of nickel-base superalloy specimens. Experience suggests that the standard deviation of the specimen logged lifetimes depends on the test stress. Table 7.32 presents the experimental lifetime data in cycles. An asterisk by the lifetime indicates a Type I-censored observation, i.e., the lifetime exceeds the stated number of cycles. Analyze these data assuming a lognormal distribution with parameters μ and σ^2 as functions of stress. Nelson (1984) centers log stress by subtracting the mean log stress calculated from all of the log stresses listed in Table 7.32 and considers both linear and quadratic models for μ, and a linear model for $\log(\sigma^2)$. Does a constant σ^2 model fit these data better? Make inferences for reliability over a range of stresses.

Table 7.32. Superalloy specimen lifetime data (in cycles at various stresses in MPa) (Nelson, 1984). An asterisk indicates a Type I-censored observation

Stress	Lifetime	Stress	Lifetime
145.9	5733	114.8	21300
85.2	13949	144.5	6705
116.4	15616	91.3	112002
87.2	56723	142.5	11865
100.1	12076	100.5	13181
85.8	152680	118.4	8489
99.8	43331	118.6	12434
113	18067	118	13030
120.4	9750	80.8	57923*
86.4	156725	87.3	121075
85.6	112968*	80.6	200027
86.7	138114*	80.3	211629
89.7	122372*	84.3	155000

7.12 When manufacturing windshield moldings, a stamping process carries debris into the die. The debris creates dents in the molding, which results in imperfect parts. Martin et al. (1987) describes an experiment performed to improve this *slugging* condition. The experiment studied four factors each at two levels: (A) poly-film thickness (0.0025, 0.00175), (B) oil mixture (1:20, 1:10), (C) gloves (cotton, nylon), and (D) metal blanks (dry underside, oily underside), with their factor levels given in parentheses. The experimenters used a 2^{4-1} fractional factorial design (i.e., a 1/2 fraction (2^{-1}) of a full factorial or 2^4 design), as given in Table 7.33. We denote the two levels by -1 and 1 in Table 7.33. Note that the A main effect is aliased with the CD interaction effect, i.e., the product of the C and D is identical to the A column. Besides the A, B, C, and D main effects, entertain the interactions AB, BC, and BD; obtain the covariates associated with the interactions by multiplying the appropriate pair of columns. For each run, the experimenters manufactured 1,000 parts and recorded the number of good parts out of the 1,000 parts. Analyze these data using a logistic regression model. What factors are the most important?

7.13 Bullington et al. (1993) reports on an experiment to improve the reliability of industrial thermostats. Corrosion-induced pinholes in the diaphragm, a key component of the thermostat, had caused an increase in early thermostat failures. Consequently, there was a need to perform an experiment to find the key factors (among a large number of possible factors) affecting the rate of corrosion. The experimenters chose 11 factors

Table 7.33. Experimental design and data for the molding experiment (number of good parts out of 1,000) (Martin et al., 1987)

	Factor				Number
Run	A	B	C	D	Good Parts
1	−1	−1	−1	−1	338
2	−1	−1	1	1	826
3	−1	1	−1	−1	350
4	−1	1	1	1	647
5	1	−1	−1	1	917
6	1	−1	1	−1	977
7	1	1	−1	1	953
8	1	1	1	−1	972

(levels in parentheses) from across a 14-stage manufacturing process: (A) diaphragm plating rinse (clean, contaminated), (B) current density (5/60, 10/15 in minutes/amps), (C) sulfuric acid cleaning (3, 30 in seconds), (D) diaphragm electroclean (2, 12 in minutes), (E) beryllium copper grain size (0.008, 0.018 in inches), (F) stress orientation (perpendicular to seam, in-line with seam), (G) humidity (wet, dry), (H) heat treatment (45 minutes at 600°F, 4 hours at 600°F), (I) brazing machine (2, 3 in seconds), (J) power element electroclean (clean, contaminated), (K) power element plating rinse (clean, contaminated). The experimenters manufactured 10 thermostats for each of the 12 factor settings (runs) and tested them up to 7,342 (×1000) cycles. Table 7.34 gives the experimental design and lifetime data; for thermostats still functioning after 7,342 (×1000) cycles, their lifetimes are Type I censored and denoted by 7,342 in the table. Fit a lognormal regression model with main effects for the 11 factors; use the columns in Table 7.34 as the covariates. Factors E and H turn out to be the most important. Try fitting the data with the E and H main effects and the EH interaction; obtain the covariate associated with the EH interaction by multiplying the E and H columns in Table 7.34. Compare this model to the model with all of the factor main effects. Would a Weibull or gamma distribution be better suited for fitting these lifetime data?

7.14 Moore and Beckman (1988) reports the failure records of 90 valve types from a pressurized water reactor as presented in Table 7.36. The five factors that may impact failure are operating system (SYS), operating method (OTY), valve type (VTY), head size (SIZ), and operating mode (OPM). Table 7.35 gives the levels of these five factors. The failure data consist of the number of failures in the stated time period (in hundreds of hours).

a) Fit an appropriate Poisson regression model.

b) Provide a 90% credible upper bound on the predicted number of failures in the next 10 years (87,600 hours) for a normally closed 2- to 10-inch air-driven globe valve in a power conversion system.

7.15 Chapter 5 considered multilevel data to assess the reliability of a system. Some of the component data may involve covariates, which we can analyze using appropriate regression models. Suppose that we have a three-component series system as given in Fig. 7.18. The system level data consist of the number of successes and failures by age of the system given in Table 7.37. The component 1 data are also success/failure counts at various ages as given in Table 7.38. The component 2 data are lifetimes, as given in Table 7.39. The component 3 data are destructive degradation measurements at various ages, as given in Table 7.40.

a) Model the component 1 data y_{1i} at times t_{1i} as $Binomial(n_{1i}, p_{1i})$, where $\log[p_{1i}/(1-p_{1i})] = \alpha_0 + \alpha_1 t_{1i}$.

b) Model the component 2 data y_{2i} as having $Weibull(\lambda, \beta)$ distributions with scale λ and shape β.

c) Model the component 3 data y_{3i} at times t_{3i} as $LogNormal(\mu_{3i}, \sigma^2)$, where $\mu_{3i} = \gamma_0 + \gamma t_{3i}$.

d) Further, assume the following prior distributions:

$$\begin{aligned}
\alpha_0 &\sim Normal(0, 100),\\
\alpha_1 &\sim Normal(0, 1),\\
\lambda &\sim Gamma(0.1, 0.1),\\
\beta &\sim Gamma(1, 1),\\
\gamma_0 &\sim Normal(0, 100),\\
\gamma &\sim Normal(0, 1), \text{ and}\\
\sigma &\sim Gamma(1, 1).
\end{aligned}$$

e) Derive the expression for system reliability. Use this to specify the likelihood for the system data.

f) Write out the likelihoods for the three sets of component data.

g) Write out an expression that is proportional to the joint posterior distribution of the model parameters.

h) Fit the model and plot the median of the posterior distribution of reliability as a function of time with corresponding 90% credible intervals for ages 1 to 25 years.

7.16 Compare deviance residual plots for binary data from a linear covariate (on the logit scale) with that when the data come from a model with a quadratic covariate or cubic covariate. Does this study suggest that binary data deviance residuals are useful for model assessment?

7.17 For the gamma regression model, verify that Cox-Snell residuals do not exist. Instead, develop deviance residuals (McCullagh and Nelder, 1989) and show how to use them.

7.18 Perform a residual analysis for the solder experiment in Example 7.11.

Table 7.34. Experimental plan and lifetime data (in 1,000s of cycles) for the thermostat experiment (Bullington et al., 1993)

	Factor											Failure Time									
Run	A	B	C	D	E	F	G	H	I	J	K										
1	−1	−1	−1	−1	−1	−1	−1	−1	−1	−1	−1	957	2846	7342	7342	7342	7342	7342	7342	7342	7342
2	−1	−1	−1	−1	−1	+1	+1	+1	+1	+1	+1	206	284	296	305	313	343	364	420	422	543
3	−1	−1	+1	+1	+1	−1	−1	−1	+1	+1	+1	63	113	129	138	149	153	217	272	311	402
4	−1	+1	−1	+1	+1	−1	+1	+1	−1	−1	+1	75	104	113	234	270	364	398	481	517	611
5	−1	+1	+1	−1	+1	+1	−1	+1	−1	+1	−1	97	126	245	250	390	390	479	487	533	573
6	−1	+1	+1	+1	−1	+1	+1	−1	+1	−1	−1	490	971	1615	6768	7342	7342	7342	7342	7342	7342
7	+1	−1	+1	+1	−1	−1	+1	+1	−1	+1	−1	232	326	326	351	372	446	459	590	597	732
8	+1	−1	+1	−1	+1	+1	+1	−1	−1	−1	+1	56	71	92	104	126	156	161	167	216	263
9	+1	−1	−1	+1	+1	+1	−1	+1	+1	−1	−1	142	142	238	247	310	318	420	482	663	672
10	+1	+1	+1	−1	−1	−1	−1	+1	+1	−1	+1	259	266	306	337	347	368	372	426	451	510
11	+1	+1	−1	+1	−1	+1	−1	−1	−1	+1	+1	381	420	7342	7342	7342	7342	7342	7342	7342	7342
12	+1	+1	−1	−1	+1	−1	+1	−1	+1	+1	−1	56	62	92	104	113	121	164	232	258	731

Table 7.35. Valve data factors and levels

Factor	Levels 1	2	3	4	5	6
SYS	containment	nuclear	power conversion	safety	process auxiliary	
OTY	air	solenoid	motor driven	manual		
VTY	ball	butterfly	diaphragm	gate	glove	directional control
SIZ	$\leq$ 2 inches	2-10 inches	10-30 inches			
OPM	normally closed	normally open				

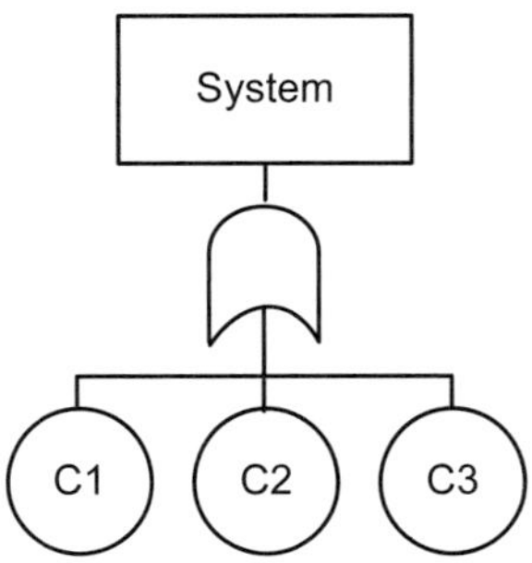

Fig. 7.18. Three-component series system.

7.19 Perform a residual analysis for the capacitor experiment in Example 7.12. Note that for each run, there are four observed lifetimes and the other six lifetimes are Type I censored at the largest observed lifetime.

7.20 Perform a residual analysis for the batteries experiment in Example 7.13.

7.21 Whitman (2003) provides data for an ALT performed in the microelectronics industry as given in Table 7.41. (Whitman (2003) does not provide the units of the lifetimes.) The experiment varied temperature at three levels (200, 215, and 230°C), where the use condition is 25°C. Note that all of the data are either interval censored or Type I censored. Whitman (2003) assumes a lognormal Arrhenius model in which the Arrhenius relationship holds for the median lifetime, i.e., for lifetimes distributed $LogNormal(\mu, \sigma^2)$, $\log(\mu) = \gamma_0 + \gamma_1(1/T)$ for temperature T. Analyze these data and assess the reliability at the use condition. Also perform a residual analysis.

7.22 The inverse power or simply power relationship is another model relating lifetime to an accelerating stress, such as elevated voltage. For a positive

Table 7.36. Valve data, factors, and number failures in time period in hundreds of hours (Moore and Beckman, 1988)

SYS	OTY	VITY	SIZ	OPM	Number Failed	Time Period	SYS	OPM	VTY	SIZ	OPM	Number Failed	Time Period
1	3	4	3	1	2	1752	3	3	4	1	1	0	3066
1	3	4	3	2	2	1752	3	3	4	1	2	0	1752
1	3	5	1	1	1	876	3	3	4	2	1	8	3504
2	1	2	2	2	0	876	3	3	4	2	2	0	1314
2	1	3	2	1	0	876	3	3	4	3	1	13	876
2	1	3	2	2	0	438	3	3	4	3	2	3	1314
2	1	5	1	1	2	1752	3	3	5	1	2	0	1314
2	1	5	1	2	4	2628	3	3	5	2	2	0	2190
2	1	5	2	1	1	438	3	4	4	2	2	1	1752
2	1	5	2	2	2	438	3	4	4	3	2	1	4380
2	2	5	2	2	3	876	3	4	5	2	2	0	1752
2	3	4	2	1	0	876	4	3	3	3	2	2	438
2	3	4	2	2	0	1752	4	3	4	2	1	2	3504
2	3	4	3	1	0	1314	4	3	4	2	2	0	1752
2	3	4	3	2	0	438	4	3	4	3	2	7	1314
2	3	5	1	1	1	876	4	3	5	1	2	0	438
2	3	5	2	2	0	1752	5	1	2	2	1	0	1314
2	3	5	3	2	0	876	5	1	2	2	2	0	876
2	4	3	1	2	0	438	5	1	2	3	1	0	438
2	4	3	2	1	0	438	5	1	2	3	2	0	2190
2	4	4	1	1	2	438	5	1	3	1	1	0	438
2	4	5	2	1	0	876	5	1	3	1	2	0	1314
3	1	1	2	1	1	15768	5	1	3	2	2	0	876
3	1	1	2	2	2	1752	5	1	4	2	1	3	1752
3	1	1	3	2	0	876	5	1	4	2	2	0	1752
3	1	2	2	1	0	876	5	1	5	1	1	3	438
3	1	2	3	1	3	3504	5	1	5	1	2	2	1314
3	1	3	2	1	1	6570	5	1	5	2	2	0	3504
3	1	3	2	2	0	1752	5	1	6	1	1	0	438
3	1	4	1	1	0	438	5	1	6	2	2	0	876
3	1	4	1	2	0	876	5	2	3	2	2	0	4818
3	1	4	2	1	5	4818	5	2	4	1	1	0	438
3	1	4	2	2	23	2628	5	3	2	2	1	0	438
3	1	4	3	2	21	1752	5	3	2	2	2	0	876
3	1	5	1	1	0	1752	5	3	2	3	1	2	1752
3	1	5	1	2	0	1752	5	3	2	3	2	0	876
3	1	5	2	1	11	13578	5	3	4	2	1	2	2190
3	1	5	2	2	3	13578	5	3	4	2	2	1	6132
3	1	5	3	2	2	438	5	3	5	2	2	0	876
3	1	6	2	1	1	876	5	4	3	1	1	1	2190
3	1	6	2	2	0	438	5	4	3	1	2	0	876
3	1	6	3	2	0	438	5	4	3	2	1	0	1314
3	2	6	2	2	1	876	5	4	4	1	2	0	438
3	3	2	2	1	0	438	5	4	4	2	1	0	438
3	3	2	3	2	0	438	5	4	5	2	2	0	438

Table 7.37. System data for three-component series system

Age (years)	Successes	Failures
0	14	15
5	15	15
10	15	15
15	15	15
20	12	15

Table 7.38. Component 1 data for three-component series system

Age (years)	Successes	Failures
0	25	25
2	25	25
4	24	25
6	25	25
8	25	25
10	25	25
15	23	25
20	19	25

Table 7.39. Component 2 data for three-component series system

Lifetime (years)									
54.95	24.41	102.50	18.75	55.53	86.13	59.49	48.25	49.45	69.76
30.64	32.42	35.25	76.33	40.09	62.48	57.42	36.41	44.72	61.03
26.73	42.23	57.64	35.86	31.50					

Table 7.40. Component 3 data for three-component series system

Age (years)	Destructive Measurement									
0.0	4.60	6.74	4.17	10.30	7.60	2.69	3.88	5.69	5.82	5.77
2.5	4.15	17.97	1.87	3.25	3.00	1.56	4.59	1.76	1.42	2.76
5.0	3.50	2.41	4.49	4.23	2.98	5.25	1.93	3.96	1.46	1.97
7.5	6.79	1.15	1.84	1.63	1.13	3.89	5.57	1.91	2.27	2.67
10.0	4.77	2.19	2.84	9.39	1.94	3.87	2.45	4.49	4.73	4.11
15.0	6.58	2.38	9.05	9.99	2.89	2.61	3.05	2.10	3.17	2.86
20.0	1.62	6.90	5.12	1.28	0.88	3.38	1.73	1.65	2.01	2.60

Table 7.41. Data for a microelectronics ALT (Whitman, 2003)

Temperature (°C)	(Lifetime Interval) Count				
200	(0, 1094)	(1094, 1521)	(1521, 1948)	(1948, 2338)	(2338, 2886)
	3	2	4	2	5
	(2886, 3469)	(3469, 4185)	(6797, ∞)		
	2	3	9		
215	(0,344)	(344, 478)	(478, 612)	(612, 735)	(735, 907)
	2	1	3	2	3
	(907, 1090)	(1090, 1315)	(1315, 1624)	(1624, 2136)	(2136, ∞)
	1	1	3	1	3
230	(0, 115)	(115, 160)	(160, 205)	(246, 304)	(304, 365)
	2	2	2	3	3
	(365, 441)	(441, 544)	(544, 716)	(716, ∞)	
	3	2	1	2	

stress variable V, the nominal lifetime takes the form A/V^{γ_1}. For Weibull lifetimes, we can express the characteristic life ψ (third parameterization of the Weibull distribution in Appendix B) as $\log(\psi) = \gamma_0 + \gamma_1[-\log(V)]$. The data displayed in Table 7.42 from Nelson (1990) are breakdown times of an insulating fluid at seven high voltages.

a) Analyze the lifetime data using the Weibull power relationship model.
b) Predict the mean lifetime and lifetime distribution at 20 kV.
c) Perform a residual analysis.

Table 7.42. Data for an insulating fluid ALT (Nelson, 1990)

Voltage (kV)	Lifetime (minutes)
26	5.79, 1579.52, 2323.70
28	68.85, 108.29, 110.29, 426.07, 1067.60
30	7.74, 17.05, 20.46, 21.02, 22.66, 43.40, 47.30, 139.07, 144.12, 175.88, 194.90
32	0.27, 0.40, 0.69, 0.79, 2.75, 3.91, 9.88, 13.95, 15.93, 27.80, 53.24, 82.85, 89.29, 100.58, 215.10
34	0.19, 0.78, 0.96, 1.31, 2.78, 3.16, 4.15, 4.67, 4.85, 6.50, 7.35, 8.01, 8.27, 12.06, 31.75, 32.52, 33.91, 36.71, 72.89
36	0.35, 0.59, 0.96, 0.99, 1.69, 1.97, 2.07, 2.58, 2.71, 2.90, 3.67, 3.99, 5.35, 13.77, 25.50
38	0.09, 0.39, 0.47, 0.73, 0.74, 1.13, 1.40, 2.38

7.23 See Example 10.5, which analyzes Weibull accelerated life test data on pressure vessels. The regression model for the data involves random effects because the pressure vessels are wrapped in Kevlar-49 fibers from eight different spools. Fit the model presented in Example 10.5 and assess the model fit using a Bayesian χ^2 goodness-of-fit test.

7.24 In Example 7.11, use the posterior draws for the model parameters as summarized in Table 7.21 to obtain predictive distributions at each of the nine PCB type and temperature factor level combinations. Based on these results, what is the recommended factor-level combination?

7.25 In Example 7.12, use the posterior draws for the model parameters as summarized in Table 7.24 to obtain predictive distributions at each of the eight temperature and voltage factor-level combinations. Based on these results, what is the recommended factor-level combination?

8

Using Degradation Data to Assess Reliability

While reliability analysts have long used lifetime data for product/process reliability assessments, they began to employ degradation data for the same purpose in the 1990s. Assessing reliability with degradation data has a number of advantages. The analyst does not have to wait for failures to occur and can use less acceleration to collect degradation data. This chapter explains how to assess reliability using degradation data and also discusses how to accommodate covariates such as acceleration factors that speed up degradation and experimental factors that impact reliability in reliability improvement experiments. We also consider situations in which degradation measurements are destructive and conclude by introducing alternative stochastic models for degradation data.

8.1 Introduction

Using lifetime data to assess the reliability of highly reliable products is often problematic. For a practical testing duration (which there is always pressure to reduce), few or perhaps no failures may occur. If most or all of the observations are censored, then these observations provide little information about reliability for a warranty period that may be orders of magnitude longer than the testing duration. We may use accelerated life tests as discussed in Sect. 7.7. However, predicting failure times at normal-use conditions, which requires extrapolation, depends critically on identifying an appropriate relationship between the accelerating variable and the failure time distribution.

Degradation data may provide a superior alternative to highly censored data, which provide little information, or accelerated data for which identifying an accelerating relationship can be difficult. All failures likely arise from a degradation mechanism at work, such as the progression of a chemical reaction or the propagation of a crack, which may or may not be observable. When there are several observed characteristics that degrade (or grow) over time, the analyst must choose one of the degrading characteristics, and then relate it to failure.

When we can define failure directly in terms of a particular observable characteristic, however, the issues of choosing a characteristic and relating it to failure vanish. For example, a crack grows over time, and failure is defined as occurring when the crack reaches a specified length. In another example, the brightness of a fluorescent light decreases over time, and industry defines failure as occurring when the fluorescent light's luminosity degrades to 60% of its luminosity after 100 hours of use (i.e., 0.6 × luminosity at 100 hours).

Failures defined in terms of observable characteristics are called *soft failures*, because the products are still working, albeit at a reduced level of performance. *Hard failures* occur when products fail completely. For hard failures, an analyst may model the probability of failure as a function of an observable characteristic. In addition to the challenge of developing a model, we may need to account for measurement error in the observable characteristic.

In the remainder of this chapter, we focus on the modeling of soft failures. In this case, degradation data often provide more information than lifetime data, and in a shorter time, demonstrated later in Example 8.1. The advantage of having more information depends on how much error the observable characteristic is measured with, however.

We begin to model degradation data by considering the simple situation in which a linear degradation curve exists and starts at 0. For the ith unit, assume that the degradation at time t is

$$D_i(t) = D(t, \theta_i) = (1/\theta_i)t, \tag{8.1}$$

i.e., it has intercept 0 and slope $1/\theta_i$, where θ_i is a unit specific parameter. When the unit's degradation reaches a critical *threshold* D_f, declare the unit as having failed. Consequently, the lifetime t_i of the ith unit is $D_f\theta_i$. Figure 8.1 illustrates the degradation for a single unit. For this particular unit, $\theta_i = 10.45$, and with $D_f = 50$ (dotted line), its lifetime is 523 (dashed line).

Figure 8.2 presents degradation curves from a sample of 100 units. How do these degradation curves provide information about their lifetime distribution? In order for lifetimes to follow a Weibull distribution, the reciprocal slopes θ_i must follow a Weibull distribution. It follows that if the θ_i have a $Weibull(\lambda, \beta)$ distribution, then the lifetimes have a $Weibull(\lambda/D_f^\beta, \beta)$ distribution, a Weibull distribution with scale λ/D_f^β and shape β (see the first form in Appendix B, without a location parameter). Consequently, these degradation curves provide information about their lifetime distribution through their reciprocal slopes. Figure 8.3 presents a histogram of the resulting 100 lifetimes from the degradation curves in Fig. 8.2, which exhibits the characteristic skewness of the Weibull distribution.

For the $Weibull(\lambda/D_f^\beta, \beta)$ distribution, an expression for the reliability function is

$$R(t) = \exp[-(\lambda/D_f^\beta)t^\beta],$$

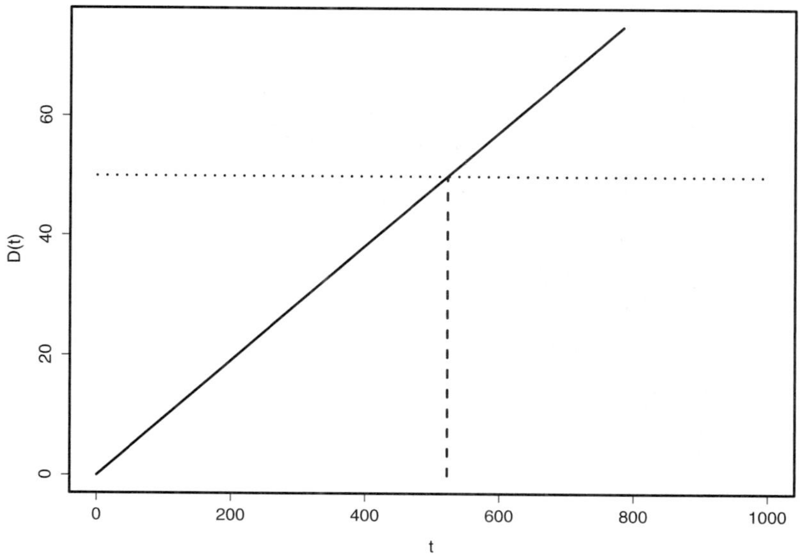

Fig. 8.1. Linear degradation over time t for a single unit. The solid line displays the unit's degradation. The dotted and dashed lines indicate the threshold D_f and the unit's lifetime, respectively.

and the α quantile of the lifetime distribution as

$$t_\alpha = [-(D_f^\beta/\lambda)\log(1-\alpha)]^{1/\beta}.$$

Recall that the α quantile is the time by which $\alpha \times 100\%$ of the population modeled by the lifetime distribution has failed.

Besides the form of the degradation curve, a degradation data model must account for measurement error. That is, degradation data obtained by sampling the degradation curve over time are usually measured with error, which we assume to be additive. Denote the degradation of the ith unit at the jth time t_{ij} as $D_i(t_{ij})$ and observe it with measurement error ε_{ij}. These assumptions lead to the following model for the observed degradation y_{ij}:

$$Y_{ij} = D_i(t_{ij}) + \varepsilon_{ij}, \tag{8.2}$$

where the measurement errors ε_{ij} follow a $Normal(0, \sigma_\varepsilon^2)$ distribution and σ_ε^2 is the measurement error variance. Note that $D_i(t_{ij})$ from Eq. 8.1 depends on the unit specific effect θ_i, which has a distribution, so that the degradation data model in Eq. 8.2 is a random effects model as discussed in Sect. 7.1.1. Whereas Fig. 8.1 presented the true degradation curve for a unit, Fig. 8.4 illustrates the observed degradation with normal measurement error for the same unit.

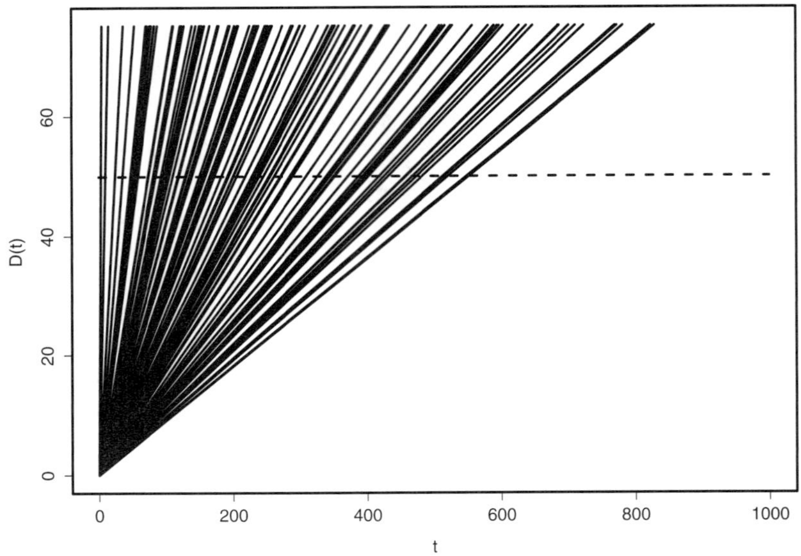

Fig. 8.2. Linear degradation over time t for a sample of 100 units for which the reciprocal slopes have a Weibull distribution. The dashed line indicates the threshold D_f. The resulting lifetimes have a Weibull distribution.

Example 8.1 Drug potency. Developing a new drug involves performing a stability study to determine the drug's shelf life. Because the potency of a drug degrades over time, define its shelf life (or lifetime) as the length of time it takes for the drug's potency to decrease to 90% of its original stated potency. Consequently, the threshold D_f is 90. Consider the degradation data in Table 8.1, as plotted in Fig. 8.5; we adapted these data from a stability study reported by Chow and Shao (1991), which followed 24 batches of a drug over a 36-month period. Here, assume that the 24 batches are a random sample of batches. Note that we adjusted the data so that their initial potency is 100% (but measured with error). However, we leave the analysis of the original data as Exercise 8.7.

Because the degradation curves in Fig. 8.5 are nearly linear, use the following linear degradation model:

$$D_i(t) = D(t, \theta_i) = 100 - (1/\theta_i)t, \tag{8.3}$$

with intercept 100 and slope $-1/\theta_i$. For the ith unit, Eq. 8.3 gives the potency at time t. For the degradation data or degradation observations y_{ij}, use Eq. 8.3 with normal measurement errors as presented in Eq. 8.2, that is, $Y_{ij} = 100 - (1/\theta_i)t_{ij} + \varepsilon_{ij}$.

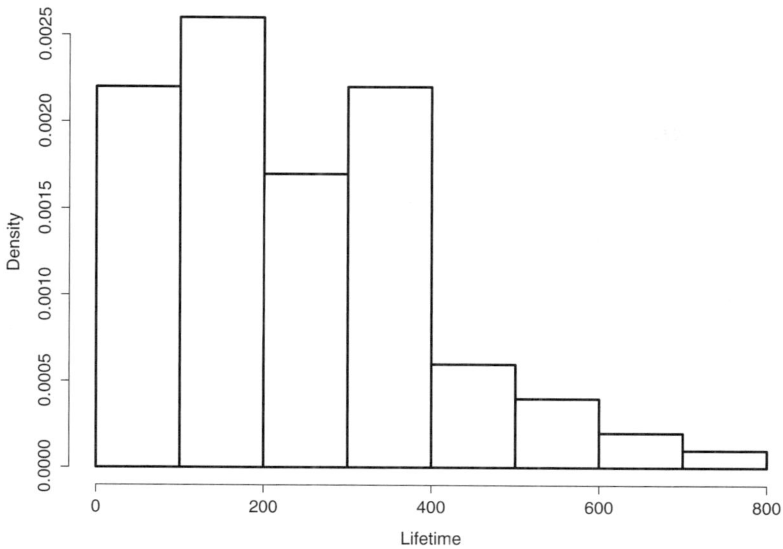

Fig. 8.3. Scaled histogram of 100 lifetimes, resulting from linear degradation.

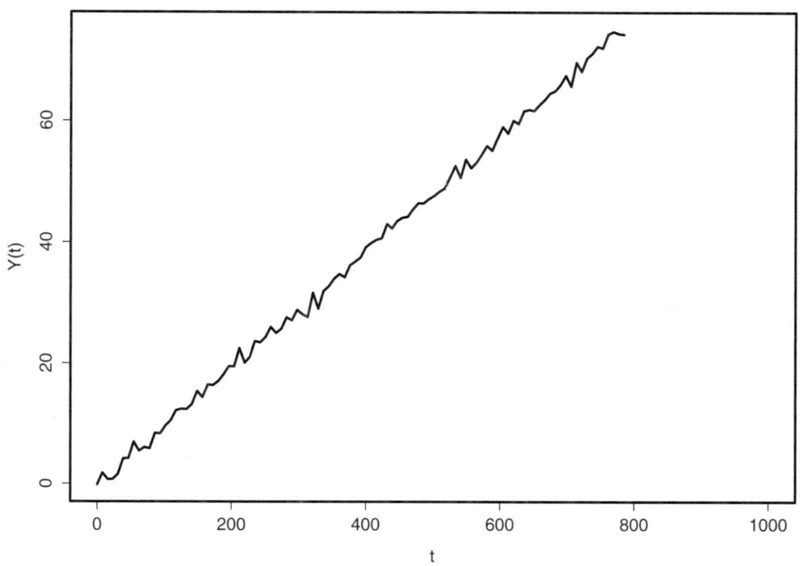

Fig. 8.4. Linear degradation over time t for a single unit, observed with measurement error.

Having specified the form of the degradation curve and measurement error, we need to determine a reasonable distribution for the reciprocal slopes θ_i. Here, the θ_i of the 24 batches are conditionally independent following some distribution, because the 24 batches are a random sample of batches. One way to determine a suitable distribution for the θ_i is by inspecting pseudo lifetimes. Obtain *pseudo lifetimes* for the 24 batches by first fitting Eq. 8.3 using least-squares separately to the data for each of the batches. From these fits, obtain the $\hat{\theta}_i$, estimates of θ_i, to calculate their corresponding pseudo lifetimes using $\hat{t}_i = (100 - 90)\hat{\theta}_i$; these times are when the fitted lines reach the threshold D_f. A lognormal probability plot of the pseudo lifetimes $\hat{t}_i$ (or equivalently, a normal probability plot of the logged pseudo lifetimes) suggests that the lifetimes follow a lognormal distribution instead of the Weibull distribution assumed in Sect. 8.1. Consequently, if the θ_i follow a $LogNormal(\mu_\theta, \sigma_\theta^2)$ distribution, then the lifetimes are $(100 - 90)\theta_i = 10\theta_i$ and have a $LogNormal[\log(10) + \mu_\theta, \sigma_\theta^2]$ distribution.

For the $LogNormal[\log(10) + \mu_\theta, \sigma_\theta^2]$ distribution, the reliability function is

$$R(t) = 1 - \Phi[(\log(t) - \log(10) - \mu_\theta)/\sigma_\theta], \tag{8.4}$$

and the α quantile of the lifetime distribution is

$$t_\alpha = \exp[z_\alpha \sigma_\theta + \log(10) + \mu_\theta], \tag{8.5}$$

where Φ is the standard normal cumulative distribution function, and z_α is the α quantile of the standard normal distribution.

Next, consider an analysis of the drug potency degradation data. A summary of the model for these data in Table 8.1 is

$$Y_{ij} \sim Normal[100 - (1/\theta_i)t_{ij}, \sigma_\varepsilon^2],$$

where

$$\theta_i \sim LogNormal(\mu_\theta, \sigma_\theta^2).$$

The likelihood function consists of a normal probability density contribution for each observed degradation y_{ij}; from the assumption that the Y_{ij} are conditionally independent, the overall likelihood function is a product of these contributions. A contribution of θ_i to the prior density function is the lognormal probability density function for θ_i. From the assumption that the θ_i are conditionally independent, the total contribution of the θ_i to the prior density function is the products of these contributions. We complete the model by specifying independent and diffuse prior distributions for μ_θ, σ_θ^2, and σ_ε^2:

$$\begin{aligned} \mu_\theta &\sim Normal(0, 10000), \\ \sigma_\theta^2 &\sim InverseGamma(0.01, 0.01), \text{ and} \\ \sigma_\varepsilon^2 &\sim InverseGamma(0.01, 0.01). \end{aligned}$$

This is an appropriate choice of prior distributions because μ_θ takes on real values, but σ_θ^2 and σ_ε^2 are both variances, which take on only positive

Table 8.1. Drug potency degradation data (in percent of original stated potency)

	Time (months)					Time (months)			
Batch	0	12	24	36	Batch	0	12	24	36
1	99.9	98.9	95.9	92.9	13	99.8	98.8	93.8	89.8
2	101.1	97.1	94.1	91.1	14	100.1	99.1	93.1	90.1
3	100.3	98.3	95.3	92.3	15	100.7	98.7	93.7	91.7
4	100.8	96.8	94.8	90.8	16	100.3	98.3	96.3	93.3
5	100.0	98.0	96.0	92.0	17	100.2	98.2	97.2	94.2
6	100.1	98.1	98.1	95.1	18	99.8	97.8	95.8	90.8
7	99.6	98.6	96.6	92.6	19	100.8	98.8	95.8	94.8
8	100.4	99.4	96.4	95.4	20	100.0	98.0	96.0	92.0
9	100.9	98.9	96.9	96.9	21	99.6	99.6	92.6	88.6
10	100.5	99.5	94.5	93.5	22	100.2	98.2	97.2	94.2
11	101.1	98.1	93.1	91.1	23	99.8	97.8	95.8	90.8
12	100.9	97.9	95.9	93.9	24	100.0	99.0	95.0	92.0

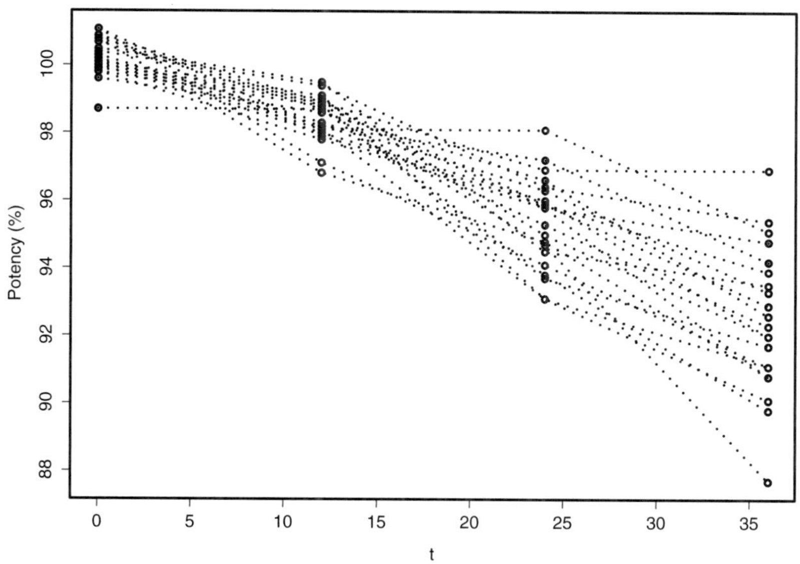

Fig. 8.5. Plot of drug potency degradation data over time t in months.

real values. If the number of random effects is small, also consider using a $Uniform(0, U)$ (large U) distribution as a diffuse prior for σ_θ (Gelman, 2006).

We obtain draws from the joint posterior distribution of $(\mu_\theta, \sigma_\theta^2, \sigma_\varepsilon^2)$ using Markov chain Monte Carlo (MCMC). See Table 8.2, which presents marginal posterior distribution summaries for these parameters.

Table 8.2. Posterior distribution summaries of the drug potency degradation model parameters (based on degradation data)

			Quantiles				
Parameter	Mean	Std Dev	0.025	0.050	0.500	0.950	0.975
μ_θ	1.646	0.057	1.537	1.555	1.645	1.741	1.761
σ_θ	0.2482	0.0481	0.1690	0.1790	0.2430	0.3339	0.3563
σ_ε	0.9221	0.0790	0.7835	0.8030	0.9162	1.0600	1.0930
$R(36)$	0.9249	0.0468	0.8102	0.8348	0.9343	0.9828	0.9875
$t_{0.1}$	37.87	2.94	31.61	32.74	38.05	42.36	43.12

By using a Bayesian approach, we can easily obtain the posterior distribution for the reliability function $R(t)$ at time t; for each draw from the joint posterior distribution of $(\mu_\theta, \sigma_\theta^2, \sigma_\varepsilon^2)$, simply evaluate $R(t)$ given in Eq. 8.4 to obtain a draw from the reliability posterior distribution. Figure 8.6 displays the posterior medians and 90% credible intervals of the drug potency reliability $R(t)$ based on Eq. 8.4, where the 0.05 and 0.95 posterior quantiles determine the 90% credible intervals. Note how in Fig. 8.6 the drug potency reliability starts to dramatically drop after 30 months, as well as the associated uncertainty that increases with age. As an example, Table 8.2 provides the posterior distribution summaries for $R(36)$, the reliability at 36 months, which shows that the 90% credible interval of $R(36)$ is $(0.835, 0.923)$; Fig. 8.6 plots this 90% credible interval as a dotted line. Similarly, we can assess the α quantile of the shelf life denoted by t_α through its posterior distribution; easily obtain draws from the posterior distribution of t_α, by evaluating Eq. 8.5 with draws from the joint posterior distribution of $(\mu_\theta, \sigma_\theta^2)$, as displayed in Fig. 8.7. As an example, Table 8.2 presents the posterior distribution summaries for the 0.1 quantile, denoted by $t_{0.1}$ (i.e., the age by which 10% of the drugs will have failed); the 95% credible interval of $t_{0.1}$, based on the 0.025 and 0.975 posterior quantiles, is $(31.61, 43.12)$ months, as displayed in Fig. 8.7 as a dashed line.

8.1.1 Comparison with Lifetime Data

Using Example 8.1, let us consider the claim that degradation data are advantageous because they generally provide more information than do lifetime data. Reviewing the degradation data in Table 8.1 reveals that the lifetime

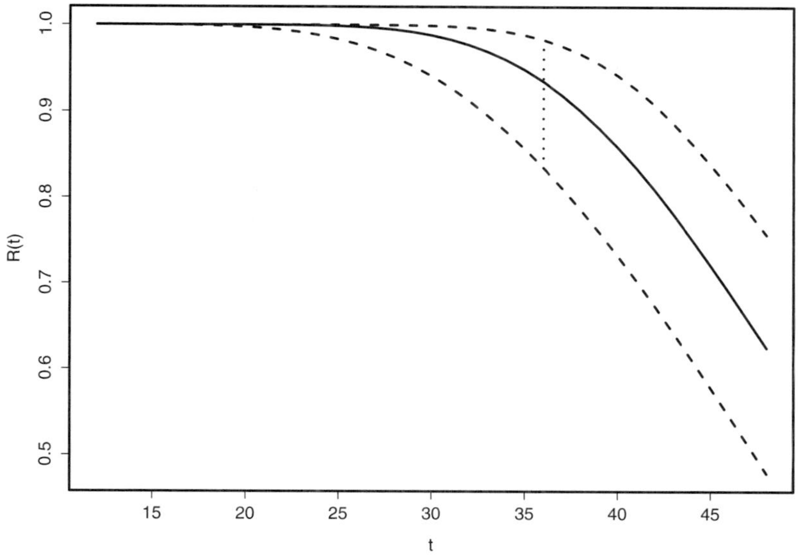

Fig. 8.6. Drug potency reliability over time t in months and 90% credible intervals. The solid line gives the posterior medians. The dotted lines give the 0.05 and 0.95 posterior quantiles. The dotted line shows the 90% credible interval for $R(36)$, the reliability at 36 months.

data consist of two interval-censored observations, both (24, 36) for the 13th and 21st batches, and 22 Type I-censored observations (at 36) for the remaining batches. Recall that the lifetimes follow a $LogNormal[\log(10) + \mu_\theta, \sigma_\theta^2]$ distribution. In analyzing these lifetime data, we use the same prior distributions for μ_θ and σ_θ^2, and obtain draws from their joint posterior distribution with MCMC. See Table 8.3, which presents posterior distribution summaries for μ_θ and σ_θ^2, as well as those for $R(36)$ and $t_{0.1}$. Note the decrease in precision of the results based on the lifetime data as compared with the results obtained from degradation data in Table 8.2. In particular, the 90% credible interval for $R(36)$ is now $(0.791, 0.978)$, which is wider than the interval obtained using the degradation data (i.e., $(0.835, 0.923)$). Also, the 95% credible interval for $t_{0.1}$ is $(30.95, 42.98)$ months, which is wider than the interval obtained using the degradation data (i.e., $(31.61, 43.12)$ months).

8.2 More Complex Degradation Data Models

Up to this point, we have considered only simple degradation data models, such as Eq. 8.2, with the true degradation curve being linear in time t as given in Eq. 8.1. This section considers more complex models for degradation data.

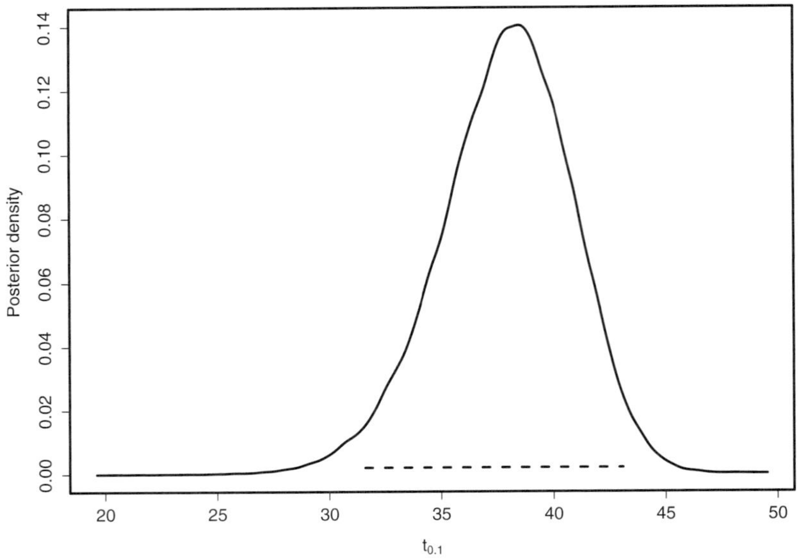

Fig. 8.7. Posterior distribution of the 0.1 quantile of the drug shelf life distribution $t_{0.1}$. The dashed line shows the 95% credible interval.

Table 8.3. Posterior distribution summaries of drug potency model parameters (based on lifetime data)

			Quantiles				
Parameter	Mean	Std Dev	0.025	0.050	0.500	0.950	0.975
μ_θ	1.569	0.114	1.379	1.397	1.561	1.769	1.806
σ_θ	0.2128	0.0767	0.0828	0.0954	0.2095	0.3438	0.3715
$R(36)$	0.9029	0.0586	0.7617	0.7910	0.9138	0.9781	0.9844
$t_{0.1}$	36.68	2.93	30.95	32.06	36.57	41.70	42.98

First, the form of the true degradation curve $D(\cdot)$ may be a nonlinear function of time t like

$$a_0/(1 - a_0^{\theta_2}\theta_1\theta_2 t)^{1/\theta_2} ,$$

as used in Example 8.2. For the ith unit, express its true degradation curve as

$$a_0/(1 - a_0^{\theta_{2i}}\theta_{1i}\theta_{2i} t)^{1/\theta_{2i}} .$$

Here, there is one parameter common to all units, a_0, but there can be more. Also, there are two parameters that are specific to the ith unit, $(\theta_{1i}, \theta_{2i})$. Further, we assume that the units are a random sample of units so that the $(\theta_{1i}, \theta_{2i})$ are conditionally independent with a specified distribution. In summary, an expression for a more complex degradation curve is

$$D(t, \boldsymbol{\theta}_i, \boldsymbol{\nu}), \tag{8.6}$$

where $D(\cdot)$ is some function of t, $\boldsymbol{\theta}_i$ is a vector of random effects, and $\boldsymbol{\nu}$ is a vector of parameters common to all units. Consequently, the distribution assumed for $\boldsymbol{\theta}_i$ describes a population of true degradation curves $D(t, \boldsymbol{\theta}_i, \boldsymbol{\nu})$ given $\boldsymbol{\nu}$. For example, we may assume that the $\boldsymbol{\theta}_i$ or possibly transformed $\boldsymbol{\theta}_i$, say $g(\boldsymbol{\theta}_i)$ for some function $g(\cdot)$, follow a multivariate normal distribution with parameters $(\boldsymbol{\mu}, \boldsymbol{\Sigma})$. A more general description is

$$g(\boldsymbol{\theta}_i) \sim \mathbf{H}(\boldsymbol{\eta}), \tag{8.7}$$

where the distribution $\mathbf{H}$ has parameters $\boldsymbol{\eta}$.

Now, let y_{ij} be the observed degradation for the ith unit at the jth time t_{ij}. Then, one possible model for degradation data is

$$Y_{ij} = D(t_{ij}, \boldsymbol{\theta}_i, \boldsymbol{\nu}) + \varepsilon_{ij}. \tag{8.8}$$

Typically, we assume that the measurement errors ε_{ij} are conditionally independent and distributed as $Normal(0, \sigma_\varepsilon^2)$. But the measurement errors need not have a normal distribution. Neither do the measurement errors need to be additive. For example, a multiplicative measurement error model has the form $Y_{ij} = D(t_{ij}, \boldsymbol{\theta}_i, \boldsymbol{\nu})\varepsilon_{ij}$, and the ε_{ij} might have a lognormal distribution.

In an analysis of these more complex degradation data models, we need to specify prior distributions for $\boldsymbol{\nu}$, $\boldsymbol{\eta}$, and σ_ε^2. We then can obtain draws from the corresponding joint posterior distribution of $(\boldsymbol{\nu}, \boldsymbol{\eta}, \sigma_\varepsilon^2)$ by MCMC.

8.2.1 Reliability Function

For the more complex degradation data model in Eq. 8.8, the reliability function at time t for lifetime T takes the form

$$R(t) = \mathbf{P}(T \geq t) = \mathbf{P}_{\boldsymbol{\theta}}[D(t, \boldsymbol{\theta}, \boldsymbol{\nu}) \leq D_f | \boldsymbol{\nu}, \boldsymbol{\eta}]. \tag{8.9}$$

Note that the reliability function depends on the true degradation and does not involve the measurement error distribution. Moreover, $\boldsymbol{\nu}$ and the assumed probability distribution of $\boldsymbol{\theta}$ determines the probability statement in Eq. 8.9; the probability distribution of $\boldsymbol{\theta}$ depends on $\boldsymbol{\eta}$ through Eq. 8.7.

While $R(t)$ in Eq. 8.9 may not have a closed form, it is a function of $\boldsymbol{\nu}$ and $\boldsymbol{\eta}$. Consequently, an easily calculated approximation of the posterior distribution of $R(t)$ is as follows with suitably large integers n_{inner} and n_{outer}:

1. Take a draw from the joint posterior distribution of $(\boldsymbol{\nu}, \boldsymbol{\eta})$.
2. Draw n_{inner} values of $\boldsymbol{\theta}$ from $\mathbf{H}(\boldsymbol{\eta})$.
3. Estimate $R(t)$ by the observed proportion of n_{inner} times that $D(t, \boldsymbol{\theta}, \boldsymbol{\nu}) \leq D_f$.
4. Repeat Steps 1–3 n_{outer} times.

We can calculate point and interval estimates (i.e., a $(1-\alpha) \times 100\%$ credible interval) of $R(t)$ by taking the median and the $\alpha/2$ and $1-\alpha/2$ quantiles of the $R(t)$ posterior distribution draws from Step 4. To obtain the α quantile posterior distribution, revise Step 3 by computing the lifetimes t by solving $D(t, \boldsymbol{\theta}, \boldsymbol{\nu}) = D_f$ for t and taking the α quantile of these lifetimes. We use this four-step algorithm, next in Example 8.2, to make inferences on $R(t)$ and the α quantile of the lifetime distribution.

Example 8.2 Crack growth. To illustrate the use of more complex degradation data models, let us analyze the fatigue crack growth data introduced by Hudak et al. (1978). We consider the version of the data that Lu and Meeker (1993) captured from published plots of these data in Bagdonov and Kozin (1985). In this study, the investigators monitored the crack lengths of 21 test units over 0.12 million cycles of fatigue testing. All cracks started at 0.9 inch, and the investigators defined failures as occurring when the crack lengths reach 1.6 inches, i.e., $D_f = 1.6$. Table 8.4 presents the crack growth data, which Fig. 8.8 plots.

The crack growth rate $\frac{\mathrm{da}}{\mathrm{dt}}$ determines the true crack length $a(t)$ according to the Paris Law, which has the form $\theta_1 k(a)^{\theta_2+1}$, for some function $k(\cdot)$. When $k(a) = a$, the solution for $a(t)$ is

$$a(t) = a_0/(1 - a_0^{\theta_2}\theta_1\theta_2 t)^{1/\theta_2}, \tag{8.10}$$

where $a_0 = 0.9$. Note that from Eq. 8.10, the crack length $a(t)$ is nonlinear in time t. By taking logarithms in Eq. 8.10, the transformed true crack length is

$$\log[a(t)/a_0] = -(1/\theta_2)\log(1 - {a_0}^{\theta_2}\theta_1\theta_2 t).$$

For the observed transformed crack length, let

$$Y_{ij} = \log[a(t)/a_0] + \varepsilon_{ij} = -(1/\theta_{2i})\log(1 - {a_0}^{\theta_{2i}}\theta_{1i}\theta_{2i}t_{ij}) + \varepsilon_{ij},$$

assuming that the measurement errors ε_{ij} are conditionally independent and distributed as $Normal(0, \sigma_\varepsilon^2)$. Furthermore, in this example, we assume that the $\boldsymbol{\theta} = (\theta_1, \theta_2)$ have a $MultivariateNormal(\boldsymbol{\mu}, \boldsymbol{\Sigma})$ distribution. Note that in this example, there are no additional parameters $\boldsymbol{\nu}$, as in Eq. 8.8.

For an analysis of the crack growth data, the likelihood consists of a normal density contribution for each y_{ij}, $y_{ij} \sim Normal(-[1/\theta_{2i}]\log[1 - {a_0}^{\theta_{2i}}\theta_{1i}\theta_{2i}t_{ij}], \sigma_\varepsilon^2)$, and a multivariate normal density contribution for each $\boldsymbol{\theta}_i$, $\boldsymbol{\theta}_i \sim MultivariateNormal(\boldsymbol{\mu}, \boldsymbol{\Sigma})$. Moreover, we use the following diffuse prior distributions for $(\boldsymbol{\mu}, \boldsymbol{\Sigma})$ and σ_ε^2:

$\boldsymbol{\mu} \sim MultivariateNormal(\boldsymbol{\mu}_0, \boldsymbol{\Sigma}_{\boldsymbol{\mu}_0})$, with

$$\boldsymbol{\mu}_0 = \begin{pmatrix} 0 \\ 0 \end{pmatrix} \quad \text{and} \quad \boldsymbol{\Sigma}_{\boldsymbol{\mu}_0} = \begin{pmatrix} 1000 & 0 \\ 0 & 1000 \end{pmatrix},$$

$\boldsymbol{\Sigma} \sim InverseWishart(\boldsymbol{\Sigma}_0, 2)$, with

$$\boldsymbol{\Sigma}_0 = \begin{pmatrix} 10 & 0 \\ 0 & 10 \end{pmatrix},$$

and $\sigma_\varepsilon^2 \sim InverseGamma(3, 0.0001)$. Note that $\boldsymbol{\Sigma}$ is a 2×2 symmetric matrix, where Σ_{ij} denotes the ith row and jth column entry, so that $\Sigma_{ij} = \Sigma_{ji}$.

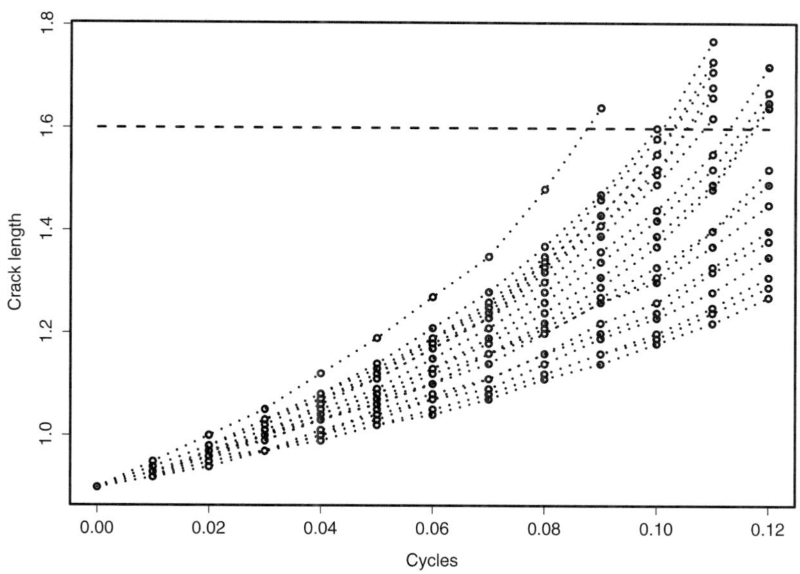

Fig. 8.8. Plot of crack growth data (crack length in inches) over millions of cycles.

We obtain draws from the joint posterior distribution of $(\boldsymbol{\mu}, \boldsymbol{\Sigma}, \sigma_\varepsilon^2)$ by MCMC. See Table 8.5, which summarizes the posterior distributions of the crack growth data model parameters.

We can then make inferences about the reliability function $R(t)$ using the four-step algorithm, presented above. Figure 8.9 displays the resulting crack growth reliability posterior median and 90% credible intervals.

8.3 Diagnostics for Degradation Data Models

In this section, we discuss diagnostics for assessing how well a model fits the degradation data. As discussed in Sects. 8.1 and 8.2, degradation data models,

Table 8.4. Crack growth data (crack length in inches) (Lu and Meeker, 1993)

Cycles	Unit																				
(millions)	1	2	3	4	5	6	7	8	9	10	11	12	13	14	15	16	17	18	19	20	21
0.00	0.90	0.90	0.90	0.90	0.90	0.90	0.90	0.90	0.90	0.90	0.90	0.90	0.90	0.90	0.90	0.90	0.90	0.90	0.90	0.90	0.90
0.01	0.95	0.94	0.94	0.94	0.94	0.94	0.94	0.93	0.92	0.92	0.93	0.93	0.92	0.93	0.92	0.92	0.93	0.92	0.92	0.92	0.92
0.02	1.00	0.98	0.98	0.98	0.98	0.98	0.98	0.97	0.97	0.96	0.96	0.97	0.97	0.96	0.96	0.95	0.96	0.94	0.94	0.94	0.94
0.03	1.05	1.03	1.03	1.03	1.03	1.03	1.02	1.00	1.01	1.00	1.00	1.00	0.99	1.00	0.99	0.97	0.97	0.97	0.97	0.97	0.97
0.04	1.12	1.08	1.08	1.07	1.07	1.07	1.07	1.06	1.05	1.04	1.04	1.03	1.03	1.03	1.03	1.00	1.00	1.01	0.99	0.99	0.99
0.05	1.19	1.14	1.13	1.12	1.12	1.12	1.11	1.11	1.09	1.08	1.08	1.07	1.06	1.07	1.06	1.03	1.05	1.04	1.02	1.02	1.02
0.06	1.27	1.21	1.19	1.19	1.19	1.18	1.17	1.17	1.15	1.13	1.13	1.10	1.10	1.12	1.10	1.07	1.08	1.07	1.05	1.05	1.04
0.07	1.35	1.28	1.26	1.25	1.24	1.23	1.23	1.23	1.21	1.19	1.18	1.16	1.14	1.16	1.16	1.11	1.11	1.09	1.08	1.08	1.07
0.08	1.48	1.37	1.35	1.34	1.34	1.33	1.32	1.30	1.28	1.26	1.24	1.22	1.20	1.20	1.21	1.16	1.16	1.14	1.12	1.12	1.11
0.09	1.64	1.47	1.46	1.43	1.43	1.41	1.41	1.39	1.36	1.34	1.31	1.29	1.26	1.26	1.27	1.22	1.20	1.19	1.16	1.16	1.14
0.10		1.60	1.58	1.55	1.55	1.51	1.52	1.49	1.44	1.42	1.39	1.37	1.31	1.30	1.33	1.26	1.24	1.23	1.20	1.19	1.18
0.11			1.77	1.73	1.71	1.68	1.66	1.62	1.55	1.52	1.49	1.48	1.40	1.37	1.40	1.33	1.32	1.28	1.25	1.24	1.22
0.12									1.72	1.67	1.65	1.64	1.52	1.45	1.49	1.40	1.38	1.35	1.31	1.29	1.27

Table 8.5. Posterior distribution summaries of crack growth data model parameters

Parameter	Mean	Std Dev	Quantiles				
			0.025	0.050	0.500	0.950	0.975
μ_1	3.732	0.162	3.411	3.467	3.733	3.997	4.053
μ_2	1.577	0.064	1.451	1.472	1.577	1.683	1.704
Σ_{11}	0.5773	0.1961	0.3092	0.3372	0.5383	0.9378	1.0630
Σ_{12}	−0.1049	0.0606	−0.2476	−0.2151	−0.0962	−0.0258	−0.0131
Σ_{22}	0.0738	0.0289	0.0344	0.0383	0.0683	0.1289	0.1447
σ_ε	0.0062	0.0003	0.0056	0.0057	0.0061	0.0067	0.0068

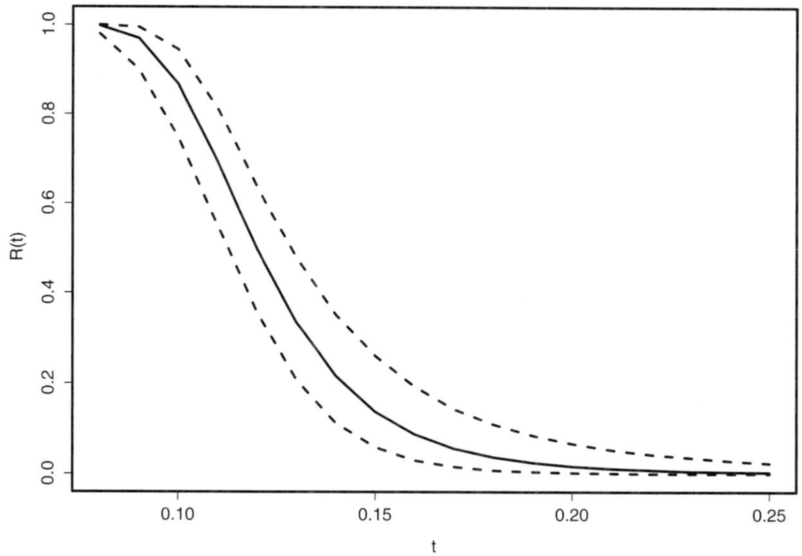

Fig. 8.9. Crack growth median posterior reliability over millions of cycles and 90% credible intervals. The solid line gives the posterior medians. The dashed lines give the 0.05 and 0.95 posterior quantiles.

such as the one in Eq. 8.2, are often random effects models; that is, they have parameters that are modeled hierarchically. Consequently, we can apply the deviance information criterion (DIC) diagnostic of Sect. 4.6 for model selection. For example, the analyst may want to entertain different distributions for the random reciprocal slopes because their distributions determine the lifetime distribution; recall that lognormal and Weibull distributed reciprocal slopes θ_i in Eq. 8.1 imply that the lifetimes follow a lognormal or Weibull distribution, respectively. It is also possible to assess the adequacy of a model numerically using a Bayesian χ^2 goodness-of-fit test presented in Sect. 3.4; in fact, Example 3.5 showed how to apply a Bayesian χ^2 goodness-of-fit test to a normal random effects model, which is similar to the degradation data model given in Eq. 8.1.

We can also apply graphical diagnostics to assess degradation data models, such as residual analysis, presented in Sect. 7.6. To apply residual analysis to degradation data models, evaluate the observed residuals

$$\varepsilon_{ij} = [y_{ij} - D(t_{ij}, \boldsymbol{\theta}_i, \boldsymbol{\nu})]/\sigma_\varepsilon \,, \tag{8.11}$$

where $D(t_{ij}, \boldsymbol{\theta}_i, \boldsymbol{\nu})$ is the more complex degradation data model given in Eq. 8.8. For normally distributed measurement errors, these residuals follow a standard normal distribution. As discussed in Sect. 7.6, obtain the posterior distribution of a residual by propagating the posterior distribution of $(\boldsymbol{\theta}_i, \boldsymbol{\nu}, \sigma_\varepsilon)$ through Eq. 8.11. That is, evaluate Eq. 8.11 with draws from the joint posterior distribution of $(\boldsymbol{\theta}_i, \boldsymbol{\nu}, \sigma_\varepsilon)$ to obtain draws from the residual posterior distribution. For example, we can assess the normality assumption of the measurement errors ε_{ij} by a normal plot of the medians of the residual posterior distributions; if normality holds, the plotted points appear as a straight line.

We may also look for patterns in plots of the medians of the residual posterior distributions against the times t_{ij}. For example, a systematic pattern on either side of zero may suggest a departure from the form of the degradation curve $D(\cdot)$. If the model incorporates covariates as discussed in Sect. 8.4, look for patterns in the plots of the medians of the residual posterior distributions against the covariates, as done in Sect. 7.6. When the number of observations per unit is small, plot the residuals from all the units together to have a basis to identify any unusual points.

Finally, besides using DIC to assess the assumed distribution of the $\boldsymbol{\theta}_i$, we can use appropriate plots of the $\boldsymbol{\theta}_i$ posterior distribution summaries for the same purpose. For example, for univariate θ_i, the logged medians of the θ_i posterior distributions will look normal when the θ_i have a lognormal distribution.

Next, we return to the drug potency example to illustrate the application of these model diagnostics.

Example 8.3 Drug potency degradation model diagnostics. To illustrate residual analysis, goodness of fit, and model selection using DIC, let us revisit the drug potency degradation model used in Example 8.1.

Figure 8.10 plots medians of the residual posterior distributions against time for the drug potency degradation data; the plot shows some lack of fit because the degradation curve is not strictly linear. Figure 8.11 presents a normal plot of all the medians of the residual posterior distributions, which is consistent with the assumed normality of the measurement errors. Moreover, Fig. 8.12 displays a normal plot of logged medians of the θ_i posterior distributions; this plot is also consistent with the assumed lognormal distribution of the θ_i.

As a demonstration of the numerical diagnostics, for a Bayesian χ^2 goodness-of-fit test based on five equally spaced bins, about 53% of the R^B values exceed the 0.95 quantile of the $ChiSquared(4)$ reference distribution, which indicates a lack of fit and confirms what appears in Fig. 8.10. (See Exercise 8.18, which suggests an alternative model for the drug potency degradation data by first transforming the data.) As an illustration of model selection, we investigate whether a Weibull distribution fits the θ_i better than a lognormal distribution. Using the DIC diagnostic, the lognormal DIC is 277.43 versus 280.07 for the Weibull DIC. Consequently, the lognormal distribution is preferable because the lognormal DIC is lower.

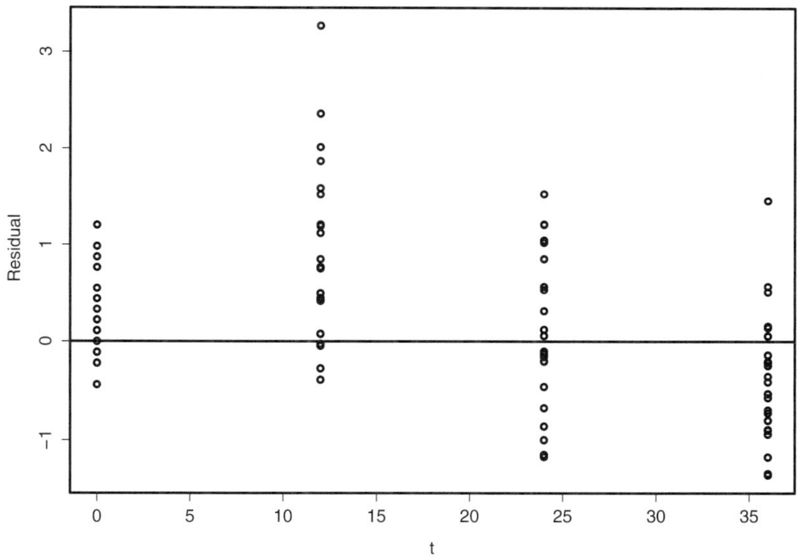

Fig. 8.10. Plot of drug potency residuals (medians of posterior distributions) over time t in months.

Next, we consider how to analyze degradation data when covariate information is available.

8.4 Incorporating Covariates

Like Chap. 7, which presented reliability data regression models, we can incorporate covariates in degradation data models. Covariates arise in a number

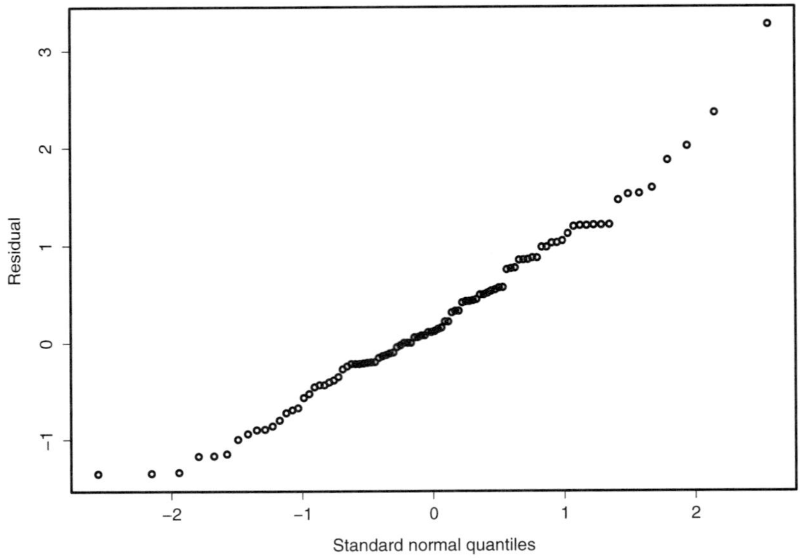

Fig. 8.11. Normal plot of drug potency residuals (medians of posterior distributions).

of ways. For example, an analyst can compare two or more populations by using the special covariates called dummy variables as discussed in Sect. 7.1.1. In this section, however, we focus on continuous variables that impact the degradation curve distributions. That is, consider degradation curves that are functions of these continuous variables. Accelerated degradation testing, which accelerates degradation via accelerating factors, collects data on this type of degradation curve. An analyst also encounters this type of degradation curve in experiments to improve reliability, which vary a number of factors simultaneously. We consider these two situations in turn in Sects. 8.4.1 and 8.4.2.

8.4.1 Accelerated Degradation Testing

Accelerated degradation testing, much like accelerated life testing, accelerates degradation by setting covariates called accelerating factors at higher values than those at normal use conditions. That is, the degradation is more severe than that observed under normal use conditions. Consequently, from accelerated degradation tests, we obtain *accelerated degradation data.*

One example of an accelerating factor is temperature. In Example 8.4, the relative luminosity of light-emitting diodes (LEDs) degrades faster at higher temperatures. Its true degradation curve is a function of temperature T and time t as follows:

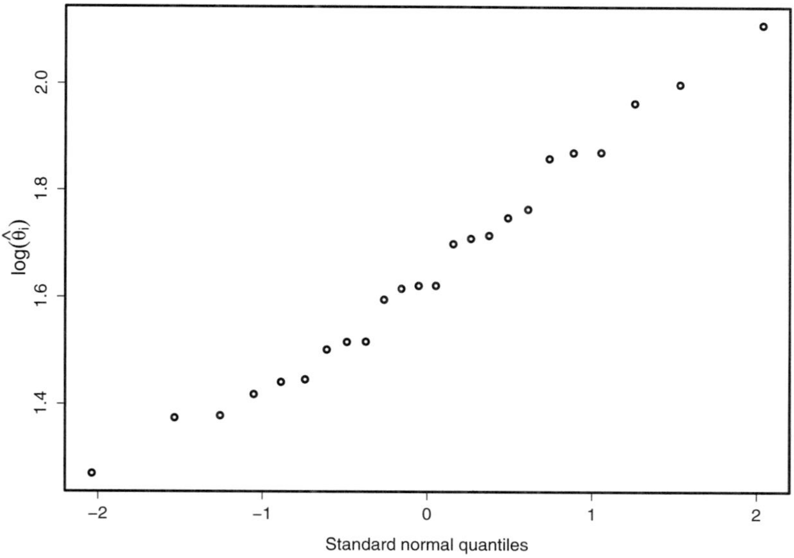

Fig. 8.12. Normal plot of logged estimated batch specific effects $\hat{\boldsymbol{\theta}}_i$ (logged medians of posterior distributions) for the drug potency degradation model.

$$D(t, T, \boldsymbol{\theta}, \boldsymbol{\nu}) = 1/\{1 + \exp(\theta_1) a(T, \nu) t^{\exp(\theta_2)}\},$$

for some acceleration factor function $a(\cdot)$, where $\boldsymbol{\theta} = (\theta_1, \theta_2)$ and $\boldsymbol{\nu}$ is a scalar ν. We have generalized the degradation curve in Eq. 8.6 to $D(t, \mathbf{x}, \boldsymbol{\theta}, \boldsymbol{\nu})$ by incorporating covariates $\mathbf{x}$, which in this case is a single covariate, temperature T.

Similarly, we can generalize the degradation data model in Eq. 8.8 as

$$Y_{ijk} = D[t_{ijk}, \mathbf{x}_i, \boldsymbol{\theta}_{ij}, \boldsymbol{\nu}] + \varepsilon_{ijk}, \tag{8.12}$$

where y_{ijk} is the observed degradation associated for the jth unit at the ith value $\mathbf{x}_i$ of the accelerating factor vector and the kth time t_{ijk}. Assume that the measurement errors ε_{ijk} are conditionally independent and distributed as $Normal(0, \sigma_\varepsilon^2)$. Further assume that the unit effects $\boldsymbol{\theta}_{ij}$ (possibly transformed, say $g(\boldsymbol{\theta}_{ij})$ for some function $g(\cdot)$) follow a distribution described by

$$g(\boldsymbol{\theta}_{ij}) \sim \mathbf{H}(\boldsymbol{\eta}). \tag{8.13}$$

For example, (θ_1, θ_2) in Example 8.4 have an assumed multivariate normal distribution. In more complex situations, the distribution of $\boldsymbol{\theta}_{ij}$ may depend on the acceleration factor vector values $\mathbf{x}_i$, but we do not consider such situations further.

Next, we consider an LED luminosity example, which uses the accelerated degradation data model given in Eq. 8.12.

Example 8.4 LED luminosity. Consider the degradation of relative luminosity (proportion of initial luminosity) for LEDs. Luminosity degrades slowly at 20°C, the standard operating temperature for LEDs; for this reason, it is impractical to test at this temperature. Instead, the tester must accelerate degradation. Let us define an LED failure as occurring when the LED relative luminosity drops to 0.5, i.e., 50% of initial luminosity.

Consider an accelerated degradation test that involves testing 25 units each at 25°C, 65°C, and 105°C. Tables 8.6, 8.7, and 8.8 present simulated luminosity data at each temperature and Fig. 8.13 plots these data.

The LED degradation data model follows the form given in Eq. 8.12, where an expression for the true degradation of luminosity at time t and temperature T (in degrees Celsius) is

$$D(t, T) = 1/\{1 + \beta_1[AF(T, T_U, \beta_3)t]^{\beta_2}\}. \tag{8.14}$$

The acceleration factor $AF(T, T_U, \beta_3)$ takes the form

$$\exp\left\{\beta_3\left[\frac{11605}{T_U + 273.15} - \frac{11605}{T + 273.15}\right]\right\}, \tag{8.15}$$

where T_U is the normal use temperature of 20°C. That is, the true degradation follows an Arrhenius relationship in temperature as discussed in Sect. 7.7.1. Note that the accelerating factor exceeds one for $T > T_U$, which by Eq. 8.14 means that higher than normal use temperatures accelerate the true degradation. Therefore, we can express the model for the luminosity degradation data as

$$Y_{ijk} = D(t_{ijk}, T_i) + \epsilon_{ijk}$$

for the kth time t_{ijk} of the jth unit at the ith temperature That is, we observe the true degradation $D(t_{ijk}, T_i)$ with measurement error ϵ_{ijk} distributed as $Normal(0, \sigma_\varepsilon^2)$. Note that $\theta_1 = \log(\beta_1)$ and $\theta_2 = \log(\beta_2)$ model the LED degradation curve population by assuming distributions for them. Also, modeling the logarithms of β_1 and β_2 ensures that β_1 and β_2 are positive. For the $\boldsymbol{\theta}_{ij}$ $(= (\theta_{1ij}, \theta_{2ij}))$, assume that they are distributed as

$$\boldsymbol{\theta}_{ij} \sim MultivariateNormal(\boldsymbol{\mu}, \boldsymbol{\Sigma})\,.$$

In Eq. 8.14, the only additional parameter is $\nu = \log(\beta_3)$. Consequently, the LED degradation data model has the form of Eq. 8.12, in which $\boldsymbol{\theta}$ consists of two parameters and $\boldsymbol{\nu}$ consists of one parameter.

In an analysis of the LED degradation data, the likelihood consists of a normal density contribution for each of the degradation data y_{ijk} (i.e., $Y_{ijk} \sim Normal(D(t_{ijk}, T_i), \sigma_\varepsilon^2)$ for $D(t_{ijk}, T_i)$ given in Eqs. 8.14 and 8.15 in which $\beta_1 = \exp(\theta_1)$, $\beta_2 = \exp(\theta_2)$, and $\beta_3 = \exp(\nu)$) and a multivariate

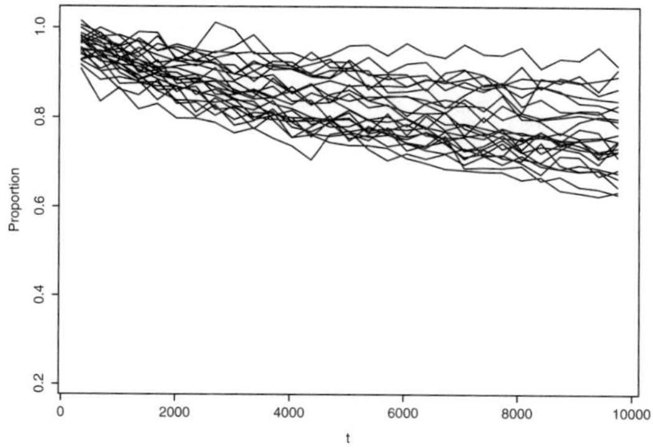

(a)

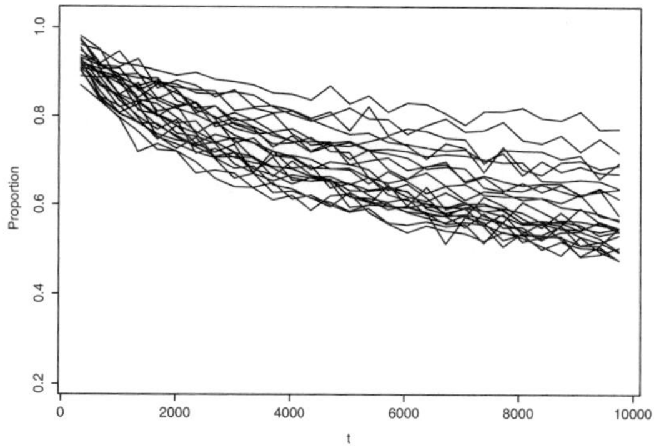

(b)

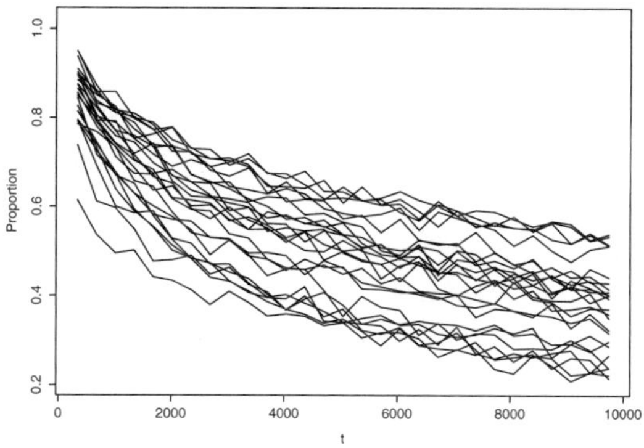

(c)

Fig. 8.13. Plot of LED luminosity data over time t in hours at (a) 25°C, (b) 65°C, (c) 105°C.

Table 8.6. LED luminosity data at 25°C (proportion of initial luminosity)

Time	Unit												
(hours)	1	2	3	4	5	6	7	8	9	10	11	12	13
336	0.9704	0.9081	0.9483	0.9171	0.9872	1.0151	0.9814	0.9642	0.9262	0.9532	1.0055	0.9793	0.9680
672	0.9439	0.8351	0.9485	0.8832	0.9688	0.9798	0.9571	0.9784	0.9030	0.9450	0.9882	0.9648	0.9883
1008	0.9614	0.8663	0.9348	0.8893	0.9264	0.9834	0.9439	0.9514	0.8627	0.9524	0.9329	0.9585	0.9527
1344	0.9008	0.8183	0.9193	0.8805	0.9547	0.9678	0.9761	0.8990	0.8498	0.9275	0.9453	0.9705	0.9343
1680	0.9273	0.8304	0.8926	0.8388	0.9385	0.9219	0.9480	0.9057	0.8833	0.9125	0.9809	0.9489	0.9640
2016	0.8753	0.7983	0.8888	0.8706	0.9248	0.9269	0.9572	0.9144	0.8309	0.9100	0.9303	0.9629	0.9502
2352	0.8793	0.7980	0.8957	0.8120	0.9276	0.9147	0.9457	0.9181	0.8339	0.8961	0.9472	0.9511	0.9273
2688	0.9106	0.7888	0.8518	0.8432	0.9221	0.9308	0.9331	0.9058	0.8048	0.8845	0.9182	0.9441	0.9469
3024	0.8572	0.7656	0.8511	0.8170	0.8826	0.8963	0.9425	0.8976	0.8265	0.8623	0.9169	0.9591	0.9442
3360	0.8572	0.7768	0.8631	0.7750	0.9039	0.9205	0.9464	0.8881	0.7952	0.8457	0.8807	0.9844	0.9038
3696	0.8698	0.7544	0.8463	0.8072	0.8799	0.9145	0.9178	0.8955	0.8049	0.8876	0.9132	0.9389	0.9254
4032	0.8369	0.7370	0.8594	0.8003	0.8552	0.9076	0.9152	0.8568	0.8164	0.8537	0.9089	0.9131	0.9236
4368	0.839	0.7056	0.8282	0.7571	0.8720	0.8913	0.8910	0.8756	0.7855	0.8735	0.8994	0.9459	0.8951
4704	0.7949	0.7628	0.8518	0.7930	0.8824	0.9049	0.8874	0.8335	0.7949	0.8481	0.9243	0.9120	0.8890
5040	0.8113	0.7706	0.8619	0.7695	0.8706	0.9048	0.9114	0.8738	0.7963	0.8376	0.9235	0.9102	0.9286
5376	0.7658	0.7300	0.8385	0.7768	0.8732	0.9052	0.8974	0.8392	0.7711	0.8466	0.8629	0.9349	0.8948
5712	0.8094	0.7047	0.8275	0.7662	0.8398	0.8567	0.8860	0.8488	0.7762	0.8462	0.8932	0.9001	0.8967
6048	0.761	0.7203	0.8128	0.7580	0.8042	0.8589	0.8920	0.8475	0.7731	0.8493	0.8700	0.9202	0.8967
6384	0.8047	0.7087	0.8097	0.7673	0.8330	0.8652	0.8954	0.8399	0.7789	0.8301	0.8864	0.9245	0.8613
6720	0.7731	0.7201	0.7861	0.7431	0.8410	0.8227	0.8627	0.8245	0.7492	0.8117	0.8780	0.8824	0.9011
7056	0.7853	0.6961	0.7874	0.7128	0.8338	0.8514	0.8810	0.8277	0.7590	0.8426	0.8859	0.9066	0.9085
7392	0.7681	0.7152	0.7777	0.7206	0.7922	0.8747	0.8649	0.8252	0.7650	0.8161	0.8951	0.8965	0.8714
7728	0.7555	0.7180	0.7682	0.7443	0.8285	0.8787	0.8713	0.7972	0.7312	0.8507	0.8551	0.8741	0.8764
8064	0.7531	0.6998	0.7684	0.7096	0.8066	0.8266	0.8476	0.8115	0.7449	0.7979	0.8120	0.8891	0.9154
8400	0.7574	0.6744	0.7575	0.7015	0.7927	0.8423	0.8712	0.7999	0.7700	0.7970	0.8884	0.8854	0.8654
8736	0.7496	0.7155	0.7683	0.7304	0.7643	0.8153	0.8605	0.8060	0.7379	0.8015	0.8781	0.8846	0.8893
9072	0.7396	0.7133	0.7549	0.7528	0.7922	0.8144	0.8333	0.8170	0.7522	0.8069	0.8954	0.8784	0.8806
9408	0.7212	0.6708	0.7572	0.7133	0.8036	0.8105	0.8306	0.8162	0.7100	0.8001	0.8601	0.8843	0.8515
9744	0.7262	0.6841	0.7486	0.7489	0.7992	0.8281	0.8166	0.7806	0.7245	0.7936	0.9070	0.8922	0.8648

Time	Unit											
(hours)	14	15	16	17	18	19	20	21	22	23	24	25
336	0.9284	0.9824	0.9396	0.9439	0.9512	0.9553	0.9747	0.9663	0.9749	0.9289	0.9704	0.9993
672	0.9360	0.9408	0.9300	0.9076	0.9025	0.9248	0.9998	0.9591	0.9402	0.9488	0.9343	0.9749
1008	0.9198	0.9302	0.8748	0.9270	0.9082	0.9177	0.9803	0.9276	0.9104	0.8989	0.9191	0.9602
1344	0.8853	0.9065	0.8764	0.9279	0.8931	0.9134	0.9429	0.9195	0.8834	0.8783	0.8936	0.9905
1680	0.9004	0.8852	0.8696	0.9023	0.8591	0.8928	0.9070	0.9087	0.8982	0.8782	0.8613	0.9885
2016	0.8572	0.8691	0.8686	0.8999	0.8387	0.8844	0.9328	0.8830	0.8892	0.8369	0.8632	0.9259
2352	0.8778	0.8580	0.8597	0.8764	0.8372	0.8851	0.9241	0.8881	0.8509	0.8019	0.8689	0.9548
2688	0.8819	0.8477	0.8230	0.8811	0.8339	0.8670	0.9212	0.8677	0.8627	0.8284	0.8180	1.0120
3024	0.8798	0.8265	0.7958	0.8534	0.8409	0.8568	0.8968	0.8423	0.8387	0.7830	0.8221	0.9959
3360	0.8555	0.8172	0.8114	0.8604	0.8021	0.8416	0.9257	0.8279	0.8366	0.7930	0.8006	0.9426
3696	0.8307	0.8378	0.8093	0.8426	0.7951	0.8449	0.9068	0.8082	0.7974	0.8194	0.8135	0.9428
4032	0.8201	0.7898	0.7928	0.8693	0.7557	0.8295	0.9085	0.8001	0.7907	0.7791	0.7560	0.9442
4368	0.8008	0.7646	0.7926	0.8320	0.7804	0.8552	0.9189	0.7993	0.7978	0.7690	0.7617	0.9278
4704	0.7932	0.7585	0.7966	0.8337	0.8041	0.8415	0.9304	0.7960	0.7795	0.7887	0.7449	0.9390
5040	0.8171	0.7643	0.7683	0.8538	0.7485	0.8188	0.8948	0.7993	0.7806	0.7537	0.7384	0.9604
5376	0.7942	0.7495	0.7573	0.8229	0.7813	0.7959	0.8694	0.7653	0.7971	0.7616	0.7381	0.9620
5712	0.7911	0.7784	0.7617	0.8292	0.7625	0.8030	0.8713	0.7726	0.7710	0.7365	0.7324	0.9385
6048	0.7989	0.7421	0.7527	0.7972	0.7450	0.7811	0.8693	0.7273	0.7860	0.7435	0.7098	0.9662
6384	0.7796	0.7334	0.7693	0.7756	0.7379	0.8106	0.8846	0.7389	0.7614	0.7282	0.6991	0.9427
6720	0.7698	0.7602	0.7524	0.8138	0.7479	0.8115	0.8828	0.7706	0.7561	0.7209	0.6862	0.9341
7056	0.7615	0.6877	0.7267	0.7611	0.7087	0.7771	0.8691	0.7285	0.7569	0.7320	0.6832	0.9641
7392	0.7541	0.6894	0.7321	0.8023	0.7304	0.7619	0.8595	0.7383	0.7193	0.7161	0.6799	0.9420
7728	0.7751	0.6884	0.7517	0.7675	0.7106	0.7772	0.8774	0.7619	0.7152	0.6967	0.6793	0.9400
8064	0.7297	0.6928	0.7239	0.7496	0.7047	0.7634	0.8671	0.7257	0.7585	0.7161	0.6618	0.9577
8400	0.7696	0.6598	0.7361	0.7691	0.7451	0.7625	0.8706	0.6993	0.7222	0.6929	0.6677	0.9106
8736	0.7431	0.6690	0.7333	0.7479	0.6951	0.7516	0.8574	0.6865	0.7426	0.6862	0.6383	0.9320
9072	0.7500	0.6460	0.7256	0.7763	0.6821	0.7493	0.8484	0.6950	0.7312	0.6807	0.6322	0.9289
9408	0.7259	0.6410	0.6974	0.7688	0.6842	0.7535	0.8430	0.6880	0.7244	0.6835	0.6252	0.9580
9744	0.7333	0.6292	0.7448	0.7128	0.6654	0.7640	0.8385	0.6773	0.7247	0.6458	0.6347	0.9166

normal density contribution for each of the $\boldsymbol{\theta}_{ij}$ (i.e., one for each of the 75 LEDs with $\boldsymbol{\theta}_{ij} \sim MultivariateNormal(\boldsymbol{\mu}, \boldsymbol{\Sigma})$). Consequently, the likelihood depends on the parameters $(\boldsymbol{\mu}, \boldsymbol{\Sigma})$, ν, and σ_ε^2. Regarding the prior distributions for the parameters $(\boldsymbol{\mu}, \boldsymbol{\Sigma})$, ν, and σ_ε^2, we use the following diffuse distributions:

$\boldsymbol{\mu} \sim MultivariateNormal(\boldsymbol{\mu}_0, \boldsymbol{\Sigma}_{\boldsymbol{\mu}_0})$, with

$$\boldsymbol{\mu}_0 = \begin{pmatrix} 0 \\ 0 \end{pmatrix} \quad \text{and} \quad \boldsymbol{\Sigma}_{\boldsymbol{\mu}_0} = \begin{pmatrix} 1000 & 0 \\ 0 & 1000 \end{pmatrix},$$

$\boldsymbol{\Sigma} \sim InverseWishart(\boldsymbol{\Sigma}_0, 3)$, with

Table 8.7. LED luminosity data at 65°C (proportion of initial luminosity)

Time (hours)	Unit												
	1	2	3	4	5	6	7	8	9	10	11	12	13
336	0.9538	0.9746	0.9007	0.9188	0.9160	0.9361	0.9206	0.9015	0.9112	0.8681	0.9060	0.9176	0.9258
672	0.8857	0.9047	0.8384	0.9049	0.8673	0.9166	0.8548	0.8415	0.8941	0.8330	0.8730	0.8979	0.8431
1008	0.8879	0.9069	0.8052	0.8478	0.8420	0.8854	0.8073	0.8061	0.8807	0.7971	0.8442	0.8885	0.8224
1344	0.8635	0.8565	0.7848	0.8298	0.7947	0.8668	0.8268	0.7869	0.8132	0.7653	0.8164	0.8314	0.7848
1680	0.8350	0.8756	0.7602	0.7993	0.7484	0.8797	0.7538	0.7553	0.8032	0.7253	0.7788	0.8411	0.7827
2016	0.8165	0.8705	0.7467	0.8238	0.7784	0.8586	0.7376	0.7280	0.7617	0.7220	0.7566	0.8232	0.7528
2352	0.8034	0.8514	0.7382	0.7612	0.7621	0.8136	0.7200	0.7525	0.7555	0.6975	0.7341	0.8285	0.7314
2688	0.7824	0.8491	0.7196	0.7867	0.7083	0.7988	0.7069	0.7047	0.7390	0.6596	0.7424	0.7765	0.7105
3024	0.7959	0.8538	0.7376	0.7578	0.7095	0.7730	0.6831	0.6601	0.6933	0.6404	0.7206	0.7849	0.7184
3360	0.7761	0.8469	0.6713	0.7410	0.7079	0.7617	0.6884	0.6715	0.6727	0.6391	0.6816	0.7730	0.7103
3696	0.7572	0.8209	0.6543	0.7223	0.6870	0.7633	0.6452	0.6274	0.7232	0.6497	0.6680	0.7314	0.6643
4032	0.7451	0.8035	0.6683	0.7104	0.6837	0.7440	0.6359	0.6307	0.6827	0.6129	0.6475	0.7568	0.7001
4368	0.7203	0.7894	0.6610	0.6849	0.6906	0.7436	0.6147	0.6160	0.6646	0.6219	0.6626	0.7227	0.6680
4704	0.7343	0.8028	0.6518	0.6829	0.6383	0.7432	0.6057	0.5953	0.6539	0.5998	0.6571	0.7404	0.6577
5040	0.7416	0.8189	0.5946	0.6805	0.6489	0.6920	0.5857	0.6282	0.6487	0.5836	0.6187	0.7044	0.6323
5376	0.7232	0.7950	0.5995	0.6480	0.6335	0.7011	0.5999	0.6057	0.6225	0.5922	0.6056	0.7257	0.6310
5712	0.7175	0.7934	0.6461	0.6765	0.6401	0.7079	0.5986	0.6032	0.5971	0.5598	0.5962	0.6816	0.6366
6048	0.7079	0.7607	0.5982	0.6536	0.6185	0.6798	0.5848	0.5836	0.5939	0.5553	0.5948	0.6590	0.6314
6384	0.6831	0.7805	0.6403	0.6266	0.6086	0.6884	0.5810	0.5712	0.6036	0.5628	0.5825	0.6890	0.5921
6720	0.6873	0.7953	0.5903	0.6432	0.6284	0.6991	0.5579	0.5604	0.5822	0.5522	0.5748	0.6933	0.5797
7056	0.6841	0.7665	0.6024	0.6116	0.6058	0.6514	0.5615	0.5776	0.5926	0.5569	0.5791	0.6551	0.5742
7392	0.6402	0.7694	0.6039	0.6314	0.5864	0.6647	0.5564	0.5537	0.5565	0.5149	0.5697	0.6678	0.6188
7728	0.6320	0.7883	0.5880	0.6283	0.5902	0.6346	0.5388	0.5072	0.5801	0.5630	0.5594	0.6545	0.5990
8064	0.6510	0.7481	0.5608	0.6120	0.5605	0.6322	0.5330	0.5384	0.5360	0.5181	0.5437	0.6451	0.5704
8400	0.6302	0.7405	0.5650	0.6169	0.5582	0.6472	0.5321	0.5594	0.5339	0.5291	0.5221	0.6493	0.5500
8736	0.6468	0.7585	0.5609	0.6056	0.5479	0.6326	0.5286	0.5244	0.5530	0.5307	0.5545	0.6496	0.5836
9072	0.6245	0.7227	0.5438	0.6249	0.5453	0.6207	0.5109	0.4867	0.5297	0.4851	0.5103	0.6219	0.5504
9408	0.6312	0.7514	0.5628	0.5763	0.5608	0.6364	0.5181	0.5078	0.5169	0.4901	0.5098	0.6388	0.5407
9744	0.6366	0.7204	0.5491	0.5686	0.5429	0.5780	0.4956	0.4765	0.5325	0.5048	0.5487	0.6150	0.5485

Time (hours)	Unit											
	14	15	16	17	18	19	20	21	22	23	24	25
336	0.9281	0.9472	0.9805	0.9307	0.9736	0.9069	0.9029	0.9696	0.9244	0.9608	0.9175	0.8887
672	0.8747	0.9134	0.9495	0.9119	0.9178	0.8994	0.8356	0.9159	0.8951	0.9474	0.8975	0.8853
1008	0.8200	0.9101	0.8727	0.8924	0.9454	0.8854	0.7955	0.8919	0.8720	0.9184	0.8475	0.8347
1344	0.8330	0.9281	0.8615	0.8844	0.8924	0.8407	0.7202	0.8939	0.8365	0.9124	0.8739	0.7993
1680	0.8030	0.8616	0.8106	0.8849	0.8728	0.8707	0.7383	0.8384	0.8270	0.9013	0.8037	0.7703
2016	0.7785	0.8798	0.8191	0.8341	0.8642	0.8501	0.7216	0.8568	0.7967	0.8910	0.7845	0.7922
2352	0.7393	0.8680	0.8308	0.8180	0.8584	0.8293	0.6717	0.8067	0.7658	0.8973	0.7890	0.7174
2688	0.7594	0.8425	0.7938	0.8227	0.8556	0.7937	0.6784	0.8226	0.7436	0.8818	0.7422	0.7275
3024	0.7479	0.8586	0.7408	0.7901	0.8126	0.8077	0.6716	0.7818	0.7186	0.8768	0.7517	0.7079
3360	0.7053	0.8213	0.7513	0.7922	0.8183	0.7910	0.6375	0.7671	0.6744	0.8697	0.7408	0.6960
3696	0.6837	0.8235	0.7119	0.8062	0.8076	0.7656	0.6106	0.7385	0.7089	0.8519	0.7132	0.6881
4032	0.6620	0.8135	0.7077	0.7608	0.8116	0.7721	0.6211	0.7409	0.6559	0.8492	0.7128	0.6821
4368	0.6536	0.7957	0.6807	0.7361	0.7923	0.7489	0.5874	0.7531	0.6543	0.8359	0.7042	0.6639
4704	0.6591	0.8101	0.6985	0.7408	0.7752	0.7358	0.6204	0.7221	0.6565	0.8682	0.6824	0.6648
5040	0.6180	0.7682	0.6929	0.7449	0.7601	0.7336	0.6251	0.7200	0.6188	0.8278	0.6841	0.6463
5376	0.6148	0.8219	0.6810	0.7250	0.7631	0.7444	0.5834	0.6628	0.6102	0.8459	0.6620	0.6624
5712	0.6065	0.7580	0.6420	0.7273	0.7427	0.7551	0.5580	0.6499	0.5897	0.8097	0.6333	0.6235
6048	0.6150	0.7728	0.6299	0.7223	0.7378	0.7348	0.5383	0.6600	0.5977	0.8299	0.6477	0.6106
6384	0.5778	0.7671	0.5957	0.7112	0.7303	0.7058	0.5536	0.6740	0.5834	0.8271	0.6164	0.6037
6720	0.5749	0.7522	0.5959	0.7159	0.7241	0.7402	0.5425	0.6439	0.5133	0.8066	0.5768	0.5941
7056	0.5506	0.7492	0.5931	0.7131	0.7266	0.7300	0.5367	0.6474	0.5724	0.7816	0.5949	0.6069
7392	0.5814	0.6954	0.5875	0.7044	0.7006	0.6755	0.5196	0.6231	0.5622	0.8116	0.5992	0.5760
7728	0.5499	0.7259	0.5757	0.6975	0.7192	0.7324	0.5405	0.6197	0.5220	0.8116	0.5928	0.5619
8064	0.5508	0.7240	0.5574	0.6985	0.6688	0.6821	0.5271	0.6195	0.5357	0.8196	0.5832	0.6068
8400	0.5244	0.7231	0.5571	0.6560	0.7011	0.6698	0.5088	0.6161	0.5013	0.7968	0.5616	0.5855
8736	0.5398	0.7063	0.5662	0.6546	0.6771	0.6984	0.4880	0.5959	0.5185	0.7922	0.5553	0.5635
9072	0.5417	0.7140	0.5641	0.6585	0.6888	0.7020	0.5294	0.5907	0.5052	0.8114	0.5930	0.5563
9408	0.5265	0.6615	0.5560	0.6555	0.6751	0.6840	0.4964	0.5807	0.5030	0.7717	0.5614	0.5331
9744	0.4980	0.6967	0.5488	0.6388	0.6727	0.6913	0.4773	0.5649	0.4773	0.7727	0.5404	0.5420

$$\boldsymbol{\Sigma}_0 = \begin{pmatrix} 10 & 0 \\ 0 & 10 \end{pmatrix},$$

$\nu \sim Normal(0, 100)$, and $\sigma_\varepsilon^2 \sim InverseGamma(3, 0.0001)$.

We then obtain draws from the joint posterior distribution of $(\boldsymbol{\mu}, \boldsymbol{\Sigma}, \eta, \sigma_\varepsilon^2)$ by MCMC. Table 8.9 summarizes the marginal posterior distributions of these parameters.

Subsequently, we can make inferences for the reliability function $R(t)$ using the four-step algorithm described in Sect. 8.2.1. Figure 8.14 presents the LED reliability posterior median and 90% credible intervals at the normal use temperature of 20°C.

Table 8.8. LED luminosity data at 105°C (proportion of initial luminosity)

Time (hours)	Unit 1	2	3	4	5	6	7	8	9	10	11	12	13
336	0.9104	0.7385	0.7912	0.8984	0.8754	0.8949	0.7856	0.8267	0.7953	0.8985	0.8844	0.9382	0.9029
672	0.8549	0.6122	0.7257	0.8616	0.8214	0.8067	0.7681	0.7318	0.7409	0.8275	0.8511	0.8322	0.8382
1008	0.8196	0.5970	0.6741	0.8123	0.7701	0.7586	0.7341	0.6397	0.6774	0.7711	0.8251	0.8188	0.8111
1344	0.7986	0.5503	0.6302	0.7894	0.7190	0.7449	0.6872	0.5905	0.6299	0.7598	0.7538	0.7745	0.8024
1680	0.7731	0.4778	0.5822	0.7334	0.7423	0.6984	0.6364	0.5557	0.6014	0.7352	0.7633	0.7398	0.7742
2016	0.7795	0.4806	0.5711	0.7381	0.6691	0.6710	0.6460	0.5136	0.5775	0.6577	0.7779	0.7342	0.7243
2352	0.7310	0.4910	0.5672	0.6725	0.6556	0.6370	0.6013	0.4839	0.5478	0.6235	0.7187	0.6712	0.7229
2688	0.7287	0.4485	0.5304	0.6534	0.6257	0.6232	0.6038	0.4414	0.4922	0.6204	0.7078	0.6561	0.6894
3024	0.6932	0.4641	0.5268	0.6556	0.5835	0.6174	0.5715	0.4323	0.5253	0.6084	0.7034	0.6404	0.6999
3360	0.7082	0.4258	0.4963	0.6115	0.5487	0.6096	0.5487	0.4110	0.4900	0.5994	0.6899	0.6176	0.7191
3696	0.6763	0.4084	0.4973	0.6223	0.5885	0.5623	0.5495	0.3864	0.4826	0.5446	0.6303	0.6007	0.6747
4032	0.6632	0.3677	0.4905	0.6195	0.5781	0.5607	0.5263	0.3693	0.4571	0.5527	0.6560	0.5717	0.6740
4368	0.6797	0.3922	0.4474	0.5819	0.5627	0.5213	0.5020	0.3582	0.4824	0.5453	0.6188	0.5758	0.6492
4704	0.6347	0.4072	0.4795	0.5852	0.5561	0.5459	0.4993	0.3386	0.4643	0.5495	0.6156	0.5545	0.6182
5040	0.6336	0.3789	0.4628	0.5654	0.5375	0.5360	0.5245	0.3479	0.4651	0.4969	0.6447	0.5638	0.6199
5376	0.6140	0.3364	0.4293	0.5556	0.5069	0.5091	0.5005	0.3208	0.4156	0.4782	0.6144	0.5183	0.6146
5712	0.6200	0.3477	0.4203	0.5018	0.5163	0.5155	0.4887	0.2886	0.4138	0.4714	0.6041	0.4934	0.6164
6048	0.6346	0.3357	0.4108	0.4920	0.4821	0.4854	0.4633	0.3260	0.4080	0.4325	0.6065	0.4877	0.5877
6384	0.6151	0.3432	0.4030	0.5262	0.5158	0.5105	0.4900	0.2891	0.4079	0.4757	0.5725	0.4886	0.5690
6720	0.6021	0.3374	0.3930	0.5149	0.4470	0.5020	0.4351	0.2822	0.3972	0.4533	0.6134	0.4561	0.6056
7056	0.5802	0.3347	0.4031	0.5185	0.4820	0.4698	0.4728	0.2788	0.3787	0.4488	0.5801	0.5024	0.5918
7392	0.5924	0.3225	0.3917	0.4782	0.4765	0.4540	0.4357	0.2967	0.3737	0.4077	0.5572	0.4942	0.5547
7728	0.5891	0.3356	0.3822	0.4976	0.4237	0.4660	0.4616	0.2574	0.3835	0.4320	0.5590	0.4597	0.5653
8064	0.5722	0.3110	0.4315	0.4709	0.4504	0.4418	0.4546	0.2627	0.3658	0.3936	0.5532	0.4293	0.5527
8400	0.5564	0.3154	0.3922	0.4801	0.4359	0.4569	0.4250	0.2693	0.3527	0.4121	0.5488	0.4503	0.5514
8736	0.5657	0.2915	0.3857	0.4414	0.4444	0.4084	0.4474	0.2438	0.3824	0.3674	0.5680	0.3950	0.5269
9072	0.5560	0.2927	0.3552	0.4581	0.4260	0.4218	0.4141	0.2095	0.3276	0.3698	0.5621	0.4350	0.5182
9408	0.5240	0.3112	0.3718	0.4311	0.4342	0.4095	0.4162	0.2291	0.3602	0.3765	0.5264	0.4184	0.5413
9744	0.5334	0.2871	0.3703	0.4293	0.4004	0.4085	0.4026	0.2226	0.3229	0.3588	0.5386	0.3939	0.5152

Time (hours)	Unit 14	15	16	17	18	19	20	21	22	23	24	25
336	0.8442	0.8740	0.8505	0.9508	0.8493	0.6145	0.8559	0.9502	0.8971	0.8144	0.8649	0.7940
672	0.7170	0.7895	0.7940	0.8582	0.7966	0.5373	0.7804	0.8723	0.7984	0.7456	0.8030	0.6921
1008	0.6734	0.7407	0.7928	0.8583	0.7428	0.4958	0.7333	0.8175	0.7122	0.6705	0.7890	0.6009
1344	0.5946	0.7036	0.7070	0.8001	0.7163	0.5025	0.6581	0.8097	0.6745	0.6525	0.7529	0.5858
1680	0.5535	0.6830	0.6747	0.7899	0.6666	0.4421	0.6449	0.7840	0.6723	0.5796	0.7178	0.5901
2016	0.5023	0.6200	0.6831	0.7506	0.6530	0.4341	0.5878	0.7357	0.6180	0.5267	0.6906	0.5289
2352	0.4847	0.6149	0.6554	0.7327	0.5959	0.4132	0.5416	0.7231	0.5937	0.4844	0.7052	0.5178
2688	0.4741	0.6079	0.6525	0.7074	0.5728	0.3794	0.5313	0.6817	0.5902	0.4736	0.6907	0.4710
3024	0.4335	0.5681	0.5911	0.7094	0.5987	0.4095	0.5280	0.6684	0.5388	0.4533	0.6452	0.4715
3360	0.4073	0.5767	0.5982	0.6947	0.5702	0.3849	0.4927	0.6865	0.5296	0.4392	0.6377	0.4289
3696	0.3957	0.5587	0.5749	0.6458	0.5524	0.3545	0.4449	0.6510	0.4828	0.4034	0.6290	0.3967
4032	0.3927	0.5304	0.5813	0.6448	0.5200	0.3589	0.4608	0.6115	0.4626	0.3674	0.6407	0.4025
4368	0.4200	0.5439	0.5619	0.6419	0.5435	0.3539	0.4597	0.6169	0.4595	0.3994	0.6287	0.3832
4704	0.3428	0.5154	0.5700	0.6594	0.4803	0.3331	0.4121	0.6138	0.4695	0.3528	0.5960	0.3647
5040	0.3403	0.5010	0.5378	0.6088	0.5002	0.3392	0.3538	0.5730	0.4488	0.3698	0.5852	0.3412
5376	0.3058	0.5055	0.5314	0.6449	0.4812	0.3491	0.3984	0.5809	0.4622	0.3224	0.5780	0.3370
5712	0.3176	0.4382	0.5085	0.6055	0.4715	0.3564	0.3740	0.5424	0.4273	0.3139	0.5836	0.3329
6048	0.3174	0.4683	0.5025	0.6176	0.4848	0.3340	0.3674	0.5726	0.4413	0.3288	0.5978	0.2989
6384	0.3046	0.4776	0.5067	0.5903	0.4692	0.3365	0.3121	0.5249	0.4103	0.2903	0.5479	0.3251
6720	0.2634	0.4162	0.4704	0.5559	0.4525	0.3147	0.3382	0.5188	0.4055	0.2954	0.5894	0.2910
7056	0.2800	0.4002	0.5011	0.5917	0.4542	0.3077	0.3020	0.5138	0.4029	0.3176	0.5746	0.3306
7392	0.2683	0.4228	0.4698	0.5753	0.4382	0.3127	0.3156	0.5437	0.3523	0.2804	0.5626	0.2717
7728	0.2873	0.4363	0.4795	0.5581	0.4297	0.3276	0.3136	0.4987	0.3706	0.2365	0.5488	0.2597
8064	0.2530	0.4172	0.4626	0.5782	0.4396	0.3157	0.2816	0.4715	0.3669	0.2263	0.5129	0.2716
8400	0.2781	0.4242	0.4836	0.5449	0.3982	0.3285	0.2829	0.4526	0.3588	0.2667	0.5366	0.2629
8736	0.2377	0.4043	0.4392	0.5638	0.3894	0.2619	0.2807	0.4650	0.3537	0.2606	0.5022	0.2721
9072	0.2611	0.4212	0.4160	0.5179	0.3884	0.2927	0.3013	0.4424	0.3401	0.2231	0.4777	0.2782
9408	0.2577	0.3846	0.4615	0.5373	0.4057	0.2760	0.2791	0.4448	0.3497	0.2275	0.5048	0.2294
9744	0.2153	0.4013	0.4420	0.5134	0.3495	0.2980	0.2387	0.4156	0.3165	0.2670	0.5126	0.2385

Table 8.9. Posterior distribution summaries of LED data model parameters

Parameter	Mean	Std Dev	Quantiles 0.025	0.050	0.500	0.950	0.975
μ_1	−7.640	0.150	−7.944	−7.896	−7.637	−7.400	−7.356
μ_2	−0.4110	0.0265	−0.4613	−0.4538	−0.4116	−0.3662	−0.3558
Σ_{11}	1.405	0.248	1.003	1.053	1.378	1.849	1.969
Σ_{21}	−0.1906	0.0385	−0.2773	−0.2598	−0.1865	−0.1351	−0.1276
Σ_{22}	0.0433	0.0075	0.0311	0.0327	0.0425	0.0567	0.0603
ν	−1.107	0.021	−1.161	−1.143	−1.104	−1.079	−1.077
σ_ε	0.0161	0.0003	0.0155	0.0156	0.0160	0.0166	0.0167

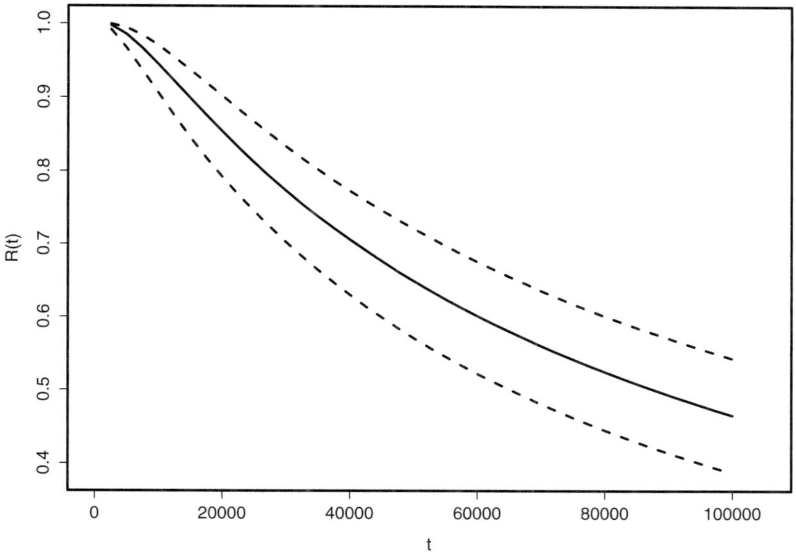

Fig. 8.14. LED reliability over time t in hours and 90% credible intervals at 20°C. The solid line shows the posterior medians. The dashed lines show the 0.05 and 0.95 posterior quantiles.

8.4.2 Improving Reliability Using Designed Experiments

In Sect. 7.8, we saw how covariates arise in reliability improvement experiments that collect lifetime data. Similarly, designed experiments can collect degradation data instead of lifetimes. As discussed in Sect. 7.8, such designed experiments simultaneously vary multiple factors; first we identify the factors that impact degradation, and then recommend levels for these factors that lead to reduced degradation or, in other words, reliability improvement. See Sect. 7.8 and Wu and Hamada (2000) for a discussion of these types of experimental plans.

Take, for example, the experiment considered in Example 8.5, which varies three factors involved in producing fluorescent lights. The experiment measured degradation of fluorescent light luminosity and the rate of degradation depends on the values of the three factors. We can generalize the degradation data model in Eq. 8.8 as

$$Y_{ijk} = D[t_{ijk}, \boldsymbol{\theta}_{ij}(\mathbf{x}_i), \boldsymbol{\nu}] + \varepsilon_{ijk}\,, \tag{8.16}$$

for the ith covariate values $\mathbf{x}_i$ associated with the factors and jth unit at the kth time t_{ijk}, where the measurement errors ε_{ijk} are conditionally independent and distributed as $Normal(0, \sigma_\varepsilon^2)$.

We see in Eq. 8.16 that the distribution of the unit effects $\boldsymbol{\theta}_{ij}$ depends on the covariate values $\mathbf{x}_i$. An expression for the distribution of the $\boldsymbol{\theta}_{ij}$ (possibly transformed, say $g(\boldsymbol{\theta}_{ij})$ for some function $g(\cdot)$) is

$$g[\boldsymbol{\theta}_{ij}(\mathbf{x}_i)] \sim \mathbf{H}[\boldsymbol{\eta}(\mathbf{x}_i)]\,. \tag{8.17}$$

For example, let

$$g[\boldsymbol{\theta}_{ij}(\mathbf{x}_i)] \sim MultivariateNormal[\boldsymbol{\mu}(\mathbf{x}_i), \boldsymbol{\Sigma}(\mathbf{x}_i)]\,,$$

where

$$\boldsymbol{\mu}(\mathbf{x}_i) = \mathbf{x}_i^T \boldsymbol{\beta}. \tag{8.18}$$

In Eq. 8.18, $\mathbf{x}_i$ denotes the p covariate values associated with the ith unit. If $\boldsymbol{\theta}_{ij}$ consists of m effects, then $\boldsymbol{\beta}$ is an $m \times p$ matrix of parameters. Regarding $\boldsymbol{\Sigma}(\mathbf{x}_i)$, we use simpler forms that involve few parameters. We may develop more complex models for $\boldsymbol{\Sigma}(\mathbf{x}_i)$, but this is beyond the scope of the book.

Next, we consider a fluorescent lamp experiment to illustrate the analysis of degradation data from a designed experiment.

Example 8.5 Fluorescent lamp brightness. The key quality attribute of a fluorescent lamp is its brightness, which decreases over time. Define lamp failure as occurring when the luminosity degrades to 60% of the luminosity that the lamp had at 100 hours of use, thus, $D_f = 0.6$. Let $L(t)$ denote the luminosity at time t. From Lin (1976), the log relative luminosity for the ith lamp takes the form

$$D(t, \theta) = \log[L(t)/L(100)] = -(1/\theta_i)(t - 100)\,.$$

To complete the true degradation model, we assume

$$\theta_i \sim LogNormal[\mu_\theta(\mathbf{x}_i), \sigma_\theta^2(\mathbf{x}_i)], \tag{8.19}$$

where

$$\mu_\theta(\mathbf{x}_i) = \mathbf{x}_i^T \boldsymbol{\beta} \quad \text{and}$$
$$\log[\sigma_\theta^2(\mathbf{x}_i)] = \mathbf{x}_i^T \boldsymbol{\gamma}.$$

Note that this degradation model is a simple version of $D[t, \boldsymbol{\theta}_i(\mathbf{x}_i), \boldsymbol{\nu}]$ from Eqs. 8.16 and 8.17, where $\boldsymbol{\theta}_i(\mathbf{x}_i) = \theta_i(\mathbf{x}_i)$. Also note that the θ_i have a lognormal distribution, which implies that the lifetimes have a lognormal distribution. Specifically, we can calculate the lifetime t as

$$t = 100 - [\log(0.6)]\theta\,.$$

Then, from Eq. 8.19,

$$T - 100 \sim LogNormal\{\log[-\log(0.6)] + \mu_\theta(\mathbf{x}_i), \sigma_\theta^2(\mathbf{x}_i)\}. \tag{8.20}$$

The experiment to improve fluorescent lamp reliability involved three factors chosen from a seven-step manufacturing process: factor A, the amount of electric current in the exhaustive process; factor B, the concentration of the mercury dispenser in the coating process; and factor C, the concentration of argon in the argon filling process. The experiment studied each factor at two levels, denoted by $(-,+)$, using a 2^{3-1} fractional factorial design. Table 8.10 displays the four-run experimental plan. Table 8.11 presents and Fig. 8.15 plots the fluorescent light degradation data for experimental runs 1–4.

Table 8.10. Fluorescent lamp experimental plan

	Factor		
Run	A	B	C
1	$-$	$-$	$-$
2	$-$	$+$	$+$
3	$+$	$-$	$+$
4	$+$	$+$	$-$

In an analysis of the fluorescent lamp degradation data, the likelihood consists of a normal density contribution for each observed degradation y_{ijk} and a multivariate normal density contribution for each $\boldsymbol{\theta}_i$. We also use the following diffuse prior distributions for $\boldsymbol{\beta}$, $\boldsymbol{\gamma}$, and σ_ε^2:

$\boldsymbol{\beta} \sim MultivariateNormal(\boldsymbol{\mu}_0, \boldsymbol{\Sigma}_0)$ and $\boldsymbol{\gamma} \sim MultivariateNormal(\boldsymbol{\mu}_0, \boldsymbol{\Sigma}_0)$ with

$$\boldsymbol{\mu}_0 = \begin{pmatrix} 0 \\ 0 \\ 0 \\ 0 \end{pmatrix} \quad \text{and} \quad \boldsymbol{\Sigma}_0 = \begin{pmatrix} 10 & 0 & 0 & 0 \\ 0 & 10 & 0 & 0 \\ 0 & 0 & 10 & 0 \\ 0 & 0 & 0 & 10 \end{pmatrix},$$

$\boldsymbol{\beta} \sim MultivariateNormal(\boldsymbol{\mu}_0, \boldsymbol{\Sigma}_0)$, and $\sigma_\varepsilon^2 \sim InverseGamma(0.1, 0.1)$. In this case, the prior distributions for β_i and γ_i, $i = 1, \ldots, 4$, are independent $Normal(0, 10)$ distributions.

We obtain draws from the joint posterior distribution of $(\boldsymbol{\beta}, \boldsymbol{\gamma}, \sigma_\varepsilon^2)$ by MCMC. Table 8.12 summarizes the marginal posterior distributions of the degradation data model parameters. The results suggest that only factors A and C impact μ_θ; only the posterior distributions of β_2 and β_4 are concentrated away from zero and correspond to the factor A and C main effects.

We can use these results to recommend factor levels at which the fluorescent lamp reliability is the best. Because the $\boldsymbol{\beta}$ factor effects for factors A and C are positive, $(A, C) = (+,+)$ are the recommended levels. Figure 8.16 displays the predictive lifetime distributions at the four runs of the experiment. We obtain the predictive distributions by taking the posterior draws

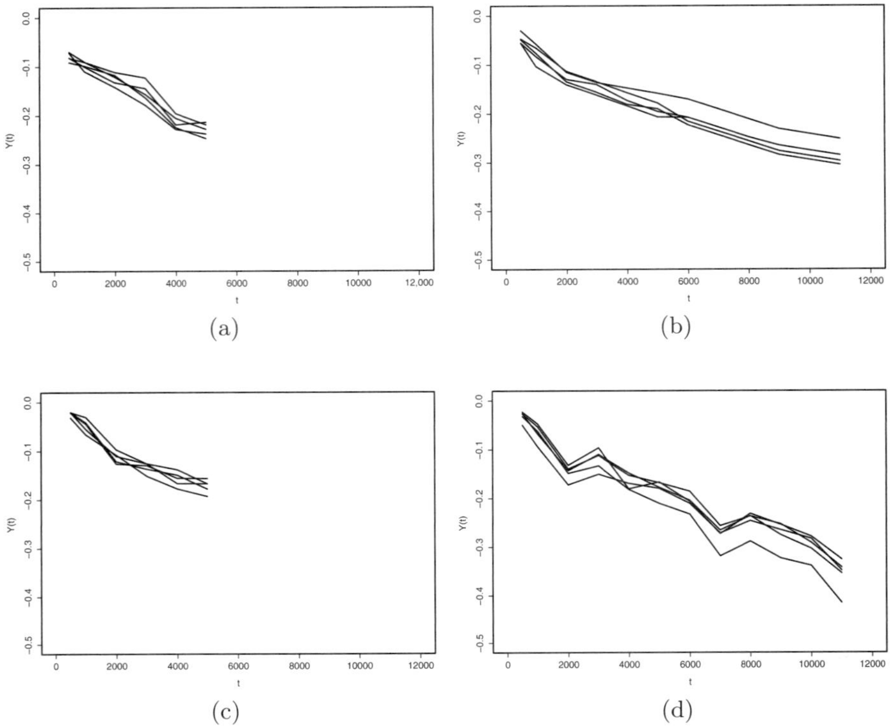

Fig. 8.15. Plot of fluorescent lamp degradation data (log relative luminosity) over time t in hours at (a) run 1, (b) run 2, (c) run 3, and (d) run 4.

of the model parameters, evaluating $\mu_\theta(\mathbf{x}_i)$ and $\sigma_\theta^2(\mathbf{x}_i)$, $i = 1, \ldots, 4$, and drawing a lognormal time using Eq. 8.20 and adding 100. Note that run 3 has $(A, C) = (+, +)$, the recommended factor levels, and indeed has the best lifetime distribution, i.e., the run with the longest predicted lifetimes.

Thus far, we have considered nondestructive measurements, so that we can observe a unit's degradation over time. Next, consider the destructive measurements case, in which a unit's degradation can be measured only once.

8.5 Destructive Degradation Data

There are situations in which measuring degradation is necessarily destructive, such as in testing the dielectric breakdown strength of insulation. When a measurement is destructive, only one measurement per unit is possible. We refer to such data obtained via destructive measurements as *destructive degradation data*. This section also applies to situations where it is logistically too

Table 8.11. Fluorescent lamp degradation data (log relative luminosity)

Time	Run 1 (−,−,−) Unit					Run 3 (−,+,+) Unit				
(hours)	1	2	3	4	5	1	2	3	4	5
500	−0.0822	−0.0817	−0.0702	−0.0719	−0.0912	−0.0205	−0.0215	−0.0315	−0.0205	−0.0203
1000	−0.0903	−0.0999	−0.0898	−0.1094	−0.0983	−0.0304	−0.0442	−0.0660	−0.0550	−0.0414
2000	−0.1112	−0.1322	−0.1209	−0.1417	−0.1172	−0.0968	−0.1263	−0.1078	−0.1103	−0.1213
3000	−0.1225	−0.1444	−0.1564	−0.1785	−0.1634	−0.1257	−0.1293	−0.1511	−0.1257	−0.1367
4000	−0.1958	−0.2186	−0.2054	−0.2282	−0.2244	−0.1663	−0.1555	−0.1773	−0.1379	−0.1489
5000	−0.2187	−0.2136	−0.2279	−0.2374	−0.2468	−0.1667	−0.1560	−0.1927	−0.1667	−0.1777
6000	−0.2285	−0.2237	−0.2522	−0.2761	−0.2712	−0.2099	−0.2302	−0.2519	−0.2099	−0.1913

Time	Run 2 (+,−,+) Unit					Run 4 (+,+,−) Unit				
(hours)	1	2	3	4	5	1	2	3	4	5
500	−0.0302	−0.0556	−0.0556	−0.0486	−0.0473	−0.0496	−0.0319	−0.0261	−0.0229	−0.0264
1000	−0.0575	−0.0829	−0.1031	−0.0762	−0.0654	−0.0938	−0.0621	−0.0676	−0.0470	−0.0528
2000	−0.1152	−0.1297	−0.1407	−0.1346	−0.1131	−0.1721	−0.1486	−0.1429	−0.1320	−0.1396
3000	−0.1362	−0.1403	−0.1621	−0.1562	−0.1345	−0.1504	−0.1333	−0.1098	−0.0963	−0.1120
4000	−0.1475	−0.1730	−0.1840	−0.1809	−0.1572	−0.1692	−0.1817	−0.1480	−0.1804	−0.1527
5000	−0.1585	−0.1955	−0.2067	−0.1900	−0.1783	−0.1796	−0.2104	−0.1775	−0.1666	−0.1680
6000	−0.1705	−0.2076	−0.2076	−0.2231	−0.2159	−0.2105	−0.2325	−0.2041	−0.2062	−0.1856
7000	−0.1905	−0.2278	−0.2278	−0.2436	−0.2360	−0.2712	−0.3185	−0.2653	−0.2727	−0.2562
8000	−0.2109	−0.2480	−0.2480	−0.2640	−0.2559	−0.2460	−0.2881	−0.2358	−0.2313	−0.2364
9000	−0.2312	−0.2649	−0.2649	−0.2843	−0.2764	−0.2648	−0.3231	−0.2751	−0.2540	−0.2526
10000	−0.2414	−0.2751	−0.2751	−0.2944	−0.2865	−0.2827	−0.3385	−0.3032	−0.2781	−0.2909
11000	−0.2516	−0.2850	−0.2850	−0.3046	−0.2967	−0.3481	−0.4149	−0.3540	−0.3254	−0.3417
12000	−0.2567	−0.2902	−0.2902	−0.3095	−0.3016	−0.3320	−0.4038	−0.3710	−0.3235	−0.3403

Table 8.12. Posterior distribution summaries of fluorescent lamp degradation data model parameters

			Quantiles				
Parameter	Mean	Std Dev	0.025	0.050	0.500	0.950	0.975
β_1	10.21	0.03	10.16	10.17	10.21	10.26	10.27
β_2	0.1693	0.0280	0.1135	0.1232	0.1697	0.2145	0.2228
β_3	0.0276	0.0280	−0.0267	−0.0177	0.0275	0.0742	0.0819
β_4	0.0814	0.0282	0.0253	0.0353	0.0814	0.1260	0.1360
γ_1	−7.768	1.577	−11.240	−10.590	−7.622	−5.464	−5.133
γ_2	0.267	1.591	−2.844	−2.307	0.252	2.916	3.491
γ_3	0.359	1.566	−2.742	−2.207	0.350	2.923	3.450
γ_4	−0.135	1.582	−3.214	−2.705	−0.141	2.496	3.091
σ_ε	0.0514	0.0027	0.0465	0.0472	0.0513	0.0559	0.0569

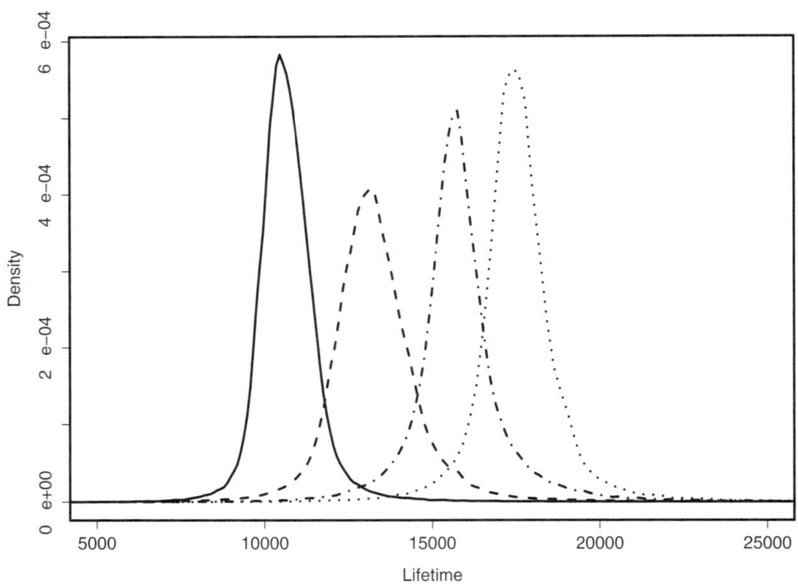

Fig. 8.16. Fluorescent lamp predictive lifetime distributions at runs 1, 2, 4, and 3 from left to right for lifetimes in hours. The recommended levels are $(A, C) = (+, +)$ at which the experimenters performed run 3.

expensive to return a unit to service after testing; in such situations, we also measure a unit's degradation only once.

For destructive degradation data, consider a model like those presented in Sects. 8.1 and 8.2 that implies a known lifetime distribution. The analyst can, under such a model, derive the probability density function for a destructive measurement at time t as follows. Let the true degradation $D(t)$ for the ith unit at time t be

$$D_i(t) = \beta_0 - \beta_1(1/\theta_i)t, \tag{8.21}$$

where the θ_i are assumed conditionally independent with a specified probability distribution. By setting the degradation $D(t)$ equal to D_f, calculate the lifetime t as

$$t = [(\beta_0 - D_f)/\beta_1]\theta = c\theta, \tag{8.22}$$

where

$$c = (\beta_0 - D_f)/\beta_1.$$

Equation 8.22, for example, implies that lifetime T has a $LogNormal(\log(c)+\mu, \sigma^2)$ distribution if θ has a $LogNormal(\mu, \sigma^2)$ distribution. Under this model, a derivation of the probability density function for the destructive degradation $z = D(t)$ yields

$$f(z) = [\sqrt{2\pi}(\beta_0 - z)\sigma]^{-1} \exp\{-0.5[\frac{\log[\beta_1 t/(\beta_0 - z)] - \mu}{\sigma}]^2\}. \tag{8.23}$$

In an analysis of the destructive degradation data, the likelihood consists of a contribution from Eq. 8.23 for each destructive measurement. Note that we have assumed no measurement error; when there is measurement error, a destructive degradation observation has a more complicated probability density function. We leave the derivation as Exercise 8.17. Prior distributions for the parameters β_0, β_1, μ, and σ^2 also need specification.

Based on the lifetimes having a $LogNormal(\log(c) + \mu, \sigma^2)$ distribution, the reliability function at time t is

$$R(t) = 1 - \Phi[(\log(t) - \log(c) - \mu)/\sigma], \tag{8.24}$$

where Φ is the standard normal cumulative distribution function.

Example 8.6 Insulation aging. Nelson (1981) presents an experiment that measured the dielectric breakdown strength of insulation under various temperatures at various ages. The study measured insulation specimens under four temperatures (180, 225, 250, and 275°C) at eight times (1, 2, 4, 8, 16, 32, 48, and 64 weeks). This experiment employed acceleration because the normal use temperature is 150°C. Table 8.13 presents and Fig. 8.17 plots the breakdown strength data (in kV). Also, the experimenters defined failure as occurring when the breakdown strength reaches 2 kV, so that the threshold $D_f = 2$.

Table 8.13. Insulation strength data (breakdown voltage in kV) (Nelson, 1981)

Week	Temp	Volt	Week	Temp	Volt
1	180	15.0	16	180	18.5
1	180	17.0	16	180	17.0
1	180	15.5	16	180	15.3
1	180	16.5	16	180	16.0
1	225	15.5	16	225	13.0
1	225	15.0	16	225	14.0
1	225	16.0	16	225	12.5
1	225	14.5	16	225	11.0
1	250	15.0	16	250	12.0
1	250	14.5	16	250	12.0
1	250	12.5	16	250	11.5
1	250	11.0	16	250	12.0
1	275	14.0	16	275	6.0
1	275	13.0	16	275	6.0
1	275	14.0	16	275	5.0
1	275	11.5	16	275	5.5
2	180	14.0	32	180	12.5
2	180	16.0	32	180	13.0
2	180	13.0	32	180	16.0
2	180	13.5	32	180	12.0
2	225	13.0	32	225	11.0
2	225	13.5	32	225	9.5
2	225	12.5	32	225	11.0
2	225	12.5	32	225	11.0
2	250	12.5	32	250	11.0
2	250	12.0	32	250	10.0
2	250	11.5	32	250	10.5
2	250	12.0	32	250	10.5
2	275	13.0	32	275	2.7
2	275	11.5	32	275	2.7
2	275	13.0	32	275	2.5
2	275	12.5	32	275	2.4
4	180	13.5	48	180	13.0
4	180	17.5	48	180	13.5
4	180	17.5	48	180	16.5
4	180	13.5	48	180	13.6
4	225	12.5	48	225	11.5
4	225	12.5	48	225	10.5
4	225	15.0	48	225	13.5
4	225	13.0	48	225	12.0
4	250	12.0	48	250	7.0
4	250	13.0	48	250	6.9
4	250	12.0	48	250	8.8
4	250	13.5	48	250	7.9

Table 8.13. (cont.)

Week	Temp	Volt	Week	Temp	Volt
4	275	10.0	48	275	1.2
4	275	11.5	48	275	1.5
4	275	11.0	48	275	1.0
4	275	9.5	48	275	1.5
8	180	15.0	64	180	13.0
8	180	15.0	64	180	12.5
8	180	15.5	64	180	16.5
8	180	16.0	64	180	16.0
8	225	13.0	64	225	11.0
8	225	10.5	64	225	11.5
8	225	13.5	64	225	10.5
8	225	14.0	64	225	10.0
8	250	12.5	64	250	7.2667
8	250	12.0	64	250	7.5
8	250	11.5	64	250	6.7
8	250	11.5	64	250	7.6
8	275	6.5	64	275	1.5
8	275	5.5	64	275	1.0
8	275	6.0	64	275	1.2
8	275	6.0	64	275	1.2

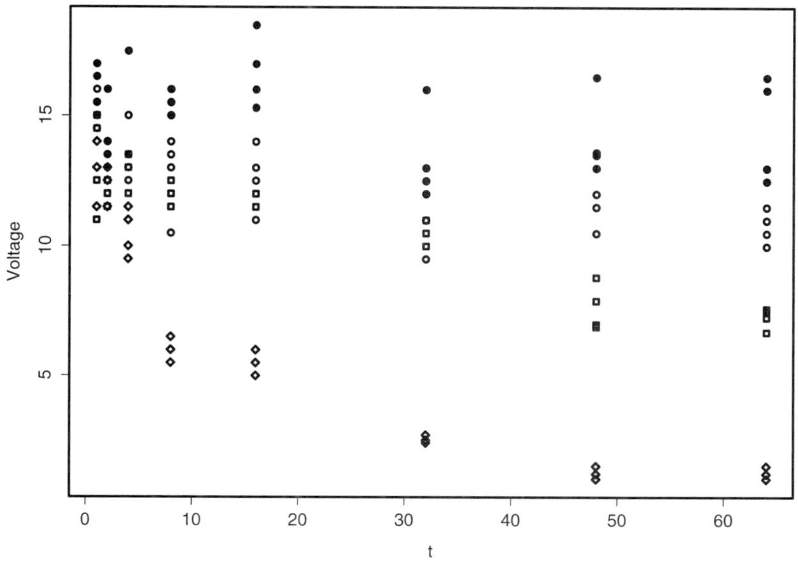

Fig. 8.17. Insulation strength data (in volts) over time t in weeks at 180°C (solid circle), 225°C (open circle), 250°C (square), and 275°C (diamond).

We use the following model proposed by Nelson (1981) for breakdown voltage V (on the log base 10 scale, which engineers still often use in practice) at absolute temperature T (in kelvins) and time t in weeks:

$$\log_{10}(V) = \log(V)/\log(10) = \alpha - \exp(-\gamma/T)(1/\theta)t. \qquad (8.25)$$

Consequently, the log of the voltage degradation follows the form given in Eq. 8.21. Solving $\log(V) = \log(D_f)$ for lifetime t, i.e., the time until the breakdown voltage reaches its threshold, yields the following expression for lifetime t:

$$t = \{[\alpha - \log(D_f)/\log(10)]/\exp(-\gamma/T)\}\theta = c\theta,$$

where $c = [\alpha - \log(D_f)/\log(10)]/\exp(-\gamma/T)$.

In an analysis of the breakdown voltage data, the likelihood consists of contributions from each of the breakdown voltage data according to Eq. 8.23, where $\beta_0 = \alpha$, $\beta_1 = \exp(-\gamma/T)$. Recall that the effects θ_i have a $LogNormal(\mu, \sigma^2)$ distribution, so that μ and σ^2 also appear in Eq. 8.23. Consequently, the destructive degradation data model parameters are (α, γ, μ, σ^2), and we use the following independent and diffuse prior distributions for them:

$$\begin{aligned}
&\alpha \sim LogNormal(0, 1000)I(\max\{\log(V_i)/\log(10)\}, \infty)\ ,\\
&\gamma \sim LogNormal(0, 1000),\\
&\mu \sim Normal(0, 1000), \text{ and}\\
&\sigma^2 \sim InverseGamma(0.1, 0.1)I(0.001, 33),
\end{aligned}$$

where the $I(\cdot, \cdot)$ notation indicates the support of the prior distribution, i.e., the interval on which we define the prior distribution. The use of a restricted prior distribution for α follows from α, the initial voltage, exceeding the largest observed degraded logged (base 10) voltage.

We used a trick in analyzing these data with a software package (e.g., WinBUGS) that does not support the specialized probability density function given in Eq. 8.23. Assuming a very small normal measurement error gives the model $Y = D(t) + \varepsilon$ for destructive degradation $D(t)$ in Eq. 8.21, where the measurement error $\varepsilon \sim Normal(0, 10^{-6})$ and $\theta \sim LogNormal(\mu, \sigma^2)$; recall that we considered this same normal error degradation data model earlier in this chapter. Consequently, in an analysis, the likelihood consists of a normal density contribution for each observed destructive degradation and a lognormal density contribution for each of the θ_i, i.e., one for each observation. We obtain draws from the joint posterior distribution of (α, γ, μ, σ^2) by MCMC. Table 8.14 presents the marginal posterior distribution summaries for the model parameters (α, γ, μ, σ^2). We obtain draws from the posterior distribution of the insulation reliability at 150°C by evaluating Eq. 8.24 with the model parameters' posterior draws. Figure 8.18 plots the insulation posterior reliability median and 90% credible intervals.

To assess the model fit, we can apply a Bayesian χ^2 goodness-of-fit test if the cumulative distribution function of $z = \log(V)$ based on Eq. 8.23 is

Table 8.14. Posterior distribution summaries of insulation strength degradation data model parameters

			Quantiles				
Parameter	Mean	Std Dev	0.025	0.050	0.500	0.950	0.975
α	2.926	0.005	2.920	2.920	2.925	2.934	2.937
γ	3.811	7.201	0.002	0.003	0.303	22.390	27.450
μ	3.103	0.120	2.864	2.902	3.104	3.300	3.339
σ	1.344	0.087	1.186	1.210	1.339	1.495	1.530

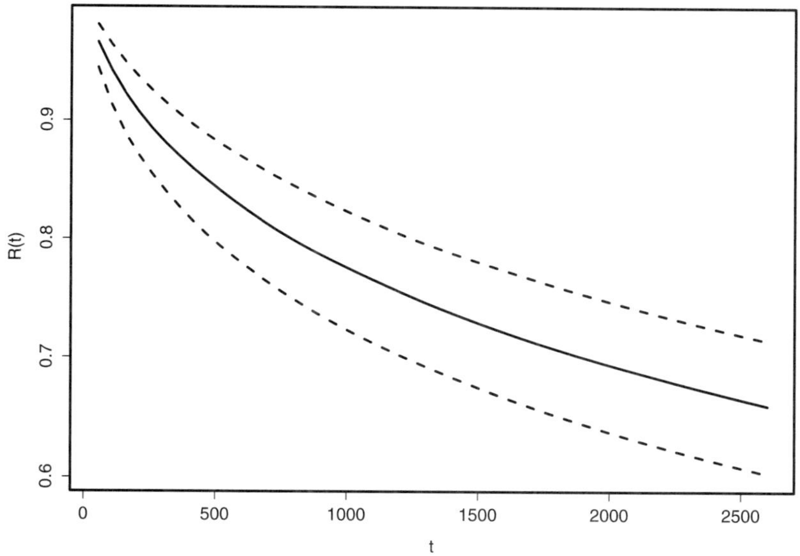

Fig. 8.18. Insulation reliability over time t in weeks and 90% credible intervals at 150°C. The solid line gives the posterior medians. The dotted lines give the 0.05 and 0.95 posterior quantiles.

available. Rather than integrating Eq. 8.23, note that simply from Eq. 8.25 we have

$$\log\left(\frac{t\exp(-\gamma/T)}{\alpha - [\log(V)/\log(10)]}\right) \sim Normal(\mu, \sigma^2).$$

We apply a Bayesian χ^2 goodness-of-fit test to the degradation data using five equally spaced bins and find that about 19% of the R^B values exceed the 0.95 quantile of the $ChiSquared(4)$ reference distribution, which indicates some lack of fit.

8.6 An Alternative Degradation Data Model Using Stochastic Processes

Thus far in this chapter, we have considered parametric models for degradation curves. This section considers an alternative for modeling degradation data based on stochastic processes.

A *Wiener process* is one type of stochastic process that has the property that the degradation W_i at time t_i follows $W_i \sim Normal(\delta t_i, \nu t_i)$. That is, there is a linear drift in the degradation curve, which has variance that increases with time. Moreover, the degradations measured at times t_i and t_j are correlated because $W_i - W_{i-1} \sim Normal(0, \nu(t_i - t_{i-1}))$. Consequently, all the measured degradations modeled by a Wiener process are correlated. We can also incorporate measurement error by assuming that $Y_i = W_i + \varepsilon_i$, for times $t_0, t_1, \ldots, t_n$ and true degradation $W_i = W(t_i)$, where the measurement errors ε_i have conditionally independent $Normal(0, \sigma^2)$ distributions, and are independent of the W_i.

Letting $\mathbf{Y} = (Y_0, Y_1, \ldots, Y_n)$ be the observed degradation measurements at times $\mathbf{t} = (t_0, t_1, \ldots, t_n)$ under a Wiener process, then we have

$$\mathbf{Y} \sim MultivariateNormal_{n+1}(\boldsymbol{\mu}, \boldsymbol{\Sigma}), \tag{8.26}$$

where $\boldsymbol{\mu} = (\mu_0, \mu_1, \ldots, \mu_n)$ and $\boldsymbol{\Sigma} = (\sigma_{ij})$. The components of $\boldsymbol{\mu}$ and $\boldsymbol{\Sigma}$ are

$$\mu_i = \mathrm{E}(Y_i) = \mathrm{E}(W_i) = W(0) + \delta t_i \tag{8.27}$$

and

$$\Sigma_{ij} = \begin{cases} \nu t_i + \sigma^2 & i = j \\ \nu \min(t_i, t_j) & i \neq j, \end{cases}$$

where σ_{ij} is the (i, j)th entry of $\boldsymbol{\Sigma}$.

Given a threshold D_f, zero initial degradation (i.e., $W(0) = 0$), and δ, the growth in degradation per unit time with $\delta > 0$, the lifetimes have an $InverseGaussian(D_f/\delta, D_f^2/\nu)$ distribution. Consequently, the reliability function takes the form

$$R(t) = 1 - \{\Phi[(\delta t - D_f)(\nu t)^{-1/2}] + \exp(2\delta/\nu)\Phi[-(\delta t + D_f)(\nu t)^{-1/2}]\}. \tag{8.28}$$

Note that the drift in the degradation curve may not be linear in clock time r as given in Eq. 8.27; in such situations, the analyst may need to first transform the clock times. For example, the linear drift may be valid for transformed time t, such as $t = 1 - \exp(-\lambda r)$ for clock time r. Consequently, analyze the degradation data in transformed time t. We can report the results in clock time r by using the appropriate inverse of the transformation function.

Example 8.7 Transistor gain. Whitmore (1995) considers the gain of transistors that declines with age, as the data in Table 8.15 (presented at various clock times r in thousands of hours) demonstrate. Whitmore (1995) did not provide the units of gain. Reviewing the plot of the transistor gain data (by clock time r) in Fig. 8.19 reveals a departure from a linear drift. For comparison, using transformed time $t = 1 - \exp(-\lambda r)$ with $\lambda = 0.333$ gives the plot displayed in Fig. 8.20, which appears as a straight line. For illustration, we define a transistor failure as occurring when its gain reaches $0.925W(0)$, i.e., 92.5% of the original gain $W(0)$. Recall that the reliability function in Eq. 8.28 is for a Wiener process model starting at zero with positive drift. Consequently, in calculating reliability, we recast the transistor gain example by starting at 0 and define failure as occurring when the degradation reaches $0.075W(0)$. Also let $\mu_i = W(0) - \delta t_i$ so that the drift δ is necessarily positive.

Table 8.15. Transistor gain versus clock time in thousands of hours (Whitmore, 1995)

Time	Gain	Time	Gain
0	90.9	1.35	87.0
0.05	90.3	1.5	87.1
0.115	90.1	1.735	86.9
0.18	89.9	1.896	86.5
0.25	89.6	2.13	86.9
0.32	89.6	2.46	85.9
0.42	89.3	2.8	85.4
0.54	89.1	3.2	85.2
0.63	89.0	3.9	84.6
0.72	89.1	4.6	83.8
0.81	88.5	5.65	83.9
0.875	88.4	7.8	82.3
0.941	88.5	8.688	82.5
1.01	88.3	10	82.3
1.1	87.7		
1.2	87.5		

In an analysis of the transistor gain data, the likelihood consists of a multivariate normal density contribution specified by Eq. 8.26. For the model parameters ($W(0)$, δ, ν, σ^2), we use the following diffuse and independent prior distributions:

$$
\begin{aligned}
&W(0) \sim InverseGamma(0.0031, 1),\\
&\delta \sim InverseGamma(0.001, 0.001)I(0.001, 33.3),\\
&\nu \sim InverseGamma(0.001, 0.001)I(0.001, 20), \text{ and}\\
&\sigma^2 \sim InverseGamma(0.001, 0.001)I(0.1, 1000),
\end{aligned}
$$

where $I(\cdot, \cdot)$ denotes the interval on which the prior distribution is defined. We chose the lower bound for the $W(0)$ prior distribution because it exceeded

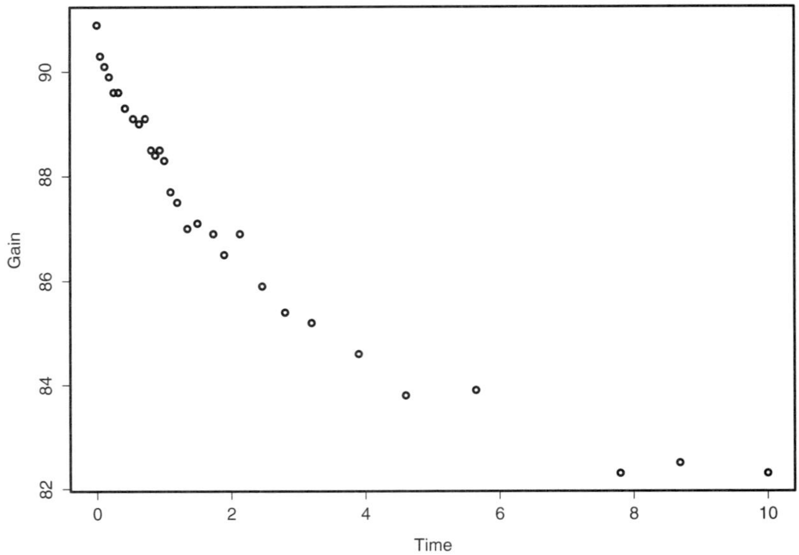

Fig. 8.19. Plot of transistor gain data by clock time in thousands of hours.

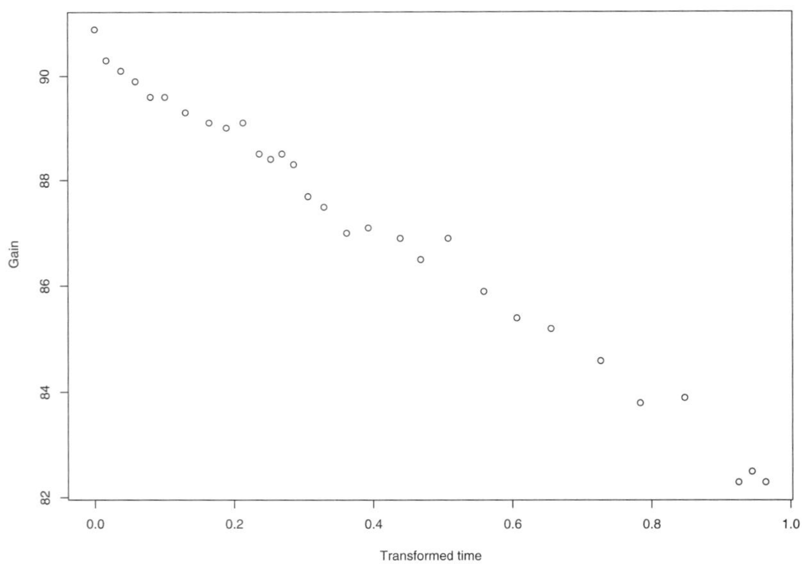

Fig. 8.20. Plot of transistor gain data by transformed time.

all the observed transistor gains and obtained draws from the joint posterior distribution of the model parameters by MCMC. Table 8.16 summarizes the marginal posterior distributions of the model parameters.

Table 8.16. Posterior distribution summaries of transistor gain data model parameters

			Quantiles				
Parameter	Mean	Std Dev	0.025	0.050	0.500	0.950	0.975
δ	8.397	0.382	7.650	7.843	8.391	8.980	9.187
ν	0.0931	0.2158	0.0013	0.0016	0.0208	0.4178	0.6395
$W(0)$	90.52	0.13	90.28	90.32	90.52	90.73	90.77
σ_m^2	0.1169	0.0179	0.1005	0.1009	0.1116	0.1511	0.1638

To assess transistor reliability using Eq. 8.28, we have $D_f = 0.075W(0)$, but notice that D_f is unknown because $W(0)$ is a parameter. Also evaluate reliability in terms of clock time r, so use the inverse function of transformed time t, which is $r = (-1/\lambda)\log(1-t)$. Evaluating reliability using Eq. 8.28 for the model parameters' posterior draws, we obtain Fig. 8.21, which plots the posterior reliability medians and 90% credible intervals.

8.7 Related Reading

The use of degradation data for assessing reliability is relatively new. Lu and Meeker (1993) is an important early paper on this topic. Regarding the incorporation of covariates, Boulanger and Escobar (1994) and Meeker et al. (1998) discuss analyzing accelerated degradation data. Tseng et al. (1995) and Chiao and Hamada (1996) consider reliability improvement experiments using degradation data. Nelson (1981) is an early paper dealing with destructive degradation data. Considering alternatives to parametric models, Whitmore (1995) explores the Wiener process for modeling degradation data with measurement error. Another alternative to the parametric modeling of degradation curves is the use of nonparametric regression, because the form of the degradation curves does not have to be specified. This is an attractive alternative when a parametric model is neither obvious from the data nor driven by the science/engineering of the problem. See Horng-Shiau and Lin (1999), which considers this topic. Excellent general references on nonparametric regression are Ramsay and Silverman (1997) and Green and Silverman (1994).

Regarding priors for degradation data model parameters, recall that if the reciprocal slopes have a lognormal distribution, then the log reciprocal slopes have a normal distribution. Gelman (2006) suggests using a $Uniform(0, U)$

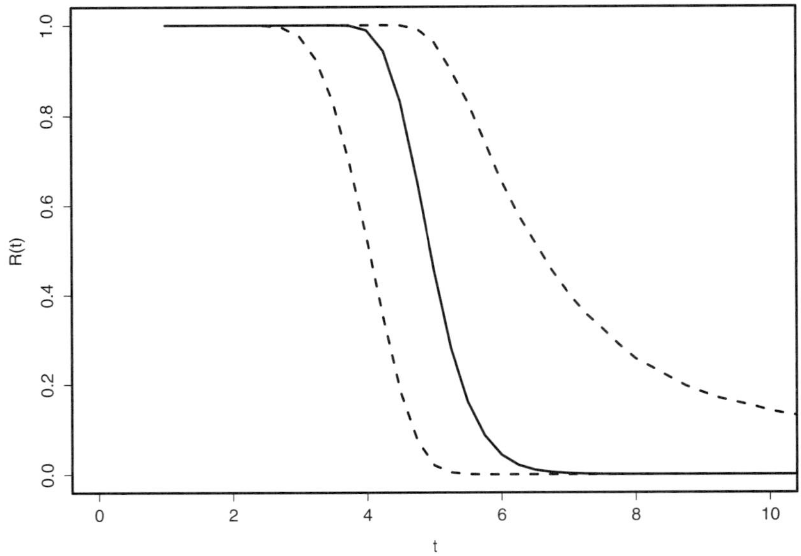

Fig. 8.21. Transistor reliability over time t in thousands of hours. The solid line is the posterior medians. The dashed lines are the 0.05 and 0.95 posterior quantiles or 90% credible intervals.

distribution (large U) as a diffuse prior distribution for the standard deviation of the random effects normal distribution if the number of random effects is small. This choice of prior distribution has little impact on the results for Example 8.1, although this dataset involves a large number of random effects.

8.8 Exercises for Chapter 8

8.1 Hamada (2006) reports the transformed light intensity in lumen/meter^2 (negative logarithm shifted to equal -5.0000 at 0 hours) of nine LEDs at 50, 100, 150, 200, and 250 hours. Table 8.17 presents these transformed data. Analyze these degradation data assuming a linear degradation model with unit-dependent slope. How well does this model fit the data? Assuming a threshold of -4.3, estimate the reliability at 300 hours and provide a 90% credible interval.

8.2 McDonald et al. (1995) provides emissions data for an experimental car. The experimenters measured HC, CO, and NO_2 in gram per mile at 0, 4,000, and 24,000 miles on 16 cars. (See the emissions data in Table 8.18.) Fit an appropriate degradation data model, assuming the data at each of the three inspections have a lognormal distribution. For an HC standard

Table 8.17. Transformed LED light intensity data (Hamada, 2006)

Part	Hours 50	100	150	200	250
1	−4.6995	−4.4568	−4.3583	−4.1734	−3.9900
2	−4.5853	−4.1105	−3.3781	−3.5268	−3.3326
3	−4.4918	−4.0063	−3.6119	−3.4022	−3.2968
4	−4.5660	−4.1605	−3.8304	−3.5544	−3.1773
5	−4.3200	−3.9120	−3.6500	−3.2970	−2.4583
6	−4.6152	−4.2759	−3.9528	−3.6652	−3.5268
7	−4.6886	−3.9686	−3.6382	−3.4022	−3.3668
8	−4.2336	−3.8077	−3.4673	−3.2189	−3.1773
9	−4.2759	−3.8491	−3.2571	−2.9957	−1.9456

of 0.41 gram per mile, would only one car in 10,000 fail the standard based on a 95% credible upper bound? If not, at what mileage would this requirement be met?

Table 8.18. Emissions data (grams per mile) (McDonald et al., 1995)

	0 miles			4,000 miles			24,000 miles		
Car	HC	CO	NO_x	HC	CO	NO_x	HC	CO	NO_x
1	0.16	2.89	2.21	0.26	1.16	1.99	0.23	2.64	2.18
2	0.38	2.17	1.75	0.48	1.75	1.90	0.41	2.43	1.59
3	0.20	1.56	1.11	0.40	1.64	1.89	0.35	2.20	1.99
4	0.18	3.49	2.55	0.38	1.54	2.45	0.26	1.88	2.29
5	0.33	3.10	1.79	0.31	1.45	1.54	0.43	2.58	1.95
6	0.34	1.61	1.88	0.49	2.59	2.01	0.48	4.08	2.21
7	0.27	1.14	2.20	0.25	1.39	1.95	0.41	2.49	2.51
8	0.30	2.50	2.46	0.23	1.26	2.17	0.36	2.23	1.88
9	0.41	2.22	1.77	0.39	2.72	1.93	0.41	4.76	2.48
10	0.31	2.33	2.60	0.21	2.23	2.58	0.26	3.73	2.70
11	0.15	2.68	2.12	0.22	3.94	2.12	0.58	2.48	2.32
12	0.36	1.63	2.34	0.45	1.88	1.80	0.70	3.10	2.18
13	0.33	1.58	1.76	0.39	1.49	1.46	0.48	2.64	1.69
14	0.19	1.54	2.07	0.36	1.81	1.89	0.33	2.99	2.35
15	0.23	1.75	1.59	0.44	2.90	1.85	0.48	3.04	1.79
16	0.16	1.47	2.25	0.22	1.16	2.21	0.45	3.78	2.03

8.3 In the preceding exercise, assume a Weibull distribution. Does the Weibull distribution provide a better fit?

8.4 Analyze the crack length data of Example 8.2 directly with expected value $a(t) = a_0/(1 - a_0^{\theta_2}\theta_1\theta_2 t)^{1/\theta_2}$ and normally distributed measurement error. How do the results change for $R(t)$ and $t_{0.1}$? Is this model better than the one used in Example 8.2?

8.5 Incorporate the transformed clock time into the Wiener process model for the transistor gain data of Example 8.7; use $t = 1 - \exp(-\lambda r)$, so that λ is an additional parameter. Reanalyze the transistor gain data and comment on Example 8.7's use of $\lambda = 0.333$.

8.6 An experiment studied the impact of five factors on the degradation of voltage drop for windshield wiper switches. The eight-run experiment displayed in Table 8.19 studied one factor (A) at four levels, denoted by 0–3, and another four factors (B–E) at two levels, denoted by $(-, +)$. There were four switches available for each of the eight runs. For each switch, the experimenters recorded the initial voltage drop (in volts) across multiple contacts and then remeasured the voltage drop every 20,000 cycles up to 180,000 cycles. (See Table 8.19 for the voltage-drop data in volts.)
   a) Analyze the voltage-drop data using a degradation model with an intercept and slope that both have distributions assuming that the four-level factor is quantitative, with evenly spaced levels. Treat a voltage drop of 120 as unacceptable. What factors impact the failure times of the switch? What levels should the experimenters set the important factors at to improve the reliability?
   b) Perform a residual analysis.

8.7 We constructed the drug potency data in Example 8.1 so that we might consider a simpler model. Table 8.20 displays the actual data from the experiment reported in Chow and Shao (1991). Analyze these data assuming that the intercept and slope follow a bivariate normal distribution. How do the results differ from those presented in Example 8.1?

8.8 Lu et al. (1997) considers the hot-carrier induced degradation of semiconductors. Table 8.21 displays their transductance degradation data in percent for five devices. The experimenters defined failure as occurring when the transductance reaches 15% of the original transductance. Lu et al. (1997) uses the model $\log(Y) = \beta_0 + \beta_1 \log(t) + \varepsilon_t$ for the observed degradation y at time t, where $\varepsilon_t \sim Normal(0, \sigma_t^2)$ and $\log(\sigma_t^2) = \alpha_0 + \alpha_1 |t - t_0|$ for $t_0 = 3.66$. Lu et al. (1997) also assumes a bivariate normal distribution for (β_0, β_1).
   a) Analyze these degradation data using this model. Do the data support the dependence of σ_t^2 on time t? Do the data support the need for assuming that (β_0, β_1) follow a distribution? That is, will a common (β_0, β_1) be sufficient?
   b) Analyze the data as lognormal failure times. How does the resulting reliability function compare with that obtained from analyzing the degradation data?
   c) Perform a residual analysis.

8.9 Analyze the drug potency data of Example 8.1 using a Wiener process. How do the results compare with those obtained in Example 8.1?

8.10 Whitmore and Schenkelberg (1997) analyzes degradation data from an accelerated test of heating cables using a Wiener process. Table 8.22 displays the data that are log resistances at various times (in thousands of hours).

Table 8.19. Wiper switch experiment, experimental plan and voltage-drop data (in volts)

Factor					Inspection (thousands of cycles)									
A	B	C	D	E	0	20	40	60	80	100	120	140	160	180
0	−	−	−	−	24	37	40	65	72	77	90	101	117	128
					22	36	47	64	71	86	99	118	127	136
					17	34	40	52	66	79	91	98	115	119
					24	30	38	46	57	71	73	91	98	104
0	+	+	+	+	45	60	79	90	113	124	141	153	176	188
					51	68	84	104	122	136	148	166	191	197
					42	58	70	82	103	119	128	143	160	175
					41	56	56	70	81	89	98	108	113	128
1	−	−	+	+	28	40	56	69	87	86	110	121	132	146
					46	50	81	95	114	130	145	161	185	202
					45	54	79	90	111	132	143	168	185	202
					37	58	81	99	123	143	166	191	202	231
1	+	+	−	−	54	51	64	66	78	84	90	93	106	109
					47	45	50	53	58	57	61	55	61	66
					47	54	63	68	70	77	88	86	91	102
					53	55	66	68	91	90	98	104	118	120
2	−	+	−	+	18	35	48	56	65	81	89	98	117	124
					20	37	52	53	67	75	85	95	112	122
					32	54	76	98	119	143	158	181	205	231
					28	39	54	73	89	98	117	127	138	157
2	+	−	+	−	44	50	48	46	55	63	65	71	68	76
					43	44	55	56	58	62	66	66	72	72
					40	46	45	49	55	62	61	61	64	66
					55	67	73	75	91	88	102	111	115	119
3	−	+	+	−	39	47	58	72	84	104	109	129	143	154
					29	42	55	67	82	91	104	117	130	136
					36	45	56	80	93	101	121	138	154	170
					31	40	60	72	82	98	103	117	130	146
3	+	−	−	+	61	67	69	86	86	88	95	103	107	118
					68	75	82	90	95	109	107	118	120	133
					60	72	85	84	87	98	99	111	113	125
					65	68	69	75	79	84	95	96	101	100

The experimenters tested five cables at each of three test temperatures, 200°C, 240°C, and 260°C. Whitmore and Schenkelberg (1997) transforms clock time using $t = 1 - \exp(-\lambda r)$ and assumes that the Wiener process parameters δ and ν, as well as the time transformation parameter λ, all depend on absolute temperature s (in kelvins) as follows: $\delta(s) = \alpha_0 + \alpha_1/s$, $\log[\nu(s)] = \beta_0 + \beta_1/s$ and $1/\lambda(s) = \gamma_0 + \gamma_1/s$. Assuming a normal use temperature of 100°C, what is the probability that a cable's resistance will double in its first 10 years of life? Analyze the heating cable data using

Table 8.20. Actual drug potency degradation data (in percent of stated potency) (Chow and Shao, 1991)

	Time (months)					Time (months)			
Batch	0	12	24	36	Batch	0	12	24	36
1	105	104	101	98	13	105	104	99	95
2	106	102	99	96	14	104	103	97	94
3	103	101	98	95	15	105	103	98	96
4	105	101	99	95	16	103	101	99	96
5	104	102	100	96	17	104	102	101	98
6	102	100	100	97	18	106	104	102	97
7	104	103	101	97	19	105	103	100	99
8	105	104	101	100	20	103	101	99	95
9	103	101	99	99	21	101	101	97	90
10	103	102	97	96	22	102	100	99	96
11	101	98	93	91	23	103	101	99	94
12	105	102	100	98	24	105	104	100	97

a parametric degradation model. How do these results compare with those reached using a Wiener process?

8.11 Suppose that the demand lifetime of a component must exceed 18.5 seconds. Table 8.23 displays the demand lifetimes that were collected at various ages. Because the degrading characteristic in this case is a lifetime, assume a lognormal distribution in which the mean of the log lifetime depends on age.
  a) Analyze these destructive degradation data.
  b) Predict the reliability at various ages and provide 95% credible intervals.
  c) Perform a residual analysis.

8.12 In the preceding exercise, assume a Weibull distribution. Does the Weibull distribution provide a better fit?

8.13 The analysis presented in Example 8.5 suggested that σ_θ^2 did not depend on the experimental factors. Assess whether this simpler model holds using the model selection DIC diagnostic.

8.14 Assess the model fit in Example 8.4.

8.15 Develop residual analysis for degradation data modeled by a Wiener process. Try out your proposal on Example 8.7. Also try out your proposal on Exercises 8.10 and 8.11.

8.16 Assess the model fit in Examples 8.2, 8.5, and 8.6.

8.17 Suppose that the degradation curves follow the model presented in Sect. 8.5 and that the degradation is measured destructively with measurement error having a $Normal(0, \sigma^2)$ distribution. Develop the probability density function needed to analyze destructive degradation data measured with error.

Table 8.21. Percent transductance degradation data (Lu et al., 1997)

Time	Item				
(seconds)	1	2	3	4	5
100	1.05	0.58	0.86	0.60	0.62
200	1.40	0.90	1.25	0.60	0.64
300	1.75	1.20	1.45	0.60	1.25
400	2.10	1.75	1.75	0.90	1.30
500	2.10	2.01	1.75	0.90	0.95
600	2.80	2.00	2.00	1.20	1.25
700	2.80	2.00	2.00	1.50	1.55
800	2.80	2.00	2.00	1.50	1.90
900	3.20	2.30	2.30	1.50	1.25
1000	3.40	2.60	2.30	1.70	1.55
1200	3.80	2.90	2.60	2.10	1.50
1400	4.20	2.90	2.80	2.10	1.55
1600	4.20	3.20	3.15	1.80	1.90
1800	4.50	3.60	3.20	2.10	1.85
2000	4.90	3.80	3.20	2.10	2.20
2500	5.60	4.20	3.80	2.40	2.20
3000	5.90	4.40	3.80	2.70	2.50
3500	6.30	4.80	4.00	2.70	2.20
4000	6.60	5.00	4.20	3.00	2.80
4500	7.00	5.60	4.40	3.00	2.80
5000	7.80	5.90	4.60	3.00	2.80
6000	8.60	6.20	4.90	3.60	3.10
7000	9.10	6.80	5.20	3.60	3.10
8000	9.50	7.40	5.80	4.20	3.10
9000	10.50	7.70	6.10	4.60	3.70
10000	11.10	8.40	6.30	4.20	4.40
12000	12.20	8.90	7.00	4.80	3.70
14000	13.00	9.50	7.20	5.10	4.40
16000	14.00	10.00	7.60	4.80	4.40
18000	15.00	10.40	7.70	5.30	4.10
20000	16.00	10.90	8.10	5.80	4.10
25000	18.50	12.60	8.90	5.70	4.70
30000	20.30	13.20	9.50	6.20	4.70
35000	22.10	15.40	11.20	8.00	6.40
40000	24.20	18.10	14.00	10.90	9.40

Table 8.22. Log resistance heating cable test data (Whitmore and Schenkelberg, 1997)

Temp	Time (1000 hours)	Unit 1	2	3	4	5
200°C	0.496	−0.120682	−0.118779	−0.123600	−0.126501	−0.124359
	0.688	−0.112403	−0.109853	−0.115186	−0.118941	−0.111966
	0.856	−0.103608	−0.101593	−0.105657	−0.110288	−0.107869
	1.024	−0.096047	−0.094567	−0.098569	−0.103419	−0.100304
	1.192	−0.085673	−0.084698	−0.088613	−0.095465	−0.085916
	1.360	−0.077677	−0.076070	−0.079332	−0.084769	−0.077947
	2.008	−0.045218	−0.040623	−0.045835	−0.052268	−0.045597
	2.992	0.000526	0.004237	0.000533	−0.008265	0.000524
	4.456	0.059261	0.063742	0.061032	0.051139	0.059544
	5.608	0.093394	0.095117	0.093612	0.082414	0.084912
240°C	0.160	−0.005152	−0.019888	−0.045961	−0.023188	−0.044267
	0.328	0.056930	0.046278	0.015198	0.040737	0.018173
	0.496	0.112631	0.101628	0.067119	0.095504	0.072214
	0.688	0.173202	0.162705	0.128670	0.156129	0.131555
	0.856	0.214266	0.202604	0.168271	0.196349	0.171394
	1.024	0.272668	0.257563	0.221611	0.250900	0.225281
	1.192	0.311422	0.297875	0.260910	0.291937	0.266314
	1.360	0.351988	0.338902	0.302126	0.332887	0.306105
	2.008	0.489847	0.461855	0.440738	0.473130	0.443941
	2.992	0.656780	0.629991	0.606275	0.638651	0.611724
	4.456	0.851985	0.798431	0.834114	0.798457	
260°C	0.160	0.123360	0.127605	0.120759	0.105206	0.120115
	0.328	0.251084	0.254944	0.247156	0.232389	0.247949
	0.496	0.393107	0.394496	0.391516	0.375789	0.388406
	0.688	0.517137	0.518485	0.513872	0.500556	0.511850
	0.856	0.598797	0.599265	0.595704	0.583362	0.595220
	1.024	0.693925	0.694445	0.688930	0.679117	0.690324
	1.192	0.774347	0.774428	0.770313	0.758314	0.770782

8.18 The diagnostics presented in Example 8.3 revealed problems with the drug potency degradation data model used in Example 8.1. Explore whether transforming the degradation data first, e.g., by taking logarithms, provides a better model.

8.19 Batra et al. (2004) reports on an experiment that measured the degradation in the resistance of electronic packaging. The experiment studied two factors, pad size and design geometry, each at two levels. The pad size levels were 12 (−) and 18 (+) mils. The design geometry factor levels were symmetric (−) and asymmetric (+). Table 8.24 presents the

Table 8.23. Component demand lifetimes (in seconds) at various ages (in months)

Age	Lifetime	Age	Lifetime	Age	Lifetime
45	125.30	120	98.50	220	62.80
45	98.00	120	102.00	220	63.00
45	96.30	120	134.30	220	74.00
46	73.50	161	131.00	241	66.00
47	93.20	162	78.00	241	60.00
64	99.90	163	81.00	263	56.03
65	96.00	163	81.00	263	3.43
72	91.80	181	60.00	263	35.20
74	77.30	181	6.05	263	46.90
85	99.60	183	41.59	263	67.50
88	111.00	200	52.00	263	55.00
88	107.00	200	55.60	263	50.40
89	78.00	207	80.00	263	65.70
110	99.80	207	110.00	264	48.00
113	71.30	208	63.00	264	50.00

Table 8.24. Percent change in resistance of electronic package (Batra et al., 2004)

Design	Pad	Thermal Cycles			
Geometry	Size	0	100	250	500
−	−	0	−3.2	2.2	4.8
−	−	0	−3.8	0.0	0.8
−	−	0	−5.7	0.7	1.0
−	+	0	3.2	4.5	8.2
−	+	0	4.9	8.7	10.1
−	+	0	5.6	5.6	9.2
+	−	0	2.5	9.4	11.0
+	−	0	2.5	6.9	9.6
+	−	0	5.4	7.0	13.5
+	+	0	2.0	9.8	12.6
+	+	0	6.7	11.8	18.3
+	+	0	5.0	6.9	16.9

measured resistance degradation as percent change after thermal cycling at cycles 0, 100, 250, 500.

a) Analyze these degradation data assuming a linear degradation model with a common slope for all the units at a factor level combination.
b) Which factor level combination has the least degradation?
c) How well does this simple model fit the data?

8.20 In the preceding exercise, analyze the resistance degradation data assuming a linear degradation model with a unit-dependent slope.

a) How well does this model fit the data?
b) Which factor level combination has the least degradation?

9

Planning for Reliability Data Collection

This chapter considers planning for reliability data collection. Data collection planning determines how to optimally collect data, given a limited amount of resources (typically, money, time, and the number of units to test). This chapter discusses various planning criteria and presents a simulation-based framework to evaluate these criteria. Depending on the situation, planning can involve single and multiple planning variables. For multiple planning variable situations, we show how to use a genetic algorithm to find a near-optimal plan. This chapter illustrates data collection planning for a number of problems involving binomial, lifetime, accelerated life test, degradation, and system reliability data.

9.1 Introduction

The preceding chapters focused on making reliability assessments from available data. This chapter considers planning for reliability data collection, which explores how much of what kind of data to collect, given specific testing constraints. In Sect. 9.5, we consider plans in a specific context known as experimental designs.

This chapter presents a simulation-based framework for data collection planning and illustrates the framework's versatility with the planning problems considered throughout the chapter.

Planning for reliability data collection determines how many resources are required to meet a specified goal and how best to allocate these (often limited) resources. To assess a plan, we must develop a *planning criterion* that evaluates how well a plan meets a specified goal; the criterion typically is related to the quality of the inferences made with the collected data. Assuming that better plans have larger criterion values, for a given amount of resources, the best data collection plan maximizes this criterion. Alternately, we may want to find the data collection plan that requires the least amount of resources, while ensuring that the criterion meets some minimum required value.

In a Bayesian approach to planning, the criterion must depend on the posterior distribution of the model parameters using the data that the plan will collect. Theoretically, the criterion depends on all possible data that the proposed plan could obtain; in using such criteria, we perform a so-called *preposterior analysis*, because we have not yet collected the actual data. In practice, the analyst can use simulation to evaluate the criterion by repeatedly making draws from the model parameters' prior distributions, generating data according to the proposed data collection plan (given these model parameter draws), and obtaining the model parameter posterior distributions with the generated data. Consequently, if a Bayesian analysis of a corresponding dataset is available, then we can use this simulation-based framework for planning.

In the next section, we consider possible criteria for reliability data collection planning and present additional details about this simulation-based framework.

9.2 Planning Criteria, Optimization, and Implementation

The main approach taken in this chapter is to use a planning criterion that directly assesses the quality of the inference resulting from the plan. In most situations, the analyst focuses on the inference for a function of the model parameters, e.g., the reliability function $R(t)$. As a planning criterion, use the β quantile of the preposterior distribution of the length (or reciprocal length) of the $(1-\alpha) \times 100\%$ credible interval of the reliability function $R(t)$ at some specified time t for specified α and β as the planning criterion. To find the plan, minimize the length of the preposterior credible interval for the β quantile (or maximize the reciprocal length). For example, the analyst might minimize the preposterior 90% credible interval length of the 0.95 quantile of $R(20)$, the reliability at 20 years. We refer to this approach as the direct approach and use it because of its interpretability; i.e., with probability 0.95, the 90% posterior credible interval length will be no larger than the planning criterion.

Another planning criterion used extensively in the Bayesian literature is the *expected Shannon information gain* (EIG) between the prior density function $p(\boldsymbol{\theta})$ and the posterior density function $p(\boldsymbol{\theta} \mid Y, X)$ (Polson, 1993), defined as

$$E_{Y \mid X} \left\{ E_{\boldsymbol{\theta} \mid Y,X} \left(\log[p(\boldsymbol{\theta} \mid Y, X)/p(\boldsymbol{\theta})] \right) \right\} , \tag{9.1}$$

where X denotes the data collection plan and Y denotes the data that the plan will collect. That is, select the data collection plan that maximizes the additional information gained for the vector of model parameters $\boldsymbol{\theta}$. We see that this planning criterion is the mean (first expectation of Eq. 9.1) of the preposterior distribution of the second expectation of Eq. 9.1. See Polson (1993), which provides details for the development of this planning criterion using the decision theoretic framework proposed by Lindley (1956).

9.2.1 Optimization in Planning

For a specified planning criterion evaluated using simulation, the challenge is to find the optimal plan. Data collection planning variables define how and what data will be collected. The number of planning variables can vary from case to case. A typical single planning variable is the number of units to test. For the single planning variable case, where the planning criterion is monotonic in the number of tested units, a simple bisection search can achieve the optimization. This situation arises, for example, in minimizing the number of tests, while maintaining at least a 0.90 probability that the 95% credible interval length does not exceed a target L_{target}; here, the criterion is the probability of the 95% credible interval length not exceeding L_{target} and the requirement is that the criterion be at least 0.90. Apply a bisection search within a range that includes a solution, so that the criterion at the high (largest) value must satisfy the requirement; otherwise, extend the range. In performing a bisection search, evaluate the criterion for the midrange value, i.e., halfway between the low and high values, and set the new high value to the midrange value if it satisfies the requirement; otherwise, set the new low value to the midrange value. Repeat this process until the low and high values converge. When the criterion is not monotonic, we can use other standard search algorithms, such as the golden section search.

For cases with multiple planning variables, some of the variables may be discrete, while other variables may be continuous. For example, for a population of units that have lognormal lifetimes, the analyst must determine a continuous test duration, i.e., censoring time, and a discrete number of units to test. Cases that include multiple planning variables may involve only discrete planning variables, however. Consider degradation data collecting where the planning variables are the number of units to test and the number of evenly spaced inspection times. When there is no requirement of equally spaced inspection times, then the inspection times become additional continuous planning variables. For the multiple planning variable case, we use a *genetic algorithm* (GA) to find the optimal plan.

A GA can handle both discrete and continuous planning variables and does not require the calculation of derivatives. Specify a GA by a population size M and number of generations G. A GA generates an initial population by

1. Randomly generating M candidate plans denoted by $P_1, \ldots, P_M$.
2. Evaluating the planning criterion for plans $P_1, \ldots, P_M$.
3. Ordering the plans $P_1, \ldots, P_M$ by increasing planning criterion values.

The GA then generates G additional populations as follows. For the gth generation, $g = 1, \ldots, G$:

1. Generating M candidate plans by crossover denoted by $P_{M+1}, \ldots, P_{2M}$.
2. Generating M candidate plans by mutation denoted by $P_{2M+1}, \ldots, P_{3M}$.
3. Evaluating the planning criterion for plans $P_{M+1}, \ldots, P_{3M}$.
4. Ordering plans $P_1, \ldots, P_{3M}$ by increasing planning criterion values.
5. The gth generation consists of the M best plans, which have the smallest planning criterion values.

We now describe a GA more fully. A GA works by constructing an initial population of M solutions (i.e., values for the planning variables) by randomly generating solutions that meet any specified constraints (such as a limit on the total required resources). The GA evaluates the criterion for each of the solutions in the initial population and ranks the solutions from smallest to largest, with the smallest value being the best solution in the initial population. If maximizing a criterion to find the best solution, the GA can rank the reciprocal of the original criterion; if the original criterion can be zero, the GA can rank the negative of the original criterion instead.

After generating an initial population of M solutions, the GA populates the second (and subsequent) GA generations using two genetic operations: crossover and mutation. The genetic crossover operation generates M additional solutions as follows. Crossover occurs when the GA randomly selects two different parent solutions from the current population of M solutions according to probabilities that are inversely proportional to their rank among the M solutions. That is, the probability of choosing the ith ranked solution is $(M - i + 1)/[M(M + 1)/2]$. The GA obtains a new or child solution from the two parent solutions by randomly picking one of the two parents and taking its value for the first planning variable, and then repeating this operation for each of the remaining planning variables. The GA continues to perform crossover operations (i.e., selecting two parent solutions from the current population of M solutions and so on) until it generates M additional solutions. Note that the GA checks the solutions to make sure they do not exceed any specified constraints, so the GA generates solutions until there are M feasible solutions. The GA then evaluates the planning criterion for each of these additional solutions. An alternative to requiring feasible solutions is having the GA penalize those solutions that do not meet the constraint.

The GA proceeds next by mutating each of the M solutions in the current population (i.e., applying genetic mutation to each of the planning variable values). Because less mutation is better as we find better solutions in the later generations, use a GA that employs relaxation, which reduces the probability of mutating in subsequent generations The GA accomplishes this relaxation by making the mutation probability an exponentially decaying function of generation. Notationally, in generation g, the GA mutates each planning variable value with probability $\exp(-\mu g)$, where μ is a user-specified mutation rate parameter; μ controls the rate at which mutations occur as the generation number g increases. For the examples presented in this chapter, we set $\mu = 0.01$, although GA performance does not seem to be overly sensitive when using a different μ value. Also, an analyst may use a version of mutation that employs "punctuated equilibrium"; after every so many generations, reset the mutation probability to the starting value and decrease it in subsequent generations until the next reset.

When the GA mutates a planning variable, it does so by drawing a new value from a distribution, which has a mean equal to the current planning variable value y and variance that decreases as g increases. For discrete (integer) planning variables restricted to an interval (L, U), the GA mutates by means of a logit transformation as described in the following steps:

1. Compute $z = (y-L)/(U-L)$ where y, L, and $U$are the current, minimum, and maximum planning variable values.
2. Calculate $a = \log[z/(1-z)] + [Uniform(0,1) - 0.5] \times \sigma \times \exp(-\mu g)$, where $Uniform(0,1)$ denotes a draw from a uniform distribution and $\log[z/(1-z)]$ is the logit transformation of z from Step 1. Here σ is a user-specified parameter that controls the variance, which decreases as g increases through $\exp(-\mu g)$.
3. Compute $u = L + (U + 1 - L) \times \exp(a)/[1 + \exp(a)]$.
4. The mutated planning variable value is floor(u), which provides the largest integer that does exceed u, and lies between L and U.

The logit transformation produces a mutated planning variable value with a mean that is approximately equal to the current value and has a variance that decreases as g increases. By repeatedly using the mutation operation, the GA generates M additional solutions satisfying any specified constraints and then evaluates the planning criterion for each solution. A GA can mutate a continuous planning variable defined on the entire real line by a similar algorithm, such as adding a normal random variable (which has a variance that decreases as g increases) to the current value. For positive planning variables, a GA can mutate the logged current value of the planning variable and then exponentiate it. For a continuous planning variable that is restricted to an interval (L, U), the GA can use the same algorithm given above for the discrete planning variable, except with the last step omitted.

The GA we use is "elitist," which means that the population in the next generation consists of the M best solutions from the $3M$ solutions currently being considered (M current solutions, M crossover solutions, and M mutation solutions). Perform the GA described above for G generations and take the best Gth generation solution as the nearly optimal data collection plan.

9.2.2 Implementing the Simulation-Based Framework

The key to implementing the simulation-based framework for reliability data collection planning is developing a high-performance Markov chain Monte Carlo (MCMC) algorithm for analyzing the data that the plan will collect.

This has led us away from using high-level languages, such as C, for anything except simple problems. For more complicated problems, developing formulas, coding the formulas, and implementing a particular sampling algorithm are time-consuming, error-prone activities. However, MCMC software offers an attractive possibility for performing these tasks. We develop a driver program that carries out the optimization algorithm, makes draws from the prior distribution, generates data according to a candidate data collection plan, calls the MCMC software (which makes draws from the appropriate posterior distribution), and finally uses the posterior draws to evaluate the planning criterion.

We have developed such an implementation with a driver program written in R (Venables et al., 2006). The driver program uses the MCMC software YADAS (Graves, 2007a,b) via a system call. R is programmable and provides easy access to random number generators, so it is simple to write a driver program by coding a bisection search or other univariate search method and by coding a GA, making draws from prior distributions, and generating data from the data collection plan under consideration. Other MCMC software, such as WinBUGS (Spiegelhalter et al., 2003; Gilks et al., 1994), may be used instead of YADAS, as long as interfaces between R and such software exist.

Now with the implementation of the simulation-based framework for reliability data collection planning, let us demonstrate the versatility of this framework by considering a few data collection planning problems in the remainder of this chapter. To illustrate the generality of this framework, we have selected these problems from the inference problems discussed in previous chapters. We believe that reliability data collection planning is a rich source of new research problems, and therefore have been purposefully selective in choosing these sample problems.

We begin with planning for binomial data and illustrate the direct and decision theoretic criteria approaches to reliability data collection planning.

9.3 Planning for Binomial Data

For binomial data, x is the number of successes in n tests where π is the probability of success and $X \sim Binomial(n, \pi)$. In this case, the sample size n determines the plan. The parameter of interest, π, may be the reliability of some component, subsystem, or system. Let's assume for illustrative purposes that there is available prior information represented by a beta distribution. The prior distribution may reflect both prior information and previous data, however. For example, for prior guess $\tilde{\pi}$ based on an equivalent "prior" sample size $\tilde{n}$, a $Beta[\tilde{n}\tilde{\pi}, \tilde{n}(1-\tilde{\pi})]$ distribution captures this prior information. By specifying $\tilde{\pi}$ and $\mathbf{P}(\pi \geq \tilde{\pi})$, the analyst can determine $\tilde{n}$. If previous data (x^*, n^*) are also available (i.e., x^* successes in n^* trials), then a $Beta[\tilde{n}\tilde{\pi} + x^*, \tilde{n}(1-\tilde{\pi}) + (n^* - x^*)]$ distribution combines the prior information and previous data. Because the sample size n determines the binomial data plan, any planning criterion improves as n increases. Consequently, the search for a plan requires a constraint. Taking the decision theoretic approach, we can use the cost per sample and determine the sample size n by maximizing

$$vEIG(n) - cn, \tag{9.2}$$

where $EIG(n)$ is the EIG from Eq. 9.1, v is the value of one unit of information gain, and c is the cost of one test. Denote the optimal sample size by $n_{optimal}$. Here, we assume a linear cost structure with no overhead; otherwise, subtract a constant c_0 from Eq. 9.2. The practical difficulties of using this approach are interpreting what one unit of information gain is, and then assigning a meaningful value to v in the same units as the cost c.

Bernardo (1997) expresses EIG in log base 2 ($\log_2$), so that

$$\log_2\left(\frac{p(\pi \mid x, n)}{p(\pi)}\right) = \log\left(\frac{p(\pi \mid x, n)}{p(\pi)}\right) / \log(2),$$

where the prior density function $p(\pi)$ is

$$p(\pi) = \frac{\Gamma(\alpha+\beta)}{\Gamma(\alpha)\Gamma(\beta)}\pi^{\alpha-1}(1-\pi)^{\beta-1},$$

and the posterior density function $p(\pi \mid x, n)$ based on the data (x successes out of n tests) is

$$p(\pi \mid x, n) = \frac{\Gamma(\alpha+\beta+n)}{\Gamma(\alpha+x)\Gamma(\beta+(n-x))}\pi^{\alpha+x-1}(1-\pi)^{\beta+(n-x)-1}.$$

We can express $EIG(n)$ as

$$\begin{aligned} EIG_2(n) &= \textstyle\int p(x|n)[\int p(\pi \mid x, n)\log_2(\frac{p(\pi \mid x,n)}{p(\pi)})\mathrm{d}\pi]\mathrm{d}x \\ &= \textstyle\int\int p(x \mid n, \pi)p(\pi)\log_2(\frac{p(\pi \mid x,n)}{p(\pi)})\mathrm{d}\pi\mathrm{d}x. \end{aligned}$$

The subscript 2 in $EIG_2(n)$ denotes the use of the log base 2 function rather than the natural log function. The form of $EIG_2(n)$ suggests the use of simulation to evaluate it in this simple case — take the average of $\log_2(\frac{p(\pi \mid x,n)}{p(\pi)})$ from repeated draws of π using its beta prior distribution and of x from its $Binomial(n, \pi)$ distribution.

Suppose that we have a best guess for π of 0.18 and choose an effective sample size of 25 to satisfy a probability of 0.99 that π is smaller than 0.39. This prior information implies a $Beta(4.5, 20.5)$ distribution for $p(\pi)$, which has a mean of 0.18. In Eq. 9.2, v is the value that the decision maker assigns the gain of one bit of information (on the log base 2 scale). For illustration, let $v = 5,000$ and $c = 10$. That is, gaining one bit of information about π is worth \$5,000 and each test costs \$10 to run. By maximizing the planning criterion given in Eq. 9.2, we find that the optimal sample size $n_{optimal}$ is 334. Figure 9.1 displays the bisection search results.

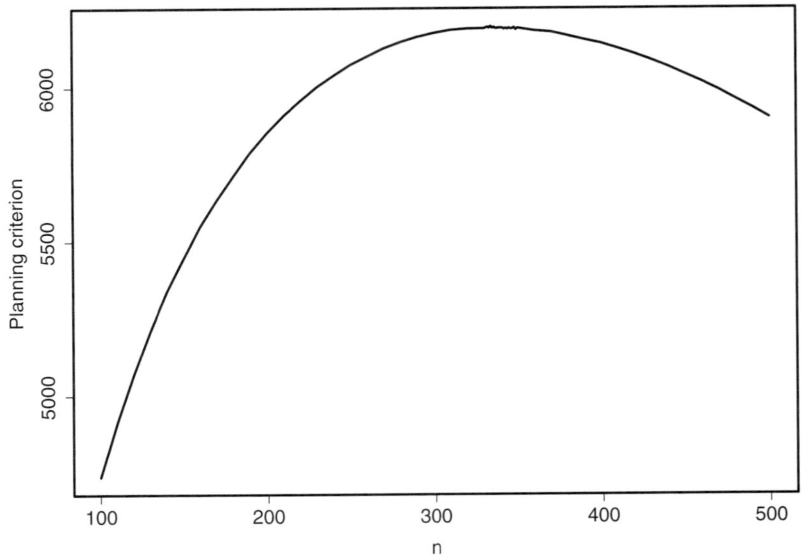

Fig. 9.1. Search results for binomial data planning criterion given in Eq. 9.2 as a function of sample size n.

Taking the direct approach to planning, based on the posterior distribution of the credible interval length, we determine the plan by finding the minimum sample size n that ensures with probability at least γ that the length of the $\alpha \times 100\%$ credible interval will be no longer than L_{target}. For example, suppose $\gamma = 0.90$ and $\alpha = 0.95$. The maximum sample n_{max}, the high end of the range of sample sizes to choose from, needs specification and checking to ensure that it meets this requirement, i.e., the probability is at least γ that the length of the $\alpha \times 100\%$ credible interval is no larger than L_{target}. If not, specify a larger n_{max}. Because the criterion is monotonic in n, we can use a bisection search to find the optimal sample size $n_{optimal}$.

Assuming a $Beta(1, 1)$ prior distribution for π, consider finding the minimum sample size n for which the probability that the $\alpha \times 100\%$ credible interval length does not exceed $L_{target} = 0.1$ is at least γ. Let $\gamma = 0.90$, $\alpha = 0.95$, $n_{max} = 500$, and verify that a sample size of 500 meets the stated requirements. We then use a bisection search, which yields $n_{optimal} = 377$. Table 9.1 displays the bisection search results, where the planning criterion is the probability that the length of the $\alpha \times 100\%$ credible interval is no bigger than L_{target}. Consequently, a sample size of 377 meets the stated requirements.

We can also consider an example in which there are existing data. Assuming a $Beta(1, 1)$ prior distribution for π and that there are existing data

Table 9.1. Bisection search results for direct binomial data planning criterion as a function of sample size n

n	Planning Criterion
500	1.000
250	0.419
375	0.895
437	1.000
406	0.999
390	0.993
382	0.955
378	0.924
376	0.899
377	0.914

consisting of 40 successes out of 50 tests, let $\gamma = 0.90$ and $\alpha = 0.95$. Find the minimum sample size n for which the probability that the $\alpha \times 100\%$ credible interval length does not exceed $L_{target} = 0.1$ is at least γ. Combining the $Beta(1, 1)$ distribution with the existing data yields a $Beta(41, 11)$ prior distribution for π. Letting $n_{max} = 500$, we verify that a sample size of 500 meets the requirements and find that a bisection search yields $n_{optimal} = 256$. Note that with the previous 50 tests, only 256 additional tests are required for a total of 306 tests and not 377 tests as in the preceding situation.

9.4 Planning for Lifetime Data

This section considers planning for a population with lifetimes that follow a $LogNormal(\mu, \sigma^2)$ distribution. We focus here on the following two characteristics of a lifetime distribution: a lifetime β quantile

$$q_\beta = \exp(\mu + z_\beta \sigma), \tag{9.3}$$

where z_β is the standard normal β quantile, and reliability

$$R(t) = 1 - \Phi((\log(t) - \mu)/\sigma), \tag{9.4}$$

at some time t, where $\Phi(\cdot)$ is the standard normal cumulative distribution function.

We take the direct approach based on the posterior distribution of the credible interval length. That is, make sure that the probability of the $\alpha \times 100\%$ credible interval length for a lifetime quantile or reliability not exceeding L_{target} is at least γ. There are two cases to consider: one with censoring and one without. For the censoring case, stop the data collection at time t_c; the lifetimes for those units still working at time t_c are Type I censored. When there is no censoring, the minimum sample size that meets the requirement (i.e., the probability of the $\alpha \times 100\%$ credible interval length not exceeding L_{target} is at least γ) determines the data collection plan. For a specified censoring time t_c, also determine a minimum sample size that meets the requirement.

For illustrative purposes, we use the following prior distributions for the model parameters μ and σ^2:

$$\mu \sim Normal(a_\mu, b_\mu) \quad \text{and}$$
$$\sigma^2 \sim InverseGamma(a_{\sigma^2}, b_{\sigma^2}).$$

Consider a data collection planning example, which focuses on reliability at time $t = 24$ months. Assuming that the lifetimes have a $LogNormal(\mu, \sigma^2)$ distribution, use the following prior distributions: $\mu \sim Normal(4, 1)$ and $\sigma^2 \sim InverseGamma(2, 1)$. Letting $\gamma = 0.90$, $\alpha = 0.95$, $L_{target} = 0.1$, we find that a sample size $n_{max} = 500$ meets the stated requirement, i.e., the probability of the $\alpha \times 100\%$ credible interval length of $R(24)$ not exceeding L_{target} is at least γ. A bisection search yields $n_{optimal} = 235$. Now consider stopping the data collection at time $t_c = 24$ (months), which yields Type I-censored lifetimes. In this case, a bisection search yields $n_{optimal} = 352$.

Next, we consider a data collection planning example, which focuses on a lifetime quantile. For the $LogNormal(\mu, \sigma^2)$ lifetime distribution, now assume the following prior distributions: $\mu \sim Normal(4, 0.25)$ and $\sigma^2 \sim InverseGamma(10, 5)$. That is, we know μ and σ^2 more precisely than in the preceding example. Suppose now that the 0.2 quantile of the population lifetime distribution is of interest, i.e., the time (in months) by which 20% of the population has failed. Let $\gamma = 0.90$, $\alpha = 0.95$, $L_{target} = 2$, and $n_{max} = 5{,}000$. We verify that a sample size of 5,000 meets the stated requirements, i.e., the probability of the $\alpha \times 100\%$ credible interval length of 0.2 quantile not exceeding L_{target} is at least γ. Then, a bisection search yields $n_{optimal} = 4{,}523$. Note the large sample size needed to meet the specified L_{target} of 2.

A different data collection planning problem determines both the censoring time t_c and sample size n that meets the requirement stated above and minimizes the total test time $t_c \times n$. This problem is left as Exercise 9.5.

9.5 Planning Accelerated Life Tests

Consider data collection planning for an *accelerated life test* as discussed in Sect. 7.7. Because testing units at different levels (or values) of the accelerating factor can be viewed as an experiment having one experimental factor, we can also call the data collection plan an experimental design. Besides determining the levels of the accelerating factor, the plan needs to specify the number of tested units at each of these levels.

We assume the following lognormal regression model for accelerated lifetimes:

$$Y_{ij} \sim LogNormal[\mu(v_i), \sigma^2], \tag{9.5}$$

for the jth lifetime at the ith level of accelerating factor denoted by v_i. That is, the lifetimes have a lognormal distribution with location parameter $\mu(v) = \beta_0 + \beta_1 v$ and common scale parameter σ^2. In planning the accelerated life test, two characteristics of the lifetime distribution at the normal use condition v_U are of interest; they are a lifetime δ quantile

$$q_\delta = \exp[\mu(v_U) + z_\delta \sigma], \tag{9.6}$$

and reliability

$$R(t) = 1 - \Phi\{[\log(t) - \mu(v_U)]/\sigma\}, \tag{9.7}$$

at some time t.

An experimental design consists of m levels of the accelerating factor, v_i, $i = 1, \ldots, m$, with $v_U < v_{low} \leq v_i \leq v_{high}$, and the corresponding number of items tested n_i, $i = 1, \ldots, m$. The experimenter needs to specify a high accelerating factor level v_{high} at which the lognormal regression model still holds. Among a number of planning problems, we consider the one that minimizes the total number of items tested $\sum_{i=1}^{m} n_i$ and meets a requirement based on the posterior distribution of the credible interval length. If the experiment stops after a specified time, then the censoring time t_c becomes another planning variable that needs determination.

For illustrative purposes, we use the following prior distributions for β_0, β_1, and σ^2:

$$\begin{aligned}
\beta_0 &\sim Normal(a_{\beta_0}, b_{\beta_0}),\\
\beta_1 &\sim Normal(a_{\beta_1}, b_{\beta_1}), \text{ and}\\
\sigma^2 &\sim InverseGamma(a_{\sigma^2}, b_{\sigma^2}).
\end{aligned}$$

Consider an accelerated life test, which uses temperature as the accelerating factor. Suppose that the normal use temperature is 180°C and that the life test employs $m = 3$ different temperature levels. In scaled temperature v, $v_U = 0$ corresponds to normal use temperature. Further, let $v_{low} = 0.5 = v_1$, $v_{high} = 1 = v_3$, and $v_{low} + 0.1(v_{high} - v_{low}) \leq v_2 \leq v_{low} + 0.9(v_{high} - v_{low})$. Consequently, consider three temperature levels, but where the first and third levels are specified, so that v_{low} corresponds to 220°C and v_{high} corresponds to 260°C. Finally, restrict the number of tests n_i at the temperature levels by $2 \leq n_i \leq 400$, $i = 1, 2, 3$.

We assume that the lifetimes Y_i have a $LogNormal[\mu(x_i), \sigma^2]$ distribution with $\mu(x_i) = \beta_0 + \beta_1 x_i$, where $x_i = 1000/((180 + v_i \times 80) + 273.15)$. Also, the following distributions capture the prior knowledge about the model parameters:

$$\begin{aligned}
\beta_0 &\sim Normal(-7.3, 0.15^2),\\
\beta_1 &\sim Normal(7.5, 0.15^2), \text{ and}\\
\sigma^2 &\sim InverseGamma(100, 0.11^2 \times 100).
\end{aligned}$$

Note that the mean of the σ^2 prior distribution is $0.11^2 \times 100/(100-1) \approx 0.11^2$. For $\gamma = 0.9$, $\alpha = 0.95$, and $L_{target} = 0.1$, let us focus on the reliability at time $t = 10{,}500$ days. First, check to make sure that the planning criterion for $n_i = 400, i = 1, 2, 3$ and $v_2 = 0.5$ meets the requirement; otherwise, increase the maximum sample sizes. We use a GA that minimizes the total sample size, $n_1 + n_2 + n_3$. Instead of discarding (n_1, n_2, n_3) cases, which have a planning criterion ρ (i.e., the probability that the $\alpha \times 100\%$ credible interval length does not exceed L_{target}) that does not exceed γ, penalize these cases by minimizing:

$$n_1 + n_2 + n_3 + [100(\gamma - \rho)/0.01]I(\rho < \gamma) + [25(\gamma - \rho)/0.01]I(\rho > \gamma).$$

That is, add 100 to $n_1+n_2+n_3$ for every 0.01 the planning criterion ρ is below the requirement γ; this penalizes large sample sizes whose planning criteria ρ are substantially below the requirement γ. Also, add 25 for every 0.01 the planning criterion ρ exceeds γ, where $I(\cdot)$ is the indicator function. A GA found the nearly optimal solution $v_{2,optimal} = 0.713$, which corresponds to 237°C, and $\mathbf{n}_{optimal} = (n_1, n_2, n_3)$, where $n_1 = 251$, $n_2 = 296$, and $n_3 = 86$. This accelerated life test plan tests 251 units at 220°C, 296 units at 237°C, and 86 units at 260°C.

9.6 Planning for Degradation Data

This section considers planning for degradation data as discussed in Chap. 8. The degradation data experiment consists of measuring n units each at m inspection times, where t_i, $i = 1, \ldots, m$, denotes the times.

For illustration, we assume the following degradation data model, which is motivated by Example 8.1:

$$\begin{aligned} Y_{ij} &\sim Normal[\beta_0 - (1/\theta_i)t_{ij}, \sigma_\varepsilon^2] \quad \text{and} \\ \theta_i &\sim LogNormal(\mu_\theta, \sigma_\theta^2), \end{aligned}$$

where y_{ij} is the observed degradation for the ith unit at the jth inspection time t_{ij}. That is, the units have a common starting value β_0 and degrade over time and have reciprocal slopes that vary from unit to unit according to a lognormal distribution. Recall that a unit fails when its degradation exceeds a threshold D_f. We can then express the lifetimes as $\theta_i(\beta_0 - D_f)$, which follow a $LogNormal[\log(\beta_0 - D_f) + \mu_\theta, \sigma_\theta^2]$ distribution. Consequently, the reliability function $R(t)$ equals $1 - \Phi[(\log(t) - \log(\beta_0 - D_f) - \mu_\theta)/\sigma_\theta]$.

In the analysis of the degradation data, we use the following prior distributions for β_0, μ_θ, σ_θ^2, and σ_ε^2:

$$\begin{aligned} \beta_0 &\sim Normal(a_{\beta_0}, b_{\beta_0}), \\ \mu_\theta &\sim Normal(a_{\mu_\theta}, b_{\mu_\theta}), \\ \sigma_\theta^2 &\sim InverseGamma(a_{\sigma_\theta^2}, b_{\sigma_\theta^2}), \text{ and} \\ \sigma_\varepsilon^2 &\sim InverseGamma(a_{\sigma_\varepsilon^2}, b_{\sigma_\varepsilon^2}). \end{aligned}$$

We focus on assessing the population reliability at a specified time t. Among many planning problems, consider the case in which the maximum time t_m is specified and use evenly spaced inspection times; further, inspect at the start so that $t_1 = 0$. Our goal is to minimize the total number of inspections mn that meets the requirement that the $\alpha \times 100\%$ credible interval length does not exceed with a posterior probability of at least γ.

We focus on estimating the reliability at $t = 48$ months, $R(48) = 1 - \Phi[\{\log(48) - (\log(\beta_0 - D_f) + \mu_\theta)\}/\sigma_\theta]$, where $D_f = 90$. Let $t_m = 36$, $\gamma = 0.90$, $\alpha = 0.95$, and $L_{target} = 0.05$; that is, for a maximum testing time t_m of 36 months, let us find a plan for which the probability that the 95% credible interval length is less than 0.05 month exceeds 0.90. Assume the following prior distributions for the model parameters β_0, μ_θ, and σ_θ:

$$
\begin{aligned}
\beta_0 &\sim Normal(100, 1), \\
\mu_\theta &\sim Normal(1.6, 0.15^2), \\
\sigma_\theta^2 &\sim InverseGamma(10, 0.25^2 \times 10), \text{ and} \\
\sigma_\varepsilon^2 &\sim InverseGamma(5, 5).
\end{aligned}
$$

Note that the mean of the σ_θ^2 prior distribution is $0.25^2 \times 10/(10-1) \approx 0.25^2$. Consider the following ranges for the number of inspections m and the number of test units n: $3 \leq m \leq 10$ and $2 \leq n \leq 100$. First, check to make sure that the planning criterion for $m = 10$ and $n = 100$ meets the requirement; otherwise, increase the maximum of test units and maximum number of inspections. We use a GA, which minimizes the total sample size, mn. Instead of discarding (m, n) cases for which the planning criterion ρ (i.e., the probability that the $\alpha \times 100\%$ credible interval length does not exceed L_{target}) does not exceed γ, penalize these cases by minimizing:

$$
mn + [100(\gamma - \rho)/0.01]I(\rho < \gamma) + [25(\gamma - \rho)/0.01]I(\rho > \gamma),
$$

i.e., add 100 to mn for every 0.01 the planning criterion ρ is below the requirement γ; this penalizes large sample sizes whose planning criteria ρ are substantially below the requirement γ. Also, add 25 for every 0.01 the planning criterion ρ exceeds γ, where $I(\cdot)$ is the indicator function. The GA found the nearly optimal solution $m_{optimal} = 3$ and $n_{optimal} = 25$. That is, the plan consists of observing the degradation of 25 units three times each (at 12, 24, and 36 months).

9.7 Planning for System Reliability Data

This section considers data collection planning for assessing the reliability of a system, as presented in Chap. 5. We illustrate system reliability data collection planning by using the simplified system shown in Fig. 9.2. In this simplified system, components, subsystems, and the system are referred to as nodes. The system is node 0, which consists of two subsystems (nodes 1 and 2) in series.

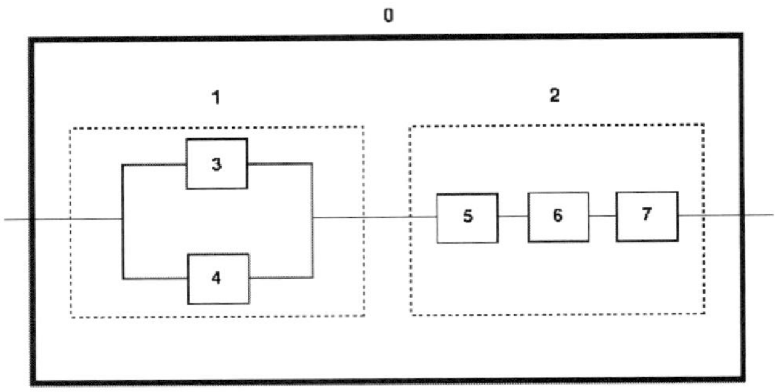

Fig. 9.2. Simplified system reliability block diagram.

The first subsystem consists of two components in parallel (nodes 3 and 4), and the second subsystem consists of three components in series (nodes 5, 6, and 7). Expressions for the subsystem and system reliabilities in terms of the component reliabilities are: $\pi_1 = 1 - (1 - \pi_3)(1 - \pi_4)$, and $\pi_2 = \pi_5\pi_6\pi_7$ for the subsystem reliabilities, and $\pi_0 = \pi_1\pi_2 = \{[1 - (1 - \pi_3)(1 - \pi_4)]\pi_5\pi_6\pi_7\}$ for the system reliability.

Suppose that there are binomial data and prior information for the simplified system as shown in Tables 9.2 and 9.3, respectively. When data are available at the ith node, there are x_i successes in n_i tests with reliability π_i. If node i is a subsystem or the full system (i.e., not a component), then there are expressions for the π_i in terms of the component reliabilities, as given above.

Next, we consider prior distributions for the node reliabilities. Use a beta prior distribution in terms of a best guess for reliability $\tilde{\pi}_i$ as given in Table 9.3 and a precision (or equivalent sample size, i.e., number of tests) $\tilde{n}_i$; that is, let $\pi_i \sim Beta[\tilde{n}_i\tilde{\pi}_i, \tilde{n}_i(1 - \tilde{\pi}_i)]$.

If information is available at the subsystem or system level, treat this prior information as binomial data. If $\tilde{\pi}_i$ is the estimated reliability, and its precision is $\tilde{n}_i$, then its contribution to the likelihood is proportional to

$$\pi_i^{\tilde{n}_i\tilde{\pi}_i}(1 - \pi_i)^{\tilde{n}_i(1-\tilde{\pi}_i)}.$$

Note that Table 9.3 provides no precisions $\tilde{n}_i$, so a prior distribution for each $\tilde{n}_i$ needs specification. We assume the same precision $\tilde{n}$, i.e., $\tilde{n}_i = \tilde{n}$, and use the following prior distribution for $\tilde{n}$:

$$\tilde{n} \sim Gamma(5, 1).$$

That is, the prior information on average is worth about a sample of size five or five tests.

Figures 9.3, 9.4, 9.5, and 9.6 display the resulting prior distributions for the node reliabilities as dashed lines. Combining these prior distributions with the node data using MCMC yields the posterior distributions displayed as the solid lines in Figs. 9.3, 9.4, 9.5, and 9.6. From these results, we calculate the 90% credible interval for the system (node 0) reliability as (0.689, 0.865), which has a length of 0.176. Note that even though there are no data for the first subsystem (node 1), the system data (node 0) and the component data (nodes 3 and 4) dramatically improve what is known about the reliability of the first subsystem.

Table 9.2. Data for simplified system (number of successes/number of tests)

Node	Data
0	15/20
1	
2	10/10
3	34/40
4	47/50
5	3/5
6	8/8
7	16/17

Table 9.3. Prior reliabilities $\tilde{\pi}$ for simplified system

Node	$\tilde{\pi}$
0	0.8
1	0.9
2	0.9
3	0.9
4	0.9
5	0.95
6	0.95
7	0.95

Now consider data collection planning for assessing system reliability. In this context, we refer to the planning problem as *resource allocation* because typically limited resources need to be allocated to the various levels of the system. When additional funding becomes available, we must determine where the tests should be done and how many tests should be performed. Consider the optimal allocation of additional tests within a fixed budget that results in the least uncertainty of system reliability for the simplified system. That is, determine how many tests should be performed at the system, subsystem, and component level (i.e., nodes 0–7 in our simplified system) under a fixed budget for specified costs at each level (system, subsystem, and component).

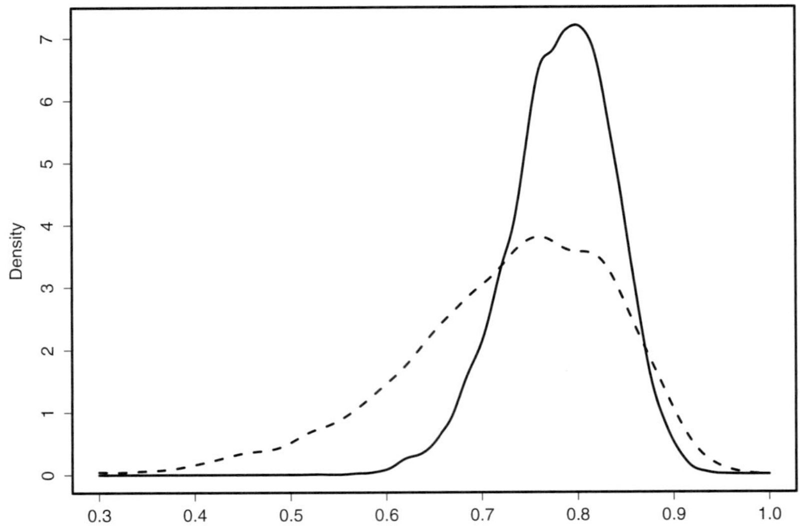

Fig. 9.3. Plot of simplified system reliability prior distribution (dashed line) and posterior distribution (solid lines) for node 0.

We assume that there is a cost for collecting additional data, with higher level data being more costly than lower level data. Consider the following costs as an example of the costs for testing at each node in the simplified system. Recall that node 0 is the system, nodes 1 and 2 are subsystems, and nodes 3–7 are components:

(0: \$5), (1: \$2), (2: \$3), (3: \$1), (4: \$1), (5: \$1), (6: \$1), (7: \$1).

We evaluate a candidate allocation (i.e., number of tests for each node) using a preposterior-based uncertainty criterion. That is, with a fixed budget, minimize the γ quantile of posterior distribution of the $\alpha \times 100\%$ credible interval length. We use a GA to find the optimal allocation, which involves eight discrete planning variables: the sample sizes (number of tests) at the eight nodes of the simplified system.

Recall that the length of the 90% credible interval of system reliability based on the existing data was 0.176. To illustrate the GA for the allocation problem described above, consider a fixed budget of \$1,000 and use populations of size $M = 20$ to generate $G = 50$ generations. Consequently, the GA generates and evaluates a total of 2,020 $(= 20 + 40 \times 50)$ candidate allocations. The uncertainty criterion is based on 500 draws from joint prior distribution of the node reliabilities (i.e., the GA generates $N_d = 500$ datasets for each evaluation and $N_p = 2,000$ draws are taken from the node reliability posterior

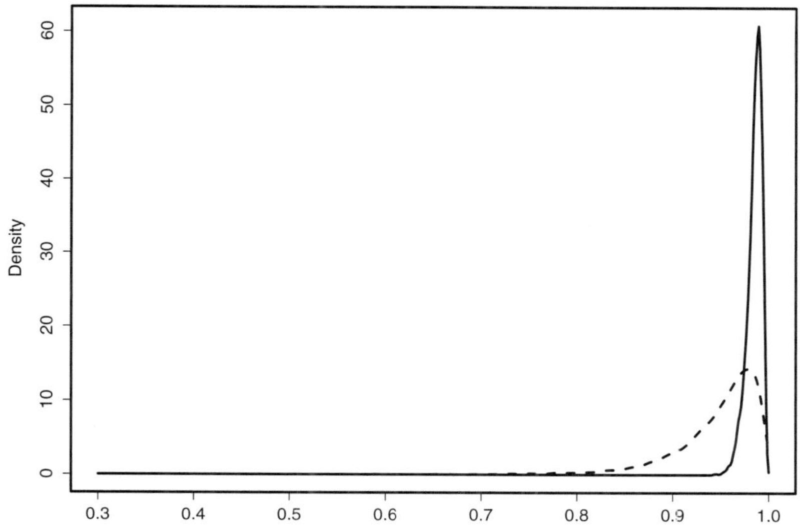

(a)

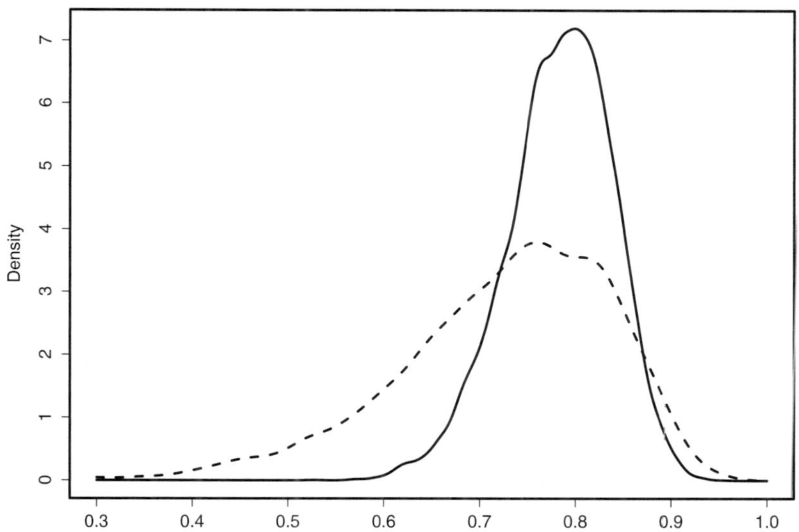

(b)

Fig. 9.4. Plot of simplified system reliability prior distribution (dashed line) and posterior distribution (solid lines) for (a) node 1 and (b) node 2.

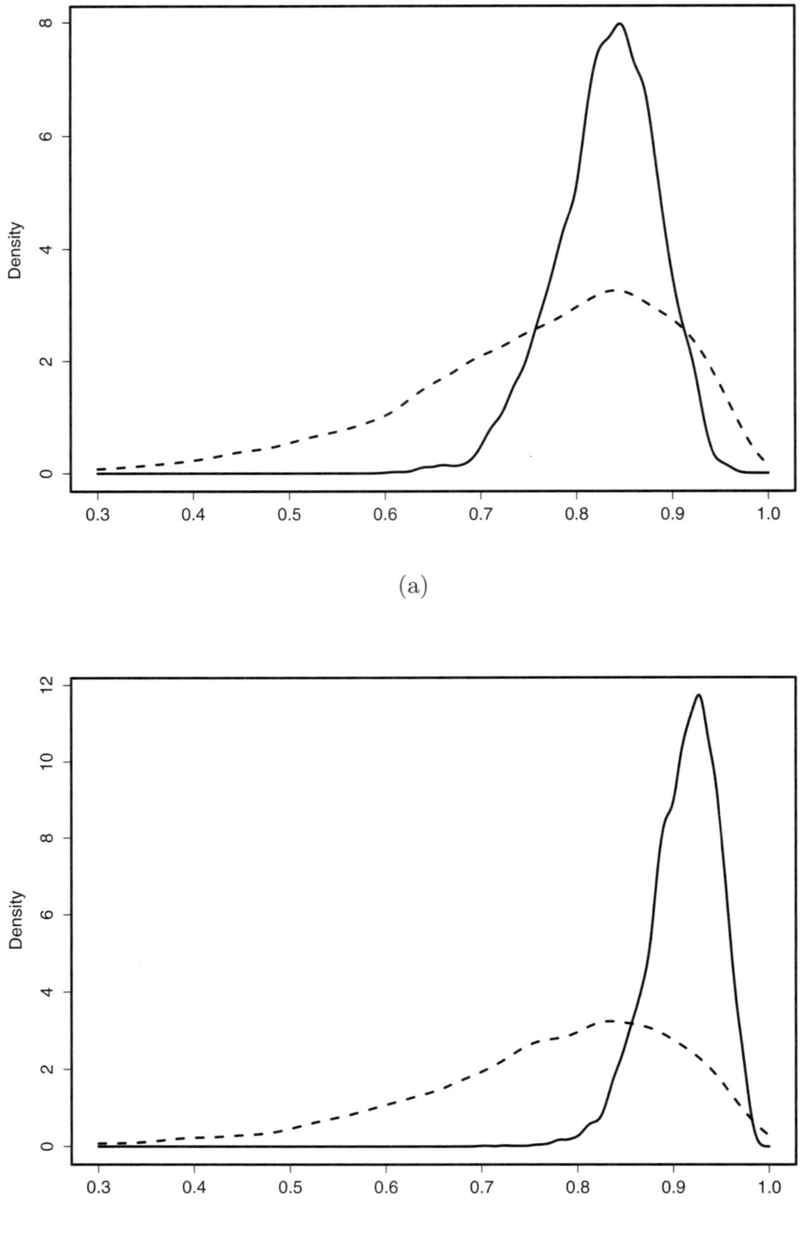

Fig. 9.5. Plot of simplified system reliability prior distribution (dashed line) and posterior distribution (solid lines) for (a) node 3 and (b) node 4.

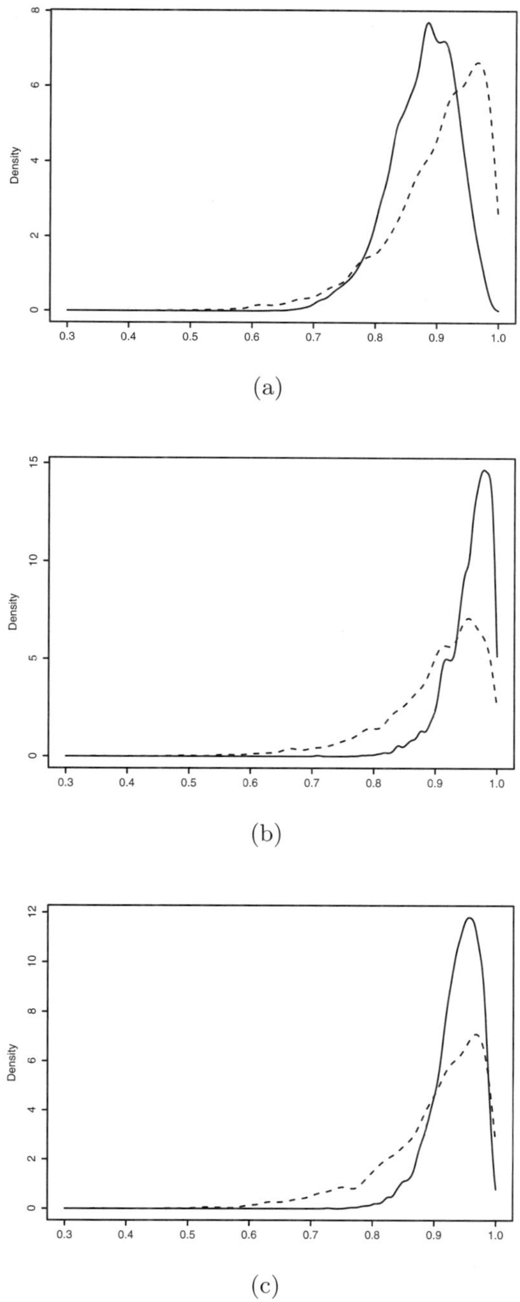

Fig. 9.6. Plot of simplified system reliability prior distribution (dashed line) and posterior distribution (solid lines) for (a) node 5, (b) node 6, and (c) node 7.

distributions for each dataset to calculate the system reliability $\alpha \times 100\%$ credible interval and its length). For a budget of \$1,000, what resource allocation yields the most reduction in the uncertainty criterion for system reliability? Figures 9.7 and 9.8 display the best uncertainty criterion and resource allocation found during each generation. The uncertainty criterion starts at 0.085 for the initial population and decreases to 0.0725 in generation 50 with an allocation of (0, 0, 175, 0, 0, 208, 137, 128) for nodes 0–7.

We evaluated this resource allocation twice with $N_p = 50,000$ and $N_d = 100,000$ and obtained uncertainty criterion values of 0.073358 and 0.073363, so take the true uncertainty criterion for this allocation as 0.0734. This resource allocation suggests that there are enough data for node 1 (the two-component parallel subsystem), and the cost structure prohibits additional system tests. (That is, the system cost equals the sum of the subsystem costs, which equals the sum of the component costs.) Because the node 2 subsystem cost equals the sum of its component costs, we tried an allocation that proportionally allocated the subsystem tests to its components giving the allocation (0, 0, 0, 0, 0, 439, 289, 270). Evaluating this allocation again with $N_p = 50,000$ and $N_d = 100,000$ gave uncertainty criterion values of 0.071439 and 0.071426, which rounded gives 0.0714. Consequently, there is some improvement by doing all component tests for the node 2 subsystem. The GA did not identify this better allocation because the uncertainty criterion difference was within the simulation variability of the uncertainty criterion evaluations.

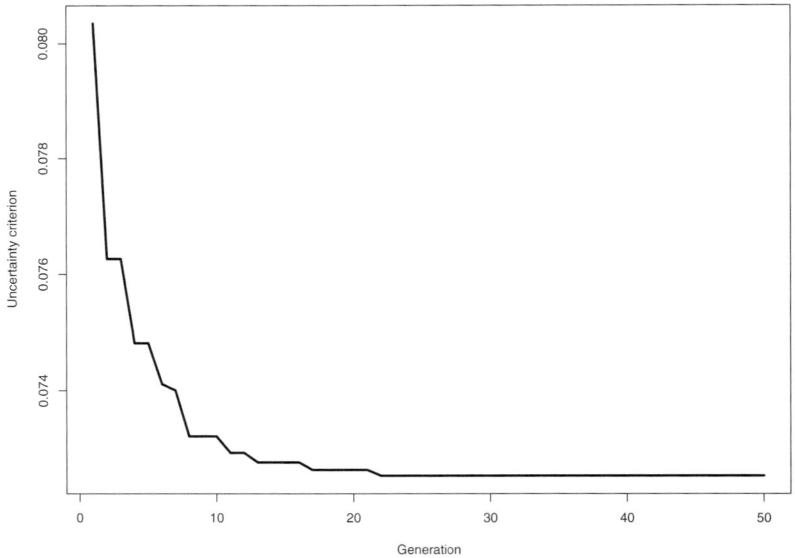

Fig. 9.7. GA evolution of best uncertainty criterion for system reliability planning.

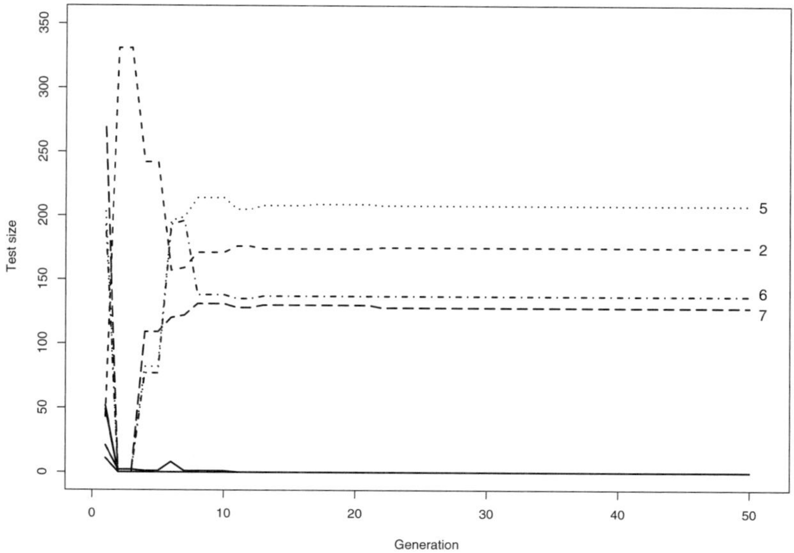

Fig. 9.8. GA evolution of best resource allocation for system reliability planning.

9.8 Related Reading

Müller and Parmigiani (1995) and Müller (1999) consider simulation-based experimental design using a Bayesian approach. Michalewicz (1992) and Goldberg (1989) present genetic algorithms. See Hamada et al. (2001) for an application of GAs to Bayesian experimental design. Lindley (1956) proposes the Bayesian approach to data collection planning using the decision theoretic framework. See also Bernardo (1997). Adcock (1997) reviews alternative planning criteria based on the posterior distribution similar to the technique used in this chapter. A number of papers discuss accelerated life testing from a Bayesian perspective: Chaloner and Larntz (1992), Polson (1993), Verdinelli et al. (1993), Zhang and Meeker (2006). Hamada et al. (2004) considers resource allocation for fault trees.

9.9 Exercises for Chapter 9

9.1 For the first planning situation in Sect. 9.3, use the direct approach and evaluate the 0.90 posterior quantile of the 95% credible interval length for the sample size of 377. That is, $\gamma = 0.90$ and $\alpha = 0.95$. What sample size would be required if we want L_{target} to be 50% of the direct criterion just calculated for a sample size of 377?

9.2 Determine the binomial data collection plan based on the posterior of the credible interval length for a $Beta(0.5, 0.5)$ prior distribution when there are existing data for the following cases: (a) 5 out of 10 successes, (b) 50 out of 100 successes, (c) 9 out of 10 successes, and (d) 90 out of 100 successes. Let $\alpha = 0.95$, $\gamma = 0.90$, and $L_{target} = 0.05$.

9.3 Assume a $Poisson(\lambda t)$ distribution for the number of restarts X in a time period of length t and a diffuse gamma prior distribution for λ, the restart rate per time unit. Suppose that you want to estimate λ. Determine the optimal Poisson data collection plan determined by t based on the posterior of the credible interval length when $\alpha = 0.95$, $\gamma = 0.90$, and $L_{target} = 0.01$.

9.4 In the first planning situation in Sect. 9.4, choose less diffuse prior distributions and study their effect on the optimal data collection plan.

9.5 In the first planning situation in Sect. 9.4, find the optimal data collection plan with planning variables, censoring time t_C, and number of units n, that meets the stated requirement and minimizes the total test-time $t_C \times n$. Another variation with censoring is to fix the censoring time t_C and determine the minimum number of units n that meets the stated requirement.

9.6 In the first planning situation in Sect. 9.4, assume a Weibull distribution. Choose comparable prior distributions for the Weibull model parameters (using prior predictive distributions as discussed in Chap. 4) and determine the optimal data collection plan. How do the two optimal data collection plans under the different distributional assumptions compare?

9.7 For the accelerated life testing planning example in Sect. 9.5, assume less diffuse prior distributions and study their effect on the optimal data collection plan.

9.8 For the accelerated life testing planning example in Sect. 9.5, study what impact on the optimal data collection plan that four and five temperature levels have. Does the optimal plan require fewer total number of units?

9.9 For the accelerated life testing planning example in Sect. 9.5, assume a Weibull distribution. Choose comparable prior distributions for the Weibull model parameters (using prior predictive distributions as discussed in Chap. 4) and determine the optimal data collection plan. How do the two optimal data collection plans under the different distributional assumptions compare?

9.10 For the logistic regression model, $\text{logit}[\pi(\text{t})] = \beta_0 + \beta_1 \text{t}$, suppose that at the ith time t_i, we test n_i units and observe success/failure data. We test at m times, $t_1, \ldots, t_m$, where $t_m = t_{max}$ for specified t_{max}. Develop a data collection plan for inference of $\pi(t_m ax + 10)$ using the direct approach for specified t_{max}, $m \geq 2$, γ, α, and L_{target}. How does the optimal data collection plan change if we require n_i to be constant? Further, how does the optimal data collection plan change if we require equally spaced testing times?

9.11 For the degradation data planning example in Sect. 9.6, assume less diffuse prior distributions and study their effect on the optimal data collection plan.

9.12 For the degradation data planning example in Sect. 9.6, study what impact on the optimal data collection plan that three and four not necessarily equally spaced inspections have. Does the optimal plan require fewer total number of inspections?

9.13 For the degradation data planning example in Sect. 9.6, assume that the intercept is random as well, say $Normal(\mu_{\beta_0}, \sigma^2_{\beta_0})$. Study the impact of the random intercept assumption on the optimal data collection plan.

9.14 Perform data collection planning for other degradation data models in Chap. 8.

9.15 Study resource allocation for a simple two-component system in series, where we may collect both system and component data. Do the same for a simple two-component system in parallel.

9.16 For the resource allocation problem considered in Sect. 9.7, explore the impact of making changes to the existing data and/or prior information on system reliability resource allocation.

9.17 Consider resource allocation for some of the system reliability applications in Chap. 5.

9.18 Consider data collection planning for other problems in the book, e.g., reliability improvement experiments with lifetime data. Do a literature search to determine how much research has been done (perhaps very little) on the problem that you choose.

10

Assurance Testing

Planning for Bayesian assurance testing involves determining a test plan that guarantees that a reliability-related quantity of interest meets or exceeds a specified requirement at a desired level of confidence. Within a Bayesian hierarchical framework, this chapter determines test plans for binomial, Poisson, and Weibull testing. Also, we develop Weibull assurance test plans using available data from an associated accelerated life testing program.

10.1 Introduction

This chapter focuses on developing a test plan for assuring (or demonstrating) that, at a desired level of confidence, a reliability-related quantity of interest meets or exceeds a specified requirement. For binomial testing, we test n devices either as a demand for successful operation or for a specified length of time and observe the total number of devices failing the test x. For example, a tester may try an emergency diesel generator (EDG) to see if it will start on demand, or a tester may place a sample of a particular nonwoven material under stress for a given length of time to see if it survives the test. The reliability-related quantity of interest for both examples is the probability π that an item survives the test, and the required binomial test plan consists of the total number of devices tested n as well as the maximum allowed number of failures c.

Although practitioners often use "assuring" and "demonstrating" synonymously, Meeker and Escobar (2004) distinguishes between reliability demonstration and reliability assurance testing. A traditional *reliability demonstration test* is essentially a *classical* (i.e., frequency based) hypothesis test, which uses only the data from the test to assess whether the reliability-related quantity of interest meets or exceeds the requirement. Consider how many modern systems, such as communication devices and transportation systems, are highly reliable. For these systems, reliability demonstration tests often require an impractical amount of testing. In response to this dilemma, Meeker and Escobar (2004) defines an alternative *reliability assurance test* as one that uses additional supplementary data and information to reduce the required amount of testing. The additional data and information may include appropriate reliability models, earlier test results on the same or similar devices, expert judgment regarding performance, knowledge of the environmental conditions under which the devices are used, benchmark design information on similar devices, prior knowledge of possible failure modes, etc. Because all of the Bayesian test plans considered in this chapter use such supplementary data and information, we refer to them as reliability assurance tests.

Life testing has many aspects in common with assurance testing. However, the primary goal in designing a life test tends to be quite different than assuring conformance to a specified reliability requirement. In designing such life tests, we often have as our primary goal improving the estimation precision of certain reliability-related quantities of interest. However, such differences notwithstanding, the basic ideas underlying life and assurance testing are similar, namely, to address such questions as "How many devices do I need to test?", "How long do I need to test each device?", or "What is the maximum number of failures permitted for a successful test?"

Because data from an assumed sampling distribution provide the basis for deciding whether the population of products being tested meets the specified requirement, there are two kinds of errors to make. A population of unreliable products (one that does not meet the requirement) may, in fact, pass the test, whereas a reliable population may fail it. This important acknowledgment makes us think about the (probabilistic) *risks* that we incur in conducting the test. The precise form of the risks is an important consideration in classical assurance tests and is an important consideration in developing Bayesian assurance tests as well. The *test criteria* are precise probabilistic statements regarding the risks we are willing to incur when developing a test plan. The following sections discuss several of the more popular criteria.

To begin our discussion of test criteria, suppose that π denotes some reliability-related quantity of interest such that large values of π are more desirable than small values. Note that reliability is one such quantity, while the mean and quantiles of a specified failure time distribution are others. It is common to base both classical and Bayesian test plans on two specified levels of π: π_0, an *acceptable reliability level* (ARL), and π_1, a *rejectable reliability level* (RRL), where $\pi_1 \leq \pi_0$. The literature sometimes refers to the region $\pi_1 \leq \pi \leq \pi_0$ as the *indifference region*. Although the precise definition of ARL and RRL differ between the classical and Bayesian test criteria, we use them in an equivalent way.

10.1.1 Classical Risk Criteria

It is quite common to use two criteria in determining classical test plans. The *producer's risk* is the probability of failing the test when $\pi = \pi_0$, whereas the *consumer's risk* is the probability of passing the test when $\pi = \pi_1$. Suppose that we specify a maximum value, α, of the producer's risk and a maximum value, β, of the consumer's risk. For binomial testing, these criteria become

$$\begin{aligned} Producer's\ Risk &= \mathbf{P}(Test\ Is\ Failed \,|\, \pi_0) \\ &= \mathbf{P}(y > c \,|\, \pi_0) \\ &= \sum_{y=c+1}^{n} \binom{n}{y} (1-\pi_0)^y \pi_0^{n-y} \leq \alpha\,, \end{aligned} \tag{10.1}$$

and

$$\begin{aligned} Consumer's\ Risk &= \mathbf{P}(Test\ Is\ Passed \,|\, \pi_1) \\ &= \mathbf{P}(y \leq c \,|\, \pi_1) \\ &= \sum_{y=0}^{c} \binom{n}{y} (1-\pi_1)^y \pi_1^{n-y} \leq \beta\,, \end{aligned} \tag{10.2}$$

where $\pi_1 \leq \pi_0, n$ is the number of test units, and c is the maximum number of failures allowed.

To choose a test plan for specified values of $(\alpha, \pi_0, \beta, \pi_1)$, we find the required binomial test plan (n, c) by simultaneously solving Eqs. 10.1 and 10.2. Numerous textbooks provide additional details of this purely classical approach, for example, see Tobias and Trindade (1995).

10.1.2 Average Risk Criteria

Easterling (1970) first proposed using average operating characteristics and corresponding risk criteria. These risk criteria are similar to the classical criteria in Sect. 10.1.1, except that now we condition on the events $\pi \geq \pi_0$ and $\pi \leq \pi_1$, respectively. To do this requires a suitable prior distribution for π, as specified by $p(\pi)$. The *average producer's risk* is the probability of failing the test when $\pi \geq \pi_0$. Choosing a maximum allowable average producer's risk α, the binomial test plan (n, c) is

$$\begin{aligned} Average\ Producer's\ Risk &= \mathbf{P}(Test\ Is\ Failed \,|\, \pi \geq \pi_0) \\ &= \frac{\mathbf{P}(y > c, \pi \geq \pi_0)}{\mathbf{P}(\pi \geq \pi_0)} \\ &= \frac{\int_{\pi_0}^{1} \left[\sum_{y=c+1}^{n} \binom{n}{y} (1-\pi)^y \pi^{n-y}\right] p(\pi) \mathrm{d}\pi}{\int_{\pi_0}^{1} p(\pi) \mathrm{d}\pi} \\ &= \frac{\int_{\pi_0}^{1} \left[1 - \sum_{y=0}^{c} \binom{n}{y} (1-\pi)^y \pi^{n-y}\right] p(\pi) \mathrm{d}\pi}{\int_{\pi_0}^{1} p(\pi) \mathrm{d}\pi} \leq \alpha\,. \end{aligned} \tag{10.3}$$

Likewise, the corresponding *average consumer's risk* is the probability of passing the test when $\pi \leq \pi_1$. Choosing a maximum allowable average consumer's risk β, the binomial test plan (n, c) is

$$\begin{aligned} \textit{Average Consumer's Risk} &= \mathbf{P}(\textit{Test Is Passed} \,|\, \pi \leq \pi_1) \\ &= \frac{\mathbf{P}(y \leq c, \pi \leq \pi_1)}{\mathbf{P}(\pi \leq \pi_1)} \\ &= \frac{\int_0^{\pi_1} \left[\sum_{y=0}^{c} \binom{n}{y} (1-\pi)^y \pi^{n-y}\right] p(\pi)\mathrm{d}\pi}{\int_0^{\pi_1} p(\pi)\mathrm{d}\pi} \leq \beta\,. \end{aligned} \tag{10.4}$$

Martz and Waller (1982) discusses the use of these risks and recommends taking care in applying these criteria. For example, the average consumer's risk may be a poor indication of what is likely really desired; namely, a maximum probability β that $\pi \leq \pi_1$ for a test that passes. The average consumer's risk given in Eq. 10.4 may be substantially larger than this desired maximum conditional probability β. The prior probability that $\pi \leq \pi_1$ may be quite small; however, if indeed $\pi \leq \pi_1$, the probability of passing the test may be large. In such a situation, using the average consumer's risk may be inappropriate, and therefore misleading.

10.1.3 Posterior Risk Criteria

We now consider fully Bayesian posterior risks that convey a completely different outlook from the corresponding classical or average risks. While the classical or average risks provide assurance that satisfactory devices will pass the test and that unsatisfactory devices will fail it, posterior risks provide precisely the assurance that practitioners often desire: if the test is passed, then the consumer desires a maximum probability β that $\pi \leq \pi_1$. On the other hand, if the test is failed, then the producer desires a maximum probability α that $\pi \geq \pi_0$. Unlike the average risks, these posterior risks are fully Bayesian in the sense that they are subjective probability statements about π.

For a test that fails, the *posterior producer's risk* is the probability that $\pi \geq \pi_0$, or $\mathbf{P}(\pi \geq \pi_0 \,|\, \textit{Test Is Failed})$. Notice that this is simply the posterior probability that $\pi \geq \pi_0$ given that we have observed more than c failures. Using Bayes' Theorem, and assuming a maximum allowable posterior producer's risk α, an expression for the posterior producer's risk for the binomial test plan (n, c) is

$$\begin{aligned} \textit{Posterior Producer's Risk} &= \mathbf{P}(\pi \geq \pi_0 \,|\, \textit{Test Is Failed}) \\ &= \int_{\pi_0}^{1} p(\pi \,|\, y > c)\mathrm{d}\pi \\ &= \int_{\pi_0}^{1} \frac{f(y > c \,|\, \pi)p(\pi)}{\int_0^1 f(y > c \,|\, \pi)p(\pi)\mathrm{d}\pi}\, \mathrm{d}\pi \end{aligned} \tag{10.5}$$

$$= \frac{\int_{\pi_0}^1 \left[\sum_{y=c+1}^n \binom{n}{y}(1-\pi)^y \pi^{n-y}\right] p(\pi)\mathrm{d}\pi}{\int_0^1 \left[\sum_{y=c+1}^n \binom{n}{y}(1-\pi)^y \pi^{n-y}\right] p(\pi)\mathrm{d}\pi}$$

$$= \frac{\int_{\pi_0}^1 \left[1 - \sum_{y=0}^c \binom{n}{y}(1-\pi)^y \pi^{n-y}\right] p(\pi)\mathrm{d}\pi}{1 - \int_0^1 \left[\sum_{y=0}^c \binom{n}{y}(1-\pi)^y \pi^{n-y}\right] p(\pi)\mathrm{d}\pi} \leq \alpha.$$

Similarly, given that the test is passed, the *posterior consumer's risk* is the probability that $\pi \leq \pi_1$, or $\mathbf{P}(\pi \leq \pi_1 \,|\, \textit{Test Is Passed})$. Notice that this is simply the posterior probability that $\pi \leq \pi_1$ given that we have observed no more than c failures. Using Bayes' Theorem, and assuming a maximum allowable posterior consumer's risk β, an expression for the posterior consumer's risk for the binomial test plan (n, c) is

$$\begin{aligned} \textit{Posterior Consumer's Risk} &= \mathbf{P}(\pi \leq \pi_1 \,|\, \textit{Test Is Passed}) \qquad (10.6)\\ &= \int_0^{\pi_1} p(\pi \,|\, y \leq c)\mathrm{d}\pi \\ &= \int_0^{\pi_1} \frac{f(y \leq c \,|\, \pi)p(\pi)}{\int_0^1 f(y \leq c \,|\, \pi)p(\pi)\mathrm{d}\pi}\mathrm{d}\pi \\ &= \frac{\int_0^{\pi_1} \left[\sum_{y=0}^c \binom{n}{y}(1-\pi)^y \pi^{n-y}\right] p(\pi)\mathrm{d}\pi}{\int_0^1 \left[\sum_{y=0}^c \binom{n}{y}(1-\pi)^y \pi^{n-y}\right] p(\pi)\mathrm{d}\pi} \leq \beta. \end{aligned}$$

Example 10.1 Binomial test plan for new modems. Consider finding a binomial test plan using the posterior consumer's risk criterion. Hart (1990) develops a reliability assurance test for a new modem, denoted by B, that is similar to an earlier modem, denoted by A. Modem A is currently in production and is very reliable. The major difference between the two modems is that B operates at a different frequency than A. Also, the same production line that builds A will produce B and both modems use most of the same components. Further, Hart (1990) reports that a binomial assurance test for modem A on 150 units yielded 6 failures.

One of the test objectives is to show that, after successful testing, the 0.1 quantile of the posterior reliability distribution for B is at least 0.938, the 0.1 quantile of A's posterior reliability distribution. Similar to Hart (1990), we use a $Beta[86.4, 3.6] = Beta[(0.6 \times 150)(144/150), (0.6 \times 150)(6/150)]$ prior distribution for π. This prior distribution arises from treating an A test as "worth" 60% of a B test or $90 = 0.6 \times 150$ total tests. Note that the 0.1 quantile of this prior distribution is 0.932, which is only slightly smaller than the requirement. Therefore, we anticipate that the test plan will require only a small sample of B modems.

A *minimum sample size* (or *zero-failure*) test plan is one in which we test n modems and state that the test is passed if there are no failures, that is,

$c = 0$. Given our $Beta(86.4, 3.6)$ prior distribution, $\pi_1 = 0.938, \beta = 0.10$, and $c = 0$, we find the desired Bayesian zero-failure test plan by solving Eq. 10.6 for sample size or number of tests n. Using Eq. 10.6 yields the expression

$$\begin{aligned}
&\mathbf{P}(\pi \leq 0.938 \,|\, \mathit{Test\ Is\ Passed}) \qquad (10.7)\\
&= \frac{\int_0^{0.938} \binom{n}{0} (1-\pi)^0 \pi^n p(\pi) \mathrm{d}\pi}{\int_0^1 \binom{n}{0} (1-\pi)^0 \pi^n p(\pi) \mathrm{d}\pi} \\
&= \frac{\int_0^{0.938} \pi^n \frac{\Gamma(86.4+3.6)}{\Gamma(86.4)\Gamma(3.6)} \pi^{86.4-1} (1-\pi)^{3.6-1} \mathrm{d}\pi}{\int_0^1 \pi^n \frac{\Gamma(86.4+3.6)}{\Gamma(86.4)\Gamma(3.6)} \pi^{86.4-1} (1-\pi)^{3.6-1} \mathrm{d}\pi} \\
&= I(0.938; 86.4+n, 3.6) \leq 0.10\,,
\end{aligned}$$

where $I(z; \alpha, \beta)$ denotes the *incomplete beta function ratio*. Upon evaluating the incomplete beta function ratio in Eq. 10.7 for increasing values of n, we find that $n = 9$ is the smallest integer that satisfies the inequality. Consequently, the plan consists of testing 9 B modems. If none fail, we can then claim that $\mathbf{P}(\pi \leq 0.938 \,|\, \mathit{No\ Failures\ in\ 9\ Tests}) = 0.097 < 0.10$, as required. In this case, given no failures in the 9 B modem tests, the 0.1 quantile of the posterior distribution of π is 0.9384. Finally, the unconditional probability of passing the test is simply

$$\begin{aligned}
\mathbf{P}[\mathit{Test\ Is\ Passed}] &= \int_0^1 \binom{n}{0} (1-\pi)^0 \pi^n p(\pi) \mathrm{d}\pi \\
&= \frac{\Gamma(86.4+n)\Gamma(3.6)}{\Gamma(86.4+3.6+n)} \frac{\Gamma(86.4+3.6)}{\Gamma(86.4)\Gamma(3.6)} \\
&= \frac{\Gamma(95.4)\Gamma(90)}{\Gamma(99)\Gamma(86.4)} = 0.70.
\end{aligned}$$

10.2 Binomial Testing

Now consider both the average and posterior risks for the binomial sampling distribution within a hierarchical framework. Suppose that we have failure count data from $m > 1$ situations, such as m different plants. Let x_i denote the observed number of failures in a sample of size n_i for $i = 1, \ldots, m$, and let $\mathbf{x}$ represent all the observed failure count data. Then, conditional on the success probability π_i, assume that the X_i are conditionally independent and that $X_i \,|\, \pi_i \sim Binomial(n_i, 1-\pi_i)$. Also assume that the π_i can be modeled hierarchically — specifically, that given δ and γ they are independent and identically distributed (i.i.d.) with a common $Beta(\delta, \gamma)$ distribution. Finally, we specify a prior distribution for the hyperparameters (δ, γ), denoted by $p(\delta, \gamma)$, which is a proper (but usually diffuse) joint distribution.

10.2.1 Binomial Posterior Consumer's and Producer's Risks

Given that there are observed data $X_i \sim Binomial(n_i, 1 - \pi_i)$, suppose that we are interested in developing a binomial test plan (n, c) using the posterior risk criteria as our test criteria. Our test plan is for a situation "similar" to those previously observed, where we describe similarity by assuming that for the new situation, an item has probability π of surviving the test, where given δ and γ, π and the π_i are i.i.d. $Beta(\delta, \gamma)$.

Recall that both the posterior producer's risk and posterior consumer's risk specify criteria on the posterior distribution for the binomial probability of success π. Since there are now observed data $\mathbf{x}$, we condition on that data and now use $p(\pi \,|\, \mathbf{x})$ in place of $p(\pi)$ in Eqs. 10.5 and 10.6. In particular, for a binomial test plan (n, c), an expression for the posterior producer's risk is

$$\mathbf{P}(\pi \geq \pi_0 \mid Test\ Is\ Failed, \mathbf{x}) \tag{10.8}$$

$$= \frac{\int_{\pi_0}^{1} \left[1 - \sum_{y=0}^{c} \binom{n}{y} (1-\pi)^y \pi^{n-y}\right] p(\pi \,|\, \mathbf{x}) \mathrm{d}\pi}{1 - \int_0^1 \left[\sum_{y=0}^{c} \binom{n}{y} (1-\pi)^y \pi^{n-y}\right] p(\pi \,|\, \mathbf{x}) \mathrm{d}\pi}.$$

Similarly, the posterior consumer's risk is

$$\mathbf{P}(\pi \leq \pi_0 \mid Test\ Is\ Passed, \mathbf{x}) \tag{10.9}$$

$$= \frac{\int_0^{\pi_1} \left[\sum_{y=0}^{c} \binom{n}{y} (1-\pi)^y \pi^{n-y}\right] p(\pi \,|\, \mathbf{x}) \mathrm{d}\pi}{\int_0^1 \left[\sum_{y=0}^{c} \binom{n}{y} (1-\pi)^y \pi^{n-y}\right] p(\pi \,|\, \mathbf{x}) \mathrm{d}\pi}.$$

Now we must determine how to evaluate these criteria. There are two equivalent ways to approach the problem. First, notice that both of the posterior risk criteria are probability statements about the posterior distribution of π given different data. For a given choice of (n, c), the posterior producer's risk can be calculated by using Markov chain Monte Carlo (MCMC) to find the posterior distribution of π given $\mathbf{x}$ and $y > c$ and then calculating the proportion of posterior draws with $\pi \geq \pi_0$. Similarly, for a given choice of (n, c), the posterior consumer's risk can be calculated by using MCMC to find the posterior distribution of π given $\mathbf{x}$ and $y \leq c$ and then calculating the proportion of posterior draws with $\pi \leq \pi_1$. Notice, however, that for each choice of (n, c), this requires using MCMC to calculate two posterior distributions for π.

Suppose instead that we condition only the data $X_i \sim Binomial(n_i, 1-\pi_i)$ with $\pi_i \sim Beta(\delta, \gamma)$ and use MCMC to obtain the posterior predictive distribution $p(\pi \,|\, \mathbf{x})$. We can obtain draws from the posterior predictive distribution $p(\pi \,|\, \mathbf{x})$ using the N posterior draws for (δ, γ) by drawing $\pi^{(j)} \sim Beta(\delta^{(j)}, \gamma^{(j)})$, and then use these samples to evaluate Eqs. 10.8 and 10.9 using Monte Carlo integration. In general, to evaluate $\mathrm{E}[g(x)] = \int g(x)p(x)dx$, obtain a random sample $x_1, \ldots, x_N$ from $p(x)$ and approximate the expectation as $\frac{1}{N} \sum_{i=1}^{N} g(x_i)$.

We evaluate the posterior producer's risk as

$$\mathbf{P}[\pi \geq \pi_1 \,|\, Test\ Is\ Failed, \mathbf{x}] \approx \frac{\frac{1}{N}\sum_{j=1}^{N}\left[1 - \sum_{y=0}^{c}\binom{n}{y}(1-\pi^{(j)})^y(\pi^{(j)})^{n-y}\right] I(\pi^{(j)} \geq \pi_0)}{1 - \frac{1}{N}\sum_{j=1}^{N}\left[\sum_{y=0}^{c}\binom{n}{y}(1-\pi^{(j)})^y(\pi^{(j)})^{n-y}\right]},$$

and the posterior consumer's risk as

$$\mathbf{P}[\pi \leq \pi_1 \,|\, Test\ Is\ Passed, \mathbf{x}] \approx \frac{\sum_{j=1}^{N}\left[\sum_{y=0}^{c}\binom{n}{y}(1-\pi^{(j)})^y(\pi^{(j)})^{n-y}\right] I(\pi^{(j)} \leq \pi_1)}{\sum_{j=1}^{N}\left[\sum_{y=0}^{c}\binom{n}{y}(1-\pi^{(j)})^y(\pi^{(j)})^{n-y}\right]}.$$

The expression for the unconditional probability of passing the test is

$$\mathbf{P}(Test\ Is\ Passed \,|\, \mathbf{x}) \approx \frac{1}{N}\sum_{j=1}^{N}\left[\sum_{y=0}^{c}\binom{n}{y}(1-\pi^{(j)})^y(\pi^{(j)})^{n-y}\right]. \tag{10.10}$$

Let

$$b^{(j)}(y) = \frac{\binom{n}{y}B(n-y+\delta^{(j)}, y+\gamma^{(j)})}{B(\delta^{(j)}, \gamma^{(j)})},$$

where $B(\alpha, \beta)$ is the beta function. We can also evaluate the posterior producer's risk using the posterior draws $\delta^{(j)}, \gamma^{(j)} \,|\, \mathbf{x}$ as

$$\mathbf{P}[\pi \geq \pi_0 \,|\, Test\ Is\ Failed, \mathbf{x}] \approx \frac{\sum_{j=1}^{N}\left[1 - I(\pi_0; \delta^{(j)}, \gamma^{(j)}) - \sum_{y=0}^{c} b^{(j)}(y)[1 - I(\pi_0; n-y+\delta^{(j)}, y+\gamma^{(j)})]\right]}{\sum_{j=1}^{N}\left[1 - \sum_{y=0}^{c} b^{(j)}(y)\right]},$$

and the posterior consumer's risk as

$$\mathbf{P}[\pi \leq \pi_1 \,|\, Test\ Is\ Passed, \mathbf{x}] \approx \frac{\sum_{j=1}^{N}\left[\sum_{y=0}^{c} b^{(j)}(y) I(\pi_1; n-y+\delta^{(j)}, y+\gamma^{(j)})\right]}{\sum_{j=1}^{N}\sum_{y=0}^{c} b^{(j)}(y)},$$

where $I(z; \alpha, \beta)$ is the incomplete beta function ratio. An additional expression for the unconditional probability of passing the test is

$$\begin{aligned}\mathbf{P}(Test\ Is\ Passed \,|\, \mathbf{x}) &\approx \frac{1}{N}\sum_{j=1}^{N}\sum_{y=0}^{c} b^{(j)}(y)\\ &\approx \frac{1}{N}\sum_{j=1}^{N}\left[\sum_{y=0}^{c}\frac{\binom{n}{y}B(n-y+\delta^{(j)}, y+\gamma^{(j)})}{B(\delta^{(j)}, \gamma^{(j)})}\right].\end{aligned}$$

To obtain a test plan, simultaneously solve the pair of inequalities given by

$$\mathbf{P}[\pi \geq \pi_0 \,|\, Test\ Is\ Failed, \mathbf{x}] \leq \alpha \tag{10.11}$$

and

$$\mathbf{P}[\pi \leq \pi_1 \,|\, Test\ Is\ Passed, \mathbf{x}] \leq \beta, \tag{10.12}$$

for the pair of integers (n, c), where $0 \leq c < n$, and where α and β are the desired maximum posterior producer's and consumer's risks.

We can find such test plans because Eqs. 10.11 and 10.12 have opposite effects. For fixed c, as n increases, $\mathbf{P}[\pi \leq \pi_1 \,|\, Test\ Is\ Passed, \mathbf{x}]$ decreases, whereas $\mathbf{P}[\pi \geq \pi_0 \,|\, Test\ Is\ Failed, \mathbf{x}]$ increases. On the other hand, for fixed n, as c increases, the opposite is true. Consequently, we can use the algorithm in Fig. 10.1 to find the required test plan.

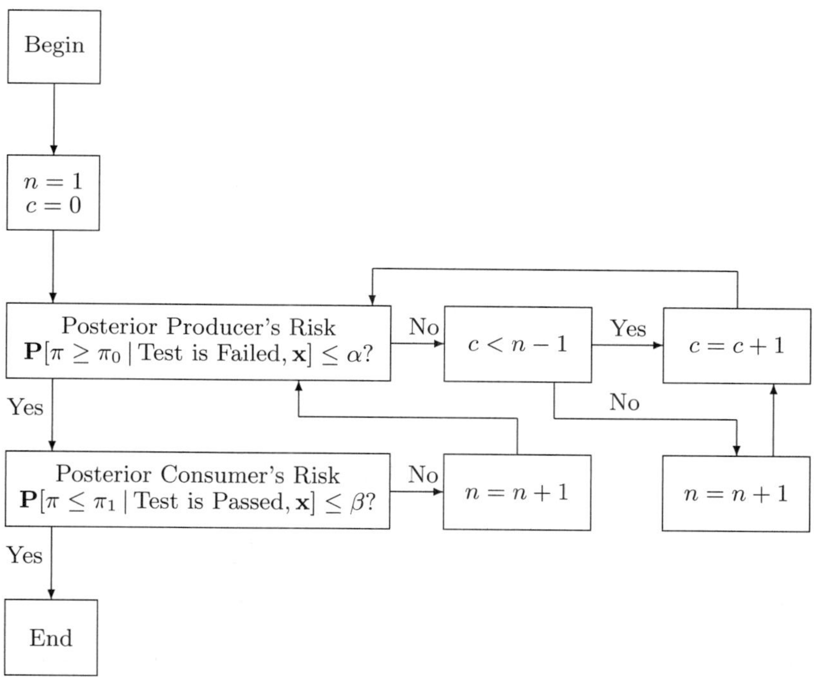

Fig. 10.1. An algorithm for finding Bayesian hierarchical binomial test plans.

Example 10.2 Hierarchical binomial test plan for EDGs. Martz et al. (1996) analyzes failure count data from EDGs in 63 U.S. commercial nuclear power plants to assess their reliability of accepting the electrical load and run (*load-run*) on demand. See Table 10.1, which summarizes the data collected over the four-year period 1988-1991.

Table 10.1. EDG load-run demand data (x failures out of n demands) (Martz et al., 1996)

Plant	x	n	Plant	x	n	Plant	x	n
1	11	854	22	0	238	43	5	216
2	2	373	23	1	370	44	0	252
3	5	618	24	2	302	45	1	419
4	5	157	25	3	152	46	0	136
5	3	542	26	2	294	47	3	185
6	2	202	27	0	101	48	7	382
7	0	65	28	0	283	49	6	304
8	2	166	29	2	117	50	2	130
9	4	574	30	0	115	51	1	121
10	2	201	31	1	196	52	2	295
11	8	388	32	0	310	53	1	289
12	2	287	33	4	134	54	0	181
13	3	431	34	2	132	55	2	150
14	5	358	35	0	242	56	0	334
15	14	1120	36	4	132	57	2	263
16	1	321	37	4	320	58	0	92
17	3	225	38	0	996	59	2	466
18	8	433	39	1	253	60	2	387
19	7	468	40	13	704	61	1	183
20	1	218	41	2	151	62	5	278
21	9	317	42	2	385	63	0	212

We model the failure counts X_i as conditionally independent given π_i with $Binomial(n_i, 1-\pi_i)$ distributions, where π_i is the ith plant reliability. Because the plants have all been built and operated to the same Nuclear Regulatory Commission (NRC)-controlled safety standards, model the 63 $\pi_i \,|\, \delta, \gamma$ as i.i.d. with a common $Beta(\delta, \gamma)$ distribution. In our analysis of these data, we use independent and diffuse $InverseGamma(0.1, 0.1)$ prior distributions for δ and γ. See Table 10.2, which summarizes the marginal posterior distributions for δ and γ and the predictive distribution for π. The hierarchical binomial model fits the load-run demand data well (see Exercise 4.13).

Using the posterior risk criteria, suppose now that we want to find the test plan having the posterior consumer's and producer's risks $\beta = 0.05$ for $\pi_1 = 0.985$, and $\alpha = 0.10$ for $\pi_0 = 0.999$, respectively.

Using the algorithm in Fig. 10.1, we obtain the required test plan $n = 42$ and $c = 0$. For this test plan, the actual posterior consumer's risk $\mathbf{P}[\pi \leq 0.985 \,|\, \mathit{Test\ Is\ Passed}, \mathbf{x}] = 0.0014$ and the actual posterior producer's risk $\mathbf{P}[\pi \geq 0.999 \,|\, \mathit{Test\ Is\ Failed}, \mathbf{x}] = 0.0992$. Therefore, at the new plant, load and run the EDGs 42 times. (Note that we are also assuming here that

Table 10.2. Posterior distribution summaries for the hyperparameters δ and γ and the predictive distribution for π

Parameter	Mean	Std Dev	Quantiles				
			0.025	0.05	0.50	0.95	0.975
δ	286.4	135.6	121.5	136.9	253.2	557.4	652.8
γ	2.710	1.243	1.226	1.360	2.402	5.171	6.204
π	0.9903	0.0067	0.9735	0.9778	0.9917	0.9981	0.9987

all the EDGs at the new plant have the same load-run demand reliability.) If there are no failures, then we can assure an EDG load-run demand reliability of 0.985 with a 0.95 probability. On the other hand, if the test is failed, we have the assurance that the EDG load-run demand reliability is not greater than 0.999 with a 0.90 probability. Finally, from Eq. 10.10, the unconditional probability of passing this test is approximately 0.69.

10.2.2 Hybrid Risk Criterion

Fitzgerald et al. (1999) uses the posterior consumer's risk criterion in Eq. 10.6 and the average producer's risk criterion in Eq. 10.3 to determine alternative binomial test plans. These "hybrid" criteria are appropriate in situations where we are interested in a single specified level of reliability, which Fitzgerald et al. (1999) refers to as the *target reliability level*. Fitzgerald et al. (1999) reasons that, in many applications of reliability assurance testing, the consumer and producer are one and the same. Consequently, in such situations, there may be less adversarial tension in determining appropriate risk criteria, and therefore, little motivation to specify two different (and acceptable) levels of reliability (π_0 and π_1) for the required test plan.

Likewise, we use this hybrid approach and seek a Bayesian binomial test plan for which there is a probability of at most β that $\pi \le \pi^*$ for a passing test, and simultaneously, there is a probability of at most α that a test is failed given $\pi \ge \pi^*$.

Notationally, we want to find the test plan (n, c) that satisfies the hybrid risk criteria given by

$$\mathbf{P}[\mathit{Test\ Is\ Failed} \,|\, \pi \ge \pi^*, \mathbf{x}] \le \alpha \tag{10.13}$$

and

$$\mathbf{P}[\pi \le \pi^* \,|\, \mathit{Test\ Is\ Passed}, \mathbf{x}] \le \beta. \tag{10.14}$$

For test plan (n, c), we can approximate the average producer's risk by

$$\mathbf{P}[\mathit{Test\ Is\ Failed} \,|\, \pi \ge \pi^*, \mathbf{x}] \approx \tag{10.15}$$
$$\frac{\sum_{j=1}^{N}\left[1 - I(\pi^*; \delta^{(j)}, \gamma^{(j)}) - \sum_{y=0}^{c} b^{(j)}(y)[1 - I(\pi^*; n - y + \delta^{(j)}, y + \gamma^{(j)})]\right]}{\sum_{j=1}^{N}\left[1 - I(\pi^*; \delta^{(j)}, \gamma^{(j)})\right]}.$$

We find the desired Bayesian test plan by employing the algorithm in Fig. 10.1, where Eq. 10.15 replaces the producer's risk inequality statement used in the algorithm.

Example 10.3 Hierarchical binomial test plan using hybrid criteria. Again consider Example 10.2. Consider finding a Bayesian test plan having the posterior consumer's and average producer's risks $\beta = 0.05$ and $\alpha = 0.05$ for $\pi^* = 0.98$, respectively. Using the algorithm in Fig. 10.1 with the modification described above, we obtain the required test plan $n = 71$ and $c = 2$. For this test plan, the actual posterior consumer's risk is $\mathbf{P}[\pi \leq 0.98 \,|\, Test\ Is\ Passed, \mathbf{x}] = 0.0497$ and the actual average producer's risk $\mathbf{P}[Test\ Is\ Failed \,|\, \pi \geq 0.98, \mathbf{x}] = 0.0333$. Because we now only consider a single target reliability level, this test plan requires significantly more testing than that found in Example 10.2, in which two different reliability values were specified. From Eq. 10.10, the unconditional probability of passing this test is approximately 0.951.

10.3 Poisson Testing

We now move from binomial testing within a hierarchical framework to Poisson testing. Recall from Chap. 6 that a homogeneous Poisson process generates a sequence of events for which the times between successive failures (the *inter-failure times*) are independently and identically $Exponential(\lambda)$ distributed. For a particular Poisson testing situation, let us develop a Bayesian test plan for assuring that the *failure rate* λ does not exceed a specified requirement. From Poisson process theory, the number of failures X occurring in fixed total test or operating time T has a $Poisson(\lambda T)$ distribution. Let T represent the total operating or exposure time of the devices during which we assume that devices are either repaired or replaced when they fail.

Suppose that the plan tests n devices for the length of time t_0, replacing devices as they fail, so that the total operating time is $T = nt_0$. In determining a Bayesian test plan, we seek (T, c), where c is the maximum allowed number of failures. The test is passed if no more than c failures occur in total test time T. Note that any combination of n and t_0 satisfying $T = nt_0$ provides an acceptable test plan.

As with binomial testing, let us assume in Poisson testing that there are data available from $m > 1$ situations. Let x_i denote the observed number of failures in total operating time T_i for the ith situation, and let $\mathbf{x}$ represent all the observed failure data. Then, conditioning on $\lambda_i, X_i \sim Poisson(\lambda_i T_i)$, where the X_i are conditionally independent. We model the λ_i hierarchically, assuming they are i.i.d. $Gamma(\eta, \kappa)$, given η and κ, and specify a prior distribution for the hyperparameters (η, κ), denoted by $p(\eta, \kappa)$.

We can write an expression for the posterior producer's risk for the Poisson test plan (T, c), where λ_0 is the rejectable failure rate and there is a maximum allowable posterior producer's risk α:

$$\begin{aligned} \textit{Posterior Producer's Risk} &= \mathbf{P}(\lambda \le \lambda_0 \,|\, \textit{Test Is Failed}) \qquad (10.16)\\ &= \int_0^{\lambda_0} p(\lambda \,|\, y > c)\mathrm{d}\lambda \\ &= \int_0^{\lambda_0} \frac{f(y > c \,|\, \lambda)p(\lambda)}{\int_0^\infty f(y > c \,|\, \lambda)p(\lambda)\mathrm{d}\lambda}\,\mathrm{d}\lambda \\ &= \frac{\int_0^{\lambda_0}\left[1 - \sum_{y=0}^{c} \frac{(\lambda T)^y \exp(-\lambda T)}{y!}\right] p(\lambda)\mathrm{d}\lambda}{\int_0^{\infty}\left[1 - \sum_{y=0}^{c} \frac{(\lambda T)^y \exp(-\lambda T)}{y!}\right] p(\lambda)\mathrm{d}\lambda} \le \alpha. \end{aligned}$$

Similarly, given that the test is passed, we can write an expression for the posterior consumer's risk for the Poisson test plan (T, c), with acceptable failure rate λ_1 and maximum allowable posterior consumer's risk β, as

$$\begin{aligned} \textit{Posterior Consumer's Risk} &= \mathbf{P}(\lambda \ge \lambda_1 \,|\, \textit{Test Is Passed}) \qquad (10.17)\\ &= \int_{\lambda_1}^{\infty} p(\lambda \,|\, y \le c)\mathrm{d}\lambda \\ &= \int_{\lambda_1}^{\infty} \frac{f(y \le c \,|\, \lambda)p(\lambda)}{\int_0^\infty f(y \le c \,|\, \lambda)p(\lambda)\mathrm{d}\lambda}\,\mathrm{d}\lambda \\ &= \frac{\int_{\lambda_1}^{\infty}\left[\sum_{y=0}^{c} \frac{(\lambda T)^y \exp(-\lambda T)}{y!}\right] p(\lambda)\mathrm{d}\lambda}{\int_0^{\infty}\left[\sum_{y=0}^{c} \frac{(\lambda T)^y \exp(-\lambda T)}{y!}\right] p(\lambda)\mathrm{d}\lambda} \le \beta. \end{aligned}$$

Since there are available data $\mathbf{x}$, we use the posterior distribution for $\lambda, p(\lambda \,|\, \mathbf{x})$, in Eqs. 10.16 and 10.17 to construct our test plan for a new situation. We can calculate the posterior producer's risk either using posterior predictive draws $\lambda^{(j)}$ or using posterior draws $(\eta^{(j)}, \kappa^{(j)})$. Let

$$g^{(j)}(y) = \frac{(\kappa^{(j)})^{\eta^{(j)}} T^y}{y!\Gamma(\eta^{(j)})(T + \kappa^{(j)})^{y+\eta^{(j)}}}.$$

$$\begin{aligned} &\mathbf{P}[\lambda \le \lambda_0 \,|\, \textit{Test Is Failed}, \mathbf{x}] \\ &\approx \frac{\sum_{j=1}^{N}\left[1 - \sum_{y=0}^{c} \frac{(\lambda^{(j)}T)^y \exp(-\lambda^{(j)}T)}{y!}\right] I(\lambda^{(j)} \le \lambda_0)}{\sum_{j=1}^{N}\left[1 - \sum_{y=0}^{c} \frac{(\lambda^{(j)}T)^y \exp(-\lambda^{(j)}T)}{y!}\right]} \\ &\approx \frac{\sum_{j=1}^{N}\left[\gamma(\eta^{(j)}, \kappa^{(j)}\lambda_0)/\Gamma(\eta^{(j)}) - \sum_{y=0}^{c} g^{(j)}(y)\gamma[y + \eta^{(j)}, (T + \kappa^{(j)})\lambda_0]\right]}{\sum_{j=1}^{N}\left[1 - \sum_{y=0}^{c} g^{(j)}(y)\Gamma(y + \eta^{(j)})\right]}, \end{aligned}$$

where $\gamma(q, z)$ denotes the *lower incomplete gamma function.*

The expressions for the posterior consumer's risk are

$$\begin{aligned}
&\mathbf{P}[\lambda \geq \lambda_1 \,|\, \mathit{Test\ Is\ Passed}, \mathbf{x}] \\
&\approx \frac{\sum_{j=1}^{N}\left[\sum_{y=0}^{c}\frac{(\lambda^{(j)}T)^y\exp(-\lambda^{(j)}T)}{y!}\right] I(\lambda^{(j)} \geq \lambda_1)}{\sum_{j=1}^{N}\left[\sum_{y=0}^{c}\frac{(\lambda^{(j)}T)^y\exp(-\lambda^{(j)}T)}{y!}\right]} \\
&\approx \frac{\sum_{j=1}^{N}\left[\sum_{y=0}^{c} g^{(j)}(y)\{\Gamma(y+\eta^{(j)}) - \gamma[y+\eta^{(j)}, (T_0+\kappa^{(j)})\lambda_1]\}\right]}{\sum_{j=1}^{N}\left[\sum_{y=0}^{c} g^{(j)}(y)\Gamma(y+\eta^{(j)})\right]} .
\end{aligned}$$

We can also write the unconditional probability of passing the test as

$$\begin{aligned}
\mathbf{P}[\mathit{Test\ Is\ Passed}|\mathbf{x}] &\approx \frac{1}{N}\sum_{j=1}^{N}\left[\sum_{y=0}^{c}\frac{(\lambda^{(j)}T)^y\exp(-\lambda^{(j)}T)}{y!}\right] \\
&\approx \frac{1}{N}\sum_{j=1}^{N}\sum_{y=0}^{c} g^{(j)}(y)\Gamma(y+\eta^{(j)}) \\
&\approx \frac{1}{N}\sum_{j=1}^{N}\left[\sum_{y=0}^{c}\frac{(\kappa^{(j)})^{\eta^{(j)}}T^y\Gamma(y+\eta^{(j)})}{y!\Gamma(\eta^{(j)})(T+\kappa^{(j)})^{y+\eta^{(j)}}}\right] .
\end{aligned}$$

To obtain a test plan, simultaneously solve the following pair of nonlinear inequalities:

$$\mathbf{P}[\lambda \leq \lambda_0 \,|\, \mathit{Test\ Is\ Failed}, \mathbf{x}] \leq \alpha \tag{10.18}$$

and

$$\mathbf{P}[\lambda \geq \lambda_1 \,|\, \mathit{Test\ Is\ Passed}, \mathbf{x}] \leq \beta, \tag{10.19}$$

where $\lambda_0 \leq \lambda_1$.

Because T is continuous, we can hold either of the risks in Eqs. 10.18 and 10.19 at its precise value. Holding the posterior producer's risk at exactly α, use the algorithm in Fig. 10.2 to obtain the desired test plan. On the other hand, holding the posterior consumer's risk at precisely β, simply reverse the two main steps in the procedure.

Example 10.4 Hierarchical Poisson test plan for pumps. Gaver and O'Muircheartaigh (1987) provides the data shown in Table 10.3 on the number of pump failures x observed in t thousands of operating hours for $m = 10$ different systems at the Farley 1 U.S. commercial nuclear power plant.

Note that we have listed the data in increasing order of the corresponding maximum likelihood estimates (MLEs) $\widehat{\lambda}$.

We model the failures as conditionally independent given their individual failure rates λ_i with $\mathit{Poisson}(\lambda_i t_i)$ distributions. Given η and κ, we model the λ_i as i.i.d. $\mathit{Gamma}(\eta, \kappa)$ and use independent and diffuse

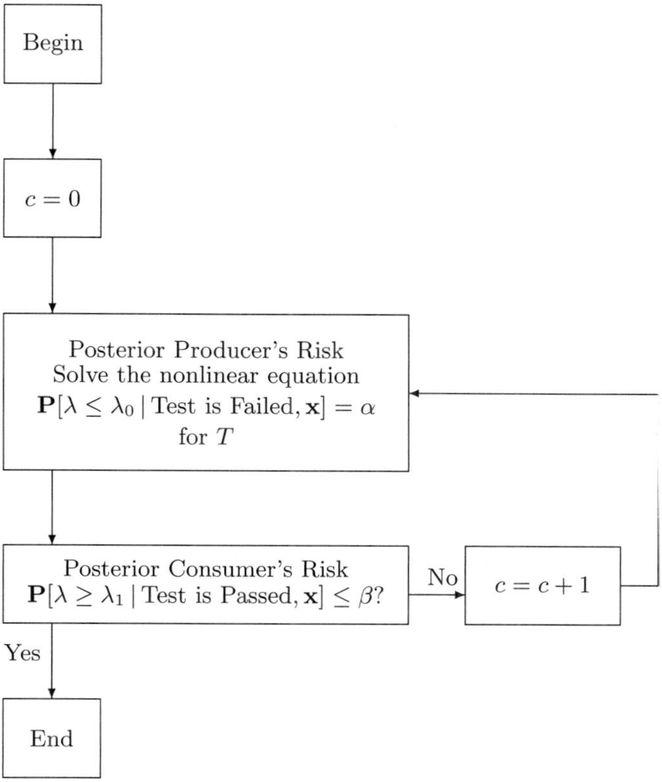

Fig. 10.2. An algorithm for finding Bayesian Poisson test plans.

Table 10.3. Pump failure count data from Farley 1 U.S. nuclear power plant (number of failures x in t thousands of operating hours) (Gaver and O'Muircheartaigh, 1987)

System	x_i (failures)	t_i (thousand hours)	$\widehat{\lambda}$ (MLE)
1	5	94.320	5.3 x 10^{-2}
2	1	15.720	6.4 x 10^{-2}
3	5	62.880	8.0 x 10^{-2}
4	14	125.760	11.1 x 10^{-2}
5	3	5.240	57.3 x 10^{-2}
6	19	31.440	60.4 x 10^{-2}
7	1	1.048	95.4 x 10^{-2}
8	1	1.048	95.4 x 10^{-2}
9	4	2.096	191.0 x 10^{-2}
10	22	10.480	209.9 x 10^{-2}

$InverseGamma(0.001, 0.001)$ prior distributions for η and κ. Table 10.4 summarizes the marginal posterior distributions of η and κ as well as the predictive distribution of λ. The hierarchical Poisson model fits the pump failure count data well (see Exercise 4.17).

Table 10.4. Posterior distribution summaries for the gamma distribution hyperparameters (η, κ) given $\mathbf{x}$ and of the predictive distribution for λ for pump example

			Quantiles				
Parameter	Mean	Std Dev	0.025	0.05	0.50	0.95	0.975
η	0.7981	0.3635	0.2995	0.3419	0.7272	1.4840	1.6950
κ	1.284	0.855	0.229	0.306	1.087	2.971	3.460
λ	0.796	1.306	0.002	0.007	0.390	2.939	4.037

Using the posterior risk criteria in Eqs. 10.18 and 10.19, suppose that we want to find the Poisson test plan with risk parameters $\lambda_0 = 0.2, \alpha = 0.05, \lambda_1 = 0.7$, and $\beta = 0.05$. Using the algorithm in Fig. 10.2 with $N = 10,000$ joint posterior draws of (η, κ) given $\mathbf{x}$, we find the required test plan $T_0 = 5.43$ and $c = 1$. The actual risks for this test plan are $\mathbf{P}[\lambda \geq 0.7 \,|\, Test\ Is\ Passed, \mathbf{x}] = 0.0171$ and $\mathbf{P}[\lambda \leq 0.2 \,|\, Test\ Is\ Failed, \mathbf{x}] = 0.0500$, and the unconditional probability of passing this test is approximately 0.44. To implement this test plan for new systems, accumulate 5,430 hours of pump operating time with repair (or replacement) of the failed pumps. If no more than one failure occurs, then the test is passed, otherwise, the test is failed.

Although it is inappropriate for Example 10.4, in many cases we are free to accumulate the required total test time T by choosing any desired combination of n devices and time on test t_0 satisfying $T = nt_0$. To illustrate this trade-off, Fig. 10.3 shows selected combinations of the number of test devices n and required test time t_0 satisfying $nt_0 = 5,430$ hours. For example, if $n = 5$, then test each of these pumps (with repair or replacement) for $t_0 = 1,000$ hours each.

10.4 Weibull Testing

The previous sections on binomial and Poisson testing considered attribute test data, which capture the survival/nonsurvival of each device on test. This section focuses on lifetime data, where testers record the actual failure times. We assume that the failure times t follow a $Weibull(\lambda, \beta)$ distribution with scale parameter λ and shape parameter β, with probability density function

$$f(t|\lambda, \beta) = \lambda\beta t^{\beta-1} \exp(-\lambda t^{\beta}), \;\; t > 0, \, \lambda > 0, \, \beta > 0. \tag{10.20}$$

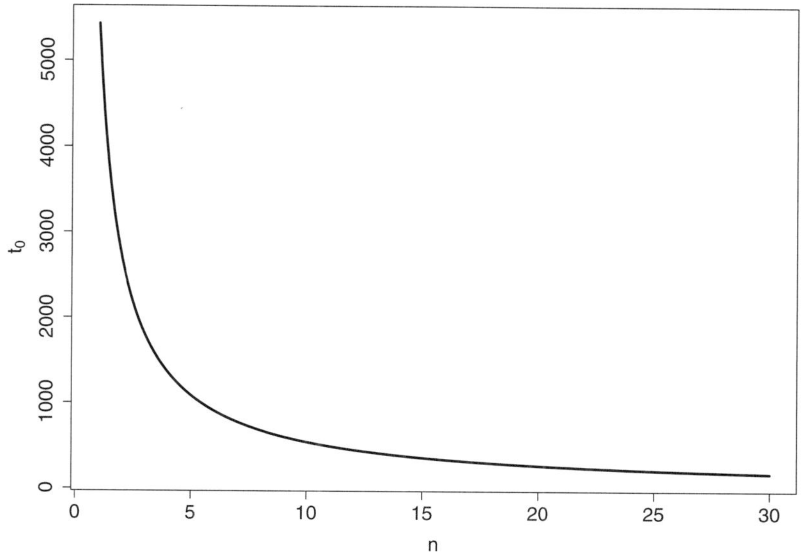

Fig. 10.3. The required test time t_0 in hours versus the number of test devices n for a Poisson test plan.

Suppose that we would like to use the posterior risk criteria to develop a Weibull test plan (n, t_0, c), where we put n units on test for t_0 time units and the test passes if no more than c units fail. To define the risk criteria, we specify requirements on reliability at time t_*, $R(t_*)$. Let

$$\begin{aligned} m(y) &= (1 - \exp[-\lambda t_0^\beta])^y \exp[-(n-y)\lambda t_0^\beta]\,, \\ k_0 &= -\log(\pi_0) t_*^{-\beta} \quad , \text{ and} \\ k_1 &= -\log(\pi_1) t_*^{-\beta}\,. \end{aligned}$$

For a Weibull test plan (n, t_0, c), we calculate the posterior producer's risk as

$$\begin{aligned} &\mathbf{P}(R(t_*) \geq \pi_0 \,|\, Test\ Is\ Failed) \qquad (10.21) \\ &\quad = \mathbf{P}(\exp(-\lambda t_*^\beta) \geq \pi_0 \,|\, Test\ Is\ Failed) \\ &\quad = \mathbf{P}(\lambda \leq -\log(\pi_0) t_*^{-\beta} \,|\, Test\ Is\ Failed) \\ &\quad = \int_0^\infty \int_0^{k_0} f(\lambda, \beta \,|\, Test\ Is\ Failed)\, \mathrm{d}\lambda \mathrm{d}\beta \\ &\quad = \int_0^\infty \int_0^{k_0} \frac{\mathbf{P}(Test\ Is\ Failed \,|\, \lambda, \beta) p(\lambda, \beta)}{\int_0^\infty \int_0^\infty \mathbf{P}(Test\ Is\ Failed \,|\, \lambda, \beta) p(\lambda, \beta)\, \mathrm{d}\lambda \mathrm{d}\beta}\, \mathrm{d}\lambda \mathrm{d}\beta \end{aligned}$$

$$= \frac{\int_0^\infty \int_0^{k_0} \left[1 - \sum_{y=0}^{c} m(y)\right] p(\lambda, \beta) \; \mathrm{d}\lambda \mathrm{d}\beta}{\int_0^\infty \int_0^\infty \left[1 - \sum_{y=0}^{c} m(y)\right] p(\lambda, \beta) \; \mathrm{d}\lambda \mathrm{d}\beta} .$$

We calculate the posterior consumer's risk as

$$\begin{aligned} &\mathbf{P}(R(t_*) \leq \pi_1 \,|\, Test\ Is\ Passed) \qquad (10.22)\\ &= \mathbf{P}(\exp(-\lambda t_*^\beta) \leq \pi_1 \,|\, Test\ Is\ Passed)\\ &= \mathbf{P}(\lambda \geq -\log(\pi_1) t_*^{-\beta} \,|\, Test\ Is\ Passed)\\ &= \int_0^\infty \int_{k_1}^\infty f(\lambda, \beta \,|\, Test\ Is\ Passed) \; \mathrm{d}\lambda \mathrm{d}\beta\\ &= \int_0^\infty \int_{k_1}^\infty \frac{\mathbf{P}(Test\ Is\ Passed \,|\, \lambda, \beta) p(\lambda, \beta)}{\int_0^\infty \int_0^\infty \mathbf{P}(Test\ Is\ Passed \,|\, \lambda, \beta) p(\lambda, \beta) \; \mathrm{d}\lambda \mathrm{d}\beta} \mathrm{d}\lambda \mathrm{d}\beta\\ &= \frac{\int_0^\infty \int_{k_1}^\infty \left[\sum_{y=0}^{c} m(y))\right] p(\lambda, \beta) \; \mathrm{d}\lambda \mathrm{d}\beta}{\int_0^\infty \int_0^\infty \left[\sum_{y=0}^{c} m(y)\right] p(\lambda, \beta) \; \mathrm{d}\lambda \mathrm{d}\beta} . \end{aligned}$$

10.4.1 Single Weibull Population Testing

One description of a failure time distribution is its reliable life. The *reliable life*, t_R, for specified R, is the time beyond which $100 \times R\%$ of the population will survive. In other words, t_R is the $(1 - R)$th quantile of the failure time distribution. For the Weibull distribution described by Eq. 10.20, $t_R = \lambda^{-1/\beta}[-\log(R)]^{1/\beta}$.

Among a variety of testing schemes, we focus here on a minimum sample size (or zero-failure) test plan $(n, t_0, c = 0)$. For a zero-failure test plan, test n devices each for a length of time t_0, and the test is passed if we observe no failures. To use such a test plan, we must determine appropriate values for both n and t_0. Meeker and Escobar (1998) considers such classical test plans in situations where the Weibull shape parameter β is known. We relax this restriction here by considering plans within a Bayesian hierarchical framework. In turn, there are two such cases to study: (1) an assurance test plan based on available data from a single Weibull population, and (2) an assurance test plan based on available data from a Weibull accelerated life test program.

Consider developing a test criterion to assure that $t_R > t_{R^*}$. For example, a manufacturer may want to assure that 99% of a certain expensive electronic product will survive a one-year warranty period; in this case, $R = 0.99$ and $t_{R^*} = 8,760$ hours. We use the posterior risk criterion $P[t_R > t_{R^*} \,|\, Test\ Is\ Passed] \geq 1 - \alpha$. If the test is passed, we would like a high probability $(1 - \alpha)$ that $t_R > t_{R^*}$ — a high probability that the 0.99 quantile of the lifetime of the electronic products is greater than 8,760 hours. This leads to the expression

$$\mathbf{P}(t_R > t_{R^*} \,|\, Test\ Is\ Passed) \qquad (10.23)$$

$$
\begin{aligned}
&= \mathbf{P}(\lambda^{-1/\beta}[-\log(R)]^{1/\beta} > t_{R^*} \,|\, \textit{Test Is Passed}) \\
&= \mathbf{P}(\lambda < -\log(R)t_{R^*}^{-\beta} \,|\, \textit{Test Is Passed}) \\
&= \int_0^\infty \int_0^{-\log(R)t_{R^*}^{-\beta}} f(\lambda, \beta \,|\, \textit{Test Is Passed})\, \mathrm{d}\lambda \mathrm{d}\beta \\
&= \int_0^\infty \int_0^{-\log(R)t_{R^*}^{-\beta}} \frac{\mathbf{P}(\textit{Test Is Passed} \,|\, \lambda, \beta) p(\lambda, \beta)}{\int_0^\infty \int_0^\infty \mathbf{P}(\textit{Test Is Passed} \,|\, \lambda, \beta) p(\lambda, \beta) \mathrm{d}\lambda \mathrm{d}\beta} \mathrm{d}\lambda \mathrm{d}\beta \\
&= \frac{\int_0^\infty \int_0^{-\log(R)t_{R^*}^{-\beta}} \exp(-n\lambda t_0^\beta) p(\lambda, \beta) \mathrm{d}\lambda \mathrm{d}\beta}{\int_0^\infty \int_0^\infty \exp(-n\lambda t_0^\beta) p(\lambda, \beta) \mathrm{d}\lambda \mathrm{d}\beta} \geq 1 - \alpha \,.
\end{aligned}
$$

Notice the similarities of this risk formulation to the posterior consumer's risk. As with the Poisson test plan, for a fixed n, we can solve for t_0 to meet the desired risk criteria. With the chosen $c = 0$, this also specifies the level of posterior producer's risk.

Suppose now that we have failure time data from $m > 1$ situations. Note that some of the available failure time data may be censored. Let $t_{ij}, i = 1, \ldots, m, j = 1, \ldots, n_i$, denote the observed failure or censoring time for the jth device in the ith situation, and let $\mathbf{t}$ denote all the observed failure time data. Given λ_i and β, model the T_{ij} as conditionally independent with $T_{ij} \sim \textit{Weibull}(\lambda_i, \beta)$. We assume a common shape parameter, because in practice, the failure times of similar devices often (but not always) exhibit the same general Weibull shape because they share common intrinsic failure mechanisms. To complete the model, let us use the following prior distributions:

$$
\begin{aligned}
\lambda_i &\sim \textit{Gamma}(\eta, \kappa), \quad i = 1, \ldots, m, \\
(\eta, \kappa) &\sim p(\eta, \kappa), \text{ and} \\
\beta &\sim p(\beta),
\end{aligned}
$$

with known hyperparameters for $p(\eta, \kappa)$ and $p(\beta)$.

We want to develop a zero-failure test plan for a new situation where we assume the failure time data will be distributed $\textit{Weibull}(\lambda, \beta)$, with $\lambda \sim \Gamma(\eta, \kappa)$. Conditioning on the observed data $\mathbf{t}$ in Eq. 10.23,

$$
\begin{aligned}
&\mathbf{P}(t_R > t_{R^*} \,|\, \textit{Test Is Passed}, \mathbf{t}) \qquad (10.24) \\
&\quad = \frac{\int_0^\infty \int_0^{-\log(R)t_{R^*}^{-\beta}} \exp(-n\lambda t_0^\beta) p(\lambda, \beta \,|\, \mathbf{t}) \mathrm{d}\lambda \mathrm{d}\beta}{\int_0^\infty \int_0^\infty \exp(-n\lambda t_0^\beta) p(\lambda, \beta \,|\, \mathbf{t}) \mathrm{d}\lambda \mathrm{d}\beta} \,.
\end{aligned}
$$

Assuming that we have $j = 1, \ldots, N$ MCMC draws from the posterior distributions (given $\mathbf{t}$) of η, κ, and β and N draws from the predictive distribution of λ, $\lambda^{(j)} \sim \Gamma(\eta^{(j)}, \kappa^{(j)})$, we can calculate our criterion as follows:

$$
\mathbf{P}(t_R > t_{R^*} \,|\, \textit{Test Is Passed}, \mathbf{t}) \qquad (10.25)
$$

$$= \frac{\int_0^\infty \int_0^{-\log(R)t_{R^*}^{-\beta}} \exp(-n\lambda t_0^\beta)p(\lambda, \beta \mid \mathbf{t})\mathrm{d}\lambda\mathrm{d}\beta}{\int_0^\infty \int_0^\infty \exp(-n\lambda t_0^\beta)p(\lambda, \beta \mid \mathbf{t})\mathrm{d}\lambda\mathrm{d}\beta}$$

$$\approx \frac{\sum_{j=1}^N \exp(-n\lambda^{(j)} t_0^{\beta^{(j)}}) I[\lambda^{(j)} \le -\log(R) t_{R^*}^{-\beta^{(j)}}]}{\sum_{j=1}^N \exp(-n\lambda^{(j)} t_0^{\beta^{(j)}})}$$

$$\approx \frac{\sum_{j=1}^N \frac{(\kappa^{(j)})^{\eta^{(j)}} \gamma[\eta^{(j)}, [-\log(R)](\kappa^{(j)}+nt_0^{\beta^{(j)}})/t_{R*}^{\beta^{(j)}}]}{(\kappa^{(j)}+nt_0^{\beta^{(j)}})^{\eta^{(j)}} \Gamma(\eta^{(j)})}}{\sum_{j=1}^N \frac{(\kappa^{(j)})^{\eta^{(j)}}}{(\kappa^{(j)}+nt_0^{\beta^{(j)}})^{\eta^{(j)}}}},$$

where $\gamma(q, z)$ denotes the lower incomplete gamma function.

Note that we may base the choice of number of test devices n on other considerations, such as cost. It may also be interesting and useful to see how t_0 functionally depends on n, which we can examine by varying n over an appropriate range, solving for t_0, and plotting the results.

Example 10.5 Hierarchical Weibull test plan for pressure vessels. Gerstle and Kunz (1983) provides the failure times (in hours) for pressure vessels that were wrapped in Kevlar-49 fibers and subsequently tested at four different stresses: 23.4, 25.5, 27.6, and 29.7 megapascals (MPa). Crowder et al. (1991) analyzes these data assuming a constant Weibull shape parameter. The fibers came from eight different spools (numbered 1–8) of material, and both studies conclude that there is a significant spool effect. In this example, consider only the failure time data obtained at 23.4 MPa. See Table 10.5, which displays the failure time data at all the stresses; an asterisk indicates a time- or Type I-censored observation.

Because any difference in the reliability of the spools is primarily due to uncontrollable random manufacturing process variability (or noise), let us model the pressure vessel failure times corresponding to each spool as conditionally independent with $Weibull(\lambda_i, \beta)$ distributions. Given η and κ, the λ_i are i.i.d. $Gamma(\eta, \kappa)$. In our analysis of these data, we use independent and diffuse $InverseGamma(0.01, 0.01)$ prior distributions for η and κ. Also, we use an independent $Exponential(1.0)$ prior distribution for β; the motivation for this prior distribution is the analysis results of Crowder et al. (1991), which suggests values of β near 1.0. See Table 10.6, which summarizes the marginal posterior distributions for η, κ, and β given $\mathbf{t}$. The hierarchical Weibull model fits these data well (see Exercise 4.20). Table 10.6 also summarizes the predictive distribution of λ.

Now suppose that we want to find a Bayesian minimum sample size test plan at a stress of 23.4 MPa for $t_{R^*} = 2,000$ hours, $R = 0.9$, and $\alpha = 0.05$. By letting $n = 1, 2, \ldots, 30$ and solving Eq. 10.25 for the corresponding test length t_0, we obtain the graph shown in Fig. 10.4.

For example, suppose that we decide to test $n = 10$ pressure vessels all wrapped from a particular spool of Kevlar-49 fibers. Figure 10.4 indicates

Table 10.5. Failure times of Kevlar-49-wrapped pressure vessels at four stress levels (An asterisk indicates a time-censored or Type I-censored observation) (Gerstle and Kunz, 1983)

Stress (MPa)	Spool	Failure Time (hours)
29.7	1	444.4 755.2 952.2 1108.2
29.7	2	2.2 8.5 9.1 10.2 22.1 55.4 111.4 158.7
29.7	3	12.5 14.6 18.7 101.0
29.7	4	254.1 1148.5 1569.3 1750.6 1802.1
29.7	5	8.3 13.3 87.5 243.9
29.7	6	6.7 15.0 144.0
29.7	7	4.0 4.0 4.6 6.1 7.9 14.0 45.9 61.2
29.7	8	98.2 590.4 638.2
27.6	1	453.4 664.5 930.4 1755.5
27.6	2	71.2 199.1 403.7 432.2 514.1 544.9 694.1
27.6	3	19.1 24.3 69.8 136.0
27.6	4	876.7 1275.6 1536.8 6177.5
27.6	5	
27.6	6	514.2 541.6 1254.9
27.6	7	
27.6	8	554.2 2046.2
25.5	1	11487.3 14032.0 31008.0
25.5	2	1134.2 1824.3 1920.1 2383.0 3708.9 5556.0
25.5	3	1087.7 2442.5
25.5	4	13501.3 29808.0
25.5	5	11727.1
25.5	6	225.2 6271.1 7996.0
25.5	7	503.6
25.5	8	2974.6 4908.9 7332.0 7918.7 9240.3 9973.0
23.4	1	41000* 41000* 41000* 41000*
23.4	2	14400.0
23.4	3	8616.0
23.4	4	41000* 41000* 41000* 41000*
23.4	5	9120.0 20231.0 35880.0
23.4	6	7320.0 16104.0 20233.0
23.4	7	4000.0 5376.0
23.4	8	41000* 41000* 41000*

Table 10.6. Posterior distribution summaries for η, κ, and β given $\mathbf{t}$ and of the predictive distribution for λ

			Quantiles		
Parameter	Mean	Std Dev	0.025	0.50	0.975
β	2.255	0.643	1.128	2.211	3.773
η	0.2100	0.1730	0.0529	0.1666	0.6113
κ	2.313E+13	3.105E+14	7.346E+3	1.121E+8	6.463E+13
λ	5.630E−6	1.221E−4	7.304E−25	3.832E−11	1.273E−5

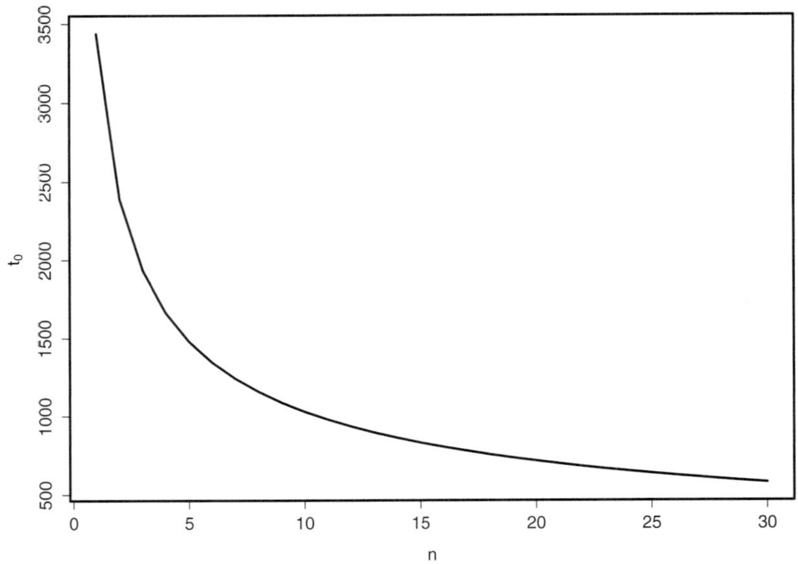

Fig. 10.4. The required Weibull test duration t_0 versus the number of test devices n for the pressure vessels example.

required testing of each of these pressure vessels for approximately $t_0 = 1,122$ hours at a stress of 23.4 MPa. If none of these fail, we can then claim, with 0.95 probability, that at least 90% of the pressure vessels wrapped from this spool will survive 2,000 hours at this stress.

10.4.2 Combined Weibull Accelerated/Assurance Testing

Now consider a Bayesian test plan based on data from a Weibull accelerated test program. Specifically, we consider again the failure time data in Table 10.5. Let $t_{ij}, i = 1, \ldots, m, j = 1, \ldots, n_i$, denote either the observed failure or censoring time for the jth device from the ith situation, with $\mathbf{t}$ denoting all of the observed failure time data. Let s_{ij} be the value of the stress s under which we obtained t_{ij}.

Conditional on λ_{ij} and β, we model the T_{ij} conditionally independent with a $Weibull(\lambda_{ij}, \beta)$ distribution. Let us define a model for λ_{ij} as

$$\lambda_{ij} = \exp(\gamma_0) s_{ij}^{\gamma_1} \omega_i \,, \tag{10.26}$$

where $\omega_i > 0$ is the random effect associated with the ith spool, and γ_0 and γ_1 are two regression parameters used to model the relationship between λ_{ij}

and s_{ij}. Recall that we presented similar regression models in Chap. 7 (see also León et al. (2007)).

We specify prior distributions $\omega_i \mid \eta, \kappa \sim Gamma(\eta, \kappa)$, $(\eta, \kappa) \sim p(\eta, \kappa)$, $\beta \sim p(\beta)$, $\gamma_0 \sim p(\gamma_0)$, $\gamma_1 \sim p(\gamma_1)$ with known hyperparameters for $p(\eta, \kappa)$, $p(\beta)$, $p(\gamma_0)$, and $p(\gamma_1)$.

To develop the Bayesian zero-failure test for a new spool, we assume the plan consists of testing n samples for time t_0 at stress s_0. We use MCMC to obtain posterior draws $\eta^{(j)}, \kappa^{(j)}, \beta^{(j)}, \gamma_0^{(j)}, \gamma_1^{(j)}$ given $\mathbf{t}$ and predictive draws $\omega^{(j)} \sim \Gamma(\eta^{(j)}, \kappa^{(j)})$. Let

$$\begin{aligned}
\boldsymbol{\theta} &= (\omega, \gamma_0, \gamma_1, \beta) \quad \text{and} \\
q^{(j)} &= \kappa^{(j)} + n \exp(\gamma_0^{(j)}) s_0^{\gamma_1^{(j)}} t_0^{\beta^{(j)}} .
\end{aligned}$$

Our test criterion is calculated as

$$\begin{aligned}
&\mathbf{P}(t_R > t_{R^*} \mid Test\ Is\ Passed, \mathbf{t}) \qquad (10.27) \\
&= \frac{\int_0^\infty \int_0^\infty \int_0^\infty \int_0^{\frac{-\log(R) t_{R^*}^{-\beta}}{\exp(\gamma_0) s_0^{\gamma_1}}} \exp(-n \exp(\gamma_0) s_0^{\gamma_1} \omega t_0^{\beta}) p(\boldsymbol{\theta} \mid \mathbf{t}) d\boldsymbol{\theta}}{\int_0^\infty \int_0^\infty \int_0^\infty \int_0^\infty \exp(-n \exp(\gamma_0) s_0^{\gamma_1} \omega t_0^{\beta}) p(\boldsymbol{\theta} \mid \mathbf{t}) d\boldsymbol{\theta}} \\
&\approx \frac{\sum_{j=1}^N \exp(-n \exp(\gamma_0^{(j)}) s_0^{\gamma_1^{(j)}} \omega^{(j)} t_0^{\beta^{(j)}}) I \left[\omega^{(j)} \leq \frac{-\log(R) t_{R^*}^{-\beta^{(j)}}}{\exp(\gamma_0^{(j)}) s_0^{\gamma_1^{(j)}}} \right]}{\sum_{j=1}^N \exp(-n \exp(\gamma_0^{(j)}) s_0^{\gamma_1^{(j)}} \omega^{(j)} t_0^{\beta^{(j)}})} \\
&\approx \frac{\sum_{j=1}^N \frac{(\kappa^{(j)})^{\eta^{(j)}} \gamma \left[\eta^{(j)}, [-\log(R)] \exp(-\gamma_0^{(j)}) s_0^{-\gamma_1^{(j)}} q^{(j)} / t_{R^*}^{\beta^{(j)}} \right]}{\Gamma(\eta^{(j)}) (q^{(j)})^{\eta^{(j)}}}}{\sum_{j=1}^N \frac{(\kappa^{(j)})^{\eta^{(j)}}}{(q^{(j)})^{\eta^{(j)}}}},
\end{aligned}$$

where $\gamma(q, z)$ denotes the lower incomplete gamma function.

More specifically, to analyze the data in Table 10.5, we use the following independent prior distributions:

$$\begin{aligned}
\beta &\sim Exponential(1.0), \\
\eta &\sim InverseGamma(0.01, 0.01), \\
\kappa &\sim InverseGamma(0.01, 0.01), \\
\gamma_0 &\sim Normal(0, 10^6), \text{ and} \\
\gamma_1 &\sim Normal(0, 10^6).
\end{aligned}$$

See Table 10.7, which summarizes the marginal posterior distributions for these five parameters. The hierarchical Weibull regression model fits these data well (see Exercise 7.23).

Table 10.7. Posterior distribution summaries for $\eta, \kappa, \beta, \gamma_0$, and γ_1 given $\mathbf{t}$ and predictive distribution for ω for pressure vessels example

Parameter	Mean	Std Dev	Quantiles 0.025	0.50	0.975
β	1.199	0.085	1.038	1.199	1.3650
η	0.6522	0.3068	0.2198	0.5977	1.391
γ_0	−84.81	11.31	−105.50	−83.40	−66.32
γ_1	27.84	1.81	24.56	27.85	31.31
κ	3.704E+15	5.666E+16	2.403E−2	4.873E+6	7.071E+15
ω	2.512	19.42	1.34E-17	3.900E-08	19.73

We can see the effect that stress has on failure time by computing the posterior distribution of the Weibull mean time to failure (MTTF) as a function of s for a randomly selected spool of Kevlar 49. Recall that the MTTF of the Weibull distribution in Eq. 10.26 is $\lambda^{-1/\beta}\Gamma(1+1/\beta)$, which, substituting using Eq. 10.26, becomes $\exp(-\gamma_0/\beta)s^{-\gamma_1/\beta}\omega^{-1/\beta}\Gamma(1+1/\beta)$.

In Fig. 10.5, we plot the median log MTTF and a 90% central credible interval as a function of stress s. Note the decreasing trend in the Weibull log MTTF as s increases, as well as the extremely heavy right tail of the posterior log MTTF distribution for a given value of s.

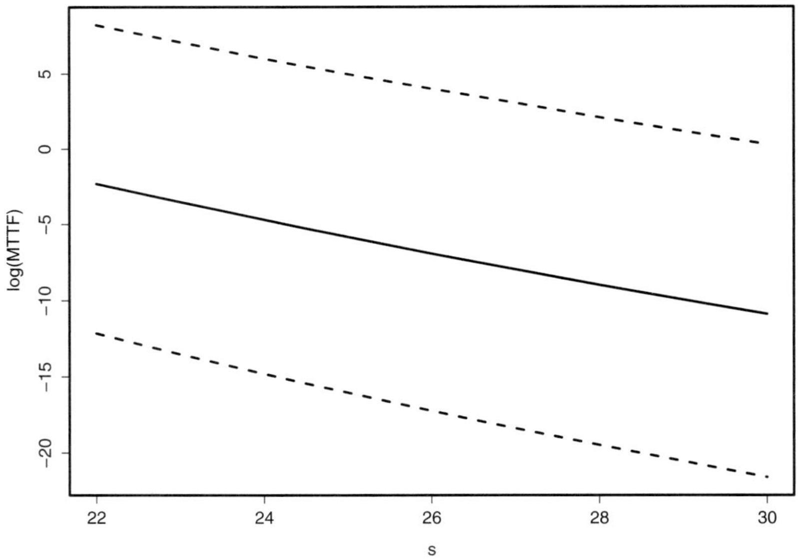

Fig. 10.5. The posterior median and 90% credible interval of the Weibull log MTTF as a function of stress s for pressure vessels example.

We can also see the effect of stress on pressure vessel reliability by computing the posterior distribution of reliability as a function of s for a randomly selected Kevlar-49 spool. Substituting using Eq. 10.26, the reliability at $t = 2,000$ hours is $\exp[-\exp(\gamma_0)s^{\gamma_1}\omega(2,000)^{\beta}]$. In Fig. 10.6, we plot the posterior median and 90% credible interval for pressure vessel reliability at $2,000$ hours as a function of stress s. Note the significant increase in the width of the 90% credible intervals as the stress s increases.

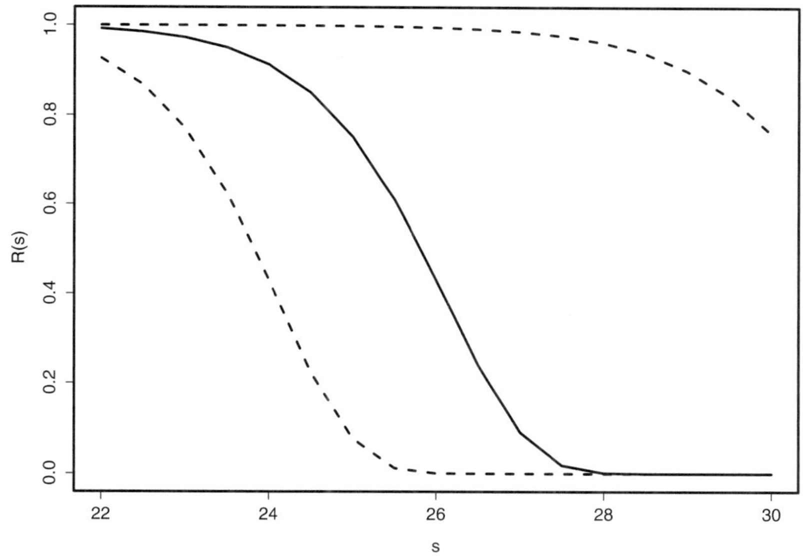

Fig. 10.6. The posterior median and 90% credible interval for $R(2000)$ as a function of stress s for pressure vessels example.

Now suppose that we want to find a Bayesian zero-failure test plan for $t_{R^*} = 2,000$ hours, $R = 0.9, \alpha = 0.05$, and $s_0 = 23$ MPa. By letting $n = 1, 2, \ldots, 15$ and solving Eq. 10.27 for the corresponding test length t_0, we obtain the solid curve in Fig. 10.7. For example, in testing $n = 10$ pressure vessels all wrapped from a spool of Kevlar-49 fiber, Fig. 10.7 indicates testing each of these vessels for approximately $t_0 = 1,900$ hours. Note that in Example 10.5, a shorter time of $t_0 = 1,132$ was required because the reliability at 2,000 hours was assessed to be higher based on the only 23 MPa data. If there are no failures, then we can claim with 0.95 probability that at least 90% of the pressure vessels wrapped from this spool will survive 2,000 hours at a stress of 23 MPa. Figure 10.7 also presents the test length t_0 as a function of n for stresses $s_0 = 24$ MPa and $s_0 = 27$ MPa.

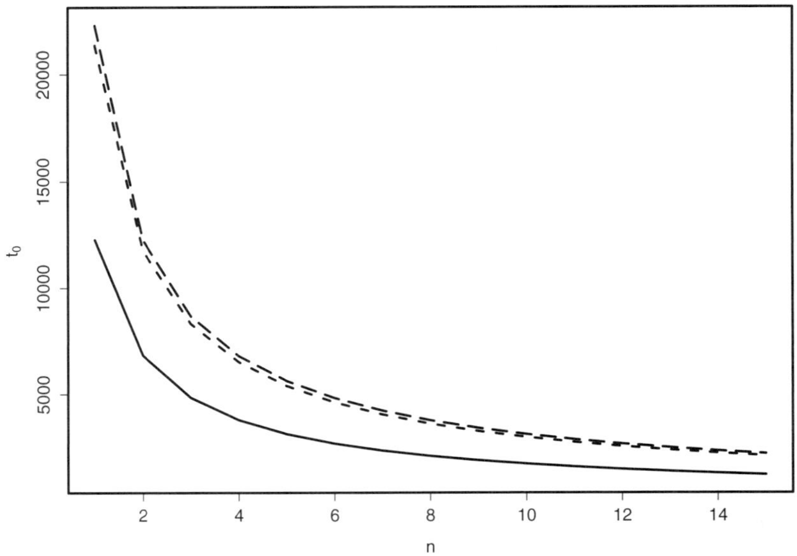

Fig. 10.7. The required test duration t_0 versus the number of test devices n for three different values of stress s_0 in MPa for pressure vessels example. The three stress values are 23 MPa (solid line), 24 MPa (short dashed line), and 27 MPa (long dashed line).

From Fig. 10.7, we can make an important observation; namely, for a given number of devices n, as the stress s_0 decreases, the required test time t_0 also decreases. This desirable situation arises because the failure time data model using the original accelerated life test data predicts that the probability of surviving the required 2,000 hours of operation increases dramatically as the stress decreases. In other words, because the fitted model predicts long failure times at low stress, achieving the required assurance needs very little additional data. For those cases, in which there is high reliability at low stress, the required assurance is already embedded in the fitted model. On the other hand, if the accelerated life test results indicate low reliability at low stress, then significant assurance testing would be required to overcome this situation.

10.5 Related Reading

There is extensive literature on Bayesian assurance testing dating back to the late 1960s, and Martz and Waller (1982) describes this early research. Brush (1986) clarifies the distinction between a posterior Bayes' and a modified classical (or average) producer's risk and compares these two criteria for various test plans. Brush (1986) also highlights the importance of calculating a Bayes' producer's risk as a supplement to the modified classical producer's risk. Sharma and Bhutani (1992) analyzes the performance of Bayes' and classical assurance test plans for simultaneously specified consumer's and producer's risks. Both Brush (1986) and Sharma and Bhutani (1992) conclude that Bayes' and classical risks need consideration.

Since 1990 there have been several more articles concerned with either Bayesian acceptance sampling or assurance test plans. Hart (1990), with comments by Ganter et al. (1990), uses Bayesian methods to determine a test plan for qualifying the reliability of some industrial products, whereas Guess and Usher (1990) considers a Bayesian approach to the assurance testing of highly reliable devices. Fan (1991) proposes a Bayesian acceptance test plan for binomial testing, while Sheng and Fan (1992) presents a method for choosing a prior distribution for binomial testing. In a Master's thesis, Jin (1991) develops Bayesian acceptance test plans based on failure-free life tests and compares these with other test plans. Pham and Turkkan (1992) considers a four-parameter beta prior in binomial testing. In a somewhat different approach, Moskowitz and Tang (1992) uses both quadratic and step-function loss functions in determining Bayesian acceptance test plans. Berger and Sun (1993) develops Bayesian sequential reliability demonstration tests using two different approaches: posterior loss and predictive loss. In addition, Berger and Sun (1993) considers three test data models. Whitmore et al. (1994) proposes two different approaches for integrating life test data into a Bayesian analysis based on the exponential distribution. For the exponential distribution, Deely and Keats (1994) develops Bayesian stopping rules for use in terminating a sequential assurance test. Vintr (1999) considers the optimization of reliability requirements from a manufacturer's point of view. Tobias and Poore (2003) proposes Bayesian reliability testing for a new generation of semiconductor processing equipment, while Kleyner et al. (2004) develops reliability demonstration test plans based on minimizing life cycle costs.

10.6 Exercises for Chapter 10

10.1 In determining a Bayesian reliability assurance test plan, what happens if the prior distribution is especially strong and satisfies the desired criteria prior to testing?
   a) How does one know when this is the case?
   b) What calculations should one perform to check whether or not this is the case?

10.2 a) What is the (classical) producer's risk for a binomial test plan with $n = 15, c = 1$, and $\pi_0 = 0.9$?
   b) What is the (classical) consumer's risk for a binomial test plan with $n = 15, c = 1$, and $\pi_1 = 0.6$?

10.3 a) What is the average producer's risk for a binomial test plan with $n = 15, c = 1, \pi_0 = 0.9$, and $p(\pi) \sim Beta(10, 1)$?
b) What is the average consumer's risk for a binomial test plan with $n = 15, c = 1, \pi_1 = 0.6$, and $p(\pi) \sim Beta(10, 1)$?

10.4 a) What is the posterior producer's risk for a binomial test plan with $n = 15, c = 1, \pi_0 = 0.9$, and $p(\pi) \sim Beta(10, 1)$?
b) What is the posterior consumer's risk for a binomial test plan with $n = 15, c = 1, \pi_1 = 0.6$, and $p(\pi) \sim Beta(10, 1)$?

10.5 Discuss the similarities and differences between the producer's and consumer's risks calculated in Exercises 10.2, 10.3, and 10.4.

10.6 Calculate a binomial test plan with a $Uniform(0, 1)$ prior distribution for π, and $\pi_0 = 0.9, \pi_1 = 0.5, \alpha = \beta = 0.05$.

10.7 The auxiliary feedwater (AFW) system is an important standby safety system in a nuclear power plant (Poloski et al., 1998). The AFW system probability of starting on demand is an important indicator of its reliability. The data in Table 10.8 are the number of AFW system failures to start on demand x_i in n_i demands at 68 U.S. commercial nuclear power plants.
a) Find the Bayesian hierarchical test plan having the posterior consumer's and producer's risk values $\pi_1 = 0.985, \beta = 0.05, \pi_0 = 0.995$, and $\alpha = 0.10$.
b) What are the actual posterior risks when using this test plan?
c) What is the unconditional probability of passing the test when using this test plan?
d) Is there anything unusual about this problem?

10.8 Derive Eq. 10.15.

10.9 Borg (1962) provides data that apparently originated at the U.S. Bureau of Naval Weapons regarding the number of minor, major, and critical defectives in successive MIL-STD-105B samples of some material. The data in Table 10.9 consist of the observed frequencies of the number of minor defectives x in samples of size $n = 150$ from $m = 205$ lots each of size 2016 items of this material.
a) Using the "hybrid" posterior consumer's and average producer's risk criteria, find the Bayesian hierarchical test plan having the risk values $\beta = 0.10$ and $\alpha = 0.05$ for $\pi^* = 0.975$.
b) What are the actual risks when using this test plan?
c) What is the unconditional probability of passing this test?

10.10 Using the data in Exercise 10.9, find the Bayesian hierarchical test plan having the posterior consumer's and producer's risk values $\pi_1 = 0.96, \beta = 0.10, \pi_0 = 0.975$, and $\alpha = 0.05$.
a) What are the actual risks when using this test plan?
b) What is the unconditional probability of passing this test?

10.11 a) What is the posterior producer's risk for a Poisson test plan with $T = 10, c = 3, \lambda_0 = 3.0$, and $p(\lambda) \sim Gamma(5, 1)$?

Table 10.8. Number of AFW system failures to start on demand x in n demands at 68 U.S. commercial nuclear power plants (Poloski et al., 1998)

Plant	x	n	Plant	x	n
Arkansas 1	0	14	North Anna 2	0	18
Arkansas 2	0	9	Oconee 1	0	18
Beaver Valley 1	0	24	Oconee 2	0	18
Beaver Valley 2	0	43	Oconee 3	0	12
Braidwood 1	0	13	Palisades	0	13
Braidwood 2	0	24	Palo Verde 1	0	7
Byron 1	0	11	Palo Verde 2	0	12
Byron 2	0	26	Palo Verde 3	0	9
Callaway	0	57	Point Beach 1	0	8
Calvert Cliffs 1	0	12	Point Beach 2	0	16
Calvert Cliffs 2	0	15	Prairie Island 1	0	3
Catawba 1	0	41	Prairie Island 2	0	7
Catawba 2	0	89	Robinson 2	1	28
Comanche Pk 1	0	66	Salem 1	0	24
Comanche Pk 2	0	14	Salem 2	0	32
Cook 1	0	18	San Onofre 2	0	13
Cook 2	0	36	San Onofre 3	0	17
Crystal River 3	1	16	Seabrook	0	17
Diablo Canyon 1	0	46	Sequoyah 1	0	30
Diablo Canyon 2	0	30	Sequoyah 2	0	41
Farley 1	0	34	South Texas 1	0	69
Farley 2	0	54	South Texas 2	0	87
Fort Calhoun	0	5	St. Lucie 1	0	35
Ginna	0	28	St. Lucie 2	0	21
Harris	0	98	Summer	0	24
Indian Point 2	1	24	Surry 1	0	26
Indian Point 3	2	32	Surry 2	0	32
Kewaunee	0	26	Three Mile Isl 1	0	6
Maine Yankee	0	23	Vogtle 1	0	103
McGuire 1	0	45	Vogtle 2	0	45
McGuire 2	0	44	Waterford 3	0	38
Millstone 2	1	11	Wolf Creek	0	51
Millstone 3	0	54	Zion 1	0	13
North Anna 1	0	20	Zion 2	0	8

b) What is the posterior consumer's risk for a Poisson test plan with $T = 10, c = 3.0, \lambda_1 = 7.0$, and $p(\lambda) \sim Gamma(5, 1)$?

10.12 For Poisson testing presented in Sect. 10.3, show that

$$\mathbf{P}(Test\ Is\ Failed|\lambda \leq \lambda_0, \eta, \kappa) = 1 - \frac{\kappa^\eta \sum_{x=0}^{c} \frac{T_0^x \gamma[x+\eta,(T_0+\kappa)\lambda_0]}{x!(T_0+\kappa)^{x+\eta}}}{\gamma(\eta, \kappa\lambda_0)}.$$

10.13 Using the expression given in Exercise 10.12 and the pump failure data in Table 10.3, find the Bayesian hierarchical test plan having the "hybrid"

Table 10.9. Minor defectives from MIL-STD-105B sampling of material (Borg, 1962)

x	Frequency	x	Frequency
0	68	9	2
1	45	10	1
2	24	12	1
3	20	13	1
4	8	15	4
5	7	18	1
6	8	20	1
7	10	22	1
8	3		

posterior consumer's and average producer's risk values $\lambda_1 = 0.3, \beta = 0.10, \lambda_0 = 0.2$, and $\alpha = 0.05$.
a) What are the actual risks for this test plan?
b) What is the unconditional probability of passing this test?

10.14 Using the pump failure data in Table 10.3, find the Bayesian hierarchical test plan having the posterior consumer's and producer's risk values $\lambda_1 = 0.1, \beta = 0.10, \lambda_0 = 0.05$, and $\alpha = 0.05$.
a) What are the actual risks for this test plan?
b) What is the unconditional probability of passing this test?
c) Is this a good test plan to use?

10.15 For the model in Sect. 10.4.1, show that we may approximate the unconditional probability of passing the Bayesian minimum sample size test plan by

$$\mathbf{P}[Test\ Is\ Passed \,|\, \mathbf{t}] \approx \frac{1}{N} \sum_{j=1}^{N} \left(\frac{\kappa^{(j)}}{\kappa^{(j)} + n t_0^{\beta^{(j)}}} \right)^{\eta^{(j)}} .$$

10.16 Using the expression in Exercise 10.15, what is the approximate unconditional probability of passing the Bayesian test plan ($n = 10, t_0 = 700$) given in Example 10.5?

10.17 Gerstle and Kunz (1983) gives the failure times for Kevlar-49-wrapped pressure vessels at a stress of 25.5 MPa. Table 10.10 displays these data. For $t_{R^*} = 300$ hours, $R = 0.9, \alpha = 0.05$, and these Weibull distributed data, find the Bayesian minimum sample size test plan time t_0 that we must test each of $n = 5$ items. What is the unconditional probability of passing this test? Is this a satisfactory test plan?

10.18 For the Weibull testing described in Sect. 10.4, suppose that we want to find a Bayesian minimum sample size test plan to assure that, at some specified time t_R, the Weibull reliability R is at least as large as a requirement R^* at the $100 \times (1 - \alpha)\%$ credible level. How does this test plan compare to the one based on the reliable life criterion?

Table 10.10. Failure times of Kevlar-49-wrapped pressure vessels at a stress of 25.5 MPa (Gerstle and Kunz, 1983)

Spool	Failure Time (hours)
1	11487.3, 14032.0, 31008.0
2	1134.3, 1824.3, 1920.1, 2383.0, 3708.9, 5556.0
3	1087.7, 2442.5
4	13501.3, 29808.0
5	11727.1
6	225.2, 6271.1, 7996.0
7	503.6
8	2974.6, 4908.9, 7332.0, 7918.7, 9240.3, 9973.0

A

Acronyms and Abbreviations

AFW	auxiliary feedwater
AHR	average hazard rate
AIC	Akaike information criterion
ALT	accelerated life test
ARL	acceptable reliability level
BFR	bathtub failure rate
BIC	Bayesian information criterion
BN	Bayesian networks
BWR	boiling water reactor
CFR	constant failure rate
CMCM	Critical Measurements and Counter Measures
DFR	decreasing failure rate
DIC	deviance information criterion
EDG	emergency diesel generator
EIG	expected Shannon information gain
FAA	Federal Aviation Administration
FPGA	field programmable gate arrays
GA	genetic algorithm
GLM	generalized linear model
HPCI	high-pressure coolant injection
HPP	homogeneous Poisson process
IEEE	Institute of Electrical and Electronics Engineers
IFR	increasing failure rate
i.i.d.	independent and identically distributed
ISO	International Organization for Standardization
LCD	liquid crystal display
LED	light-emitting diode
LHD	"load-haul-dump" (machine)
MAP	maximum a posteriori
MCMC	Markov chain Monte Carlo
MLE	maximum likelihood estimator (or estimate)

MPa	Megapascals
MPLP	modulated power law process
MSTF	mean strength to failure
MTBF	mean time between failures
MTTF	mean time to failure
NHPP	nonhomogeneous Poisson process
NRC	Nuclear Regulatory Commission
NSSS	nuclear steam supply system
PCB	printed circuit board
PEXP	piecewise exponential (model)
PLP	power law process
PWR	pressurized water reactor
ROCOF	rate of occurrence of failures
RRL	rejectable reliability level
se	standard error
SMP	shared memory processor
Std Dev	standard deviation
TAAF	"test, analyze, and fix"
USAF	United States Air Force

B

Special Functions and Probability Distributions

B.1 Greek Alphabet

Table B.1. Greek alphabet

Name	Lower Case	Upper Case	Name	Lower Case	Upper Case
Alpha	α	A	Nu	ν	N
Beta	β	B	Xi	ξ	Ξ
Gamma	γ	Γ	Omicron	o	O
Delta	δ	Δ	Pi	π	Π
Epsilon	ϵ	E	Rho	ρ	P
Zeta	ζ	Z	Sigma	σ	Σ
Eta	η	H	Tau	τ	T
Theta	θ	Θ	Upsilon	υ	Υ
Iota	ι	I	Phi	ϕ	Φ
Kappa	κ	K	Chi	χ	X
Lambda	λ	Λ	Psi	ψ	Ψ
Mu	μ	M	Omega	ω	Ω

B.2 Special Functions

B.2.1 Beta Function

$$
\begin{aligned}
B(\alpha, \beta) &= \int_0^1 s^{\alpha-1}(1-s)^{\beta-1}\mathrm{d}s \qquad \alpha > 0, \quad \beta > 0 \\
&= \frac{\Gamma(\alpha)\Gamma(\beta)}{\Gamma(\alpha+\beta)}.
\end{aligned}
$$

B.2.2 Binomial Coefficient

$$\binom{n}{x} = \frac{n!}{x!(n-x)!}.$$

B.2.3 Determinant

The determinant is defined for a $k \times k$ square matrix A:

$$\det(A) = |A| = \sum_{i=1}^{k} a_{ij}(-1)^{i+j} M_{ij},$$

where M_{ij} is the minor of matrix A, which is formed by eliminating row i and column j of matrix A.

For a 2×2 matrix,

$$\det \begin{bmatrix} a & b \\ c & d \end{bmatrix} = ad - bc.$$

B.2.4 Factorial

$$\begin{aligned} n! &= 1 \cdot 2 \cdot 3 \cdot \ldots \cdot n \\ 0! &= 1. \end{aligned}$$

B.2.5 Gamma Function

$$\Gamma(\alpha) = \int_0^\infty s^{\alpha-1} \mathrm{e}^{-s} \mathrm{d}s \qquad \alpha > 0.$$

The recursion formula for the gamma function is

$$\begin{aligned} \Gamma(\alpha+1) &= \alpha\Gamma(\alpha), \quad \text{with} \\ \Gamma(n+1) &= n! \qquad n = 0, 1, 2, \ldots. \end{aligned}$$

B.2.6 Incomplete Beta Function

$$B(z; \alpha, \beta) = \int_0^z s^{\alpha-1}(1-s)^{\beta-1} \mathrm{d}s \qquad \alpha > 0, \quad \beta > 0.$$

B.2.7 Incomplete Beta Function Ratio

$$I(z; \alpha, \beta) = \frac{1}{B(\alpha, \beta)} \int_0^z s^{\alpha-1}(1-s)^{\beta-1} \mathrm{d}s \qquad \alpha > 0, \quad \beta > 0.$$

B.2.8 Indicator Function

$$\begin{aligned} I(x \in A) &= 1 \text{ if } x \in A \\ &= 0 \text{ if } x \notin A. \end{aligned}$$

B.2.9 Logarithm

If $a^p = N$, where $a \neq 0, 1$, then $p = \log_a(N)$ is the logarithm of N to the base a. We use the notation $\log(N) = \log_e(N)$.

B.2.10 Lower Incomplete Gamma Function

$$\begin{aligned} \gamma(\alpha, z) &= \int_0^z s^{\alpha-1} \mathrm{e}^{-s} \mathrm{d}s \qquad \alpha > 0 \\ &= \Gamma(\alpha) - \Gamma(\alpha, z). \end{aligned}$$

B.2.11 Standard Normal Cumulative Density Function

$$\begin{aligned} \Phi(z) &= \int_{-\infty}^z \phi(s) \mathrm{d}s \\ &= \int_{-\infty}^z \frac{1}{\sqrt{2\pi}} \exp(-\frac{1}{2}s^2) \mathrm{d}s. \end{aligned}$$

B.2.12 Standard Normal Probability Density Function

$$\phi(x) = \frac{1}{\sqrt{2\pi}} \exp(-\frac{1}{2}x^2).$$

B.2.13 Trace

The trace is defined for a $k \times k$ square matrix A:

$$Tr(A) = \sum_{i=1}^k a_{ii}.$$

B.2.14 Upper Incomplete Gamma Function

$$\begin{aligned} \Gamma(\alpha, z) &= \int_z^\infty s^{\alpha-1} \mathrm{e}^{-s} \mathrm{d}s \qquad \alpha > 0 \\ &= \Gamma(\alpha) - \gamma(\alpha, z). \end{aligned}$$

B.3 Probability Distributions

B.3.1 Bernoulli

$X \sim Bernoulli(\pi)$.

Probability Mass Function

$$f(x \mid \pi) = \pi^x (1-\pi)^{1-x} \qquad x = 0, 1, \quad 0 \le \pi \le 1.$$

$\mathrm{E}(X) = \pi$, $\mathrm{Var}(X) = \pi(1-\pi)$.

B.3.2 Beta

$X \sim Beta(\alpha, \beta)$.

Probability Density Function

$$f(x \mid \alpha, \beta) = \frac{\Gamma(\alpha+\beta)}{\Gamma(\alpha)\Gamma(\beta)} x^{\alpha-1}(1-x)^{\beta-1} \qquad 0 \le x \le 1, \quad \alpha > 0, \quad \beta > 0.$$

$\mathrm{E}(X) = \frac{\alpha}{\alpha+\beta}$, $\mathrm{Var}(X) = \frac{\alpha\beta}{(\alpha+\beta)^2(\alpha+\beta+1)}$.

Parameters α and β are shape parameters. They are symmetrically related by

$$f(x \mid \alpha, \beta) = f(1-x \mid \beta, \alpha).$$

One interpretation of the beta distribution is as a distribution that captures the information of x_0 successes in n_0 trials. The parameterization of the beta distribution that reflects this idea is

$$f(y \mid n_0, x_0) = \frac{\Gamma(n_0)}{\Gamma(x_0)\Gamma(n_0-x_0)} y^{x_0-1}(1-y)^{n_0-x_0-1}$$
$$0 \le y \le 1, \quad n_0 > x_0 > 0.$$

$\mathrm{E}(Y) = \frac{x_0}{n_0}$, $\mathrm{Var}(Y) = \frac{x_0(n_0-x_0)}{n_0^2(n_0+1)}$.

A third parameterization of the beta distribution uses the mean (here, π) as one of the parameters.

$$f(z \mid \pi, \nu) = \frac{\Gamma(\nu)}{\Gamma(\nu\pi)\Gamma(\nu(1-\pi))} z^{\nu\pi-1}(1-z)^{\nu(1-\pi)-1}$$
$$0 \le z \le 1, \quad \pi > 0, \quad \nu > 0.$$

$\mathrm{E}(Z) = \pi$, $\mathrm{Var}(Z) = \frac{\pi(1-\pi)}{\nu+1}$.

The uniform distribution is a special case of the beta distribution with $\alpha = 1$ and $\beta = 1$. The kth order statistic from a sample of n independent, identically distributed $Uniform(0,1)$ random variables has a $Beta(k, n-k+1)$ distribution.

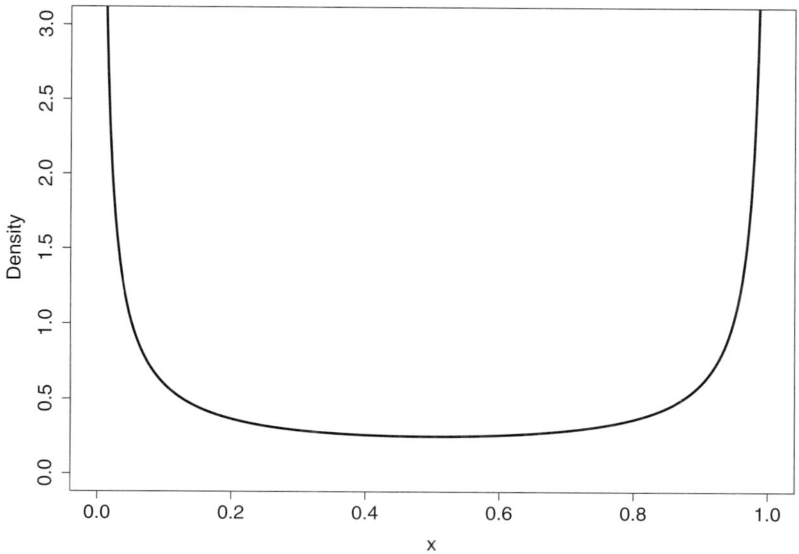

(a)

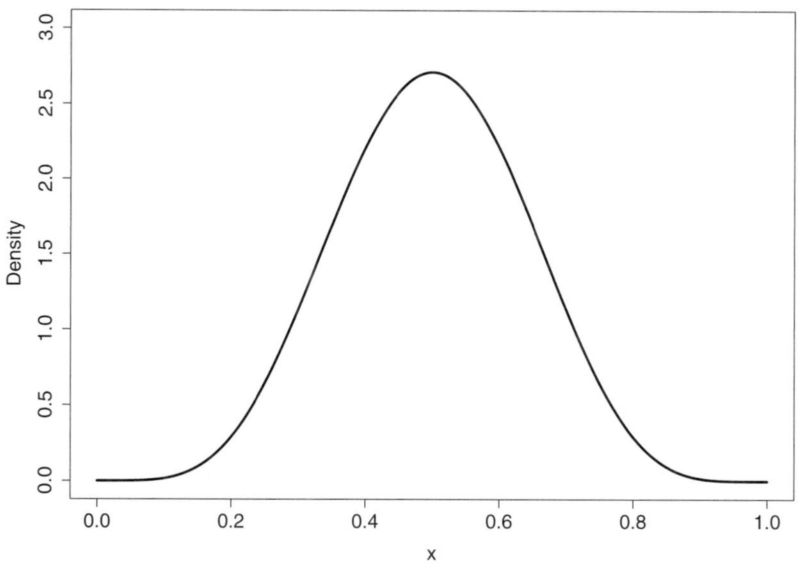

(b)

Fig. B.1. Beta distribution probability density functions with (a) $\alpha = \beta = 0.15$ and (b) $\alpha = \beta = 6$.

B.3.3 Binomial

$X \sim Binomial(n, \pi)$.

Probability Mass Function

$$f(x \mid n, \pi) = \binom{n}{x} \pi^x (1-\pi)^{n-x} \qquad x = 0, 1, 2, \ldots, n, \quad 0 \le \pi \le 1.$$

$\mathrm{E}(X) = n\pi$, $\mathrm{Var}(X) = n\pi(1-\pi)$.

The Bernoulli distribution is a special case of the binomial distribution with $n = 1$.

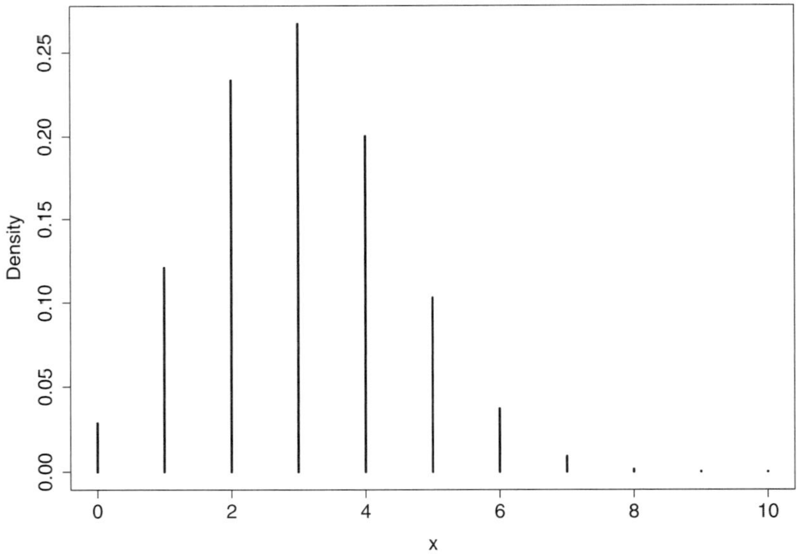

Fig. B.2. Binomial distribution probability density function with $n = 10$ and $\pi = 0.3$.

B.3.4 Bivariate Exponential

$(X, Y) \sim BivariateExponential(\lambda_1, \lambda_2, \lambda_{12})$.

Probability Density Function

$$f(x, y \mid \lambda_1, \lambda_2, \lambda_{12}) = \exp\{-\lambda_1 x - \lambda_2 y - \lambda_{12}\max(x, y)\}$$
$$x \ge 0, \quad y \ge 0, \quad \lambda_1 > 0, \quad \lambda_2 > 0, \quad \lambda_{12} > 0.$$

$\mathrm{E}(X) = \frac{1}{\lambda_1+\lambda_{12}}$, $\mathrm{E}(Y) = \frac{1}{\lambda_2+\lambda_{12}}$, $\mathrm{Var}(X) = \frac{1}{(\lambda_1+\lambda_{12})^2}$, $\mathrm{Var}(Y) = \frac{1}{(\lambda_2+\lambda_{12})^2}$, $\mathrm{Cor}(X,Y) = \frac{\lambda_{12}}{\lambda_1+\lambda_2+\lambda_{12}}$.

This is the Marshall-Olkin bivariate exponential distribution (Marshall and Olkin, 1967). Johnson and Kotz (1972) discusses additional bivariate exponential distributions. The bivariate exponential is the bivariate generalization of the exponential distribution.

B.3.5 Chi-squared

$X \sim ChiSquared(\nu)$.

Probability Density Function

$$f(x \mid \nu) = \frac{2^{-\frac{\nu}{2}}}{\Gamma(\frac{\nu}{2})} x^{\frac{\nu}{2}-1} e^{-\frac{x}{2}} \qquad x > 0, \quad \nu > 0.$$

$\mathrm{E}(X) = \nu$, $\mathrm{Var}(X) = 2\nu$.

The parameter ν is called the *degrees of freedom* of the chi-squared distribution.

Hazard Function

$$h(t \mid \nu) = \frac{\frac{1}{2}^{\frac{\nu}{2}}}{\Gamma(\frac{\nu}{2}, \frac{t}{2})} t^{\frac{\nu}{2}-1} e^{-\frac{t}{2}} \qquad t \geq 0, \quad \nu > 0,$$

where $\Gamma(\alpha, z)$ is the upper incomplete gamma function.

A chi-squared distribution is a special case of the gamma distribution with $\alpha = \frac{\nu}{2}$ and $\lambda = \frac{1}{2}$.

B.3.6 Dirichlet

$(X_1, \ldots, X_k) \sim Dirichlet(\alpha_1, \ldots, \alpha_k)$.

Probability Density Function

$$f(\mathbf{x} \mid \alpha) = \frac{\Gamma(\sum_{i=1}^k \alpha_i)}{\prod_{i=1}^k \Gamma(\alpha_i)} x_1^{\alpha_1 - 1} \cdots x_k^{\alpha_k - 1}$$

$$0 \leq x_i \leq 1, \quad \sum_{i=1}^k x_i = 1, \quad \alpha_i \geq 0.$$

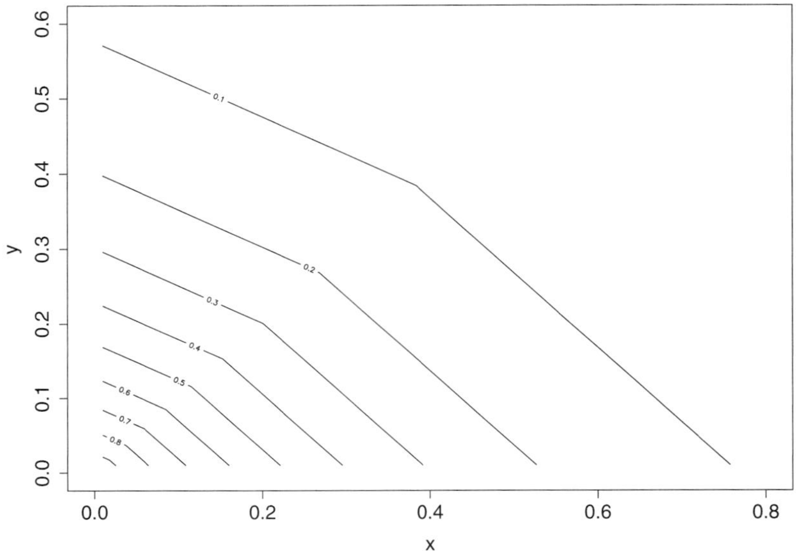

(a)

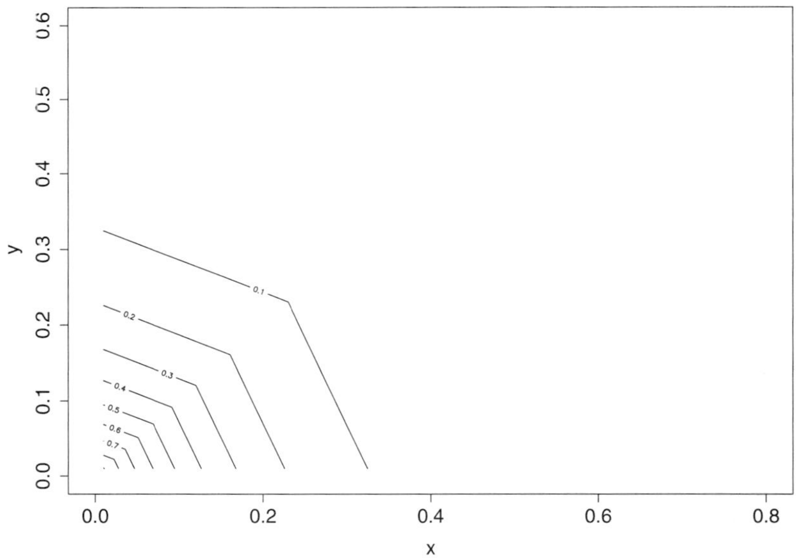

(b)

Fig. B.3. Contour plots of bivariate exponential distribution probability density functions with (a) $\lambda_1 = 2$, $\lambda_2 = 3$, $\lambda_{12} = 1$ and (b) $\lambda_1 = 3$, $\lambda_2 = 3$, $\lambda_{12} = 4$.

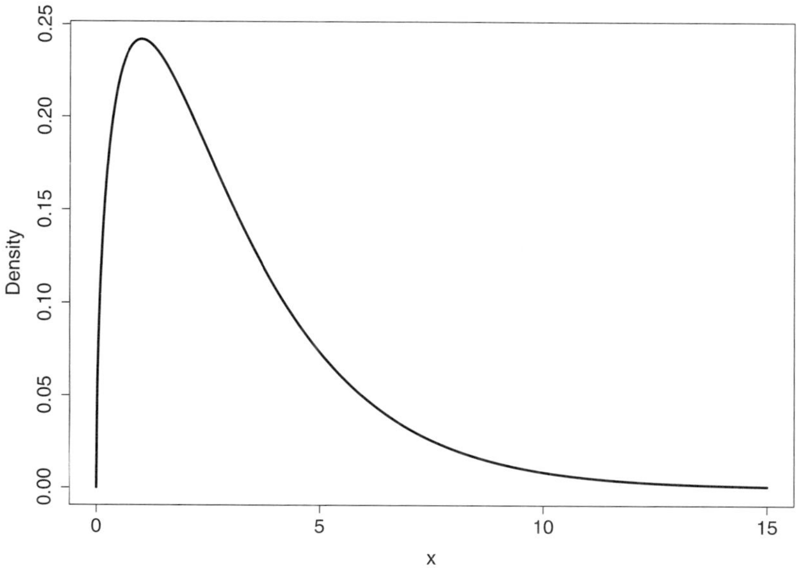

(a)

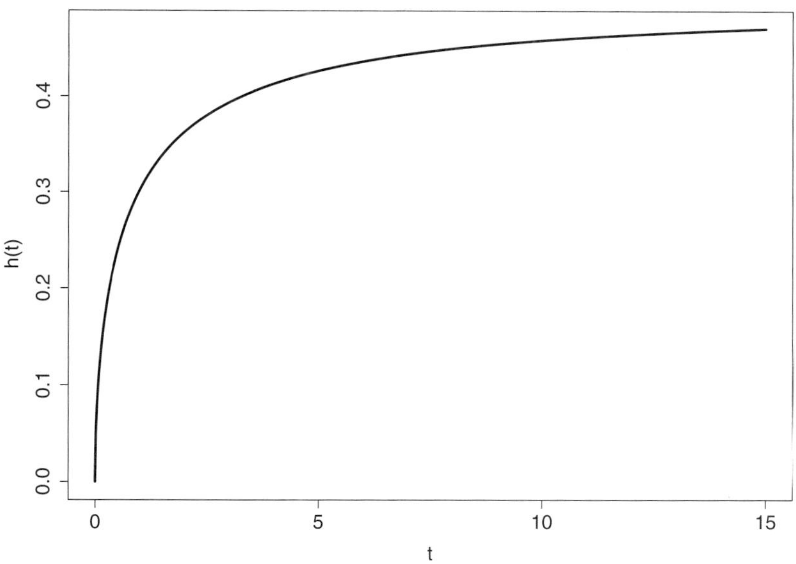

(b)

Fig. B.4. Chi-squared distribution (a) probability density function with $\nu = 3$ and (b) hazard function with $\nu = 3$.

Let $\alpha_0 = \sum_{i=1}^{k} \alpha_i$. $\mathrm{E}(X_i) = \frac{\alpha_i}{\alpha_0}$, $\mathrm{Var}(X_i) = \frac{\alpha_i(\alpha_0 - \alpha_i)}{\alpha_0^2(\alpha_0+1)}$, $\mathrm{Cov}(X_i, X_j) = \frac{-\alpha_i \alpha_j}{\alpha_0^2(\alpha_0+1)}$.

Since $\sum_{i=1}^{k} x_i = 1$, $f(\mathbf{x} \mid \alpha)$ is not a k-dimensional probability density function. Instead, it gives the joint probability density function of any subcollection of the $k-1$ random variables in $(X_1, \ldots, X_k)$.

The Dirichlet is the multivariate generalization of the beta distribution. The marginal distribution of a single X_i is $Beta(\alpha_i, \alpha_0 - \alpha_i)$.

B.3.7 Exponential

$X \sim Exponential(\lambda)$.

Probability Density Function

$$f(x \mid \lambda) = \lambda e^{-\lambda x} \qquad x > 0, \quad \lambda > 0.$$

$\mathrm{E}(X) = \frac{1}{\lambda}$, $\mathrm{Var}(X) = \frac{1}{\lambda^2}$.

Hazard Function

$$h(t \mid \lambda) = \lambda.$$

The exponential distribution is a special case of the gamma distribution with $\alpha = 1$.

For the two-parameter exponential distribution

$$f(x \mid \lambda, \mu) = \lambda e^{-\lambda(x-\mu)} \qquad x > \mu \geq 0, \quad \lambda > 0.$$

$\mathrm{E}(X) = \frac{1}{\lambda} + \mu$, $\mathrm{Var}(X) = \frac{1}{\lambda^2}$.

B.3.8 Extreme Value

$X \sim ExtremeValue(\mu, \sigma)$.

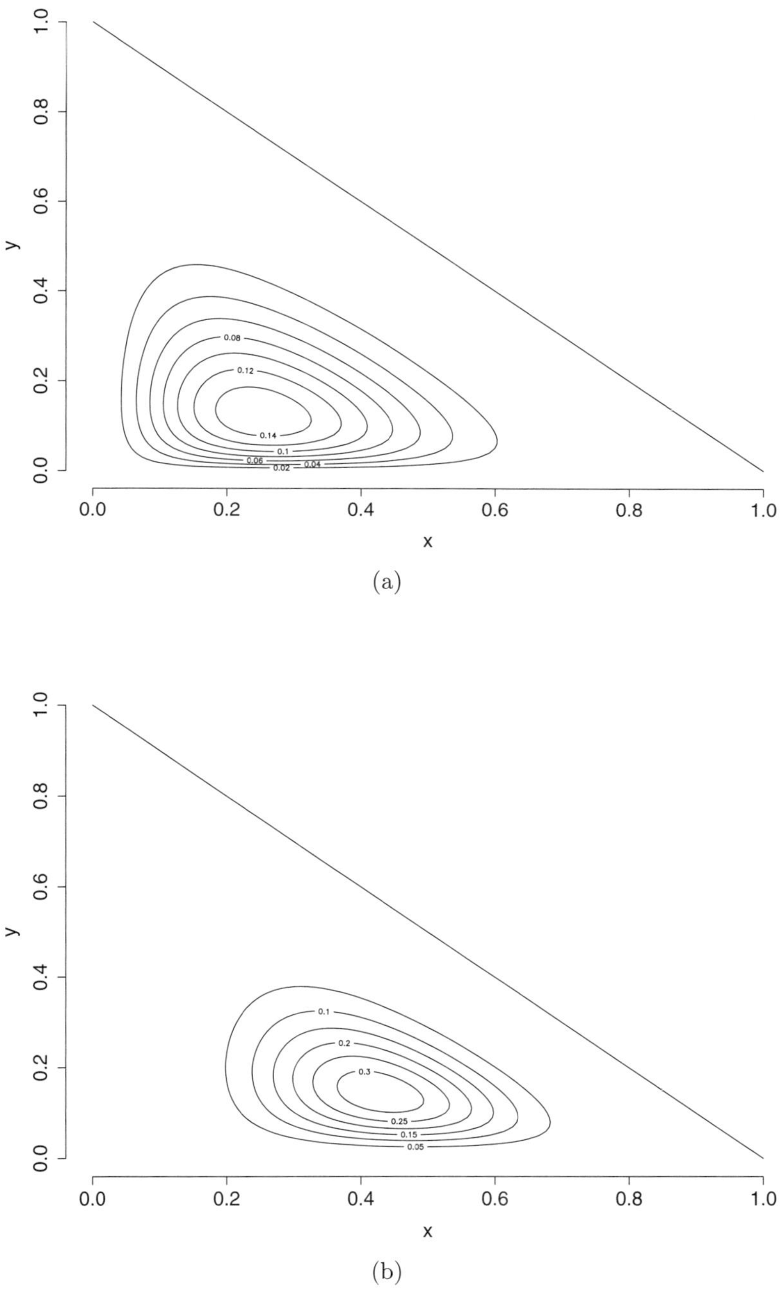

Fig. B.5. Contour plots of Dirichlet distribution probability density function with (a) $\alpha_1 = 2$, $\alpha_2 = 1$, $\alpha_3 = 5$ and (b) $\alpha_1 = 6$, $\alpha_2 = 2$, $\alpha_3 = 6$.

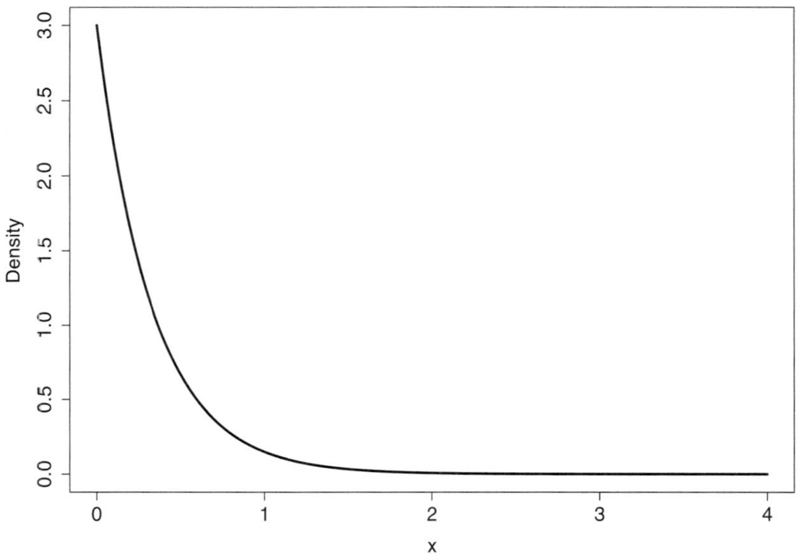

(a)

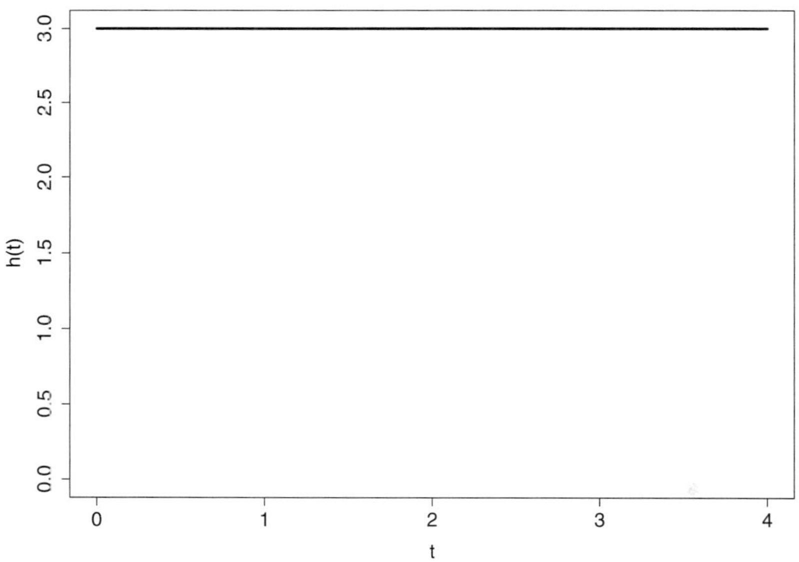

(b)

Fig. B.6. Exponential distribution (a) probability density function with $\lambda = 3$ and (b) hazard function with $\lambda = 3$.

Probability Density Function

$$f(x \mid \mu, \sigma) = \frac{1}{\sigma} \exp\left(-\frac{x-\mu}{\sigma}\right) \exp\left[-\exp\left(-\frac{x-\mu}{\sigma}\right)\right]$$

$$-\infty < x < \infty, \quad \sigma > 0, \quad -\infty < \mu < \infty.$$

$\mathrm{E}(X) = \mu + 0.57722\sigma$, $\mathrm{Var}(X) = \frac{1}{6}\pi^2\sigma^2$, where $0.57722\ldots$ is Euler's constant.

This distribution is also called the Type I or Gumbel-type extreme value distribution (maximum). The Gumbel-type extreme value distribution has cumulative distribution function

$$F(x \mid \mu, \sigma) = \exp\left[-\exp\left(-\frac{x-\mu}{\sigma}\right)\right].$$

B.3.9 Gamma

$X \sim Gamma(\alpha, \lambda)$.

Probability Density Function

$$f(x \mid \alpha, \lambda) = \frac{\lambda^\alpha}{\Gamma(\alpha)} x^{\alpha-1} \exp(-\lambda x) \qquad x > 0, \quad \alpha > 0, \quad \lambda > 0.$$

$\mathrm{E}(X) = \frac{\alpha}{\lambda}$, $\mathrm{Var}(X) = \frac{\alpha}{\lambda^2}$.

Parameter α is the *shape parameter* and λ is the *scale parameter* of the gamma distribution.

Hazard Function

$$h(t \mid \alpha, \lambda) = \frac{\lambda^\alpha}{\Gamma(\alpha, \lambda t)} t^{\alpha-1} e^{-\lambda t} \qquad t \geq 0, \quad \alpha > 0, \quad \lambda > 0.$$

The exponential distribution is a special case when $\alpha = 1$. The chi-squared distribution is a special case when $\lambda = \frac{1}{2}$ and $\alpha = \frac{\nu}{2}$.

B.3.10 Inverse Chi-squared

$X \sim InverseChisquared(\nu)$.

Probability Density Function

$$f(x \mid \nu) = \frac{2^{-\nu/2}}{\Gamma(\nu/2)} x^{-\nu/2-1} \exp(-\nu/2) \qquad x > 0, \quad \nu > 0.$$

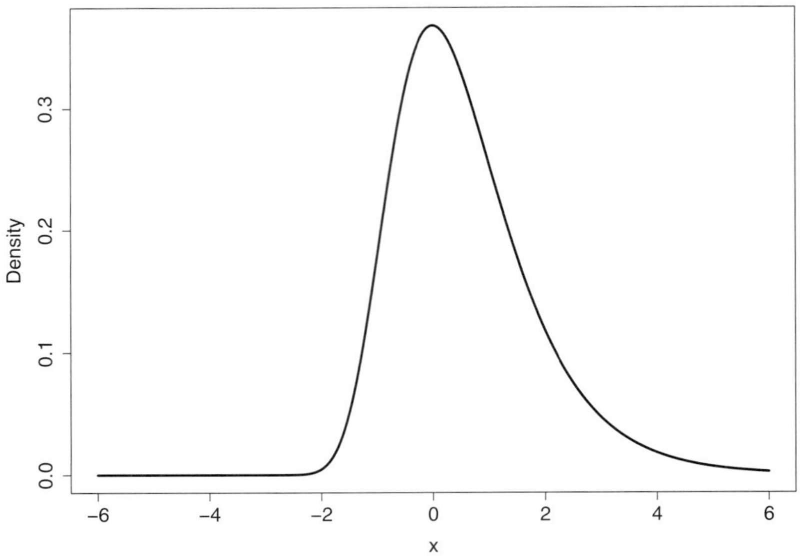

(a)

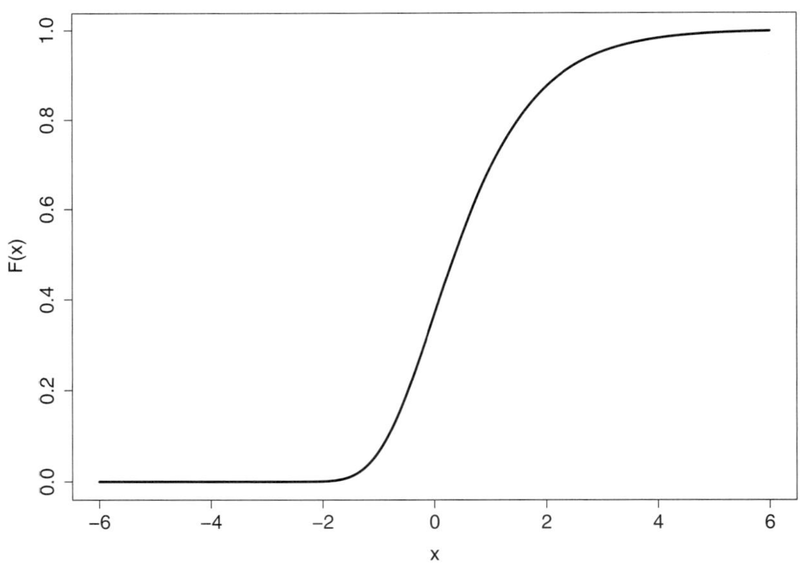

(b)

Fig. B.7. Gumble-type extreme value distribution (a) probability density functions with $\mu = 0$ and $\sigma = 1$ and (b) cumulative distribution function with $\mu = 0$ and $\sigma = 1$.

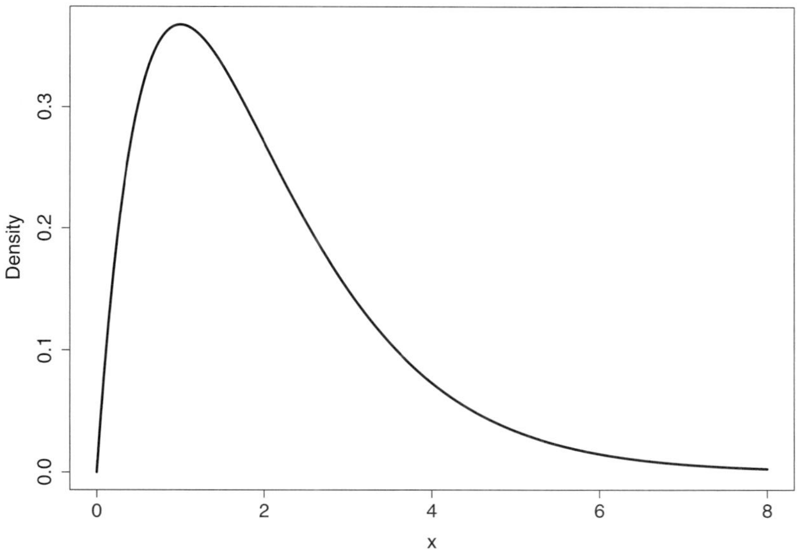

(a)

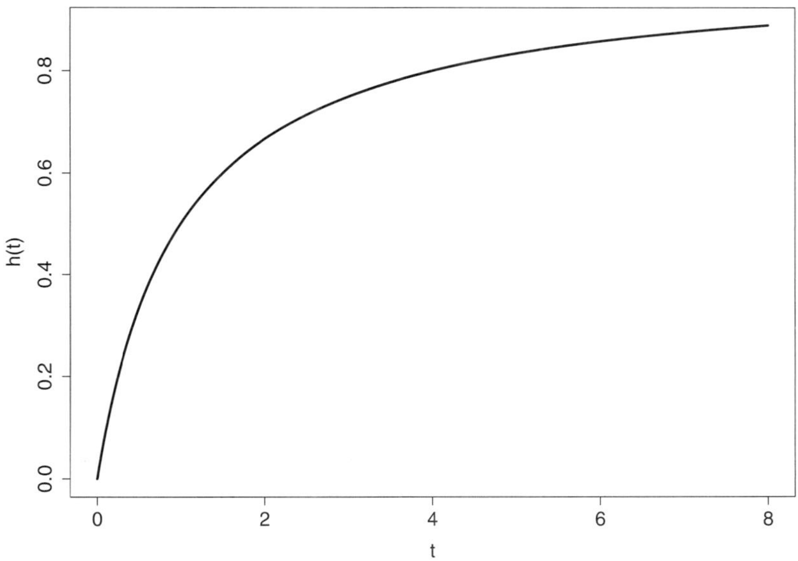

(b)

Fig. B.8. Gamma distribution (a) probability density function with $\alpha = 2$ and $\lambda = 1$ and (b) hazard function with $\alpha = 2$ and $\lambda = 1$.

$\mathrm{E}(X) = \frac{1}{\nu-2}$ for $\nu > 2$, $\mathrm{Var}(X) = \frac{2}{(\nu-2)^2(\nu-4)}$ for $\nu > 4$.

The inverse chi-squared distribution is the distribution of $\frac{1}{X}$ when $X \sim ChiSquared(\nu)$.

B.3.11 Inverse Gamma

$X \sim InverseGamma(\alpha, \lambda)$.

Probability Density Function

$$f(x \mid \alpha, \lambda) = \frac{\lambda^\alpha}{\Gamma(\alpha)} x^{-(\alpha+1)} e^{-\lambda/x} \qquad x > 0, \quad \alpha > 0, \quad \lambda > 0.$$

$\mathrm{E}(X) = \frac{\lambda}{\alpha-1}$ for $\alpha > 1$, $\mathrm{Var}(X) = \frac{\lambda^2}{(\alpha-1)^2(\alpha-2)}$ for $\alpha > 2$.

The inverse gamma is the distribution of $\frac{1}{X}$ when $X \sim Gamma(\alpha, \lambda)$.

The inverse chi-squared distribution is a special case when $\alpha = \frac{\nu}{2}$ and $\lambda = \frac{1}{2}$.

B.3.12 Inverse Gaussian

$X \sim InverseGaussian(\mu, \lambda)$.

Probability Density Function

$$f(x \mid \mu, \lambda) = \sqrt{\frac{\lambda}{2\pi x^3}} \exp\left(\frac{-\lambda(x-\mu)^2}{2x\mu^2}\right) \qquad x > 0, \quad \mu > 0, \quad \lambda > 0.$$

$\mathrm{E}(X) = \mu$, $\mathrm{Var}(X) = \frac{\mu^3}{\lambda}$.

The Wald distribution is a special case when $\mu = 1$.

B.3.13 Inverse Wishart

$\mathbf{X} \sim InverseWishart(\nu, \mathbf{\Omega})$.

Probability Density Function

$$f(\mathbf{X} \mid \nu, \mathbf{\Omega}) = \left(2^{\frac{\nu d}{2}} \pi^{\frac{d(d-1)}{4}} \prod_{i=1}^{d} \Gamma\left(\frac{\nu+1-i}{2}\right)\right)^{-1} \times \mid \mathbf{\Omega} \mid^{\frac{\nu}{2}} \mid \mathbf{X} \mid^{\frac{-(\nu+d+1)}{2}} \exp\left[-\frac{1}{2} tr(\mathbf{\Omega}\mathbf{X}^{-1})\right],$$

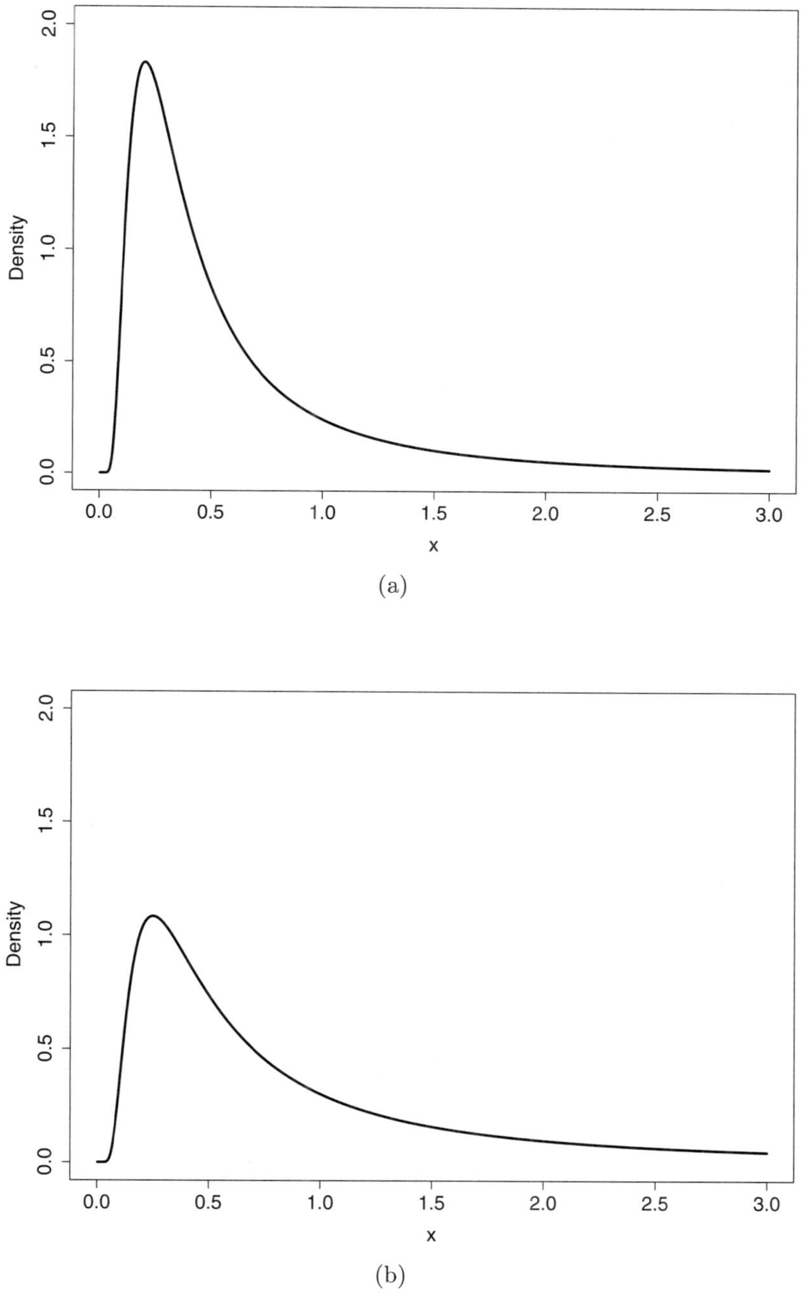

Fig. B.9. Inverse chi-squared distribution probability density function with (a) $\nu = 3$ and (b) $\nu = 2$.

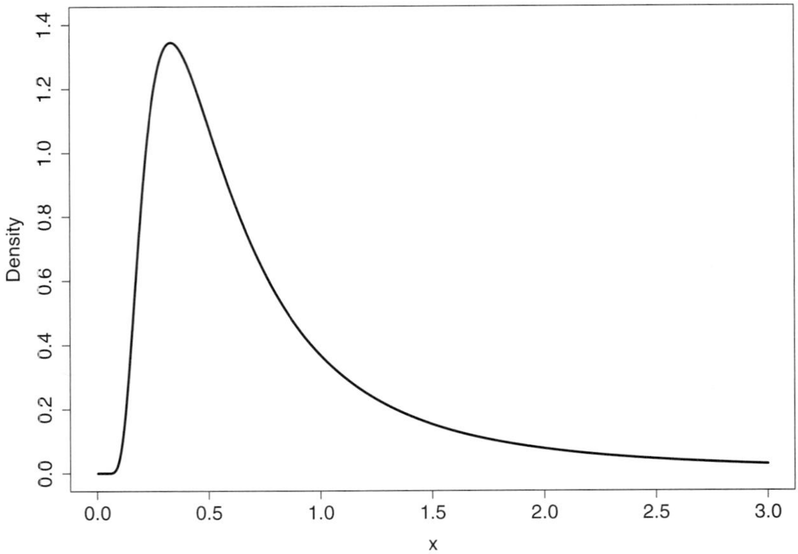

(a)

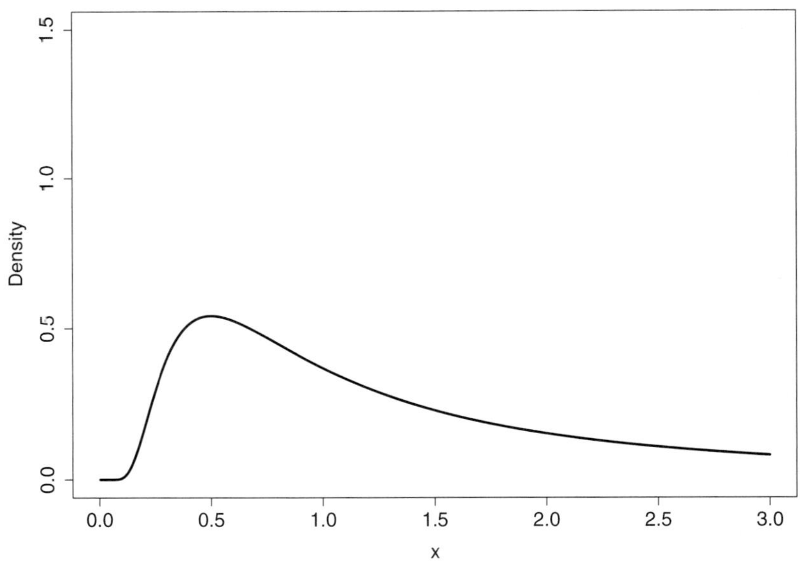

(b)

Fig. B.10. Inverse gamma distribution probability density function with (a) $\alpha = 2$ and $\lambda = 1$ and (b) $\alpha = 1$ and $\lambda = 1$.

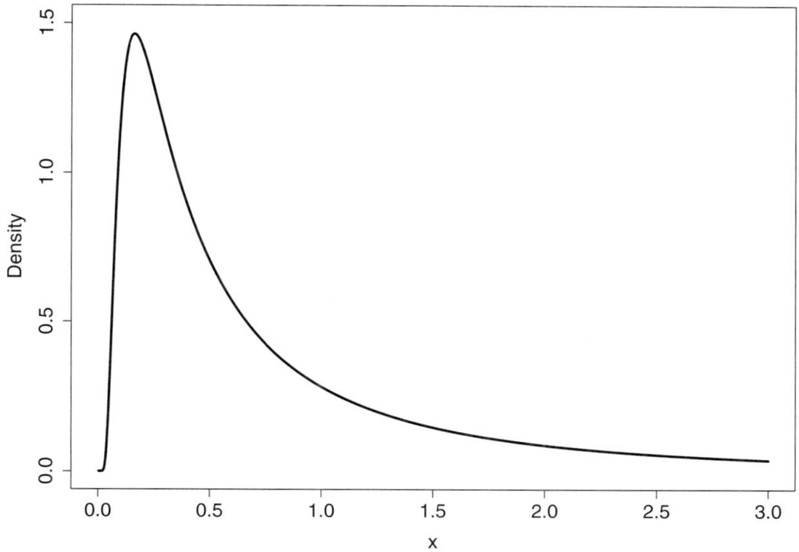

(a)

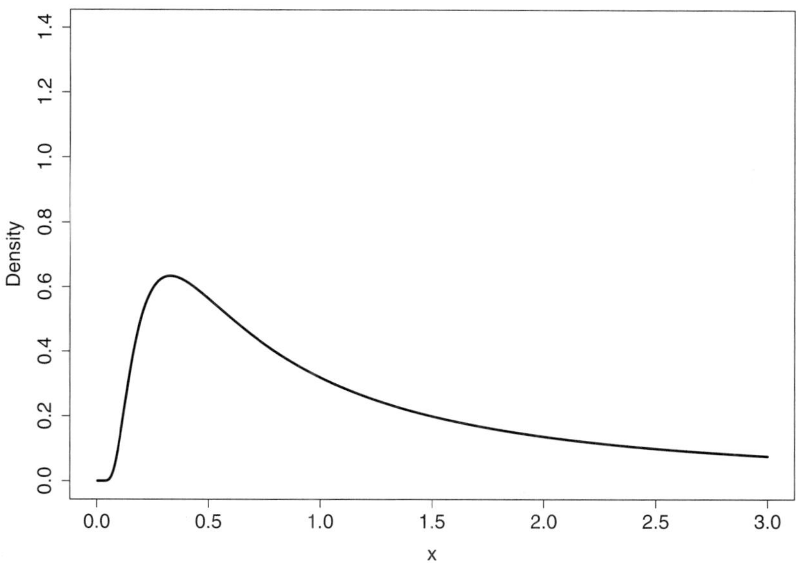

(b)

Fig. B.11. Inverse Gaussian distribution probability density function with (a) $\mu = 1$ and $\lambda = 0.5$ and (b) $\mu = 3$ and $\lambda = 1$.

where $\mathbf{X}$ and $\mathbf{\Omega}$ are $d \times d$ symmetric, positive definite matrices, $\nu \geq d$, $d > 0$.

$\mathrm{E}(\mathbf{X}) = (\nu - d - 1)^{-1}\mathbf{\Omega}$.

For additional information, see Press (1972) and Eaton (1983).

B.3.14 Logistic

$X \sim \mathit{Logistic}(\mu, \lambda)$.

Probability Density Function

$$f(x \mid \mu, \lambda) = \frac{\mathrm{e}^{-(x-\mu)/\lambda}}{\lambda(1+\mathrm{e}^{-(x-\mu)/\lambda})^2} \qquad -\infty < x < \infty, \quad -\infty < \mu < \infty, \quad \lambda > 0.$$

$\mathrm{E}(X) = \mu$, $\mathrm{Var}(X) = \frac{1}{3}\pi^2\lambda^2$.

The cumulative distribution function of the logistic distribution is

$$F(x \mid \mu, \lambda) = \frac{1}{1+\mathrm{e}^{-(x-\mu)/\lambda}}.$$

B.3.15 Lognormal

$X \sim \mathit{LogNormal}(\mu, \sigma^2)$.

Probability Density Function

$$f(x \mid \mu, \sigma^2) = \frac{1}{x\sqrt{2\pi\sigma^2}} \exp\left[-\frac{1}{2\sigma^2}(\log(x) - \mu)^2\right]$$
$$x > 0, \quad -\infty < \mu < \infty, \quad \sigma > 0.$$

$\mathrm{E}(X) = \exp(\mu + \frac{\sigma^2}{2})$, $\mathrm{Var}(X) = \exp(2\mu + 2\sigma^2) - \exp(2\mu + \sigma^2)$.

Hazard Function

$$h(t \mid \mu, \sigma) = \frac{\phi\left(\frac{\log(t)-\mu}{\sigma}\right)}{\sigma t - \sigma t \Phi\left(\frac{\log(t)-\mu}{\sigma}\right)},$$

where $\phi(\cdot)$ is the probability density function of the standard normal distribution and $\Phi(\cdot)$ is the cumulative distribution function of the standard normal distribution.

If $X > 0$ is a random variable with $\log(X) \sim N(\mu, \sigma^2)$, then X has a lognormal distribution. The mode of the lognormal distribution is $\exp(\mu - \sigma^2)$.

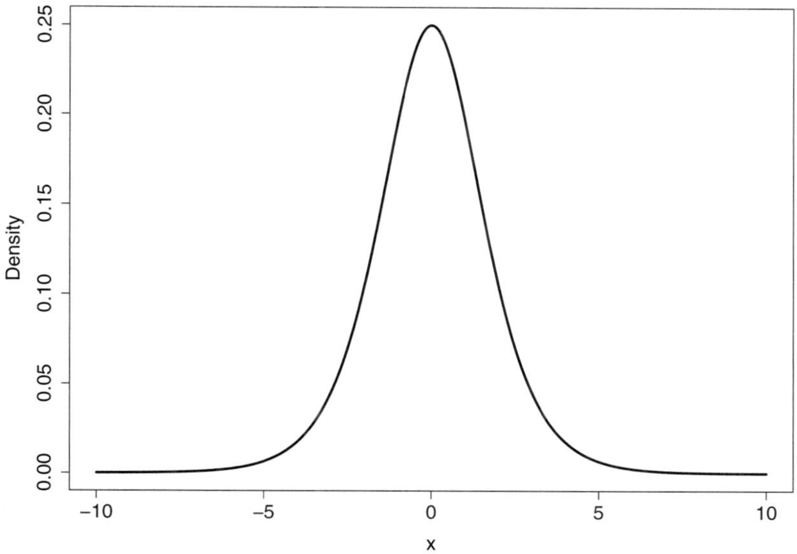

(a)

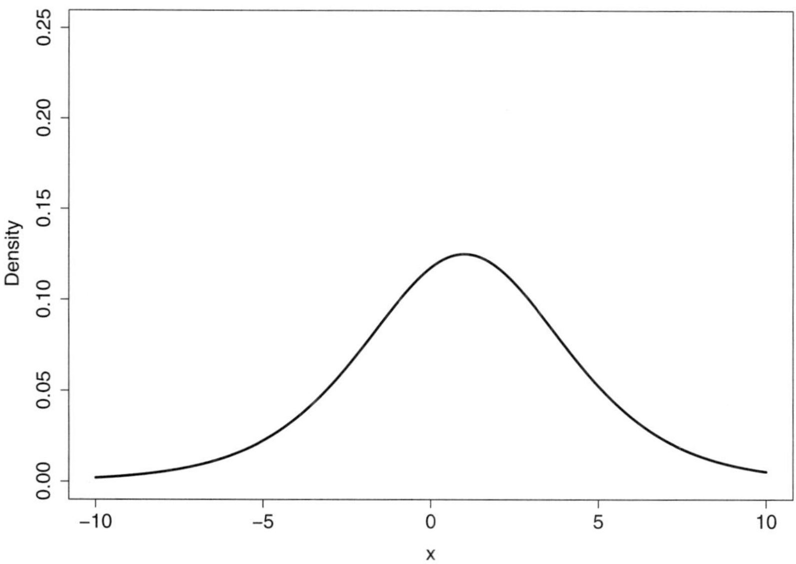

(b)

Fig. B.12. Logistic distribution probability density function with (a) $\mu = 0$ and $\lambda = 1$ and (b) $\mu = 1$ and $\lambda = 2$.

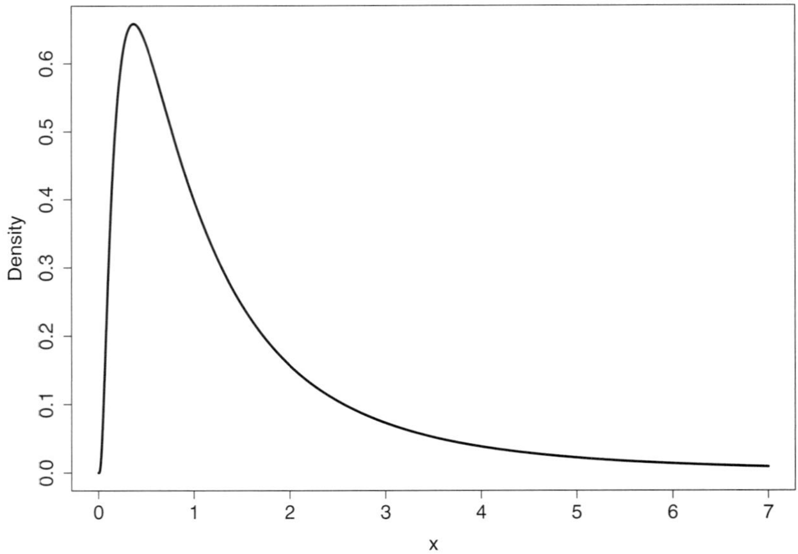

(a)

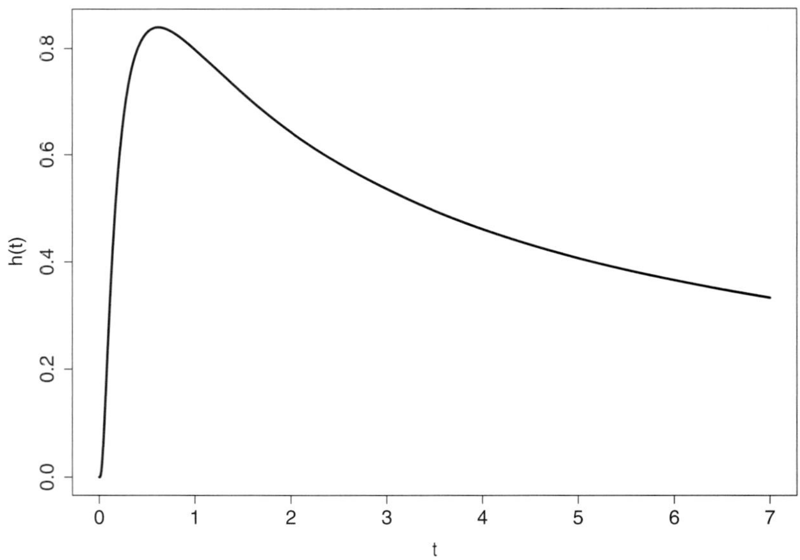

(b)

Fig. B.13. Lognormal distribution (a) probability density function with $\mu = 0$ and $\sigma^2 = 1$ and (b) hazard function with $\mu = 0$ and $\sigma^2 = 1$.

B.3.16 Multinomial

$\mathbf{X} \sim Multinomial(n, \pi_1, \ldots, \pi_k)$.

Probability Mass Function

$$f(\mathbf{X} \mid n, \boldsymbol{\pi}) = \frac{n!}{\prod_{i=1}^{k} x_i!} \pi_1^{x_1} \cdots \pi_k^{x_k}$$

$$x_i = 0, 1, 2, ..., n, \quad 0 \le \pi_i \le 1, \quad \sum_{i=1}^{k} \pi_i = 1.$$

$\mathrm{E}(X_i) = n\pi_i$, $\mathrm{Var}(X_i) = n\pi_i(1 - \pi_i)$, $\mathrm{Cov}(X_i, X_j) = -n\pi_i\pi_j$.

The multinomial distribution is a multivariate generalization of the binomial distribution — when $k = 2$, the multinomial distribution reduces to the binomial distribution. The marginal distribution of a single X_i is $Binomial(n, \pi_i)$.

B.3.17 Multivariate Normal

$\mathbf{X} \sim MultivariateNormal(\boldsymbol{\mu}, \boldsymbol{\Sigma})$.

Probability Density Function

$$f(\mathbf{X} \mid \boldsymbol{\mu}, \boldsymbol{\Sigma}) = (2\pi)^{-\frac{d}{2}} \det(\boldsymbol{\Sigma})^{-\frac{1}{2}} \exp\left[-\frac{1}{2}(\mathbf{x} - \boldsymbol{\mu})^T \boldsymbol{\Sigma}^{-1} (\mathbf{x} - \boldsymbol{\mu}) \right],$$

where $-\infty < \mathbf{x} < \infty$, $-\infty < \mu < \infty$, Σ is a $d \times d$ positive-definite, symmetric matrix.

$\mathrm{E}(\mathbf{X}) = \boldsymbol{\mu}$, $\mathrm{Var}(X) = \Sigma$.

The multivariate normal distribution is a multivariate generalization of the normal distribution. The marginal distribution of a single X_i is $Normal(\mu_i, \Sigma_{ii})$.

B.3.18 Negative Binomial

$X \sim NegativeBinomial(r, \pi)$.

Probability Mass Function

$$f(x \mid r, \theta) = \binom{x + r - 1}{r - 1} \pi^r (1 - \pi)^x \qquad x = 0, 1, 2, \ldots, \quad 0 \le \pi \le 1.$$

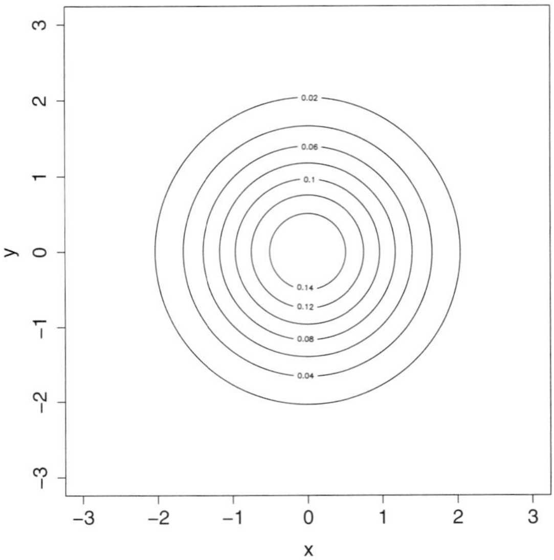

(a)

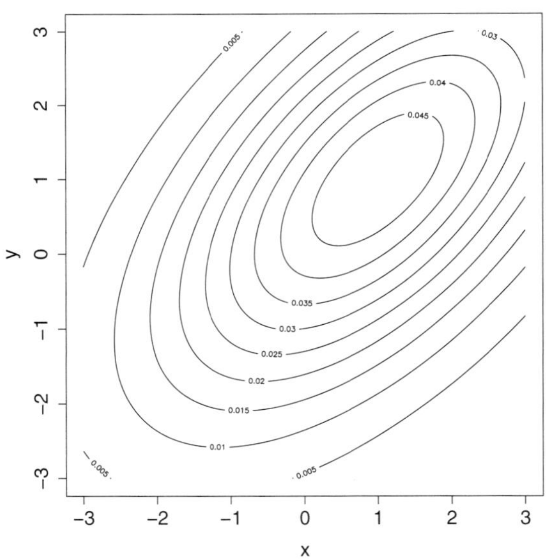

(b)

Fig. B.14. Contour plots of bivariate normal distribution probability density function with (a) $\mu_1 = \mu_2 = 0$, $\sigma_{1,1} = \sigma_{2,2} = 1$, and correlation = 0 and (b) $\mu_1 = \mu_2 = 1$, $\sigma_{1,1} = \sigma_{2,2} = 2$, and correlation = 0.6.

$\mathrm{E}(X) = \frac{r(1-\pi)}{\pi}$, $\mathrm{Var}(X) = \frac{r(1-\pi)}{\pi^2}$.

The negative binomial distribution is used to model the number of failures x observed before the rth success. The geometric distribution is a special case of the negative binomial distribution when $r = 1$.

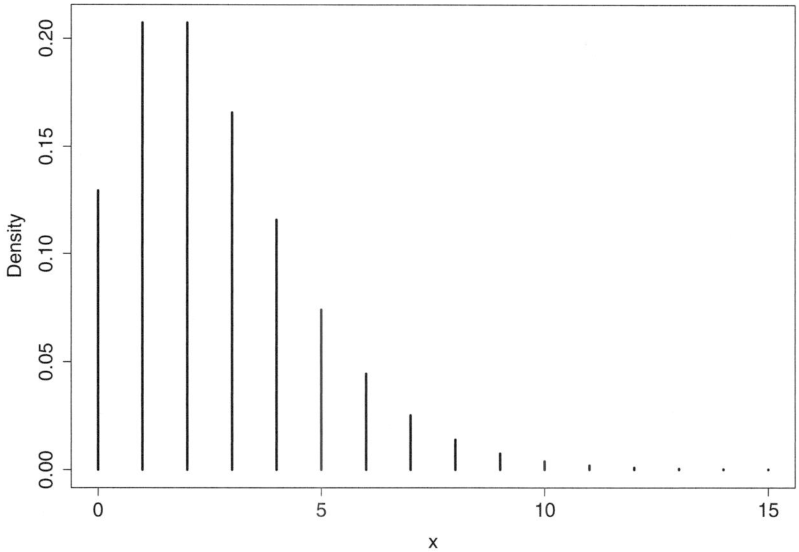

Fig. B.15. Negative binomial distribution probability density function with $r = 4$ and $\pi = 0.6$.

B.3.19 Negative Log-Gamma

$X \sim NegativeLogGamma(\alpha, \gamma)$.

Probability Density Function

$$f(x \mid \alpha, \gamma) = \frac{\gamma^\alpha}{\Gamma(\alpha)} x^{\gamma-1}[-\log(x)]^{\alpha-1} \qquad 0 \le x \le 1, \quad \alpha > 0, \quad \gamma > 0.$$

$\mathrm{E}(X) = (1 + 1/\gamma)^{-\alpha}$, $\mathrm{Var}(X) = (1 + 2/\gamma)^{-\alpha} - (1 + 1/\gamma)^{-2\alpha}$.

If $-\log(X) \sim Gamma(\alpha, \gamma)$, then $X \sim NegativeLogGamma(\alpha, \gamma)$. For more information regarding this distribution, see Martz and Waller (1982).

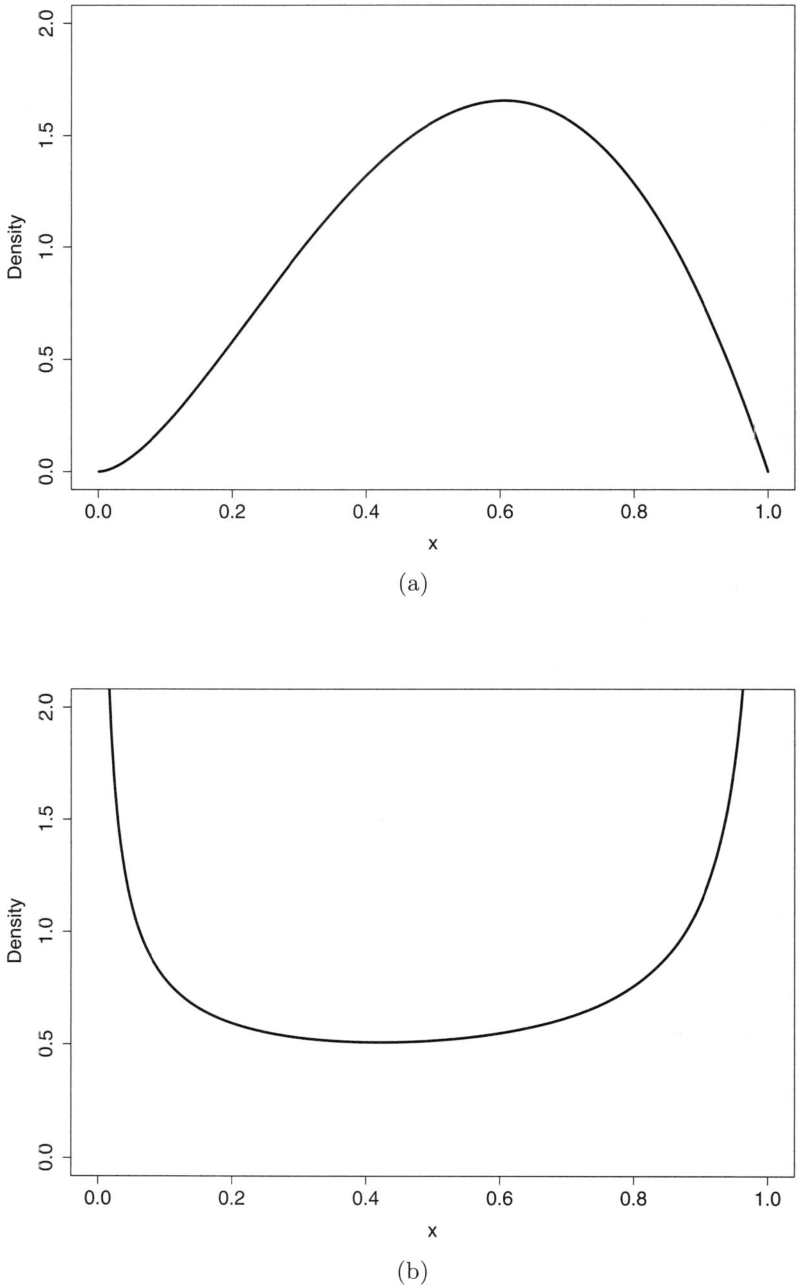

Fig. B.16. Negative log-gamma distribution probability density function with (a) $\alpha = 2$ and $\gamma = 3$ and (b) $\alpha = 0.35$ and $\gamma = 0.25$.

B.3.20 Normal

$X \sim Normal(\mu, \sigma^2)$.

Probability Density Function

$$f(x \,|\, \mu, \sigma^2) = \frac{1}{\sqrt{2\pi\sigma^2}} \exp\left[-\frac{1}{2\sigma^2}(x-\mu)^2\right]$$
$$-\infty < x < \infty, \quad -\infty < \mu < \infty, \quad \sigma > 0.$$

$\mathrm{E}(X) = \mu$, $\mathrm{Var}(X) = \sigma^2$.

The normal distribution is also known as the *Gaussian distribution*. When $\mu = 0$ and $\sigma = 1$, the distribution is called the *standard normal distribution.*

B.3.21 Pareto

$X \sim Pareto(\alpha, \beta)$.

Probability Density Function

$$f(x \,|\, \alpha, \beta) = \frac{\alpha\beta^\alpha}{x^{\alpha+1}} \qquad \alpha > 0, \quad x > \beta > 0.$$

$\mathrm{E}(X) = \frac{\alpha\beta}{\alpha-1}$ for $\alpha > 1$, $\mathrm{Var}(X) = \frac{\alpha\beta^2}{(\alpha-1)^2(\alpha-2)}$ for $\alpha > 2$.

B.3.22 Poisson

$X \sim Poisson(\lambda)$.

Probability Mass Function

$$f(x \,|\, \lambda) = \frac{\lambda^x}{x!} e^{-\lambda} \qquad x = 0, 1, \ldots, \quad \lambda > 0.$$

$\mathrm{E}(X) = \lambda$, $\mathrm{Var}(X) = \lambda$.

B.3.23 Poly-Weibull

$X \sim PolyWeibull(\boldsymbol{\beta}, \boldsymbol{\lambda})$.

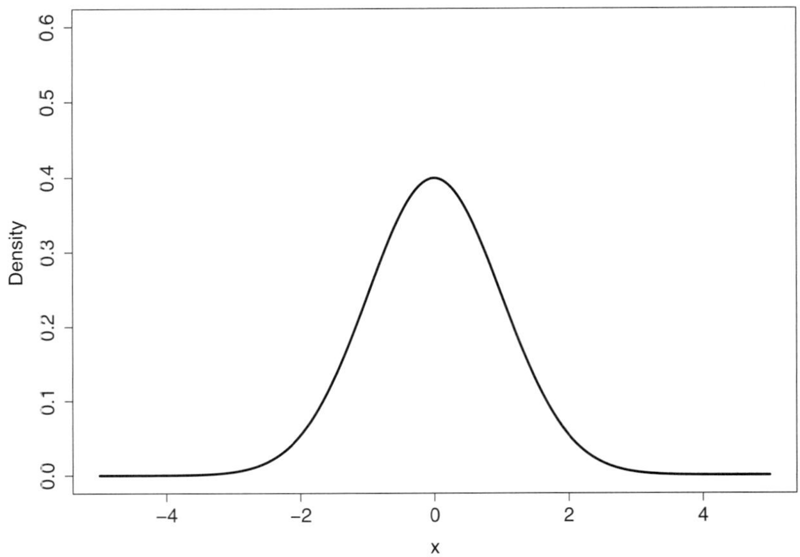

(a)

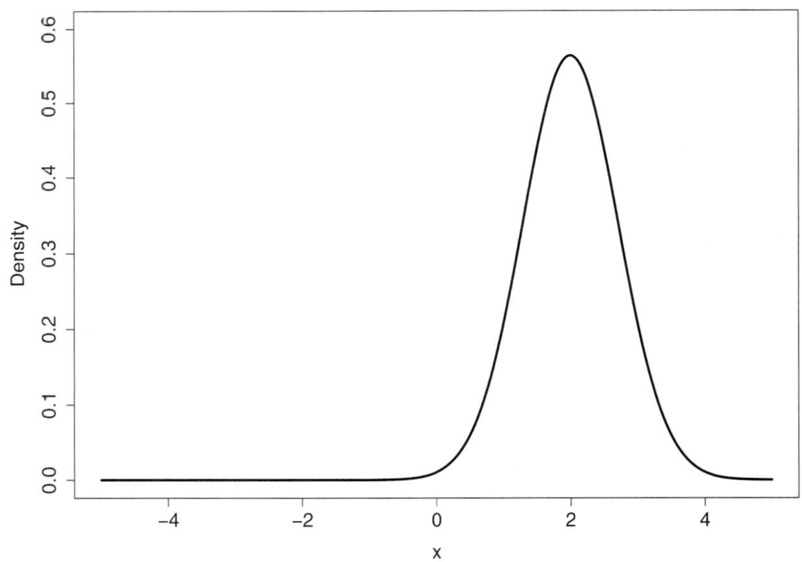

(b)

Fig. B.17. Normal distribution probability density function with (a) $\mu = 0$ and $\sigma^2 = 1$ (standard normal distribution) and (b) $\mu = 2$ and $\sigma^2 = 0.5$.

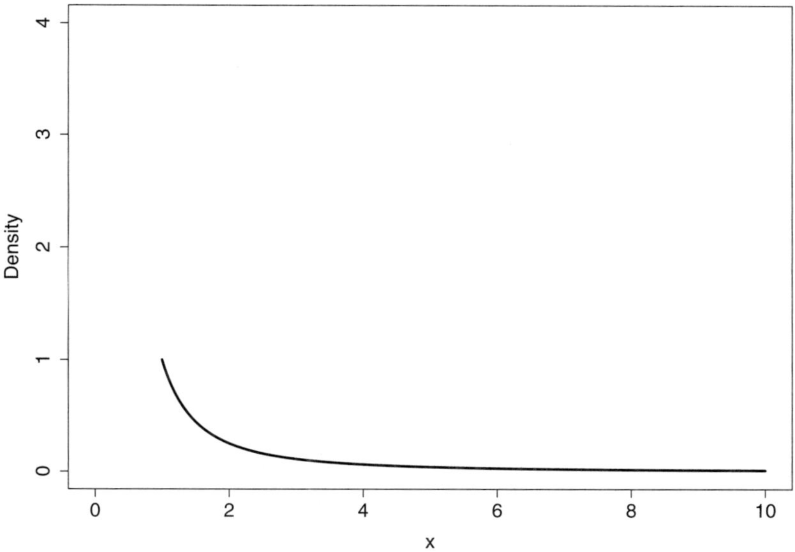

(a)

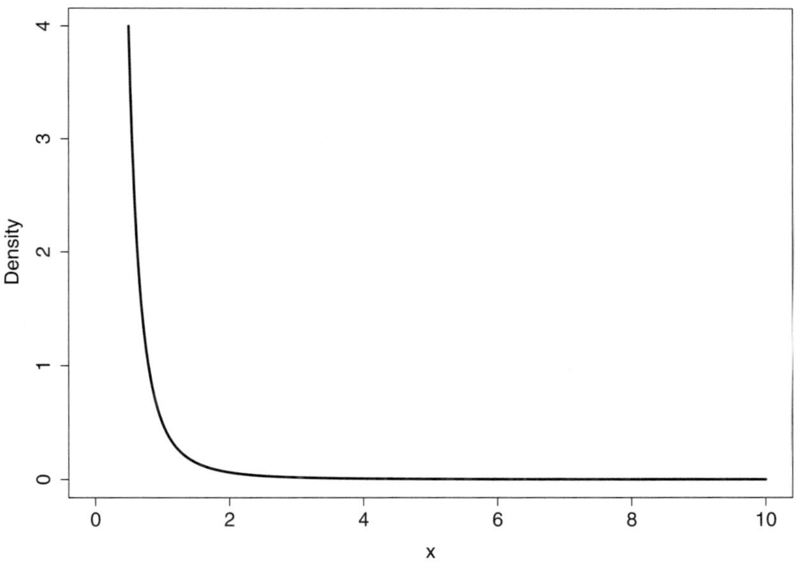

(b)

Fig. B.18. Pareto distribution probability density functions with (a) $\alpha = 1$ and $\beta = 1$ and (b) $\alpha = 2$ and $\beta = 0.5$.

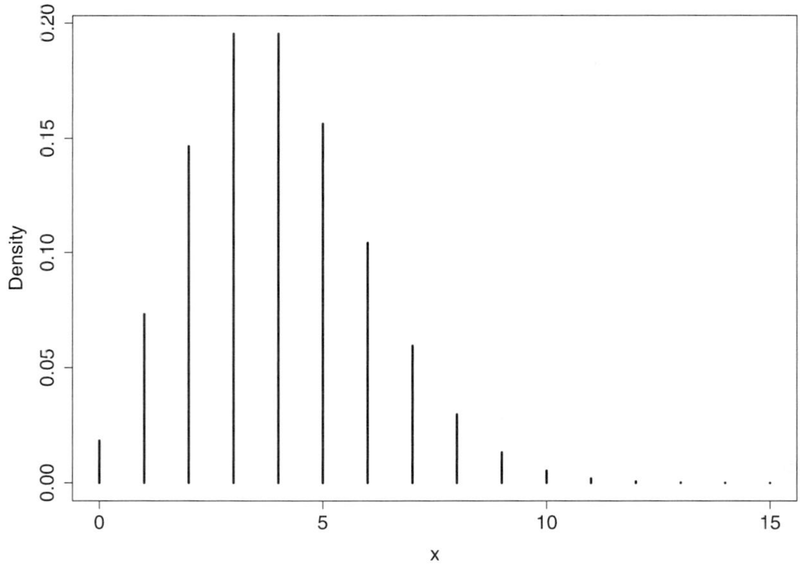

Fig. B.19. Poisson distribution probability density function with $\lambda = 4$.

Probability Density Function

$$f(x \mid \boldsymbol{\beta}, \boldsymbol{\lambda}) = \sum_{j=1}^{m} \frac{\beta_j x^{\beta_j - 1}}{\lambda_j^{\beta_j}} \exp\left[-\sum_{k=1}^{m} \left(\frac{x}{\lambda_k}\right)^{\beta_k}\right] \qquad x > 0, \quad \boldsymbol{\beta} > 0, \quad \boldsymbol{\lambda} > 0.$$

The poly-Weibull is the multiparameter generalization of the Weibull distribution. For more information regarding this distribution, see Berger and Sun (1993).

B.3.24 Student's t

$X \sim t(\nu, \mu, \sigma)$.

Probability Density Function

$$f(x \mid \nu, \mu, \sigma^2) = \frac{\Gamma[\frac{1}{2}(\nu+1)]}{\sigma\sqrt{\nu\pi}\,\Gamma(\frac{\nu}{2})} \left(1 + \frac{1}{\nu}\left(\frac{x-\mu}{\sigma}\right)^2\right)^{-\frac{\nu+1}{2}}$$
$$-\infty < x < \infty, \quad \nu = 1, 2, \ldots, \quad -\infty < \mu < \infty, \quad \sigma > 0.$$

$\mathrm{E}(X) = \mu$ for $\nu > 1$, $\mathrm{Var}(X) = \frac{\nu\sigma^2}{\nu-2}$ for $\nu > 2$.

The Cauchy distribution is a special case when $\nu = 1$.

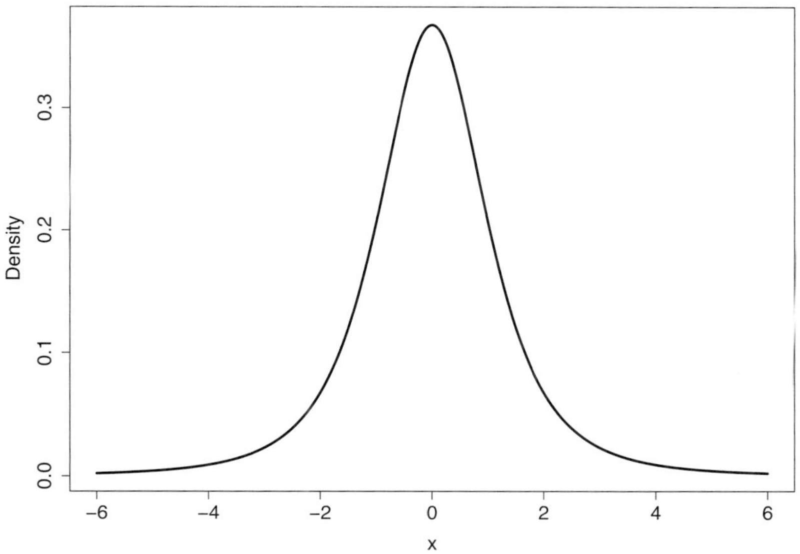

(a)

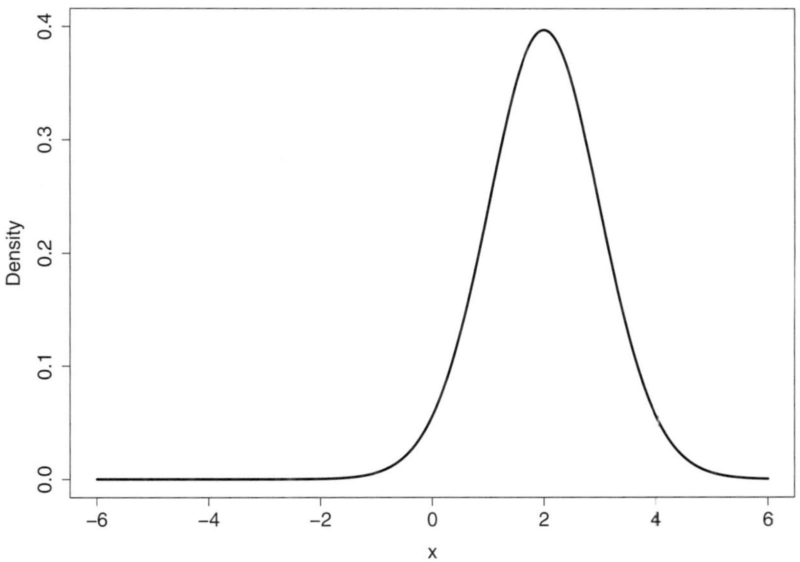

(b)

Fig. B.20. Student's t distribution probability density function with (a) $\nu = 3$, $\mu = 0$, and $\sigma = 1$ and (b) $\nu = 50$, $\mu = 2$, and $\sigma = 1$.

B.3.25 Uniform

$X \sim Uniform(\alpha, \beta)$.

Probability Density Function

$$f(x \mid \alpha, \beta) = \frac{1}{\beta - \alpha} \qquad \alpha \le x \le \beta.$$

$\mathrm{E}(X) = \frac{\alpha+\beta}{2}$, $\mathrm{Var}(X) = \frac{(\beta-\alpha)^2}{12}$.

The uniform distribution is a special case of the beta distribution, with $Uniform(0,1) = Beta(1,1)$.

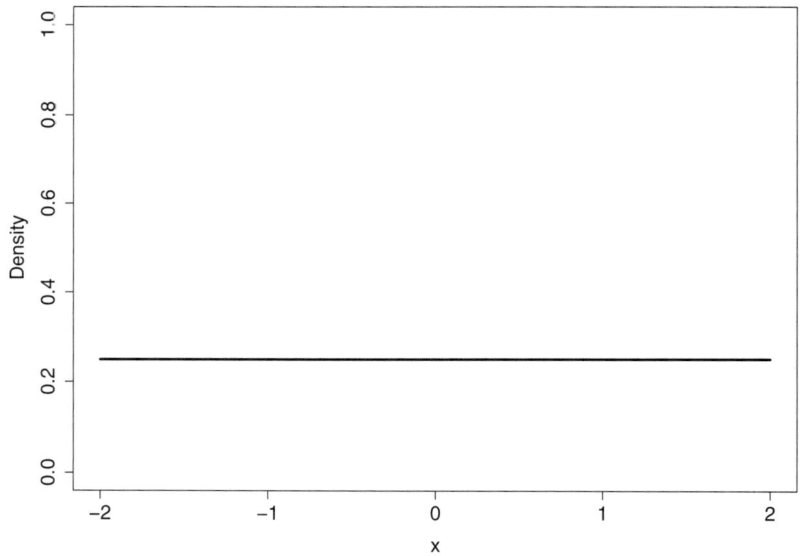

Fig. B.21. Uniform distribution probability density function with $\alpha = -2$ and $\beta = 2$.

B.3.26 Weibull

$X \sim Weibull(\lambda, \beta, \theta)$.

Parameter λ is the *scale parameter*, β is the *shape parameter*, and θ is the *location parameter*.

There are three commonly used parameterizations of the Weibull distribution.

Parameterization 1
Probability Density Function

$$f(x \mid \lambda, \beta, \theta) = \lambda\beta(x-\theta)^{\beta-1}\exp\left[-\lambda(x-\theta)^{\beta}\right] \quad 0 \le \theta < x, \quad \lambda > 0, \quad \beta > 0.$$

$$\mathrm{E}(X) = \theta + \lambda^{-\frac{1}{\beta}}\Gamma\left(\tfrac{\beta+1}{\beta}\right), \ \mathrm{Var}(X) = \lambda^{-\frac{2}{\beta}}\left[\Gamma\left(\tfrac{\beta+2}{\beta}\right) - \Gamma^2\left(\tfrac{\beta+1}{\beta}\right)\right].$$

Hazard Function

$$h(t \mid \lambda, \beta, \theta) = \lambda\beta(t-\theta)^{\beta-1} \qquad 0 < \theta \le t, \quad \lambda > 0, \quad \beta > 0.$$

Parameterization 2
The second commonly used parameterization of the Weibull distribution has $\zeta = \lambda^{\frac{1}{\beta}}$.

Probability Density Function

$$f(y \mid \zeta, \beta, \theta) = \zeta\beta\left[\zeta(y-\theta)\right]^{\beta-1}\exp\left(-\left[\zeta(y-\theta)\right]^{\beta}\right)$$
$$0 \le \theta < y, \quad \zeta > 0, \quad \beta > 0.$$

$$\mathrm{E}(Y) = \theta + \tfrac{1}{\zeta}\Gamma\left(\tfrac{\beta+1}{\beta}\right), \ \mathrm{Var}(Y) = \tfrac{1}{\zeta^2}\left[\Gamma\left(\tfrac{\beta+2}{\beta}\right) - \Gamma^2\left(\tfrac{\beta+1}{\beta}\right)\right].$$

Hazard Function

$$h(t \mid \zeta, \beta, \theta) = \zeta\beta\left[\zeta(t-\theta)\right]^{\beta-1} \qquad 0 < \theta \le t, \quad \zeta > 0, \quad \beta > 0.$$

Parameterization 3
The third commonly used parameterization of the Weibull distribution has $\psi = \lambda^{-\frac{1}{\beta}}$.

Probability Density Function

$$f(z \mid \psi, \beta, \theta) = \frac{\beta}{\psi}\left(\frac{z-\theta}{\psi}\right)^{\beta-1}\exp\left(-\left[\frac{z-\theta}{\psi}\right]^{\beta}\right)$$
$$0 \le \theta < z, \quad \psi > 0, \quad \beta > 0.$$

$$\mathrm{E}(Z) = \theta + \psi\Gamma\left(\tfrac{\beta+1}{\beta}\right), \ \mathrm{Var}(Z) = \psi^2\left[\Gamma\left(\tfrac{\beta+2}{\beta}\right) - \Gamma^2\left(\tfrac{\beta+1}{\beta}\right)\right].$$

Hazard Function

$$h(t \mid \psi, \beta, \theta) = \frac{\beta}{\psi}\left(\frac{t-\theta}{\psi}\right)^{\beta-1} \qquad 0 < \theta \le t < \infty, \quad \psi > 0, \quad \beta > 0.$$

The two-parameter exponential distribution with parameters λ and θ is a special case of the Weibull distribution with $\beta = 1$. The Raleigh distribution is a special case of the Weibull distribution with $\beta = 2$.

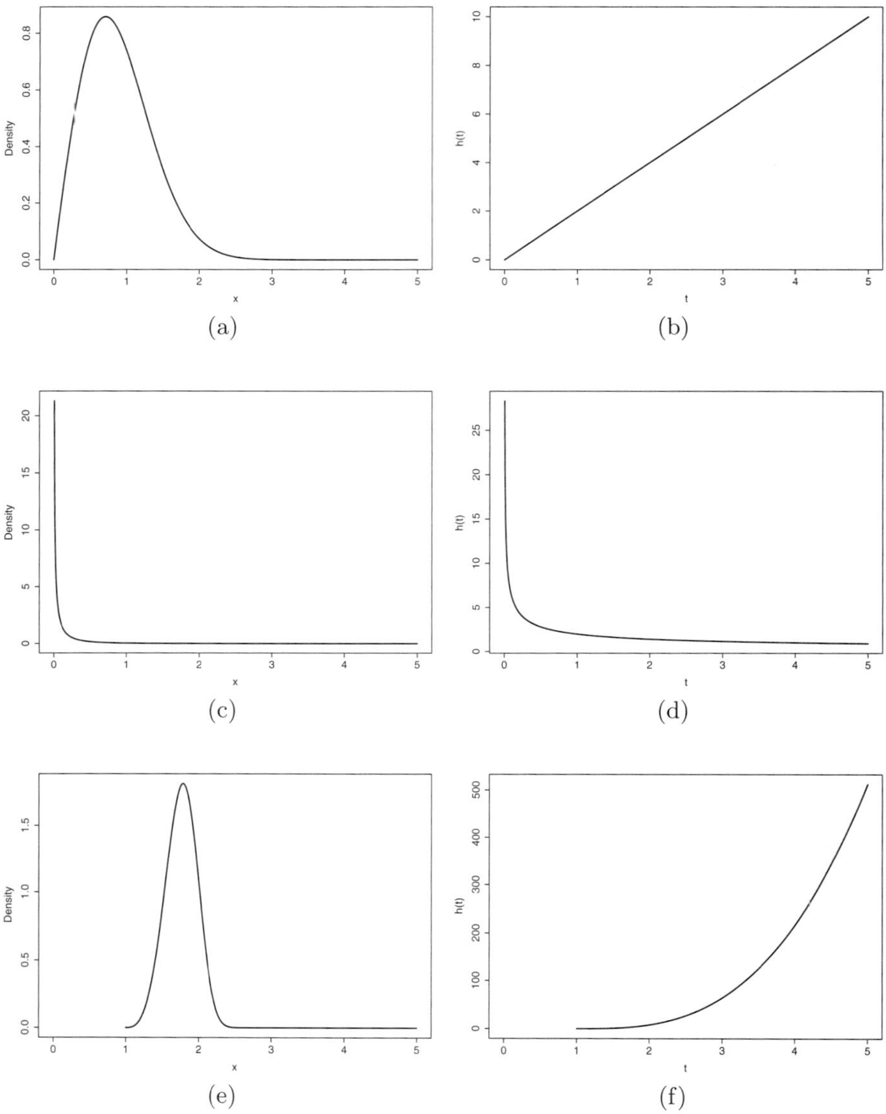

Fig. B.22. Weibull distribution (a) probability density function with $\lambda = 1$, $\beta = 2$, $\theta = 0$, (b) hazard function with $\lambda = 1$, $\beta = 2$, $\theta = 0$, (c) probability density function with $\lambda = 4$, $\beta = 0.5$, $\theta = 0$, (d) hazard function with $\lambda = 4$, $\beta = 0.5$, $\theta = 0$, (e) probability density function with $\lambda = 2$, $\beta = 4$, $\theta = 1$, and (f) hazard function with $\lambda = 2$, $\beta = 4$, $\theta = 1$.

B.3.27 Wishart

$\mathbf{X} \sim Wishart(\nu, \boldsymbol{\Omega})$.

Probability Density Function

$$f(\mathbf{X} \mid \nu, \boldsymbol{\Omega}) = \left(2^{\frac{\nu d}{2}} \pi^{\frac{d(d-1)}{4}} \prod_{i=1}^{d} \Gamma\left(\frac{\nu + 1 - i}{2}\right)\right)^{-1} \times \mid \boldsymbol{\Omega} \mid^{-\frac{\nu}{2}} \mid \mathbf{X} \mid^{\frac{\nu-d-1}{2}} \exp\left[-\frac{1}{2} tr(\boldsymbol{\Omega}^{-1}\mathbf{X})\right],$$

where $\mathbf{X}$ and $\boldsymbol{\Omega}$ are $d \times d$ symmetric, positive definite matrices, $\nu \geq d$, $d > 0$.

$\mathrm{E}(\mathbf{X}) = \nu \boldsymbol{\Omega}$.

For additional information, see Press (1972) and Eaton (1983).

References

C. J. Adcock. Sample size determination: a review. *The Statistician*, 46: 261–283, 1997.

G. Apostolakis and P. Moieni. The foundation of models of dependence in probabilistic safety assessment. *Reliability Engineering*, 18:177–195, 1987.

C. L. Atwood. The binomial failure rate common cause model. *Technometrics*, 28:139–149, 1986.

J. Bacon-Shone. Bayesian methods for system reliability and availability. *The Statistician*, 32:249–251, 1983.

J. L. Bagdonov and F. Kozin. *Probabilistic Models of Cumulative Damage*. John Wiley and Sons, New York, 1985.

R. E. Barlow and F. Proschan. *Mathematical Theory of Reliability*. John Wiley and Sons, New York, 1965.

R. E. Barlow and F. Proschan. *Statistical Theory for Reliability and Life Testing: Probability Models*. Holt, Reinhart, and Winston, New York, 1975.

A. Batra, P. Ramachandran, P. Sathyanarayanan, S. Lu, and H. Srihari. Reliability enhancement of electronic packages by design of optimal parameters. *Microelectronics Reliability*, 44:1157–1163, 2004.

J. O. Berger. *Statistical Decision Theory and Bayesian Analysis*. Springer-Verlag, New York, 1985.

J. O. Berger and D. Sun. Bayesian analysis for the poly-Weibull model. *Journal of the American Statistical Association*, 88:1412–1418, 1993.

J. Bernardo. Statistical inference as a decision problem: the choice of sample size. *The Statistician*, 46:151–153, 1997.

J. Bernardo and A. F. M. Smith. *Bayesian Theory*. John Wiley and Sons, New York, 1994.

A. Birnbaum. On the foundations of statistical inference (with discussion). *Journal of the American Statistical Association*, 57:269–306, 1962.

W. R. Blischke and D. N. P. Murthy. *Reliability: Modeling, Prediction, and Optimization*. John Wiley and Sons, New York, 2000.

A. Bobbio, L. Portinale, M. Minichino, and E. Ciancamerla. Improving the analysis of dependable systems by mapping fault trees into Bayesian networks. *Reliability Engineering and System Safety*, 71:249–260, 2001.

J. C. Borg. Sampling inspection where a continuing relationship exists between consumer and producer. Master's thesis, Lehigh University, 1962.

M. Boulanger and L. A. Escobar. Experimental design for a class of accelerated degradation tests. *Technometrics*, 36:260–272, 1994.

G. E. P. Box and G. C. Tiao. *Bayesian Inference in Statistical Analysis.* Addison-Wesley, Reading, MA, 1973.

D. M. Brender. The prediction and measurement of system availability: a Bayesian treatment. *IEEE Transactions on Reliability*, R-17:127–138, 1968a.

D. M. Brender. The Bayesian assessment of system availability: advanced applications and techniques. *IEEE Transactions on Reliability*, R-17:138–147, 1968b.

G. C. Brush. A comparison of classical and Bayes' producer's risk. *Technometrics*, 28:69–72, 1986.

R. G. Bullington, S. G. Lovin, D. M. Miller, and W. H. Woodall. Improvement of an industrial thermostat using designed experiments. *Journal of Quality Technology*, 25:262–270, 1993.

K. P. Burnham and D. R. Anderson. *Model Selection and Multimodel Inference.* Springer-Verlag, New York, 2002.

B. P. Carlin and T. A. Louis. *Bayes and Empirical Bayes Methods for Data Analysis.* Chapman & Hall, London, 1996.

G. Casella and R. Berger. *Statistical Inference.* Wadsworth & Brooks/Cole, Pacific Grove, CA, 1990.

G. Casella and E. George. Explaining the Gibbs sampler. *The American Statistician*, 46:167–174, 1992.

J. H. Cha and J. M. Kim. On availability of Bayesian imperfect repair model. *Statistics and Probability Letters*, 53:181–187, 2001.

K. Chaloner. Bayesian residual analysis in the presence of censoring. *Biometrika*, 78:637–644, 1991.

K. Chaloner and K. Larntz. Bayesian design for accelerated life testing. *Journal of Statistical Planning and Inference*, 33:245–259, 1992.

H. Chernoff and E. L. Lehmann. The use of maximum likelihood estimates in χ^2 tests for goodness of fit. *Annals of Mathematical Statistics*, 25:579–586, 1954.

R. S. Chhikara and J. L. Folks. The inverse Gaussian distribution as a lifetime model. *Technometrics*, 19:461–468, 1977.

C. H. Chiao and M. Hamada. Using degradation data from an experiment to achieve robust reliability for light emitting diodes. *Quality and Reliability Engineering International*, 12:89–94, 1996.

S. Chib and E. Greenberg. Understanding the Metropolis-Hastings algorithm. *The American Statistician*, 49:327–335, 1995.

S. C. Chow and J. Shao. Estimating drug shelf-life with random batches. *Biometrics*, 47:753–763, 1991.

D. Cochran and G. M. Cox. *Experimental Designs.* John Wiley and Sons, New York, 2nd edition, 1957.

D. W. Coit and T. Jin. Gamma distribution parameter estimation for field reliability data with missing failure times. *IIE Transactions*, 32:1161–1166, 2000.

D. Collett. *Modelling Survival Data in Medical Research.* Chapman & Hall, London, 1994.

L. W. Condra. *Reliability Improvement with Design of Experiments.* Marcel Dekker, New York, 1993.

P. Congdon. *Bayesian Statistical Modeling.* John Wiley and Sons, New York, 2001.

R. J. Cook and J. F. Lawless. *The Statistical Analysis of Recurrent Events.* Springer, New York, 2007.

D. R. Cox. *Analysis of Binary Data.* Chapman & Hall, New York, 1970.

D. R. Cox and D. Oakes. *Analysis of Survival Data.* Chapman & Hall, New York, 1984.

D. R. Cox and E. J. Snell. A general definition of residuals (with discussion). *Journal of the Royal Statistical Society, Series B*, 30:248–275, 1968.

H. Cramér. *Mathematical Methods of Statistics.* Princeton University Press, Princeton, NJ, 1946.

M. Crowder. *Classical Competing Risks.* Chapman & Hall, London, 2001.

M. J. Crowder, A. C. Kimber, R. L. Smith, and T. J. Sweeting. *Statistical Analysis of Reliability Data.* Chapman & Hall, London, 1991.

E. J. Dastrup. Estimating the discrepancy between computer model data and field data: modeling techniques for deterministic and stochastic computer simulators. Master's thesis, Brigham Young University, 2005.

J. J. Deely and J. B. Keats. Bayes' stopping rules for reliability testing with the exponential distribution. *IEEE Transactions on Reliability*, 43:288–293, 1994.

T. J. DiCiccio, R. E. Kass, A. E. Raftery, and L. Wasserman. Computing Bayes' factors by combining simulation and asymptotic approximations. *Journal of the American Statistical Association*, 92:903–915, 1997.

J. R. Van Dorp and T. A. Mazzuchi. A general Bayes Weibull inference model for accelerated life testing. *Reliability Engineering and System Safety*, 90: 140–147, 2005.

R. G. Easterling. On the use of prior distributions in acceptance sampling. *Annals of Reliability and Maintainability*, 9:31–35, 1970.

M. Eaton. *Multivariate Statistics: A Vector Space Approach.* John Wiley and Sons, New York, 1983.

M. Englehardt. Models and analyses for the reliability of a single repairable system. In N. Balakrishnan, editor, *Recent Advances in Life-Testing and Reliability: A Volume in Honor of Alonzo Clifford Cohen, Jr.*, pages 79–106. CRC Press, Boca Raton, FL, 1995.

H. Eyring, S. Gladstones, and K. J. Laidler. *The Theory of Rate Processes.* McGraw-Hill, New York, 1941.

D. Y. Fan. Bayesian acceptance sampling scheme for pass-fail components. *Communications in Statistics–Theory and Methods*, 20:2351–2355, 1991.

B. Fischoff. Debiasing. In D. Kahneman, P. Slovic, and A. Tversky, editors, *Judgment Under Uncertainty: Heuristics and Biases*, pages 422–444. Cambridge University Press, Cambridge, UK, 1982.

M. Fitzgerald, H. F. Martz, and R. L. Parker. Bayesian single-level binomial and exponential reliability demonstration test plans. *International Journal of Reliability, Quality, and Safety Engineering*, 6:123–137, 1999.

K. N. Fleming. A reliability model for common mode failures in redundant safety systems. Technical Report General Atomic Report GA-A13284, General Atomic Company, San Diego, CA, 1975.

A. Fries and A. Sen. A survey of discrete reliability-growth models. *IEEE Transactions on Reliability*, 45:582–604, 1996.

D. Gamerman. *Markov Chain Monte Carlo: Stochastic Simulation for Bayesian Inference.* Chapman & Hall, New York, 1997.

W. A. Ganter, L. Hart, H. F. Martz, and R. Fisher. Comment on reliability of modified designs. *IEEE Transactions on Reliability*, 39:520–522, 1990.

D. P. Gaver and M. Mazumdar. Some Bayes' estimates of long-run availability in a two-state system. *IEEE Transactions on Reliability*, R-18:184–189, 1969.

D. P. Gaver and I. G. O'Muircheartaigh. Robust empirical Bayes' analysis of event rates. *Technometrics*, 29:1–15, 1987.

A. Gelman. Prior distributions for variance parameters in hierarchical models. *Bayesian Analysis*, 1:515–533, 2006.

A. Gelman. Inference and monitoring convergence. In W. R. Gilks, S. Richardson, and D. J. Spiegelhalter, editors, *Markov Chain Monte Carlo in Practice*, pages 131–144. Chapman & Hall/CRC, Boca Raton, FL, 1996.

A. Gelman and D. Rubin. Inference from iterative simulation using multiple sequences. *Statistical Science*, 7:457–472, 1992.

A. Gelman, J. B. Carlin, H. Stern, and D. B. Rubin. *Bayesian Data Analysis.* Chapman & Hall, New York, 2004.

F. P. Gerstle and S. C. Kunz. Prediction of long-term failure in Kevlar 49 composites. In T. K. O'Brien, editor, *Long-Term Behavior of Composites*, number 813 in ASTM Special Technical Publication, pages 263–292. American Society for Testing and Materials, Philadelphia, 1983.

W. R. Gilks, A. Thomas, and D. J. Spiegelhalter. A language and program for complex Bayesian modeling. *The Statistician*, 43:169–178, 1994.

W. R. Gilks, S. Richardson, and D. J. Spiegelhalter, editors. *Markov Chain Monte Carlo in Practice.* Chapman & Hall, London, 1996.

J. Gill. *Bayesian Methods: A Social and Behavioral Sciences Approach.* Chapman & Hall, Boca Raton, FL, 2003.

D. E. Goldberg. *Genetic Algorithms in Search, Optimization, and Machine Learning.* Addison-Wesley, Reading, MA, 1989.

G. M. Grant, W. S. Roesener, D. G. Hall, C. L. Atwood, C. D. Gentillon, and T. R. Wolf. Reliability study: high-pressure coolant injection (HPCI) system, 1987-1993. Technical Report NUREG-CR-5500, Vol. 4, INEL-94-0158, Idaho National Engineering and Environmental Laboratory, Idaho Falls, ID, 1999.

T. L. Graves. Design ideas for Markov chain Monte Carlo software. *Journal of Computational and Graphical Statistics*, 16:24–43, 2007a.

T. L. Graves. *An Introduction to YADAS*, 2007b. http://yadas.lanl.gov.

T. L. Graves, M. Hamada, R. Klamann, A. Koehler, and H. Martz. A fully Bayesian approach for combining multi-level information in multi-state fault tree quantification. *Reliability Engineering and System Safety*, 92:1476–1483, 2007.

P. J. Green and B. W. Silverman. *Nonparametric Regression and Generalized Linear Models: A Roughness Penalty Approach.* Chapman & Hall, New York, 1994.

F. M. Guess and J. S. Usher. A Bayesian approach to demonstration testing with highly reliable devices. In *Proceedings of the 1990 Decision Sciences Institute Conference*, February 1990.

M. Guida and G. Pulcini. Bayesian analysis of repairable systems showing a bounded failure intensity. *Reliability Engineering and System Safety*, 91: 828–838, 2006.

M. Guida, R. Calabria, and G. Pulcini. Bayes' inference for a nonhomogeneous Poisson process with power intensity law. *IEEE Transactions on Reliability*, 38:603–609, 1989.

M. Hamada. *Maintenance Oriented Optimal Design of Accelerated Degradation Testing.* PhD thesis, Wichita State University, 2006.

M. Hamada, H. F. Martz, C. S. Reese, and A. G. Wilson. Finding near-optimal Bayesian experimental designs via genetic algorithms. *The American Statistician*, 55:175–181, 2001.

M. Hamada, H. F. Martz, C. S. Reese, T. Graves, V. Johnson, and A. G. Wilson. A fully Bayesian approach for combining multilevel failure information in fault tree quantification and optimal follow-on resource allocation. *Reliability Engineering and System Safety*, 86:297–305, 2004.

M. S. Hamada, H. F. Martz, A. Koehler, and E. Berg. Optimizing the availability of a buffered industrial manufacturing process. *Reliability Engineering and System Safety*, 91:1039–1048, 2006.

L. Hart. Reliability of modified designs: a Bayes' analysis of an accelerated test of electronic assemblies. *IEEE Transactions on Reliability*, 39:140–144, 1990.

P. Hokstad. A shock model for common-cause failures. *Reliability Engineering and System Safety*, 23:127–145, 1988.

J. J. Horng-Shiau and H. H. Lin. Analyzing accelerated degradation data by nonparametric regression. *IEEE Transactions on Reliability*, 48:149–158, 1999.

S. J. Hudak, A. Saxena, R. J. Bucci, and R. C. Malcolm. Development of standard methods of testing and analyzing fatigue crack growth rate data. Technical Report AFML-TR-78-40, Westinghouse R&D Center, Westinghouse Electric Corporation, Pittsburgh, PA, 1978.

F. L. Hulting and J. A. Robinson. A Bayesian approach to system reliability. Technical Report General Motors Research Publication GMR-7110, General Motors, 1990.

A. V. Huzurbazar. *Flowgraph Models for Multistate Time-to-Event Data.* John Wiley and Sons, Hoboken, NJ, 2005.

ISO 8402 International Standard: Quality Vocabulary. ISO: International Organization for Standardization, Geneva, Switzerland, 1986.

F. V. Jensen. *Bayesian Networks and Decision Graphs.* Springer-Verlag, New York, 2001.

H. C. Jin. Development of Bayesian acceptance sampling plans based on failure-free period life tests and comparisons with other plans. Master's thesis, Korea Advanced Institute of Science and Technology, 1991.

N. Johnson and S. Kotz. *Distributions in Statistics: Continuous Multivariate Distributions.* John Wiley and Sons, Hoboken, NJ, 1972.

V. E. Johnson. A Bayesian χ^2 test for goodness-of-fit. *Annals of Statistics*, 32:2361–2384, 2004.

V. E. Johnson. Bayesian model assessment using pivotal quantities. *Bayesian Analysis*, 2:719–734, 2007.

V. E. Johnson, T. L. Graves, M. S. Hamada, and C. S. Reese. A hierarchical model for estimating the reliability of complex systems. In J. M. Bernardo, M. J. Bayarri, J. O. Berger, A. P. Dawid, D. Heckerman, and M. West, editors, *Bayesian Statistics 7.* Oxford University Press, Oxford, UK, 2003.

V. E. Johnson, A. Moosman, and P. Cotter. A hierarchical model for estimating the early reliability of complex systems. *IEEE Transactions on Reliability*, 54:224–231, 2005.

R. E. Kass and D. Steffey. Approximate Bayesian inference in conditionally independent hierarchical models (parametric empirical Bayes models). *Journal of the American Statistical Association*, 84:717–726, 1989.

A. Kleyner, P. Sandborn, and J. Boyle. Minimization of life cycle costs through optimization of the validation program: a test sample size and warranty approach. Technical report, Delphi Delco Electronics, Kokomo, IL, 2004.

P. M. Ku, E. L. Anderson, and H. J. Carper. Some considerations in rolling fatigue evaluation. *ASLE Transactions*, 15:113–129, 1972.

U. Kumar and B. Klefsjö. Reliability analysis of hydraulic systems of LHD machines using the power law process model. *Reliability Engineering and System Safety*, 35:217–224, 1992.

W. Kuo. Simulation properties of the Bayesian and maximum likelihood estimators of availability. *Microelectronics Reliability*, 24:1057–1068, 1984.

W. Kuo. Bayesian availability using gamma distributed priors. *IIE Transactions*, 17:132–140, 1985.

W. Kuo. Bayes' weighted availability for a digital radio transmission system. *IEEE Transactions on Reliability*, R-35:201–207, 1986.

J. Kyparisis and N. D. Singpurwalla. Bayesian inference for the Weibull process with applications to assessing software reliability growth and predicting software failures. In L. Billard, editor, *Computer Science and Statistics: The Interface*, pages 57–64. Elsevier Science Publishers B. V., 1985.

K. B. Laskey and S. Mahoney. Network engineering for agile belief network models. *IEEE Transactions on Knowledge and Data Engineering*, 12:487–498, 2000.

J. H. Lau, G. Harkins, D. Rice, J. Kral, and B. Wells. Experimental and statistical analyses of surface-mount technology PLCC solder-joint reliability. *IEEE Transactions on Reliability*, 37:524–530, 1988.

J. F. Lawless. *Statistical Models and Methods for Lifetime Data.* John Wiley and Sons, New York, 1982.

E. Lawrence and S. Vander Wiel. Notes on system and component priors Technical Report LA-UR-05-9325, Los Alamos National Laboratory, Los Alamos, NM, 2005.

B. H. Lee. Using Bayesian belief networks in industrial FMEA modeling and analysis. In *Proceedings of the Annual Reliability and Maintainability Symposium*, pages 7–15, Philadelphia, 2001. International Symposium on Product Quality and Integrity.

M. T. L. Lee and A. J. Gross. Lifetime distributions under unknown environment. *Journal of Statistical Planning and Inference*, 29:137–143, 1991.

P. Lee. *Bayesian Statistics: An Introduction.* John Wiley and Sons, London, 1997.

R. León, R. Ramachandran, A. Ashby, and J. Thyagarajan. Bayesian modeling of accelerated life tests with random effects. *Journal of Quality Technology*, 39:3–16, 2007.

E. E. Lewis. *Introduction to Reliability Engineering.* John Wiley and Sons, New York, 2nd edition, 2001.

C. H. Lie, C. L. Hwang, and F. A. Tillman. Availability of maintained systems: a state-of-the-art survey. *AIIE Transactions*, 9:247–259, 1977.

W. Z. Lin. Fundamentals of fluorescent lamp. Technical report, Taiwan Fluorescent Lamp Co., Taiwan, 1976.

D. V. Lindley. On the measure of information provided by an experiment. *Annals of Statistics*, 27:986–1005, 1956.

D. V. Lindley and A. F. M. Smith. Bayes' estimates for the linear model (with discussion). *Journal of the Royal Statistical Society, Series B*, 34: 1–41, 1972.

C. J. Lu and W. Q. Meeker. Using degradation measurements to estimate a time-to-failure distribution. *Technometrics*, 35:161–174, 1993.

J. C. Lu, J. Park, and Q. Yang. Statistical inference of a time-to-failure distribution derived from linear degradation data. *Technometrics*, 39:391–400, 1997.

N. R. Mann, R. E. Schafer, and N. D. Singpurwalla. *Methods for Statistical Analysis of Reliability and Life Data.* John Wiley and Sons, New York, 1974.

J. Marin and C. Robert. *Bayesian Core: A Practical Approach to Computational Bayesian Statistics.* Springer, New York, 2007.

A. W. Marshall and I. Olkin. A multivariate exponential distribution. *Journal of the American Statistical Association*, 62:30–44, 1967.

B. Martin, D. Parker, and L. Zenick. Minimizing slugging by optimizing controllable factors on topaz windshield molding. In *Fifth Symposium on Taguchi Methods*, pages 519–526, Dearborn, MI, 1987. American Supplier Institute.

H. Martz and R. Waller. *Bayesian Reliability Analysis.* John Wiley and Sons, New York, 1982.

H. F. Martz and R. G. Almond. Using higher-level failure data in fault tree quantification. *Reliability Engineering and System Safety*, 56:29–42, 1997.

H. F. Martz and R. A. Waller. Bayesian reliability of complex series/parallel systems of binomial subsystems and components. *Technometrics*, 32: 407–416, 1990.

H. F. Martz, R. A. Waller, and E. T. Fickas. Bayesian reliability analysis of series systems of binomial subsystems and components. *Technometrics*, 30: 143–154, 1988.

H. F. Martz, P. H. Kvam, and L. R. Abramson. Empirical Bayes' estimation of the reliability of nuclear-power-plant emergency diesel generators. *Technometrics*, 38:11–24, 1996.

H. F. Martz, R. L. Parker, and D. M. Rasmuson. Estimation of trends in the scram rate at nuclear power plants. *Technometrics*, 41:352–364, 1999.

D. Mastran. Incorporating component and system test data into the same assessment: a Bayesian approach. *Operations Research*, 24:491–499, 1976.

D. Mastran and N. D. Singpurwalla. A Bayesian estimation of the reliability of coherent structures. *Operations Research*, 26:663–672, 1978.

E. S. McCrory. Cavity and beam excitation for the linac 7835 tube lifetimes. Technical Report Beams-Document-2326-V1, Fermi National Accelerator Laboratory, Batavia, IL, 2006.

P. McCullagh and J. A. Nelder. *Generalized Linear Models.* Chapman & Hall, London, 2nd edition, 1989.

G. C. McDonald, L. C. Vance, and D. L. Gibbons. Some tests for discriminating between lognormal and Weibull distributions: An application to emissions data. In N. Balakrishnan, editor, *Recent Advances in Life-Testing and Reliability*, pages 475–490. CRC Press, Boca Raton, FL, 1995.

W. Q. Meeker and L. A. Escobar. Reliability: the other dimension of quality. *Quality Technology and Quantitative Management*, 1:1–25, 2004.

W. Q. Meeker and L. A. Escobar. *Statistical Methods for Reliability Data.* John Wiley and Sons, New York, 1998.

W. Q. Meeker, L. A. Escobar, and J. C. Lu. Accelerated degradation tests: modeling and analysis. *Technometrics*, 40:89–99, 1998.

M. Meyer and J. M. Booker. *Eliciting and Analyzing Expert Judgment.* ASA/SIAM, Philadelphia, 2001.

Z. Michalewicz. *Genetic Algorithms + Data Structures = Evolution Programs.* Springer-Verlag, New York, 1992.

L. M. Moore and R. J. Beckman. Approximate one-sided tolerance bounds on the number of failures using Poisson regression. *Technometrics*, 30:283–290, 1988.

M. Morgan and M. Henrion. *Uncertainty: A Guide to Dealing with Uncertainty in Quantitative Risk and Policy Analysis.* Cambridge University Press, Cambridge, UK, 1991.

H. Moskowitz and K. Tang. Bayesian variables acceptance-sampling plans: quadratic loss function and step-loss function. *Technometrics*, 34:340–347, 1992.

A. Mosleh and V. Bier. On decomposition and aggregation error in estimation-some basic principles and examples. *Risk Analysis*, 12:203–214, 1992.

C. Muller. Reliability analysis of the 4.5 roller bearing. Master's thesis, Naval Postgraduate School, 2003.

P. Müller. Simulation-based optimal design. In J. M. Bernardo, J. O. Berger, A. P. Dawid, and A. F. M. Smith, editors, *Bayesian Statistics 6*, pages 459–474. Oxford University Press, London, 1999.

P. Müller and G. Parmigiani. Optimal design via curve fitting of Monte Carlo experiments. *Journal of the American Statistical Association*, 90:1322–1330, 1995.

A. Nádas. Best tests for zero drift based on first passage times in Brownian motion. *Technometrics*, 15:125–132, 1973.

M. Neil, N. Fenton, and L. Nielsen. Building large-scale Bayesian networks. *Knowledge Engineering Review*, 15:257–284, 2000.

W. Nelson. Analysis of performance-degradation data from accelerated tests. *IEEE Transactions on Reliability*, R-30:149–155, 1981.

W. Nelson. *Applied Life Data Analysis.* John Wiley and Sons, New York, 1982.

W. Nelson. Fitting of fatigue curves with nonconstant standard deviation to data with runouts. *Journal of Testing and Evaluation*, 12:69–77, 1984.

W. Nelson. *Accelerated Testing: Statistical Models, Test Plans, and Data Analyses.* John Wiley and Sons, New York, 1990.

J. Neyman. On the problem of estimating the number of schools of fish. In J. Neyman, M. Loève, and J. Yerushalmy, editors, *University of California Publications in Statistics*, volume 1(3), pages 21–36. University of California Press, Berkeley and Los Angeles, CA, 1949.

B. Ostle. *Statistics in Research.* Iowa State University Press, Ames, IA, 2nd edition, 1963.

J. Parker. Bayesian prior distributions for multi-component systems. *Naval Research Logistics Quarterly*, 19:509–515, 1972.

K. Pearson. On the criterion that a given system of deviations from the probable in the case of a correlated system of variables is such that it can be reasonably supposed to have arisen from random sampling. *Philosophy Magazine*, 50:157–172, 1900.

T. G. Pham and N. Turkkan. Bayes' binomial sampling by attributes with a general-beta prior distribution. *IEEE Transactions on Reliability*, 41: 310–316, 1992.

T. Pham-Gia and N. Turkkan. System availability in a gamma alternating renewal process. *Naval Research Logistics*, 46:822–844, 1999.

D. A. Pierce and D. W. Schafer. Residuals in generalized linear models. *Journal of the American Statistical Association*, 81:977–986, 1986.

J. P. Poloski and W. H. Sullivan. Data summaries of licensee event reports of diesel generators at U. S. commercial nuclear power plants from January 1, 1976 to December 31, 1978. Technical Report NUREG-CR-1362, EGG-EA-5092, Idaho National Engineering and Environmental Laboratory, Idaho Falls, ID, 1980.

J. P. Poloski, G. M. Grant, C. D. Gentillon, W. J. Galyean, and J. K. Knudsen. Reliability study: Auxiliary/emergency feedwater system, 1987-1995. Technical Report NUREG/CR-5500, Vol. 1, INEEL/EXT-97/00740, Idaho National Engineering and Environmental Laboratory, Idaho Falls, ID, 1998.

N. G. Polson. A Bayesian perspective on the design of accelerated life tests. In A. P. Basu, editor, *Advances in Reliability*, pages 321–330. North-Holland, Amsterdam, 1993.

L. Portinale, A. Bobbio, and S. Montani. From artificial intelligence to dependability: modeling and analysis with Bayesian networks. In A. Wilson, N. Limnios, S. Keller-McNulty, and Y. Armijo, editors, *Modern Statistical and Mathematical Methods in Reliability*. World Scientific, Singapore, 2005.

S. Press. *Applied Multivariate Analysis*. Holt, Rinehart, and Winston, New York, 1972.

F. Proschan. Theoretical explanation of observed decreasing failure rate. *Technometrics*, 18:375–383, 1963.

J. O. Ramsay and B. W. Silverman. *Functional Data Analysis*. Springer-Verlag, New York, 1997.

M. Rausand and A. Høyland. *System Reliability Theory: Models, Statistical Methods, and Applications*. Wiley-Interscience, New York, 2nd edition, 2003.

C. S. Reese, V. Johnson, M. Hamada, and A. Wilson. A hierarchical model for the reliability of an anti-aircraft missile system. Technical Report LA-UR-05-9281, Los Alamos National Laboratory, Los Alamos, NM, 2005.

S. E. Rigdon and A. P. Basu. *Statistical Methods for the Reliability of Repairable Systems*. John Wiley and Sons, New York, 2000.

C. Robert. *The Bayesian Choice: From Decision-Theoretic Foundations to Computational Implementation*. Springer-Verlag, New York, 2001.

C. Robert and G. Casella. *Monte Carlo Statistical Methods*. Springer-Verlag, New York, 2nd edition, 2004.

K. J. Ryan. Some flexible families of intensities for non-homogeneous Poisson process models and their Bayes' inference. *Quality and Reliability Engineering International*, 19:171–181, 2003.

K. J. Ryan and C. S. Reese. Estimating reliability trends for the world's fastest computer. Technical Report LA-UR-00-4201, Los Alamos National Laboratory, Los Alamos, NM, 2001.

G. Schwarz. Estimating the dimension of a model. *Annals of Statistics*, 6: 461–464, 1978.

K. K. Sharma and R. K. Bhutani. A comparison of classical and Bayes' risks when the quality varies randomly. *Microelectronics and Reliability*, 32:493–495, 1992.

K. K. Sharma and H. Krishna. Non-parametric inferences on system availability with a reference of k-out-of-m systems. *Microelectronics Reliability*, 35:289–293, 1995.

V. N. Shaw, M. B. Sattison, C. L. Atwood, A. G. Ware, and G. M. Grant. Assessment of pressurized water reactor primary system leaks. Technical Report NUREG/CR-6582, INEEL/EXT-97-01068, Idaho National Engineering and Environmental Laboratory, Idaho Falls, ID, 1998.

Z. Sheng and D. Y. Fan. Bayes' attribute acceptance-sampling plan. *IEEE Transactions on Reliability*, 41:307–309, 1992.

J. H. Sigurdsson, L. A. Walls, and J. L. Quigley. Bayesian belief nets for managing expert judgement and modelling reliability. *Quality and Reliability Engineering International*, 17:181–190, 2001.

N. Singpurwalla. *Reliability and Risk: A Bayesian Perspective*. John Wiley and Sons, New York, 2006.

N. O. Siu and D. L. Kelly. Bayesian parameter estimation in probabilistic risk assessment. *Reliability Engineering and System Safety*, 62:89–116, 1998.

D. Spiegelhalter. Bayesian graphical modeling: a case-study in monitoring health outcomes. *Applied Statistics*, 47:115–133, 1998.

D. Spiegelhalter, A. Thomas, and N. G. Best. Computation on Bayesian graphical models. In J. M. Bernardo, J. O. Berger, A. P. Dawid, and A. F. M. Smith, editors, *Bayesian Statistics 5*, pages 407–426. Oxford University Press, Oxford, UK, 1996.

D. Spiegelhalter, A. Thomas, N. Best, and D. Lunn. *WinBUGS User Manual Version 1.4*, 2003. http://www.mrc-bsu.cam.ac.uk/bugs.

M. Springer and W. Thompson. Bayesian confidence limits for the product of n binomial parameters. *Biometrika*, 53:611–613, 1966.

G. Taguchi. *Introduction to Quality Engineering*. Asian Productivity Organization, Tokyo, 1986.

Tetrahedron Inc. Reliability of blowout preventers tested under fourteen and seven days time interval. Technical report, Tetrahedron Inc., 1996.

W. E. Thompson and P. A. Palicio. Bayesian confidence limits for the availability of systems. *IEEE Transactions on Reliability*, R-24:118–120, 1975.

W. E. Thompson and M. D. Springer. A Bayes' analysis of availability for a system consisting of several independent subsystems. *IEEE Transactions on Reliability*, R-21:212–214, 1972.

F. A. Tillman, W. Kuo, C. L. Hwang, and D. L. Grosh. Bayesian reliability and availability—a review. *IEEE Transactions on Reliability*, R-31:362–372, 1982.

P. Tobias and M. Poore. Bayesian reliability testing for new generation semiconductor processing equipment. Presented at 2003 Quality and Productivity Research Conference, May 2003.

P. A. Tobias and D. C. Trindade. *Applied Reliability*. Van Nostrand Reinhold, New York, 2nd edition, 1995.

S. T. Tseng, M. Hamada, and C. H. Chiao. Using degradation data from a factorial experiment to improve fluorescent lamp reliability. *Journal of Quality Technology*, 27:363–369, 1995.

U. S. Nuclear Regulatory Commission. A review of NRC staff uses of probabilistic risk assessment. Technical Report NUREG-1489, U. S. Nuclear Regulatory Commission, Washington, DC, 1994.

W. N. Venables, D. M. Smith, and R Development Core Team. *An Introduction to R*. R Foundation for Statistical Computing, Vienna, 2006. http://www.R-project.org.

I. Verdinelli, N. Polson, and N. D. Singpurwalla. Shannon information and Bayesian design for prediction in accelerated life-testing. In R. E. Barlow, C. A. Clariotti, and R. Spizzichino, editors, *Reliability and Decision Making*, pages 247–256. Chapman & Hall, London, 1993.

W. E. Vesely. Estimating common-cause failure probabilities in reliability and risk analysis: Marshall-Olkin specializations. In J. B. Fussell and G. R. Burdick, editors, *Nuclear Systems Reliability Engineering and Risk Assessment*, pages 314–341. SIAM, Philadelphia, 1977.

W. E. Vesely, F. F. Goldberg, N. H. Roberts, and D. F. Haasl. Fault tree handbook. Technical Report NUREG-0492, U. S. Nuclear Regulatory Commission, Washington, DC, 1981.

Z. Vintr. Optimization of reliability requirements from manufacturer's point of view. In *1999 Proceedings Annual Reliability and Maintainability Symposium*, pages 183–189, 1999.

C. S. Whitman. Accelerated life test calculations using the method of maximum likelihood: an improvement over least squares. *Microelectronics Reliability*, 43:859–864, 2003.

G. A. Whitmore. Estimating degradation by a Weiner diffusion process subject to measurement error. *Lifetime Data Analysis*, 1:307–319, 1995.

G. A. Whitmore and F. Schenkelberg. Modelling accelerated degradation data using Weiner diffusion with a time scale transformation. *Lifetime Data Analysis*, 3:27–45, 1997.

G. A. Whitmore, K. D. S. Young, and A. C. Kimber. Two-state reliability tests with technological evolution: a Bayesian analysis. *Applied Statistics*, 43:295–307, 1994.

A. G. Wilson, L. A. McNamara, and G. D. Wilson. Information integration for complex systems. *Reliability Engineering and System Safety*, 92:121–130, 2007.

D. Von Winterfeldt and W. Edwards. *Decision Analysis and Behavioral Research.* Cambridge University Press, Cambridge, UK, 1986.

C. F. J. Wu and M. Hamada. *Experiments: Planning, Analysis, and Parameter Design Optimization.* John Wiley and Sons, New York, 2000.

M. Zelen. Factorial experiments in life testing. *Technometrics*, 1:269–288, 1959.

Y. Zhang and W. Q. Meeker. Bayesian methods for planning accelerated life tests. *Technometrics*, 48:49–60, 2006.

F. W. Zok, X. Chen, and C. H. Weber. Tensile strength of SiC fibers. *Journal of American Ceramic Society*, 78:1965–1968, 1995.

Author Index

Subject Index

Springer
the language of science

springer.com

Bayesian Computation with R

Jim Albert

This book introduces Bayesian modeling by the use of computation using the R language. The early chapters present the basic tenets of Bayesian thinking by use of familiar one and two-parameter inferential problems. Bayesian computational methods such as Laplace's method, rejection sampling, and the SIR algorithm are illustrated in the context of a random effects model. The construction and implementation of Markov Chain Monte Carlo (MCMC) methods is introduced.

2007. 270 pp. (Use R!) Softcover
ISBN 978-0-387-71384-7

Pattern Recognition and Machine Learning

Christopher M. Bishop

Time This is the first textbook on pattern recognition to present the Bayesian viewpoint. The book presents approximate inference algorithms that permit fast approximate answers in situations where exact answers are not feasible. It uses graphical models to describe probability distributions when no other books apply graphical models to machine learning. No previous knowledge of pattern recognition or machine learning concepts is assumed. Familiarity with multivariate calculus and basic linear algebra is required, and some experience in the use of probabilities would be helpful though not essential. The book includes an introduction to basic probability theory.

2006, 740 pp. (Information Science and Statistics) Hardcover
ISBN 978-0-387-31073-2

An Introduction to Bayesian Analysis

Jayanta K. Ghosh, Mohan Delampady and Tapas Samanta

This is a graduate-level textbook on Bayesian analysis blending modern Bayesian theory, methods, and applications. Starting from basic statistics, undergraduate calculus and linear algebra, ideas of both subjective and objective Bayesian analysis are developed to a level where real-life data can be analyzed using the current techniques of statistical computing. Advances in both low-dimensional and high-dimensional problems are covered, as well as important topics such as empirical Bayes and hierarchical Bayes methods and Markov chain Monte Carlo (MCMC) techniques.

2006. 352 pp. (Springer Texts in Statistics) Hardcover
ISBN 978-0-387-40084-6

Easy Ways to Order▶ Call: Toll-Free 1-800-SPRINGER • E-mail: orders-ny@springer.com • Write: Springer, Dept. S8113, PO Box 2485, Secaucus, NJ 07096-2485 • Visit: Your local scientific bookstore or urge your librarian to order.

Printed in the United States of America